INFRASTRUCTURE

A FIELD GUIDE TO THE INDUSTRIAL LANDSCAPE

INFRASTRUCTURE

A FIELD GUIDE TO THE INDUSTRIAL LANDSCAPE

BRIAN HAYES

W. W. NORTON & COMPANY
NEW YORK LONDON

For information about permission to reproduce selections from this book, write to
Permissions, W. W. Norton & Company, Inc., 500 Fifth Avenue, New York, NY 10110

Manufacturing by South China Printing Co. Ltd.
Book design by Brian Hayes
Production manager: Julia Druskin

Library of Congress Cataloging-in-Publication Data

Hayes, Brian.
 Infrastructure : a field guide to the industrial landscape / Brian Hayes.
 p. cm.
 Includes bibliographical references and index.
 ISBN 0-393-05997-9 (hardcover)
 1. Industrial buildings—Landscape architecture. I. Title.
 TS190.5.H39 2005
 711'.6—dc22

 2005004640

W. W. Norton & Company, Inc., 500 Fifth Avenue, New York, N.Y. 10110
 www.wwnorton.com

W. W. Norton & Company Ltd., Castle House, 75/76 Wells Street, London W1T 3QT

1 2 3 4 5 6 7 8 9 0

CONTENTS

PREFACE

A NATURE GUIDE IS A USEFUL COMPANION when you venture into the deep woods or the mountains. For better or worse, however, few of us spend much time in the wilderness; we live in an environment where most of what we see is man-made. Perhaps on your summer vacation you have a chance to hike the Appalachian Trail, but the rest of the year you drive through refinery row on the New Jersey Turnpike. Once in a lifetime, if you are incredibly lucky, you might look up a tree and spot an ivory-billed woodpecker, but on most days you are more likely to look up a utility pole and see the finned steel casing of an electrical transformer.

This book adopts the form of a nature guide, but its subject matter is everything that *isn't* nature. It is a guide to the common sights of the built environment—the power lines, water tanks, street lights, manholes, traffic signals, cellular-telephone towers—that we pass by every day and yet seldom really notice. In these pages I identify and classify some of the species that inhabit this familiar urban ecosystem. Farther afield, there are more exotic industrial habitats to explore: coal mines, oil refineries, railroad freight yards, power plants, garbage incinerators. These are places that most of us never see close up; many of us would go out of our way to *avoid* seeing them. But they are nonetheless a part of modern life—and worth a visit. There can be just as much of interest happening on a factory rooftop as there is in the forest canopy, just as much to marvel at in the operation of a strip-mining dragline as in the geological carving of a river canyon.

Some may find puzzling or distasteful the parallel I am drawing between the study of nature and the study of technology. After all, nature is good and good for you, whereas everyone knows that technology is ugly, evil, and dangerous. The mention of nature brings to mind majestic landscapes: Yosemite, Yellowstone, the Grand Canyon. The mention of industrial technology brings to mind a long list of disasters:

What is it and what does it do? We live in a world cluttered with industrial artifacts whose purpose and principles of operation are often unknown to those who see them. The curious rooftop structures on the opposite page are devices called cyclones, mounted atop a lumber yard and sawmill in Sacramento, California. Their function is to separate sawdust from a moving stream of air. The dust-laden air enters each funnel-like vessel tangentially, creating a whirlpool wind that spins the sawdust to the outside; the dust trickles down the walls of the funnel into a bin below, while the air exits through the question-mark vent at the top. The telltale shape of a cyclone is something you can expect to encounter not just at a sawmill but in many other industrial environments, from coal mines to flour mills.

Three Mile Island, Bhopal, Chernobyl, Love Canal. In the presence of nature we hold our breath in hushed reverence; in the presence of industry we hold our nose.

A few centuries ago—say, on the American western frontier—a quite different view prevailed. Nature was seen as savage, hostile, cruel. Mountains and forests were barriers, not refuges. The lights of civilization were a comforting sight. We took our charter from the book of Genesis, which grants mankind dominion over the beasts, and felt it was both our entitlement and our duty to tame the wilderness, fell the trees, plow the land, dam the rivers. In the most extreme version of this ideology, everything on the planet was put here explicitly for human use. At the opposite extreme, today, the earth-first sensibility urges us to treat the entire planet according to the campsite ethic: carry out what you carry in, and leave no trace of your passage.

It is not my mission to mediate between these strangely polarized positions. My chief aim is simply to describe and explain the technological fabric of society, not to judge whether it is good or bad, beautiful or ugly. And yet I would not argue that technology is neutral or value-free. Quite the contrary: I submit that the signs of human presence are the only elements of the landscape that have any moral or aesthetic significance at all. In nature undisturbed, a desert is not better or worse than a forest or a glacier; there is simply no scale on which to rank such things unless it is a human scale of utility or beauty. Only when people intervene in nature is there any question of right or wrong, better or worse. When we look on a pristine glade, we are mere bystanders, but when we walk down a city street, we are responsible for what we see (and what we hear and smell), and we are therefore called on to pass judgment.

The sandstone buttes of Red Rock State Park, near Gallup, New Mexico, serve as backdrop to the processing and storage facilities of El Paso Natural Gas Company and Conoco Propane. Do the tanks and towers spoil the view of the cliffs? Or does the overdramatic landscape distract from a proper appreciation of the industrial equipment?

In Cushing, Oklahoma ("pipeline crossroads of the world"), cylindrical tanks dot another bucolic landscape. Would the scene be more interesting or more attractive without them?

One winter afternoon a few years ago I stood by the side of a highway near Gallup, New Mexico, looking on a classic vista of the American West: red sandstone buttes rising from a valley floor. It was the kind of landscape made famous by films and paintings and postcards, not to mention all those Marlboro Man advertisements. But this particular vista had something more. In front of the cliffs, and towering over them, were several cylindrical spires that I recognized as petroleum fractionating columns; off to one side was a grove of gleaming white spherical tanks. The towers and tanks belonged to a plant for processing and storing propane, or liquefied petroleum gas. I suspect that most viewers of this scene would consider the industrial hardware to be an intrusion, a distraction, perhaps even a desecration of the landscape. I might offer the counterargument that the juxtaposition of natural landforms with the geometrically simpler cylinders and spheres adds visual interest to the composition, but I don't expect to win many converts to that view.

When looking on that scene west of Gallup, the obvious question is: Why did they have to build it *here*? Couldn't the gasworks have been put somewhere less conspicuous? There is an answer. The propane plant was put below the buttes because that's where the pipeline runs, carrying petroleum products from Texas to California. And the pipeline takes that path because it follows the highway and the railroad. And before the highway and railroad were built, a stagecoach line followed the same route. And before that, there was a trail used by the native peoples who have maintained an urban culture in this region for at least a millennium. In other words, this is a landscape that has been put to human use for a very long time. Today, many of us might prefer to see it put to some *other* use, but that is still a matter of impressing

Sometimes the forms of industrial hardware are full of visual interest in their own right, quite apart from how they fit into the wider landscape. At right, an insulated pipeline and a row of air-intake stacks at an installation of gas-turbines next to the Ravenswood Generating Station in Queens, New York. Below, another rooftop cyclone, in Spartanburg, South Carolina.

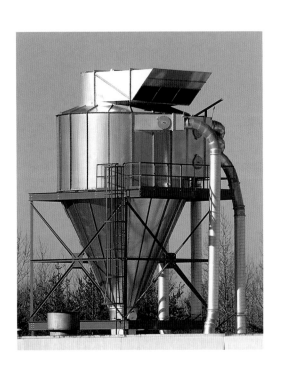

human values on the land. We might choose industry or we might choose scenery, but in either case it is our choice and responsibility.

A field guide ought to offer more than just clues to identifying species. It should also lead to an understanding of how the various elements of a landscape or an ecosystem fit together in a coherent whole. In the rain forest, trees provide shelter for some animals; other animals eat the fruits and at the same time disperse the seeds; still others are parasites on the leaves or roots. Comparable webs of interactions can be found in some industrial ecosystems. A classic example is the damming of a river, which can supply hydroelectric power and drinking water for cities, irrigation water for agriculture, and protection from flooding downstream; but the dam also permanently submerges a valley, with a consequent loss of land area, and in some cases the dam may interfere with fisheries; the effects on river navigation could be either beneficial or detrimental.

Because the networks of the industrial economy are so tangled, there is no clear starting point or finish line; it's all cycles within cycles. But we have to start somewhere, and so in this book I have imposed an overall structure that traces the flow of materials, energy, and information through the system. The story begins with basic inputs, namely, raw materials such as metal ores, coal, and petroleum, as well as water and what might be called biological raw materials—food and other products of agriculture. Then we explore the various networks that interconnect us all: the electric power grid, communication channels, and transport by road, rail, air, and water. Finally we come to the nether end of the industrial economy—the disposal and recycling of wastes.

Not everything industrial or technological will be found within these pages. My emphasis is on landscape, meaning things you can see out of doors. I have had to ignore a multitude of familiar household technologies; thus, you will have to turn elsewhere to learn how a dishwasher or a microwave oven works. With regret, I have also had to leave out many manufacturing operations. I would have greatly enjoyed learning and describing what goes on inside a pharmaceutical plant or a fabrication line for silicon chips, but all the action in such places is hidden away; from the outside, you would have a hard time telling whether a factory makes aspirin or microprocessors. The manufacturing operations that do appear in these pages are those where the machinery is more or less visible from outside the chain-link fence, as with oil refineries and steel mills. (In fact, many of the photographs in this book were taken from outside the fence—or by poking the lens of the camera through it.)

Other excluded topics may seem more or less arbitrary. I decided not to attempt a survey of military technology, partly because poking a camera lens through *those* fences could get me in trouble. I also had no room for the construction industry. Finally, I should note that this is largely a guide to the North American industrial landscape, with some attention to Europe but almost nothing about the rest of the world.

The writing of this book has been an education and an adventure. I have crisscrossed North America as a technotourist taking in the highlights of the industrial landscape. Friends looked at me strangely when I announced that I was going to New Orleans not for Mardi Gras but to look at the drainage pumps, that I was driving to Vermont not for the skiing but to see a granite quarry, that my vacation in the south of Italy was spent photographing highways and harbors. At least I avoided the crowds.

My hope is that this book will cultivate greater awareness of all the miscellaneous hardware that goes into making a civilization, and perhaps even some enthusiasm for

Some industrial activities impress a pattern directly on the landscape itself, especially in mining and agriculture. At left, ponds are prepared in the surface of a towering artificial butte, made of low-grade copper ore. The ponds will be flooded in the mining process called heap leaching. The ore pile is at the Tyrone Mine in New Mexico.

the industrial landscape. It's all around you; you might as well get to know what it's called and what it does. If you would pull off the highway to admire a mountain vista or a waterfall, you might also consider pausing for a mine or a power plant.

Welcome to the world we have made for ourselves.

ACKNOWLEDGMENTS

This is a book that springs from at least three sources. One point of departure was the question that my daughter, Amy Hayes, used to ask from the back seat: "What's *that* thing?" I thank her both for asking the question and for letting me know when my answers grew too pedantic to endure.

Looking further back, my own interest in the artifacts of the industrial world was awakened in adolescence, when I was fortunate to have an older friend, Dave Fell, an engineer and an urban-industrial counterpart of the wise woodsman. One summer we explored the back roads near Cape May Point, New Jersey. The locale is significant. Witmer Stone's *Bird Studies at Old Cape May* was a founding work of American nature writing, and Lily Pond at Cape May Point still attracts an interesting population of birds and birders. When Roger Tory Peterson, the dean of field-guide authors, made his first ornithological expedition, his destination was Cape May Point. But the area is also a rich one for the technophile. Less than a mile from Lily Pond, Dave and I found a vast, humming plant that extracts magnesium from seawater; nearby we watched a dredge deepen the canal that crosses the cape; we clambered over a World War II blockhouse and gun emplacement, abandoned by the Coast Guard; and we puzzled over rusted remnants of bygone industries, such as a steam-driven mill for crushing oyster shells into powdered lime. Dave—sadly, now deceased—taught me the pleasure of puzzling out how things work.

The third influence was Dennis Flanagan, the editor of *Scientific American* through all its best years, who educated a generation of science writers, including me. In a conversation two decades ago, Dennis pointed out that although a great deal is written about technology, almost all of it focuses on a few fashionable topics such as computers and genetic engineering; the technologies that run most of the world's industries have become almost invisible. His remark inspired this book, and his continuing guidance helped to shape it. Dennis died in January 2005.

I am grateful to many others for help and support. Diana Lutz and her parents, Josephine Lutz and the late John Lutz, were the project's first enthusiasts. The late Philip and Phylis Morrison offered encouragement and a stream of suggestions that were pure gold. Joseph Wisnovsky, a friend and colleague, helped me refine my ideas and shared many of his own; then he arranged for publication at W. W. Norton and for seven years was my editor there. After Joe's retirement, Angela von der Lippe and Alessandra Bastagli took over. Barbara Williams Flanagan offered valuable counsel. Bonnie Auslander has been my loyalest reader and critic. And, through a decade of

wandering the industrial landscape, Rosalind Reid has been my companion and fellow-traveler. She has a presence in these pages.

This book would not exist without the financial support of the Alfred P. Sloan Foundation, where Arthur L. Singer, Jr., rescued me from an early crisis and Doron Weber provided sustained support over a period of years. The grants were administered by Sigma Xi, the Scientific Research Society.

Finally I wish to acknowledge the assistance of a great many helpful and thoughtful people who invited me into their corners of the industrial landscape, showed me the sights, and helped me tell their stories: Asheville Waste Paper Company; Frank Brenner of Atlantic Scrap and Processing in Kernersville, North Carolina; Reagan Gentry, Aubrey K. McClendon, Tom Price, Jr., Ron E. Voth, and Roger Wilson of Chesapeake Energy Corporation in Oklahoma City; Ken Deffeyes of Princeton, New Jersey; R. S. Renfroe of the ChevronTexaco refinery in Pascagoula, Mississippi; Glenn Madelmeyer of Covanta Energy Corporation in Lorton, Virginia; Mike Adcock, Renee Lawrence, James Minor, Terry Rolan, David Sineath, and William Telford of the Department of Water Management in Durham, North Carolina; Duane Klabunde and Doug Stoltz of the Falkirk Mining Company in Underwood, North Dakota; Tom Spain of the Water Reclamation Facility in Henderson, North Carolina; H. L. Himes of Silver City, New Mexico; Joe Hirschi of the Illinois Clean Coal Institute; the Galatia Mine in Galatia, Illinois, operated by American Coal Company; Linda White of the Kern Wind Energy Association in Bakersfield, California; Dave Rib of the Kramer Junction Operating Company in Kramer Junction, California; Sara Moriarty and the late Judi Scioli of the Maryland Port Administration; Alan R. Blatecky and Eileen Sarro of the Microelectronics Center of North Carolina; David G. Finley of the National Radio Astronomy Observatory in Socorro, New Mexico; Joe Puglia of the New Orleans Drainage Commission; Frank Feeney, Michael Marotta, John Pampelone, Steve Violetta, and Paula Young of the New York City Department of Sanitation; Susan Terpay, C. T. Sansbury, Jr., J. L. Thomas, and Darnell Wood of Norfolk Southern Corporation; Scott A. Andrews, Giff Daughtridge, Mike D. Fox, Johnny E. Jacobs, Joe Rutkowski, and Ray Wright of Nucor Corporation; Woodrow Boyd, Sharon Hall, Robert Sisson, and Robert Yanity of Progress Energy; John Hubbard of the Roanoke Valley Resource Authority; Todd Paton of the Rock of Ages Granite Quarry in Barre, Vermont; Matt Saindon and Melvin Saindon of Zurich, Kansas; Michael B. Gwynn of Texasgulf; Michael D. O'Dell of UUNET Technologies; Bill Roser of the Wheeler Brothers Feedyard in Watonga, Oklahoma.

A big hole in the ground: This is where most of the raw materials of an industrial society come from. In this particular hole—the Rock of Ages quarry in Barre, Vermont—workers skillfully carve out immense slabs of granite for use in buildings, monuments, and gravestones. To appreciate the scale of this excavation, note that the bright blue object on a shelf near the center of the image is a Porta Potti.

OUT OF THE EARTH

THE RAW MATERIALS FOR BUILDING a civilization are mainly things we dig up out of the ground: fuels such as coal, uranium, and petroleum; the ores of iron, copper, aluminum, and dozens of other metals; stone, sand, and clay for building; and a miscellany of other minerals such as sulfur and phosphates. These things are basic inputs of the world economy—the ingredients for making everything else—and so they seem like a good place to begin an exploration of the industrial landscape.

This chapter describes what economists call the extractive industries—the various ways of scraping and drilling and blasting and digging to claw useful stuff out of the earth. We'll visit pit mines, strip mines, and underground mines, hard-rock mines, coal mines, and stone quarries. And we'll look at what happens to a few of these commodities after they are retrieved from underground: the smelting of copper, the conversion of iron ore into steel, the manufacture of cement and concrete. (Two other vital resources—water and petroleum—might have been discussed in the same context, but they are so important they deserve chapters of their own.)

To begin our tour of the industrial landscape with titanic mines and mills is to confront immediately the most spectacular ways that industry can transform the terrain. Over a period of a few decades, a big open-pit mine can blast away an entire mountain, leaving a deep crater in its place and filling nearby valleys with waste rock. Inside the mine, everything biological—the whole veneer of grass, trees, and soil that shapes and softens the contours of the human habitat—is peeled back to expose the raw, rocky heart of the planet. Critics of the industry describe such mines as wounds or scars. Miners, of course, see nothing so sinister or destructive in what they do. On the contrary, they are proud of the grandeur of their art: for the first time in human history, we have it in our power to move mountains.

Mining seems like a simple job: first you find pay dirt, and then you dig it out. But there are many ways of going about it. Different kinds of deposits call for different

mining strategies, and a single deposit might be worked in several ways over the course of its lifetime.

Suppose you strike gold. You might start making your fortune by trying to recover loose flakes of metal in sandy stream beds—a technique known as placer mining. Later you might dig a small underground mine, following a vein of ore as it meanders through the rock. With a bigger capital investment, the tunneling operation could be replaced by an open-pit mine, with enormous earthmoving machines. Then, if the ore body continues still deeper, you might return to underground methods with a vertical shaft mine. Finally, just when you think the deposit is played out at last, the discarded wastes from earlier operations become a resource worth exploiting. In some mining districts all these activities are going on at once.

PLACER MINING

The mention of placer mining evokes images of forty-niners panning for gold in the streams near Sutter's Mill in northern California. A placer (pronounced "plah-ser") is just a deposit of sand or gravel with a concentration of some useful mineral. Most placers are created by the action of rivers or glaciers (the word is Spanish for "shoal"), and they are mined by using water to separate the mineral from waste materials.

In panning for gold, the miner digs gravel from the stream bed and swirls it in a shallow pan until the lighter wastes escape and the denser flakes of "color" are left behind. As mining processes go, panning is inefficient, recovering only the most accessible minerals. On the other hand, the only equipment needed is a pie pan and a shovel.

When a placer-mining operation grows too big for panning, the next step is a sluice box or riffle box, in which larger quantities of gravel are washed over corru-

GETTING A LOOK

Mine operators have a certain reputation for wariness, secrecy, and occasionally even hostility. Think of the solitary prospectors of western legend, always chasing off "claim jumpers." Superintendents of modern mining operations have other reasons for shutting out visitors. Mines are dangerous places, and when you wander onto the property, the owners will worry about your safety—or about their legal liability for your safety. If you are snooping around with binoculars or a camera, they may think you are looking for violations of environmental or worker-safety regulations.

Then there are all the usual concerns about theft and vandalism.

Yet hostility to sightseers is by no means universal. Your chances of getting a close look are probably best at the largest quarries and open-pit mines, some of which welcome visitors and offer tours, or at least put up a sign explaining what's going on down below. Underground mines are harder to get into, but there are a few where you can strap on a hard hat and ride a hoist or an ore train into the depths of the earth. In any case, the biggest mines are so vast in scale that they can't be hidden from view. You'll see them whether you want to or not.

The photo at left shows the aptly named Lavender Pit, an abandoned mine in Bisbee, Arizona, as seen from a roadside overlook.

Hydraulicking operation uses a stream of water under high-pressure to wash away an embankment along the Rogue River in southern Oregon. The date of the photo is unknown; it comes from the collection of the Bureau of Land Management.

gations by rapidly flowing water, again leaving behind the heavier gold particles. These forms of placer mining are not economically important today, but in gold-mining districts there are plenty of roadside museums where you can see old equipment on display. For a fee, you can also try your luck at panning. (The owners of the museums have recognized that more wealth flows by on the highway out front than in the stream out back.)

A more elaborate form of placer mining is called hydraulicking. A big nozzle fed water by a high-pressure hose plays over a bank of gravel, washing it into a ravine or stream bed. Somewhere downstream a sluice box separates out the metal.

Placer mining seems like a simple, low-tech operation; in the case of panning, it's even rather picturesque. But placer mining can be more destructive to the landscape than the larger-scale methods of underground and open-pit mining. It is essentially a way of washing land into rivers, to the detriment of both. Nineteenth-century hydraulicking in the Sierra Nevada dumped such a load of silt into the Sacramento River that the riverbed was raised 20 feet above the surrounding valley floor. Hydraulicking is now outlawed almost everywhere in the United States except Alaska. It's still done in the gold and diamond fields of Brazil, Australia, and Southeast Asia.

UNDERGROUND MINING

By their very nature, underground mines are mainly out of sight. But that doesn't mean there's nothing to see on the surface. Large-scale mining entails moving millions of tons of earth and ore; it can't be done without some fairly conspicuous machinery. From topside you can see facilities for hauling ore, for ventilating and draining the mine, for processing the ore, and for disposing of wastes.

The Headframe and Hoist. In Hollywood's vision of the Old West, a mine is a horizontal tunnel carved into a hillside, with the portal outlined by a frame of rough-

A headframe of traditional style stands sentinel over a disused mine near Butte, Montana. The large pulleys, or sheaves, at the top of the frame are directly above a shaft that descends vertically into the mine. Wire rope threaded over the sheaves hauled up ore buckets, or skips, which dumped their load into the bin built along one side of the headframe.

hewn timbers. Such tunnel mines do exist, but far more common are shaft mines, where the entry is vertical rather than horizontal. Getting miners and equipment into the mine, and getting the ore out, requires a hoist, which generally turns out to be the biggest and most expensive piece of equipment on the site.

The part of the hoist you can see on the surface is called the headframe. At older and smaller mines it is likely to be a frame in the literal sense—that is, an open framework of wood or steel beams, 20 or 30 feet high. Most often it is an **A**-frame (or actually an *A*-frame, with one straight side and one sloped side). In some of the traditional mining districts of the western states—places like Butte, Montana, and Tonopah, Nevada—dozens of headframes in this style dot the hillsides.

Larger and newer mines still have a structure called a headframe, but it has lost its distinctive appearance. Instead of an open frame, it's an enclosed building or tower, typically sheathed in corrugated metal (which seems to be the mining engineer's favorite construction material).

At the top of a headframe are large pulleys called sheaves. Wire ropes pass over the sheaves to the hoist engine at ground level. Some newer hoists are built with the motors and the rest of the mechanism at the top of the tower.

Why is the hoist such a big deal in a mining operation? After all, it works just like an elevator in a building, except that the floors go down from ground level instead of up. There are two important differences. First, a mine hoist has to lift a much heavier load. A passenger elevator has a maximum load of a few thousand pounds; a mine hoist may be called on to lift 50,000 or even 100,000 pounds at a time. Second, even the tallest skyscrapers are only about a quarter of a mile high, but mine shafts sink more than two miles into the earth. Combining these two factors, a mine hoist might have to do a hundred times as much work as the elevator in a hotel or office building. That means bigger motors and thicker cables and a beefier framework.

The ore-carrying buckets of a hoist are called skips. Many mine hoists have two skips suspended from separate ropes but traveling side by side in the same shaft and driven by the same motor. When one skip is at the bottom of the shaft, the other is at the top; likewise, when one is headed downward, the other is coming up. The descending empty skip acts as a partial counterweight to the ascending loaded one.

For safety reasons, underground mines in the United States are required to have at least two separate access openings. Often they are identified as the main shaft (for ore haulage) and the man shaft (for miners). The hoist on the man shaft is generally smaller and built for lighter duty than the ore hoist, and it has cages instead of skips.

Ore Transport. The headframe lifts each skip high enough to dump its load of ore into the top of a storage bin, or hopper. The ore is drawn from the bottom of the bin and carried to the next stage in the mining operation—the mill, or ore-processing plant.

At small, isolated mines, ore might be hauled by trucks or rail cars. There are also ore-transport systems that look like rickety ski lifts going nowhere: a long line of stanchions supporting a cable with suspended ore buckets. Most modern conveyor systems are more like something at the supermarket checkout counter or the airport luggage carousel. A flexible belt supported by rollers runs in a channel or through an enclosed tube, carrying ore in a continuous stream. The conveyor belts can extend for miles, both underground and at the surface.

A headframe of more recent design stands above a coal mine in West Frankfort, Illinois, which operated from the 1970s until 1994. The orange objects visible just above the enclosed section are coal-carrying skips.

An aerial tramway threaded through a residential neighborhood in Trieste, Italy, carries buckets of limestone from a quarry to a steel mill.

Certain kinds of ores—the lightweight or soft or soluble ones, such as phosphate—can be transported even more efficiently by a slurry pipeline. The ore is carried along by a stream of water in a pipe a foot or two in diameter. If you can get close, you'll hear the gravelly solids scraping along the bottom.

Whatever the nature of the transport system, at the far end the ore is usually dumped on an outdoor storage heap, which can become a small mountain. The heap serves as a buffer between the mining and the milling operations. If the mine should have to shut down briefly, the mill could continue working on stockpiled ore; conversely, if the mill stops running, there's a place to put the mine's continuing output. This idea of buffering turns up again and again not only in mining but in many other industries as well, from bakeries to oil refineries.

Ventilation. Miners will tell you that what makes all the difference in the quality of life underground is the ventilation system. The air is not just for breathing; it also controls the temperature of the mine, and carries away dust and fumes from blasting. In coal mines the circulated air also disperses methane gas, which is an explosion hazard.

In the usual arrangement, air flows down the main shaft and comes back to the surface through auxiliary openings. If you ever get lost in the dark and twisty passages of an underground mine, the standard advice is to walk so that the air is blowing in your face. This should bring you back to the main shaft.

At some hard-rock mines the fans that keep the air moving are underground, so you won't see or hear them. Coal mines, however, are required to have the main fans on the surface so that they won't be put out of commission by an underground fire or explosion.

Ventilating fans come in two designs, centrifugal and axial. The centrifugal ones are snail-shaped, like a hair dryer; they suck air into the center and blow it out on a tan-

Conveyor belts cutting diagonally across the sky are one of the most characteristic sights in mining districts. These belts, moving through corrugated-metal housings to protect them from the weather, are at the Galatia coal mine in southern Illinois.

A ventilation fan at the West Frankfort, Illinois, coal mine is built to survive the effects of underground explosions. The blast doors at the end of the hornlike duct open automatically in response to overpressure.

gent at the periphery. Axial fans are more like household window fans or airplane propellers. They are enclosed in a long tapered duct, like the horn of a giant trumpet. At a large mine, the blowers might be 10 feet in diameter, driven by motors that consume even more power than the main hoist. It's not unusual for a ventilation system to pump 10 tons of air through the mine for every ton of ore produced.

Drainage. If you dig a hole in the ground, sooner or later it will fill up with water. The problem of removing this water from mines has a distinctive place in the history of technology. The first steam engines were not built to power ships or locomotives; they were designed to pump out mines in eighteenth-century Britain. Once those first steam-driven pumps were running, they allowed coal mines to be dug deeper, producing more coal to run larger steam engines, allowing still deeper mines, and so on. There you have the Industrial Revolution in a nutshell.

Pumps today, driven by electric motors, are more powerful but less conspicuous than the giant huffing steam engines of two centuries ago. The main pumps are usually out of sight deep underground—in the sump of the mine where the water collects. If seepage is too much for those pumps to handle, mine operators drill a pattern of wells all around the mine, in an attempt to lower the water table. At a phosphate mine I once visited in North Carolina, the pumps were drawing out some 60 million gallons a day, which is enough water for the city of Washington, D.C.

In the contest between pumps and seepage, the water always wins in the end, if only when the mine is closed and the power switched off. Both open pits and underground mines fill with water. Abandoned quarries become neighborhood swimming holes. The same thing is happening on a larger scale at the Berkeley Pit, an abandoned copper mine on the edge (literally!) of Butte, Montana. The pumps were shut

Flooding of a mine is inevitable once the pumps are shut down. Here the rising waters are gradually submerging an ore-conveyor system in the Hill Annex Mine, a former open-pit iron mine (now a state park) in Calumet, Minnesota.

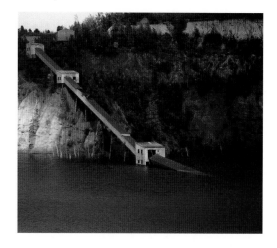

HOW LOW CAN YOU GO?

All of the world's deepest mines are in one place: South Africa, where veins of gold have lured miners to depths of more than 13,000 feet. That means they are lower than the bottom of the ocean—the sea floor of the nearby Atlantic is at about 10,000 feet.

Perhaps it should come as no surprise that conditions down there are pretty hellish. The first problem is heat. As a rule of thumb, the temperature inside the earth rises 1 degree Fahrenheit for every 300 feet of depth. Mere fans are not enough to make such a place habitable; the deep levels of the mine have to be air-conditioned, using some of the world's biggest chiller plants.

The pressure exerted by more than two miles of rock overburden also causes trouble. When a section of the tunnel fails, it's not just that the roof falls in. The surrounding rock explodes violently in an event that is likened to a miniature earthquake. These rock bursts, which come with little warning, are the main cause of death for workers in the deep mines.

Just getting down into the mine and then back up again is a major logistical challenge. No hoist can reach all the way to the bottom of a 13,000-foot shaft. Such a length of wire rope would break under its own weight, even with nothing else to lift. Thus, workers have to make the trip in stages, using multiple hoists, and ore has to be hauled up in the same way. The time and energy spent on this vertical commute has an impact on profits. Pumping water from the bottom of the mine is also expensive.

Nevertheless, there is talk of going even deeper in South Africa. The Western Ultra Deep Levels Mine (known as "Wuddles") has announced a goal of five kilometers, or more than 16,000 feet.

off in 1982, and now the mine is a shimmering lake. But it's hardly a swimming hole. The lake is toxic—so heavily contaminated with metals that it's become a resource to be exploited. A plant above the pit extracts copper from the water.

What's Going on Down There? An underground mine is like an ant colony: the little mound at the surface gives only a faint hint of the sprawling network of passages below.

The process of hollowing out the earth is done in a sequence of simple operations, repeated many times. First comes drilling, boring a series of holes 10 or 15 feet deep. The drilling was once done by a two-man crew: one man held a long rod of drilling steel and the other hit it with a sledgehammer. These days, pneumatic and hydraulic drills have replaced muscle power; the usual tool is a machine called a drilling jumbo, an octopus mounted on wheels or crawler treads that drills several holes at once.

The drilled holes are loaded with an explosive, then detonators are wired up and the rock is blasted with the ritual cry "Fire in the hole!" Once the smoke clears, the next operation is "mucking out" the debris. This work too was once done by manual labor, using shovels (miners called them muck sticks) to load the shattered rock into mule-drawn carts. Now both the loading and the hauling have been mechanized.

After the mucking out, the next priority is shoring up the walls and roof to prevent cave-ins and to hold back loose rock that might fall on a worker. (Miners talk about "getting slabbed.") Timbers were the traditional choice for this job, but now there is something called a roof bolt. It works like the expansion screw that helps hold up a heavy picture frame on a plaster wall, except that the roof bolt is 6 to 10 feet long.

More goes on underground than just digging the hole itself. There are pump rooms and fan houses, powder magazines, supply depots, machine shops, and electric-

power switching stations. Once upon a time there were barns and stables for the hundreds of animals that pulled ore cars, spending their entire lives in the dark. The modern counterparts of the stables are underground maintenance yards for mining equipment, much of which is too large to bring to the surface without disassembly. Sometimes there's a lunchroom. Everywhere there's graffiti. As for toilet facilities—yes, modern mines have those too.

SURFACE MINING

Underground mining has the advantage of surgical precision: You dig out what you want and leave everything else in place. Surface mining, in contrast, is a brute-force affair. You have to rip open the earth and shove aside vast quantities of worthless rock before you even reach the mineral you're after. But surface mines also have a not-so-

This terraced rainbow landscape is the Santa Rita mine near Silver City in southwestern New Mexico. Copper deposits in this area were known to the Apache and other early inhabitants, and in the nineteenth century the town of Santa Rita grew up to support mining activity. But later the mine completely engulfed and obliterated the town for which it was named. The mine is currently owned by the Phelps Dodge Corporation.

Blasting shatters thousands of tons of rock, sends up pillars of smoke, and sets off a minor avalanche at the Bingham Canyon mine in Utah, the world's largest copper mine. The power shovel in the foreground is unoccupied during the blast; all workers leave the pit before the explosion is detonated. The photographs on the opposite page were also made at Bingham Canyon.

secret weapon. For the immense earthmoving job, mining companies bring in immense machines, the largest on the planet. This economy of scale seems to be working in their favor: In the United States almost 85 percent of all minerals are recovered by surface mining.

There are two main types of surface mines: open pit and open cast, the latter being better known as a strip mine. The open-pit technique works best for mineral veins or ore bodies that dive deep into the earth through hard rock; most iron and copper ores are mined this way. Open-cast mining is most suitable for "bedded" sedimentary deposits that extend horizontally over large areas but lie not too far under the surface. Coal seams usually fit this description, and so do deposits of certain other soft minerals such as phosphates.

Open-Pit Mining. An open-pit mine is remarkable most of all for what is *not* there. At the Hull-Rust-Mahoning iron mine in Hibbing, Minnesota, for example, what is not there is 35 million cubic yards of rock and ore. The chasm cut into the earth is three miles long and a mile wide—so big it takes three names to cover it. And yet an open-pit mine is not just a void, a hole in the ground. It has a structure, even an architecture. It has to be designed so that the walls don't collapse, so that the maximum amount of ore can be extracted with the least amount of waste, and—the most challenging part—so that enormous ore-hauling trucks and other heavy equipment can drive into and out of the pit.

The basic design element in the architecture of an open-pit mine is the terrace, or bench. The wall of the pit is not a sheer cliff, nor is it a slope descending smoothly from the rim to the floor; instead, it is a giant stairway or cascade, formed of benches roughly 50 feet high. The nearly vertical walls between the benches are called faces. The benches help to stabilize the wall against landslides and also provide working platforms for miners and equipment. When the mine is to be expanded, work begins on the topmost bench, which is widened by excavating the face above it (a maneuver called a pushback). Then the next bench below is likewise pushed back, and then the next deepest, continuing all the way to the pit floor, where a new bench may need to be created. Because the benches generally follow lines of constant elevation, they are equivalent to contour lines on a topographic map. If you know the height of a typical bench, you can estimate the total depth of the pit just by counting benches.

The pushback of a bench or terrace, removing a complete ring of rock and ore, is nearing completion in the photograph above. One last isthmus of the bench remains to be mined; once it is gone, activity will shift to the next lower bench, which will be pushed back in turn. The tall machines on the isthmus are drilling rigs, boring holes 60 feet deep for the placement of explosives. As an indicator of scale, note that the orange object parked near one of the drilling rigs is a school bus, used for bringing crews to and from the work site. The haul road crossing the isthmus diagonally will have to be relocated.

In the aftermath of a blast *(left)*, a huge electrically powered shovel scoops up the broken rock and loads it into a series of haul trucks that carry 200 tons each. The trellis-like frame near the center of the image holds aloft the power cable for the shovel so that the trucks won't drive over it.

IT'S A BLAST

The art of earthmoving has been transformed not only by bigger trucks and shovels but also by bigger booms. In chemical explosives, black powder gave way to dynamite a century ago, and now dynamite has been replaced by materials called blasting agents, which are safer, easier to handle, and much cheaper. (But the miners still tend to call the stuff "powder.")

The most important blasting agent is ammonium nitrate, which is also a major commodity in agriculture, where it serves as a fertilizer. For explosive use, ammonium nitrate is manufactured in the form of prills, or pellets, to which a small amount of diesel fuel is added. This mixture is called ANFO, for ammonium nitrate–fuel oil. Small mines and quarries buy ANFO in 50-pound sacks; larger operations receive it in bulk and pump it into bore holes. Unfortunately, ANFO has also become the explosive of choice for terrorist bombers, a problem the explosives industry is now being asked to address.

One reason ANFO is relatively safe is that it takes a fairly serious explosion to set it off—not just a blasting cap but also a primer, or booster. In a mine blast, the charges are timed to go off in a specific sequence, a few milliseconds apart. The staggered timing helps ensure that the rock is thoroughly fractured, but without throwing boulders halfway across the mine.

Most blasting caps for mine use are triggered electrically. Signs warn motorists to turn off two-way radios because a nearby transmission might produce enough stray current in a blasting wire to cause a premature detonation. The wire is never reused. Thus, if you are ever walking through an old mine or quarry, one thing you are likely to stumble over (perhaps literally) is a tangle of abandoned blasting wire.

In the movies, explosives are set off by ramming home the plunger in a device that looks something like a bicycle pump. This "blasting machine" is an electric generator, turned by the movement of the plunger. The modern version is smaller, and it is triggered by merely pushing a couple of buttons. What's inside is not a generator but a capacitor, which is slowly charged from a battery and then rapidly discharged through the blasting caps.

Complementing the upside-down-wedding-cake structure of benches is a ribbon of haulage roads that cross the contour lines obliquely. If you think of the benches as the levels of a vast underground parking deck, then the haulage roads are the ramps that connect the levels. In some mines a single road spirals either clockwise or counterclockwise all the way from the rim to the floor, so that a vehicle making the complete trip will turn several full circles before reaching the bottom. Other mines have a road with many switchbacks, or zigzags, built mainly along one wall of the pit; this plan is usually chosen when one wall of the mine is at the edge of the ore body and will not be excavated further. If there's room, the mine will have separate roadways for empty trucks descending into the pit and laden ones climbing out.

The sequence of basic mining operations in an open pit is not much different from that at the working face of an underground mine. Again the main steps are to drill, to blast, and to remove the rubble. But each of these operations is carried out on a much larger scale. The drill holes are a foot in diameter instead of two inches, and 60 feet deep instead of 10 feet. The quantity of explosives loaded into the holes scales up accordingly, and so does the volume of rock and ore broken up by each blast. Gigantic power shovels and fleets of front-end loaders scoop up the debris. This material is hauled out of the mine by trucks that carry a load of 400,000 pounds, which is 5 or 10 times the capacity of trucks that travel the public roadways.

The world's largest open-pit mine is the Bingham Canyon copper mine, 30 miles southwest of Salt Lake City. The project was begun in 1903 by D. E. Jackling, a legendary mining engineer who is generally credited with the invention of modern large-scale surface mining. Over the years more than 5 billion tons of material have

been excavated at Bingham Canyon, leaving a pit two-and-a-half miles across at the rim and half a mile deep. Haulage roads that climb both the north and the south walls of the mine are used mainly for transporting waste rock to dump heaps in adjacent gorges. The ore exits via three large tunnels dug through the surrounding mountains. Two of the tunnels have tracks for standard rail cars; the third has a conveyor belt carrying ore directly to a mill in the nearby town of Copperton. Along the highest wall of the mine there are more than 50 benches at vertical intervals of 50 feet. Excavating a complete sequence of pushbacks, enlarging the mine from the top down, takes almost a decade. The mine has a visitor center and an overlook on the rim.

Strip Mining. In open-pit mining, nothing is ever put back in the pit: ore goes to the mill for refining, and waste rock is piled up in towering heaps and embankments whose skyward growth matches the downward progress of the pit itself. In strip min-

Strip mining, or open-cast mining, creates plateaus rather than pits. Here the artificial plateau is carved out of the hills of central Pennsylvania at a coal mine near the town of Ebensburg. The photograph was taken from atop a great heap of soil that had been set aside for later use in restoring in the mine site.

ing, on the other hand, the waste dug out of each new section of a mine is used to fill an earlier section. Only the ore (or coal) is permanently removed from the site; the waste material is retained and eventually returned to the mined terrain. In the simplest case the mining is done in long parallel strips. As the overburden is peeled off one strip, it is spread directly onto the adjacent worked-out strip.

This built-in recycling process might be expected to make strip-mining technology a favorite of environmentalists, but it has quite the opposite reputation. The trouble is that strip mining rips open truly vast tracts of countryside. Even the largest open-pit mines cover only a few square miles, but an ambitious strip mine could roll over 10 or even 100 times as much territory in the same period. Furthermore, although the old workings are filled in and much effort is put into reclamation, the land may not be "natural" again for many years.

By the 1970s, strip mining was in such ill repute that mine operators were finally moved to do something about it. What they did was change the name of the technique from strip mining to open-cast mining. This was a very helpful measure, in that it ended arguments over the original meaning of the term *strip*. (Is it a noun referring to the long strips of land or a verb describing the stripping away of overburden?) Unfortunately, the change has not entirely caught on. The latest edition of *The SME Mining Engineering Handbook* has a chapter titled "Open Cast (Strip) Mining."

A dragline is the largest of the machines used to strip away the overburden and mine the ore layer at an open-cast mine. The bucket dangling on ropes from the end of the boom is cast outward like a fishing lure and then reeled in toward the operator, scooping up earth along the way. A bucketload for this particular dragline, one of the world's largest, is 220 cubic yards. Note the school bus, which would easily fit in the bucket. The dragline is at a phosphate mine in Aurora, North Carolina. At the time the photograph was made, the Aurora mine was operated by Texasgulf Inc.; it is now owned by PotashCorp.

Strip-mining process at the Aurora mine calls for excavating 100 feet of sandy overburden in order to reach a layer of phosphate-rich ore 30 or 40 feet thick. The overburden removed from one cell or strip of the mine is used to fill in mined-out areas. The machinery in the foreground is part of a conveyor-belt system carrying overburden from an active strip to an area being reclaimed.

Whatever you choose to call the technique, and whether or not you want a strip mine in your own backyard, you cannot help being impressed by the grandeur of the operations. It is like plowing the earth on a gargantuan scale, turning over furrows a hundred yards wide, a mile long, and a hundred feet deep.

The basic steps in strip mining are somewhat different from those in hard-rock, open-pit mining. In many cases no blasting is done; the overburden is soft enough to be chewed up by earthmoving machinery without the help of explosives. When a new parcel is being opened up, the first stage is "grubbing," which is the removal of topsoil and vegetation. Then the stripping of overburden begins, along with the simultaneous casting of spoil into the worked-out adjacent strip. Often the stripping is done in layers, corresponding to different geological strata. For example, a layer of gravel or clay might be stripped off with a machine called a bucket-wheel excavator, which can handle large volumes of soft material. Harder deposits call for a stripping shovel or a dragline. The final operation is the mining of the coal or mineral layer. Depending on the size of the mine, completing a single strip can take a few weeks or a few years.

While excavating is under way in one strip, reclamation gets started in the previous ones. At most American strip mines the operator is required to restore the land to its "approximate original contour," which means that a great deal of leveling and grading needs to be done with a fleet of bulldozers and scrapers. The spoil, having been dredged up from deep underground, is biologically sterile, and so topsoil has to be laid down. (Most of it is reclaimed from the grubbing operation.) Mulch and fertilizer are also needed, followed by seeding with grasses or other cover crops.

Like other capital-intensive industries, strip mines usually operate 24 hours a day and 7 days a week. A mine at night is a spectacle. Banks of floodlights cast a harsh

glare on the working areas but leave other corners in eerie gloom. The tall booms of the shovels and draglines are also festooned with lights, and they swoop through the dark sky like the twirling neon masts of some maniacal amusement-park ride; but there is no carnival music or squealing of children, just the rhythmic thud and roar of machinery.

Surface-Mining Machinery. Looking at a dragline or a stripping shovel from a distance, you might be fooled by the scale of the machine. In basic form it looks much like equipment you see at any construction site. But then you notice a car parked nearby or a human figure climbing up to the control cab, and suddenly the true size of the machine sinks in. The control cab is an inconspicuous appendage at the front of the "house," the enclosure that forms the main body of the machine. The house is, in fact, the size of a small hotel—almost 150 feet long and 50 feet high. And the house in turn is dwarfed by the boom and its rigging, which extend a full football field out in front. Draglines are the largest self-propelled devices on land. They are big enough to be named individually, like ships.

At the end of the dragline's long boom, a bucket swings from a hoist rope, and an additional drag rope trails under the boom from the bucket to a point near the control cab. What the operator does with this apparatus is just like casting a fishing line into the surf and then reeling it in again. By adroit manipulation of the boom and the ropes, the bucket is swung forward and then allowed to plunge into the earth. Reeling in the drag rope hauls the bucket back toward the operator, scraping out ore or overburden as it goes. When the bucket is full, it is lifted by the hoist rope, then the entire dragline pivots horizontally, swinging the boom out over the spoil pile,

A VISIT TO CHIEF IRONSIDE

At the Falkirk lignite mine in North Dakota I had a chance to climb aboard a dragline called Chief Ironside, which has a 120-cubic-yard bucket and a boom 215 feet long. You enter by first climbing a few stairs onto the upper surface of one of the giant "shoes" on which the machine shuffles along, then up two more flights of stairs to the main deck. A doorway leads inside the house, which turns out to be a dim cathedral-like space filled with the hum of electrical machinery and the whoosh of ventilating fans. Half a dozen large motors turn the drums that reel in the hoist rope and the drag rope, slew the house on its pivoting axis, and propel the shoes. Along the walls, tall steel cabinets

hold switchgear to control the motors. Also conspicuous are the sheaves and reels for the two-inch-thick wire ropes. But even with all this heavy equipment, the inside of the house is still mostly empty space.

From the main area of the house you go forward through a passage lined with lockers for the crew, then through a galley kitchen (with a small fridge and microwave oven) and out into the glass-walled control cab. There the operator is enthroned 30 feet above the ground, with an unobstructed view of the bucket and the surrounding mine. The cab is heated in winter and air-conditioned in summer, and the audio system is first-rate. The operator controls the entire 13-

million-pound machine with two joysticks, positioned just in front of the armrests of the chair. The knobby handles of the joysticks were padded with slit-open tennis balls, just like trailer hitches.

The dragline Chief Ironside, which digs coal at the Falkirk mine in central North Dakota, walks backward to take up a new position. The machine walks by planting its two feet, or shoes, and then driving them forward (toward the boom) so that the rest of the body slides backward. The dragline is electrically powered through the high-voltage cable visible in the foreground. The small tractor tends to the power cord, making sure the walking dragline doesn't step on it.

where the bucket is tipped to dump the load. Under the control of a skilled operator, this immense machine can be made to move with surprising grace and agility.

The largest draglines have bucket capacities of 220 cubic yards, which is enough to hold a couple of Greyhound buses. A bucketful of dense rock could weigh a million pounds. The weight of the dragline itself can approach 30 million pounds. This is too much to be supported by wheels or by the kind of crawler treads found on bulldozers and other earthmoving machinery. Instead, when the dragline wants to move, it gets up on its feet and walks! Or perhaps I should say it waddles.

The mode of locomotion is the mechanical equivalent of sitting on the floor, planting your feet, and sliding backward on your butt. The dragline's two feet, which are actually called shoes, are broad steel platforms on either side of the machine's base. They are lifted in unison by a simple camshaft mechanism, allowing the massive body of the dragline to settle onto the earth between them. Then the shoes are moved a few feet backward, set down again with great force, and driven forward (that is, toward the boom). In response to this force, the body of the machine is pushed or dragged backward over the ground, leaving a dinosaur trail with a broad central skid mark and flanking pairs of footprints. Ahead of the slowly moving behemoth, water tankers soak the ground to lubricate the path. Even so, friction can heat the earth to the steaming point. The machine has only this one gear—reverse. To go the other way, it has to turn around and continue scooting backward.

Big draglines are powered by electricity, supplied through a cable the size of a fire hose that trails behind the machine like the cord of a vacuum cleaner. There's usual-

A spare bucket for the Falkirk dragline holds 120 cubic yards. The teeth along the cutting edge have to be replaced regularly. The metal plates welded to the back and side walls are coated with Teflon, part of a strategy for dealing with sticky materials sometimes encountered in mining lignite (a low-grade coal).

A shovel digs up-and-away, unlike a dragline's down-and-toward cycle of motion. The shovel seen here, at the Cottage Grove coal mine in southern Illinois, is scooping up 20 or 30 tons of overburden.

ly a small tractor running around near the base of the big machine to keep the cord out of the way and prevent the dragline from stepping on it. Needless to say, the cord is not plugged into the nearest wall outlet. The operating voltage is anywhere from 10,000 to 25,000 volts. Peak power demand can be close to 50,000 horsepower, or roughly 35,000 kilowatts.

Inside the house is a motor-generator set: a large electric motor that runs on the alternating current (AC) supplied by the utility company drives a direct-current (DC) generator. The DC power, which is easier to control than AC, runs the various motors that move the bucket, the boom, and the walking mechanism.

Knowing a dragline from a shovel is essential if you aspire to join the mining cognoscenti. (And whatever you do, don't call either one of them a crane!) A dragline works the way you eat ice cream, spooning down and toward you. A shovel works the way Emily Post would have you eat soup, spooning away and up. Instead of a bucket, a shovel has a "dipper"; and whereas the dragline's bucket swings freely from a rope, the shovel's dipper is rigidly mounted to a dipper stick, or handle, which in turn is hinged to the boom. After scooping up a load, the shovel pivots to swing the dipper over the spoil pile; then the operator releases a door in the bottom of the dipper, allowing the load to fall.

Shovels are smaller than draglines (the maximum dipper size is some 180 cubic yards), and they have a shorter reach. On the other hand, they have a faster working cycle, they can be controlled more precisely, and they handle somewhat harder materials. Another difference is that a dragline works from above, constantly retreating from the hole it is digging, whereas a shovel works from below, advancing into the excavated face. Like draglines, the biggest shovels are electrically powered, but smaller ones have diesel engines. Most shovels are light enough to be mounted on crawler treads.

The bucket-wheel excavator is a different kind of digging tool. Instead of taking discrete bites of soil or rock the way draglines and shovels do, it continuously gnaws through the landscape. The business end of the machine is the bucket wheel itself, which is 20 to 40 feet in diameter and has buckets protruding like sawteeth around its perimeter. The wheel is mounted on a boom that pivots up and down, slews from side to side, and in some cases extends and retracts. Material shaved out of the embankment by the spinning wheel is carried through the boom on a conveyor belt and then onto a discharge chute or another conveyor. The biggest excavators can carve out well over 10,000 cubic yards per hour.

MINE WASTES

When you drive into a mining district, what you are likely to notice first is not the mine itself or any of its machinery, but instead the towering heaps of waste rock that can overshadow whole towns. Ironically, as mining technology improves, the waste-disposal problem gets worse and worse. To see why, consider the history of the

American copper industry. Nineteenth-century copper mines exploited only the richest ores, which were 30 percent or even 60 percent metal. By 1915, miners were digging ores with only 5 percent copper, and by the 1940s just 2 percent; today the average grade is about 0.5 percent. This shift to lower-grade sources has had a dramatic impact on the economics and technology of mining, and also on the waste budget. To produce 1 ton of copper from 50 percent ore, you have to mine 2 tons of ore, and you leave 1 ton of waste. With 0.5 percent ore, you have to mine 200 tons, and you leave 199 tons of waste. Thus, even though copper production has been slowing in recent decades, the production of copper-mining wastes has been booming like never before.

Many of the wastes discarded during earlier generations of mining are richer than any of the ores coming out of the ground today. The wastes have therefore become a valuable resource for secondary mining, particularly by various methods known as solution mining.

Just as the Eskimo have many words for snow, miners have an elaborately ramified vocabulary for mine wastes. Gangue is the generic term for everything in the ore other than the desired mineral. In the coal fields, the slate and other rock separated from the coal is dumped in heaps called gob piles or culm banks. Lead and zinc mining leaves a dark, sandy material called chat, which forms steep, conical heaps. Slag from smelters, which solidifies into a glassy or cindery material, also forms steep-

Copper-mining wastes have added several new flat-topped mountains to the landscape near the Tyrone mine in southwestern New Mexico. In the photograph above, two trucks are toiling up the spiral haul road to dump more waste rock at the summit. The buildings in the foreground house the mine's ore-milling machinery. Below, a truck dumps its load of overburden at the Cottage Grove, Illinois, coal mine.

Heap leaching of copper ore, seen from overhead, looks almost like some kind of irrigated agriculture. But the acidic water that floods these furrowed fields is not suitable for growing crops; it is meant to percolate through the ore heap, dissolving some of the copper minerals so that the "pregnant liquor" can be collected in drainage ditches at the base. The leaching operation is at the Tyrone mine in New Mexico.

sided heaps. One of the wastes from phosphate mining is gypsum, a gray or white powder whose heaps are called gyp stacks. And when spoils from placer mining wash up downstream, they are called slickens.

The wastes from the milling of ore are tailings—silts and sands suspended in a large volume of dirty water. The usual strategy for getting rid of tailings is to build an earthen dam, impounding the slurry in a tailings pond. After the water has seeped away or evaporated, the solids remain behind in giant flat-topped embankments, sometimes gooey, sometimes powdery, sometimes gray, sometimes white.

Another kind of mine waste is the water that seeps through both the mine itself and the waste banks. This drainage water can become an acid strong enough to sterilize creeks and streams. The source of the problem is sulfur. Rocks brought up from deep in the earth are often rich in sulfide minerals such as iron pyrite, which start to oxidize as soon as they hit the air. The resulting sulfur oxides dissolve in water to form various acids, including sulfuric acid. The acid then leaches other substances out of the waste rock, including certain toxic metals such as cadmium and arsenic, so that

the fluid oozing out of the bottom of the hill is a rather unpleasant brew. Mines are required to remove both the acidity and the metals. The water-treatment plant looks something like a small version of a municipal sewage plant. To neutralize the acid, lime is added to the water in a mixing basin, then the water is aerated and pumped to a settling pond, or thickener, where the contaminants settle out as a sludge.

Can a mined-out landscape ever be restored to natural beauty? There are many places in the desert Southwest where mining stopped decades ago, but the landscape is still dominated by spoil piles, old pits, and rusting machinery, which leave the impression that the earth's scars will never heal. But there are also counterexamples. Consider Morris County, New Jersey, just 30 miles west of New York City. Throughout the eighteenth and nineteenth centuries this area was a thriving iron-mining district. By the 1870s there were more than 200 operating mines, producing 600,000 tons of ore per year—which was a lot in those days. The landscape was dotted with the smokestacks of ore mills and crisscrossed by the tracks of a dozen ore-hauling railroads. But if you visit the area now, you need to know where to look and what to look for if you want to find any evidence of the former mining industry. Most of the county is given over to very pricy suburban housing and corporate headquarters. There are lush lawns, woodlots, horse farms, and shopping malls, but nary a smokestack. The only telltale remnants of the mining era are place names, such as the towns of Ironia and Ferromonte.

SOLUTION MINING

The latest fad in the metals-mining industry is solution mining, a low-cost and low-labor scheme for recovering metal from low-grade ores. You might think of it as a trick for putting to good use the problem of mine drainage waters that become contaminated with metals. Some 30 percent of the copper mined in the United States, as well as 35 percent of the gold and 75 percent of the uranium, comes from solution-mining operations of various kinds.

One form of solution mining is the heap leaching of copper ore. The crushed ore is piled up into flat-topped mounds, then water and sulfuric acid are sprayed or dripped onto the top of the heap. The acidic liquid percolates through the mound, dissolving various copper compounds and forming copper sulfate, which is carried off in solution. The fluid that seeps out of the bottom of the heap is called pregnant liquor or pregnant leach solution (but it is *not* full of bloodsucking organisms about to give birth!). The pregnant fluid is run through a processing plant to extract the copper, then it passes into another pond, as barren liquor, before being pumped up onto the heap again for another cycle of leaching. The process goes on for several years.

With some copper ores, the sulfuric acid isn't needed; water alone will start the leaching process. In these mines the conversion to soluble sulfates is done by bacteria (mainly a species called *Thiobacillus ferrooxidans*) that grow on sulfides within the

A haul truck backs up to the perilous edge of a leach heap and dumps another load of low-grade ore. The scene is the Santa Rita mine in New Mexico.

Santa Rita leaching operation relies on sprinkler heads to spray the leach solution, producing a distinctive pattern of round, wet spots on the surface of the leach heap.

moist heap. It's as if the leach heap were a compost pile, where all the real work is done by microorganisms. The job of those tending the heap is merely to create a suitable habitat for the bacteria. The biologically aided process is important not only for leaching newly mined ore but also for recovering any metal remaining in gangue and mill tailings from earlier generations of mining.

The heap leaching of gold and silver is done on a smaller scale and on a quicker timetable than the solution mining of copper. Often a heap is built and then leached out within a few months. Moreover, the active chemical agent in gold and silver leaching is nothing so innocuous as mere sulfuric acid; it is sodium or potassium cyanide (yes, the stuff once used in the gas chamber at San Quentin). The presence of cyanides makes environmental protection a sticky problem in precious-metals leaching. The heap is built on an impervious pad to keep the leaching solutions out of the water table, and some provision has to be made for handling excess runoff during storms or the spring snowmelt. You may also notice elaborate fencing, which is meant not only to keep out people but also to protect wildlife.

A typical leach dump is recognizable (and distinguishable from a mere waste pile) by the plastic pipes climbing the sides and the sprinkler heads at the top. At a gold mine the processing plant will include tall canisters full of activated charcoal, where the metal is extracted from the solution. At copper mines you may notice big rusty piles of scrap iron, on which the copper is precipitated.

For uranium mining, the leaching is done without even digging the ore out of the ground. A leaching fluid is pumped down a well into the ore body, and the pregnant liquor is recovered from another well. Another mineral extracted by bore-hole mining is sulfur. In this process, invented about 1900 by Herman Frasch in southern

Louisiana, hot water is injected through a well into the sulfur-bearing layer. Molten sulfur (a beautiful tawny-purple liquid) is carried from the well in steam-heated pipes, and sprayed onto storage heaps. The solidified sulfur is a bright yellow powder that looks like some exotic spice but doesn't smell as good.

COAL MINING

Coal mining is an industry unto itself, roughly the same size as all other kinds of mining put together. It also has its own distinctive history and culture. Indeed, there are several subcultures within the American coal industry: those of the old anthracite belt in eastern Pennsylvania, the bituminous region that runs from the Appalachians west through Ohio into Illinois, and the newer strip-mining districts farther west in Wyoming and Montana.

If this book had been written a century ago, it would have described a whole infrastructure for distributing coal to homes and businesses—coal yards, coal wagons, coal bins, coal chutes, and so on. Today, far more coal comes out of the ground than ever before, and yet most of us never see a lump of the stuff. More than three-fourths of all coal goes to electric utilities, and most of the rest is sold to other big customers such as the iron-and-steel industry.

Underground Coal Mining. North American coal mines are seldom as deep as metal mines; furthermore, coal is much softer than the rock that metal miners have to batter through. But coal mining has challenges of its own. The softer ground is more vulnerable to roof falls. Worse, the coal is flammable, so that the very ground

The extraction plant at the Mesquite mine in southern California recovers gold from fluids that seep through an ore pile. In effect, the extraction process is the reverse of the leaching process. In the leach heap, liquids percolate through the ore, dissolving gold-bearing minerals. In the extraction plant, the pregnant liquor percolates through several beds of activated charcoal (contained in the large red vessels inside the framework), leaving behind the gold.

A heap of gold ore is seen in profile at the Mesquite mine. The black plastic pipe laid in a groove at the corner of the pile carries leaching fluids pumped to the top of the heap. The ditch extending along the base of the heap collects pregnant liquor. The Mesquite mine is near the town of Glamis in the desert environment of California's Imperial Valley; the evening "mist" in the distance is actually dust in the atmosphere, obscuring the base of the Chocolate Mountains.

can catch fire, and the methane gas that seeps out of the coal can explode. Coal dust is also an explosive, in addition to causing respiratory disease ("black lung").

Today, most coal from underground mines is excavated by continuous-mining machines that chew up the seam and load the coal without any need for the traditional cycle of drilling, blasting, and mucking out. The continuous miners are squat vehicles with vicious-looking augers or cutting wheels at the front end, and a pair of crab-claw arms that gather up the broken coal onto a conveyor belt. They are built low to the ground—this is a characteristic feature of all coal-mining machinery—so as to fit through passageways that may be no more than three feet high.

The traditional method of underground mining in American coal mines is called the room-and-pillar system. The rooms are the open areas where coal has been excavated; the pillars are blocks of coal left standing to support the overburden. But the term *pillar* can be misleading: Don't think of the slender, fluted columns that support the roof of a Greek temple. The pillars of a coal mine are thicker than they are high, and roughly as big as the open rooms that surround them.

An alternative to the room-and-pillar system, called long-wall mining, has been steadily gaining popularity in the United States. (It has long been the dominant technique in Europe.) A long-wall mining machine has a cutting head that moves along an exposed panel of coal that may be up to a mile long, continuously shaving off a thin layer. The machinery is protected by hydraulic roof shields that temporarily hold up the overburden. When a layer of coal has been shaved from the full length of the wall, the shields are moved forward, allowing the roof behind to cave in.

Above ground, the biggest and most distinctive structure at a coal mine is a building known traditionally as a breaker or a tipple. The nineteenth-century exemplars—

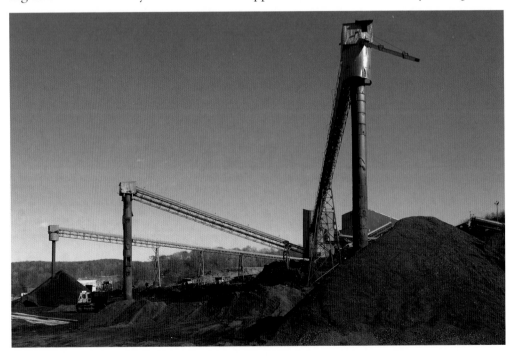

Coal ready for shipment stacks up in conical heaps at a minehead in Sonman, Pennsylvania, near Altoona. The coal is carried upward on the inclined conveyor belts, then pours down the vertical tubes and flows out of "windows" cut into these tubes. Other minerals might simply be allowed to fall off the end of an elevated conveyor belt, but the vertical tubes or towers are used with coal as a way of controlling dust.

a few of which still stand, as haunting ruins—were timber structures five or six stories high. Mine cars filled with coal were winched up a long incline, at the top of which each car tipped over to dump its load (hence the term *tipple*). The coal tumbled down a series of sloping wooden troughs, where workers picked out slate and other kinds of waste rock that inevitably get mixed up in the mined coal. For a long time these workers were called breaker boys, and with good reason: they were children as young as six.

Toward the bottom of an old breaker or tipple, the coal passed over a series of grates and screens that classified the pieces according to size. In descending sequence, the standard sizes were known as lump, stoker, nut, and egg; the still-smaller pebbles and powders were called slack. Coal in each of these sizes was intended for a different market and would fall into separate storage bins, for eventual loading onto rail cars. Sorting by size is seldom needed today, since all the coal goes to one customer.

A modern coal mine still has a tall building that looks a little like a breaker or tipple, and some people still call it by those names (although the approved term these days is coal-preparation plant). Removing slate is still the main function of the plant, but the way it is done has changed totally. Gone are the breaker boys and their Dickensian labors. The new process is called coal cleaning or coal washing, but neither name does it justice. The coal isn't merely washed; it is taken apart and put back together. Modern coal is not so much a raw material as a manufactured product.

Different kinds of coal call for different washing technologies, and even within a single plant the various sizes of lumps and particles have to be treated differently. Most of the separation methods exploit the fact that slate is somewhat heavier than coal. For example, whirling a watery mixture of materials in a funnel-shaped device called a cyclone forces the denser slate to the outside and the lighter coal to the interior. The same principle is at work in a spiral concentrator, which is essentially a helical water slide, where denser particles tend to accumulate toward the outer rim of a curving trough. Spirals are used in the washing of the smallest particles of coal.

Another technique springs from the notion that if you had a fluid of just the right density, then the coal would float in it while the slate would sink. But what fluid has the right density? Water obviously won't work; both coal and slate sink in water. The answer is to start with water and add large quantities of a heavy powder such as the iron ore magnetite, forming a slurry that is much denser than water alone. When the raw coal is dumped into a shallow, agitated tank filled with the slurry, coal can be skimmed off the surface while slate is raked up off the bottom. But a further problem remains: Having used a heavy powder to separate the coal from the slate, how do you separate the coal from the heavy powder? That's the reason for choosing magnetite as the powder: as the name suggests, it is a magnetic material. A slowly turning, magnetized drum, half-immersed in the slurry, pulls the magnetite out but leaves the coal behind.

Yet another coal-cleaning technique is froth flotation, which is also used in processing many metal ores. Bubbles rising through a frothy slurry attach themselves to

Three generations of the coalfield institution known as a breaker or a tipple show 150 years of evolution. The structure at the top is a replica of a nineteenth-century breaker, built as a movie prop (and now falling into ruins); in the middle is a breaker built in the mid-twentieth century, which closed in the 1970s; at the bottom is a modern coal-preparation plant, still in operation. All three structures are within a few miles of one another near Hazleton, Pennsylvania.

the coal, which is thereby lifted to the top of a vat; the bubbles have less affinity for the slate, which accordingly sinks.

All of these washing methods are wet processes; from the moment the coal enters the plant, it is immersed in vats of watery solutions and transported as a slurry or sludge in troughs and pipelines. But before the finished product ships to the customer, the water has to be wrung out because moist coal doesn't burn well. The drying is done in a big cylindrical centrifuge, which works like a clothes washer in the spin cycle.

THE HUMAN (AND THE INHUMAN) SIDE OF MINING

Mining is more than just a technology; it's a way of life. If you spend much time in mining communities, you begin to notice traits that set them apart from other places.

Nineteenth-century mining towns were marked by great contrasts in wealth. The mansions of the mine owners on the brow of the hill overlooked the shanties of the miners lined up near the mill or the coal breaker in the valley. Today the mine owners are even wealthier— they are multinational corporations—but they have long since left town. The miners are somewhat better off too, but fewer in number.

Many mining communities started out as company towns, where the mine owners built the housing, the stores, occasionally even the schools and churches. They named everything as well, sometimes with a singular lack of imagination. In southwestern Pennsylvania the Berwind-White Coal Mining Company simply numbered its towns from 1 through 40. In some places all the houses and mine buildings were painted the same color. (In the Southwest, "Phelps Dodge green" was a popular choice.) By now the mines have sold off most of this real estate, but company houses are still recognizable under the new vinyl siding. (In the photograph, coal-mine wastes looms over houses and former mine offices in Portage, Pennsylvania.)

In contrast to architectural uniformity, there is ethnic diversity. A stroll through the graveyards of any town in the eastern coal fields will reveal an abundance of Welsh, Irish, German, and Italian names (mostly in *separate* graveyards). My grandmother, who lived in the mining community of Hazleton, Pennsylvania, used to speak often of Hungarians; only years later did I realize that this was her generic term for all Eastern Europeans, applied indiscriminately to immigrants from Poland, Czechoslovakia, Romania, and Yugoslavia as well as Hungary.

The California gold fields were worked by many Chinese miners, who have left their own legacy on the land. African-Americans also worked underground, although until recently on a very tenuous basis. A common pattern was to

hire black miners as strike-breakers, then fire them as soon as the strike ended. Women have had an even harder time getting into the mines.

Economic cycles of boom and bust have also left distinctive marks on mining districts. The ghost towns and abandoned mining camps of the West are a familiar example, but I find even more poignant the eerily anachronistic look of mining towns that have somehow survived a long decline, but where nothing has been built or renovated since the last boom. A few towns have lasted long enough in suspended animation to capitalize on their status as antiques. For example, the borough of Jim Thorpe (the former Mauch Chunk) in Pennsylvania entices tourists with picturesque gingerbread buildings erected when the town was a prosperous anthracite center a century ago.

Another inescapable aspect of mining is its physical rigor and danger. In ancient Rome, sentencing a criminal to the mines (*damnatio ad metalla*) was considered worse punishment than being chained to an oar as a galley slave. The situation has improved only slightly. Many a mining town has a bronze plaque listing the victims of a mine fire or cave-in. American coal mines have recorded more than 120,000 fatalities since 1839. This estimate includes only deaths from accidents, not mortality from black lung and other occupational diseases. In wartime, miners have often been exempt from the draft; they might have stood a better chance of surviving if they had volunteered for combat.

Mine accidents are much reduced today, but that is in part because the mining work force is much reduced. Coal-mine employment fell from 415,000 in 1950 to 75,000 in 2002—and only a fraction of those employees work underground. In Appalachian communities where the mines have closed, the continuing economic basis of life is disability compensation. The communities live on black lung as well as die from it.

Coal washing separates coal from sulfur-bearing slate and other impurities at the Galatia mine in Illinois, owned by the American Coal Company. At left, raw coal enters the plant from a storage silo. Below, from top to bottom, are three of the washing operations carried on inside the plant. In heavy-media separation, coal floats on a slurry of water and powdered magnetite, while slate sinks; the silvery drums are magnets for recovering the magnetite powder. Froth flotation lifts coal particles on a stream of rising bubbles. In spiral concentrators a slurry of finely powdered coal and slate pours down a helical trough, where denser components are thrown to the outer rim and lighter ones slip over the inner lip.

It's worth asking why a mining company would go to so much trouble to remove a bit of slate from the coal. Part of the answer is that slate doesn't burn, and so it reduces the fuel value of the coal. But that fact alone could not justify the expense of running a coal-prep plant. The main reason for the washing operations is that much of the sulfur in coal is associated not with the coal itself but with the contaminating rock. For the coals mined in most regions of the United States, washing is the only way to get the sulfur content low enough to meet air-quality regulations.

Surface Mining of Coal. Speak of coal mines in America, and most people think of tunnels beneath the hills and hollers of Pennsylvania, West Virginia, and Kentucky. But most of the nation's coal now comes from farther west—from mines on the prairies and the Great Plains and on into Wyoming and even New Mexico. And most of that western coal is dug out by strip mining, not by underground methods.

The western strip mines have a number of advantages over the older mines back East. For starters the western seams are much thicker; 30 feet of coal is common, and some beds are 75 feet thick. The overburden can be as thin as 30 or 40 feet. Moreover, the terrain is relatively flat, which favors the use of very large equipment. The quality of the coal itself is only mediocre: much of it is ranked as subbituminous, which means it has a lower energy content than typical eastern coals. On the other hand, the western coal also has less sulfur, which can make it the cheaper fuel when the cost of pollution abatement is taken into account.

At many western coal mines, the entire output is shipped to a single customer by unit train—a set of rail cars that stay hitched together and shuttle back and forth over a fixed route. Another strategy is not to ship the coal at all but to burn it on site,

Mountaintop mining for coal, observed here from high altitude, alters the texture of the landscape of southern West Virginia, creating an island of nearly flat terrain amid crinkled hills. Within the mine, the greener, softer areas in the foreground are knobs where the coal has already been removed and the land is being reclaimed; these stumps of hills are covered with grass (but still look quite different from the forested ridges surrounding them). Elsewhere, active mining continues.

exporting electrons instead of carbon atoms. A power plant at the Four Corners (the junction of Utah, Colorado, Arizona, and New Mexico) burns coal from nearby strip mines and sends the power to southern California. Similarly, a power station at the Falkirk mine near Bismarck, North Dakota, generates electricity for Minneapolis and St. Paul. These remote generating plants can be viewed as arrangements whereby an urban area pays a less-populated region to accept a share of its air pollution.

Although strip mining of coal is usually considered a western specialty, there are also surface coal mines in the Appalachians. A technique peculiar to the area is the contour mining of steep hillsides, locally known as knobs. The basic idea is to cut benches around the perimeter of a hill, mining outcrops of coal where they appear and using the spoil to fill the bench below. A variation on the technique is called mountaintopping. Where a coal seam is reasonably close to the top of a knob, the miner pushes all the overburden over the side of the hill and then extracts the coal, leaving a plateau where a peak had been. The mining companies argue that this is an improvement to the topography, since the level ground can be put to productive use. Others vehemently disagree—particularly those who live in the valley that gets filled.

STONE QUARRIES

The material dug out of the earth in the largest quantity is not an ore or a fuel. It is the plain stuff of the earth itself—stone, sand, gravel, clay. The annual tonnage of these materials dwarfs all other mining products. The output of crushed stone alone runs to well over a billion tons per year in the United States, which is close to 10,000 pounds per person. (Where did you put your share?)

Crushed Stone. Crushed stone is one of the unappreciated marvels of modern industrial society. In highway construction it forms the foundation layer and is also a key component in the concrete or blacktop that paves the roadway. The invention of stone-crushing machines was what made modern hard-surfaced roads possible. Crushed stone is equally important in railroads and in the foundations of buildings, and it has other uses as well, such as filter beds in water-treatment plants.

Because crushed stone is a low-value, high-bulk commodity, it seldom pays to ship it very far. Quarries tend to be dug on the outskirts of cities and then later engulfed by suburban sprawl. The new neighbors dislike the noise and dust and the truck traffic, and they agitate to have the quarry shut down; the quarry operators protest that they were there first. These sporadic clashes between quarrymen and suburbanites are rich in irony, which neither side ever seems to appreciate. Without the quarries to produce stone for roads and buildings, there would be no suburbs; without suburban growth, there would be little market for quarry products. They may not get along, but they need each other.

A quarry for crushed stone looks and works much like an open-pit mine. The working face is drilled and blasted, then the broken stone is mucked out by power shovels or front-end loaders and carried to the primary crusher. This machine is the heart of the operation. It is usually the largest item of capital equipment on the site, and its capacity determines the maximum output of the quarry. Some crushers work like giant steel jaws, chewing their way through hunks and slabs of rock. Another type is more like an out-of-balance washing machine, where the stone bounces around inside a big gyrating steel drum.

The primary crusher reduces the stone to lumps no more than about a foot in diameter. These pieces are small enough to be carried by a conveyor system to secondary crushers, where the size is reduced to an inch or two. There may also be tertiary crushers for making still finer materials. Screens classify the crushed stone by size, and it is either stored in open heaps or shipped immediately by truck.

Stone pulverized by a series of crushers pours from a conveyor belt onto a conical heap at a quarry in Raleigh, North Carolina.

A stone-crushing machine—the angular yellow device at far left—is installed at the lip of a terrace in this quarry, so that blocks of stone can be pushed into the hopper, rather than hoisted up to it. The crushed stone emerges at the bottom and is lifted by conveyor belts for loading into trucks. The quarry, at Villalba, near Rome, specializes in dimension stone—specifically travertine. Crushed stone is a sideline.

Most of the crushed stone quarried in the United States is limestone. There are also quantities of granite and a material called traprock, which can be any of several hard minerals. For road paving, the stone has to meet standards for skid resistance.

Dimension Stone. Alongside the crushed-stone industry there is an older form of quarrying. Large blocks or slabs of stone are carefully carved from the earth for buildings, for statuary, and for grave markers. Slate for roofing and flagstone for paving also fall into this category. Because it is cut to measure in regular geometric shapes, it is called dimension stone.

For quite a few centuries dimension stone was civilization's most important building material. From the time of the pyramids through the age of the great cathedrals and on until the birth of the skyscraper, large structures were almost always built of jointed stones. Today, dimension stone is more often an ornamental element than a structural one; even buildings that appear to be made of stone usually turn out to have a stone veneer over a steel or concrete frame.

It's easy to tell a dimension-stone quarry from a crushed-stone pit. In the first place, the walls of the dimension-stone quarry are carved into precise, rectilinear faces and benches, like the facets of a gemstone. Whereas an open-pit mine looks like a mountain has been yanked out by its roots, a dimension-stone quarry looks like the upside-down mold for making a pyramid or a ziggurat. Second, the equipment for moving and handling dimension stone is quite different from that for crushed stone. Needless to say, there is no crusher, but in addition the stone is not hauled about in dump trucks or on conveyor belts. Instead, blocks of stone are lifted by a derrick, or jib crane—a tall mast mounted near the rim of the pit, with a hinged boom that can swivel out over the quarry. Like the hoist in an underground mine, the derrick is the center of operations for a dimension-stone quarry; it carries not only stone but also equipment and workers.

A quarry might have half a dozen derricks, positioned so that at least one of them can reach every corner of the working rock faces. Each mast is steadied by multiple guy wires that stretch out to anchorages hundreds of yards away, making for a very busy skyline. The web of cables overhead reminds me of the rigging of a sailing ship or a circus tent. Some quarries have begun using mobile cranes instead of permanently mounted derricks, but it takes a big crane to handle the great blocks of stone.

In other kinds of mining and quarrying, the main idea is to blast the rock to rubble, but that obviously won't do for dimension stone. The blocks and slabs have to be gently carved out of the earth. The traditional method of cutting stone was based on driving a long steel bit with sledgehammers; later, pneumatic and hydraulic drills took over. By drilling lines of closely spaced holes downward from the surface of a bench, workers could free the back and sides of a long slab. Only the bottom of the block remained attached to the rock matrix. Then a few more widely spaced holes were drilled inward from the face along the bottom edge of the block. When the drilling was complete, each of these holes was fitted with a device called a plug and

feathers. The feathers were two tapered steel bars, and the plug was a wedge that fit between them. Driving the plug with a sledgehammer split the block of stone away from the matrix, in much the same way that a wedge splits a log for firewood.

Derricks and their guy wires form a canopy over the Rock of Ages granite quarry near Barre, Vermont. The derricks are arrayed so that one or another of them can lift stone from any point within the quarry.

New tools have replaced these ancient techniques. Sometimes the stone is cut with a wire saw, which might be described as a cross between a chainsaw and dental floss. A wire about a quarter of an inch thick, studded with abrasive diamonds, is threaded through holes bored into the stone, forming a loop around the plane of stone to be sawn through; then a motor rotates the loop under tension.

Some kinds of stone can be cut with a torch. A device called a channeling machine burns fuel oil with pure oxygen to produce a high-temperature flame, somewhat like the oxyacetylene torch used for cutting metal, but the mechanism of cutting stone is different. The stone does not burn or melt; the torch cuts by a process called spalling. The surface of the stone expands so rapidly under the flame that chips break away from the substrate. On each pass across the stone, the channeling machine cuts a slot about two inches wide and an eighth of an inch deep. A motorized drive moves the flame steadily back and forth through the ever-deepening channel. The flame looks like the exhaust of a jet engine, and that's also what it sounds like.

Drilling has lately made a comeback as a stonecutting technology, with the development of diamond-tipped bits that last longer and bore through the stone faster. At the Rock of Ages granite quarry near Barre, Vermont, I saw all three cutting techniques at work simultaneously. None of the methods is fast. It can take six weeks to liberate a single block of stone.

The bottom edge of a cut block is still freed by splitting rather than drilling or cutting, but the splitting is no longer done with plug and feathers. The tools favored today are hydraulic wedges, inflatable air bags, or small, precise explosive charges.

BRICKS AND MORTAR, CONCRETE AND BLACKTOP

As if our rocky old planet didn't have enough stones already, people have expended a great deal of effort and ingenuity coming up with new-and-improved variations on the theme of stony building materials. Think of bricks as stones made regular and rectilinear, all the same size and shape. Concrete is moldable stone, ready to assume any shape the designer chooses. And blacktop is spreadable stone, laid down in a uniform coating, like peanut butter.

Bricks. Brickmaking is an ancient craft; according to Exodus, the Israelites did it in Egypt. There are places in the world where bricks are still made the old way: digging clay with a pick and shovel, kneading it with bare feet in a wood tub, throwing the wet clay into a wood mold, drying the bricks in the sun, and finally firing them in a wood-burning kiln.

Modern industrial brickmaking is not very different in its fundamentals, although bare feet are no longer part of the process. The digging of the clay (actually, they call it "winning" rather than digging) is done with backhoes and power shovels. Grinding mills reduce the clay to a powder, which is turned into mud again by blending and

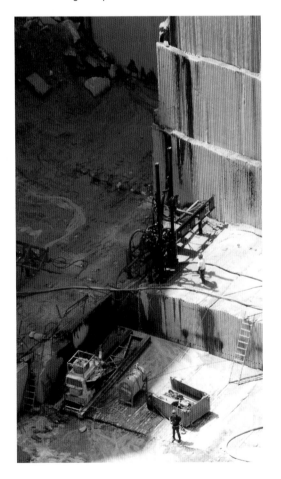

Two stonecutting technologies nibble away at the granite on the floor of the Rock of Ages quarry. The white machine mounted on a skid on the lowest level is a wire saw; it turns a diamond-studded abrasive wire threaded through holes bored in the stone. The orange device on the higher bench is a channeling machine, which cuts with a high-temperature flame.

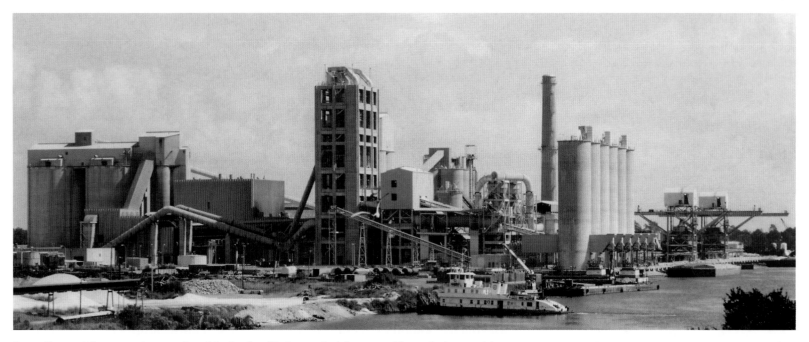

kneading with water in another kind of mill. Some bricks are still made by packing the wet clay into molds, but most of them are sliced from an extruded ribbon of clay by a fine wire, like a cheese cutter.

The heart of the brickmaking yard is the kiln, where soft clay is transformed into stony brick. The "green" (unfired) bricks are loaded onto wheeled carriages and pulled slowly through a long tunnel, which has a series of temperature zones. The bricks have to be dried for a day or two at 100 to 400 degrees Fahrenheit; then the temperature is slowly raised to 2,000 degrees or more and held at that level for two or three days; finally there's a cooling period of another day or two.

More than other building materials, bricks differ from place to place. Partly that's because the raw materials differ. Georgia clay can't be treated the same as the clay in Iowa. But there are also differences in taste and tradition that help to maintain diversity: builders in Chicago don't choose the same bricks as builders in Seattle.

Concrete and Cement. Concrete is the engineer's dream of what stone ought to be. It is heavy and hard, just like the natural product, but it molds itself to fit the straight lines and smooth curves of roadways, bridges, dams, and buildings. Gone are all the imperfections and irregularities of quarried stone. Gone too, some would say, are the charm and character of stone—but others disagree. The builders of some concrete structures—the Sydney Opera House, for example—have made concrete soar through the air in ways that stonemasons would not dare.

Concrete is not quite as ancient as bricks, but it is not new either. The dome of the Pantheon in Rome is a thin concrete shell, 140 feet across, still standing strong after almost 2,000 years. Indeed, the ancient Romans made lots of concrete, both for road paving and for building. Then came the Dark Ages, or so the story goes, and the

A portland cement plant in Mobile, Alabama, manufactures the crucial ingredient in concrete. The heart of the plant is the rotary kiln—a long, slightly inclined steel tube, rust brown in color, that extends from just behind the cylindrical silos along the waterfront almost as far as the larger silos on the inland side. It is worth noting that almost the entire plant is built of concrete.

The batch plant is the local installation where cement powder is mixed with sand, aggregate, and water to make concrete. The ingredients are weighed and assembled in the two towers, but most of the mixing happens in the conical drum of a truck as the batch is en route.

secret of concrete was lost for more than a thousand years, to be rediscovered in England in the 1820s. The story may even be true. It seems pretty clear that the modern concrete industry traces its origin to the nineteenth-century invention of portland cement by builders in the English Midlands. The name *portland cement*, by the way, is an early instance of clever marketing. The name refers to the Isle of Portland, off the Dorset coast of England, which was the source of the white limestone that Christopher Wren chose for St. Paul's Cathedral in London. Portland cement does not come from the Isle of Portland, but the name was thought to suggest quality.

Before going any further, let us clear up the perennial confusion between cement and concrete. Cement is the gray powder that comes in 50-pound sacks. Add water and you get cement paste, which is not good for much of anything. Add sand as well as water and you get mortar, the gray slop used to set tiles and bricks. Now add crushed stone—known in the trade as aggregate—and finally you get concrete.

The most familiar emblem of the concrete business is the ready-mix or transit-mix truck, with its rotating, conical drum. The biggest of these trucks hold nine cubic yards of wet concrete, which weighs some 36,000 pounds—an intimidating load when it looms up in your rearview mirror. And if the drivers of those trucks seem always to be in a hurry, they have a reason: in about 90 minutes the concrete will begin to set, and the driver will be hauling around a solidifying 36,000-pound nightmare.

Pay attention to the twirling of the drum. If it's spinning rapidly clockwise (as seen from the rear), it's busy mixing the concrete. If it's turning slowly clockwise, the concrete is already mixed; the slow rotation is just to keep the aggregate from settling out, the way that blueberries have a habit of sinking to the bottom of a muffin. At the job site, the driver will run the drum at high speed for a final few minutes of mixing. Then—and this is the cleverness in the design of the mixer—the drum is turned counterclockwise to deliver the concrete. Inside the drum is a screw-shaped blade, which forces the concrete back down into the drum when the rotation is clockwise but pumps it out when the drum spins the other way.

Home base for a fleet of ready-mix trucks is the batch plant, which in most towns is in the same neighborhood that attracts warehouses and lumberyards, often along a railroad siding. Even fairly small towns are likely to have a batch plant; given the 90-minute imperative, a single plant can't cover too much territory. For large construction projects, such as repaving a major highway or an airport runway, the contractor may set up a portable batch plant close to the action.

The batch plant is essentially a cake-mix operation. The dry ingredients are stored in open heaps or covered silos and then loaded as needed into an elevated hopper. An empty transit-mix truck drives under the outlet of this hopper, and the batch pours into a funnel above the open end of the drum, along with a measured amount of water. In many cases the whole plant is run by one person, who sits in a control room near the mixing hopper, talking with customers on the telephone, talking with drivers on the two-way radio, and meanwhile weighing out batches of ingredients with the mouse and keyboard of a computer.

Although mixing a batch of concrete is fairly straightforward, making the cement powder that goes into the concrete is a more elaborate process. The life cycle of cement is a curious one. The main idea is to take limestone and clay and drive all the water out of them; cement powder is basically dehydrated rock. Then when it's time to mix concrete, the water is added back again—and it stays in. The setting or hardening of concrete is not a drying process; the water doesn't evaporate or drain away. Instead, it combines chemically with the water-starved components of the cement, knitting the grains of powder together into a solid matrix.

At a cement plant, the water is burned out of the limestone and clay in a giant high-temperature kiln. The kiln looks nothing like the boxy oven that a potter would use to fire ceramics, or even like a big brickmaking kiln. A cement kiln is a steel tube,

A rotary kiln supplies intense heat for driving the water out of various mineral products. The heat comes from fuel burned at the lower end of the inclined tube. Raw materials enter at the upper end and tumble toward the bottom as the tube slowly turns. The shiny ring near the middle of the kiln is a bearing that supports its weight; the somewhat larger ring nearby is a sprocket for the chain that drives the rotation. This particular kiln, in Port Costa, California, processes shale to make light-weight aggregate for concrete.

10 or 12 feet in diameter and anywhere from 100 to 500 feet long; the interior is lined with firebrick to resist high temperatures. The tube is supported on rollers at intervals along its length, and a motor turns it slowly—one revolution every minute or two. The kiln is not quite horizontal but has a slope of a few degrees. Fresh ingredients continually enter the upper end of the tube, and fuel is burned at the lower end. As the materials tumble in the slowly turning kiln, they drift down the slope, getting hotter along the way as they approach the flame zone. The peak temperature is about 2,700 degrees Fahrenheit—hot enough to soften the dried limestone-clay mixture into a gooey semiliquid. When the product finally emerges from the lower end of the kiln, it cools and hardens into glassy lumps and beads called clinker. The clinker is then ground to a powder in big revolving drums full of steel balls.

The long rotary kiln is the most distinctive element of a cement-making plant, but it may not be the most conspicuous. Bins for raw materials and silos for finished products tend to dwarf everything else on the site. Somewhere near the high end of the kiln are flues and smokestacks for exhaust gases, with baghouses for trapping dust.

Asphalt. For road paving, the main rival of concrete is the substance known variously as blacktop or asphalt or macadam. Purists insist that these names don't refer to the same material, but the distinctions seem to be blurring as the years go by. The people who actually make and sell the stuff call it hot mix. Whatever the name, it consists of aggregate—crushed stone and gravel—held together by a black, tarry goo called bitumen, which is basically the dregs of the oil-refining process. The mix is kept hot so that the bitumen will flow freely enough to coat the aggregate; after the mix is spread out on the road surface, the bitumen cools and hardens.

A hot-mix plant is about the same size as a ready-mix concrete plant. Again you are likely to find one in any but the smallest town, and for roughly the same reason: like pizza, hot mix has to be delivered hot. The most common style of plant has three or four elevated metal bins that hold various sizes of aggregates, or one large bin with several internal compartments. Before the stones are loaded into the bins, they are heated and dried in a burner at the base of the tower. Then, at the top of the tower, a vibrating screen sorts the stones according to size and drops them into the appropriate bin. The stones tumbling through the drier and the shaking screen produce a distinctive noise. Below the bins—but still 15 feet above ground level—measured amounts of aggregate and hot bitumen are mixed for a few minutes in a mill. Finally a truck drives under the tower, and the mix is loaded for delivery.

ORE MILLING AND SMELTING

Unlike coal and stone, which come out of the ground ready to use, most metals need further processing before you would even recognize them for what they are. With the exception of gold nuggets and the occasional lump of native copper, metals in nature

A hot-mix plant coats crushed stone with bitumen—the sticky, bottom-of-the-barrel fraction left over from petroleum refining—to make the paving material called blacktop or macadam. The plume of vapor in this chilly-morning photograph is steam, driven out of the crushed stone as it is heated before mixing with the bitumen.

are never pure. They are combined chemically with other elements—most often sulfur or oxygen—and the chemical compounds are mixed up physically with grains of other minerals. Both the chemical and the physical mixing need to be undone. Most of the physical unscrambling happens in a process called ore milling, which separates individual grains of the sought-after mineral from everything else in the rock. The chemical isolation is done in the process of smelting; for most metals it is a fiery rite of purification.

Ore Milling. Somewhere in the neighborhood of almost any metal mine you will find a mill, also known as a concentrator. A century ago this was a fairly simple installation. The high-grade ores mined in those days needed little treatment before they went to the smelter; the ore was merely ground up into particles of the appropriate size. Grinding is still an essential part of milling, but today some degree of enrichment—called dressing or beneficiation—is also usual. If an ore includes only 1 percent of the desired mineral, there's no sense hauling the other 99 percent to the smelter.

The prototypical mineral mill is a series of buildings or sheds that cascade down a hillside in half-story steps. There is a reason for this design: it allows gravity to do the work of moving the ore from one stage of processing to the next. Even when the mill is built on flat ground, the roofline may have a stairstep profile. Another common sight is a long, sloping conveyor line, which moves material from the bottom of one processing stage to the top of another.

The first step in ore processing is crushing, which usually begins at the mine, reducing multiton boulders to lumps the size of bowling balls, baseballs, or golf balls. Then the lumps are reduced to powder in a process called grinding, done in the mill. A ball

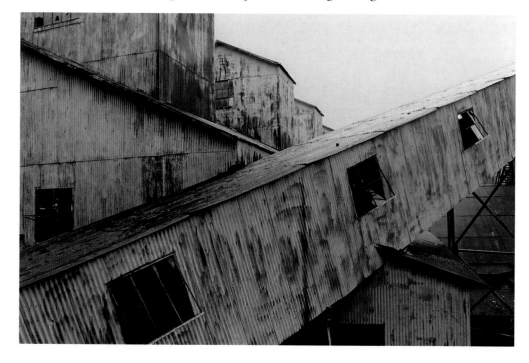

The bold diagonal is a distinctive identifying mark of ore-mill architecture, where conveyor belts and chutes move hundreds of tons of materials. The celebration of angularity shown here is part of a mill in eastern Missouri that processed lead ore until the mines closed in the 1970s. The shed roofs and the corrugated sheet-metal sheathing are also common elements of mining-industry architecture.

The world's tallest smokestack *(opposite page)* towers 1,250 feet over the INCO nickel smelter—and the little neighborhood of houses surrounding it—in Sudbury, Ontario. For scale, note the school bus.

mill is a horizontal cylinder half-filled with steel balls. Ore is poured in through a hatchway, and then the cylinder is rotated, not too rapidly, so that the balls tumble over one another, smashing the lumps of ore. There are also rod mills, which are similar in design but filled with loose steel rods or bars instead of balls. And lately there is a vogue for "autogenous mills," in which the tumbling of the ore alone does the grinding. This spares the cost of steel balls or rods, but the mills need to be larger.

Ideally, grinding breaks the ore down into individual grains that are either all metal-bearing mineral or all gangue. The next task is to separate the two kinds of particles. The most popular method is froth flotation, in which ore particles rise to the top of a water tank while gangue sinks to the bottom. This scheme sounds as if it might be based on differences in density, but that's not how it works. Both ore and gangue are too heavy to float in water; the ore rises to the surface only through the magic of bubbles—lots of tiny bubbles. Chemicals added to the water make the ore particles water-repellent, with the result that they attach themselves to air bubbles and are carried to the surface. One flotation-tank additive is pine oil, which can give a whole mill the fragrance of a freshly scrubbed public bathroom.

The bubbles in the froth are made by pumping compressed air into the bottom of a trough or by violently agitating the mixture. The term *froth* may give a somewhat misleading impression in this context. It doesn't look much like the foam atop a cup of cappuccino; it's more like pond scum.

Smelting. Whereas milling is mainly a physical process, smelting is heavy-duty chemistry. The traditional agent of smelting is fire—the metal is cooked out of the ore at

The cascading roofline is yet another common feature of mill buildings, allowing the ore to move by gravity from one stage of grinding and sorting to the next. This small mill, built directly adjacent to the mine head-frame, is in the little town of Goldfield, Nevada.

high temperature. (The similarity of the words *smelt* and *melt* is not coincidental. As for the similarity to *smell*—that *is* a coincidence, but an apt one.)

Not every mine has its own smelter; often an entire mining district shares one central facility. Finding the smelter is easy: its smokestack is the tallest structure for miles around. In the early years of the twentieth century these stacks grew steadily higher until several topped the 600-foot mark. Then, surprisingly, there was another growth spurt, and now the Kennecott smelter, around the corner from Salt Lake City, has a stack 1,200 feet high. The International Nickel Company in Ontario has gone higher still, to 1,250 feet, which happens to be almost exactly the height of the Empire State Building.

The traditional smelting process has three stages. First, the ore is roasted to drive off some of its sulfur. Then, the roasted ore is loaded into a furnace and heated beyond its melting point along with a flux—a substance such as limestone that combines with some of the contaminants. The flux and the impurities form a slag that is skimmed off the top. What remains is a mixture of metal and sulfides called matte. In the final stage of smelting, known as conversion, air blown through the matte burns away the remaining sulfur.

Roasting was once done in open heaps of ore and cordwood, which would smolder for months, emitting great fogbanks of sulfur-laden fumes. Later, the heaps were replaced by big cylindrical ovens, 40 feet tall and 30 feet in diameter. Now the ovens have given way to the fluidized-bed roaster (in which ore particles are suspended in a strong updraft as they roast) and sintering machines (in which ore is heated so intensely that the particles soften and stick together). These new roasters are much faster, and so they can be made smaller. They are usually out of sight inside one of the shedlike buildings that also house the smelting furnaces and the converters.

In picturesque terms, the smelting process is a kind of chemical seduction. At the outset, copper and sulfur atoms are locked in a tight embrace. Then, in the sultry atmosphere of the smelter, along comes the temptress oxygen, luring away the sulfur atoms from their copper mates. At the end, the copper is left in solitude, while the sulfur and oxygen run off together.

Isolating metallic copper is the aim of this whole exercise. Unfortunately, copper isn't the only product; there are also those sulfur oxides. What can be done with them? For many years they simply went up the stack, where they combined with water vapor to form sulfuric acid, which then fell back to earth as acid rain. That's why smelter stacks are built so tall—in order to disperse the fallout over a wide area. But even with 600- and 1,200-foot chimneys, smelters are notorious for creating downwind deserts—fan-shaped tracts where not even weeds will grow.

Puffing untreated sulfur fumes out the top of the stack is no longer socially or legally acceptable, so smelters have had to find other ways to cope with the stuff. Mostly they adopt the make-lemonade strategy: if you can't help producing sulfuric acid, capture it and sell it as a by-product. Most smelters now have an acid plant, where the fumes pass through a catalytic converter not too different from the one in the exhaust

pipe of your car. The converter is a tall, cylindrical vessel full of vanadium pentoxide—an exotic and expensive substance that happens to have a knack for converting sulfur dioxide into sulfur trioxide. This latter gas combines with water in an adjacent tower to form liquid sulfuric acid. The acid plant is easy to spot. It looks almost like a dairy: all the equipment is either gleaming stainless steel or sparkling white.

Even if a smelter has a sulfuric acid plant, it will also have scrubbers and baghouses to trap any remaining sulfur oxides and other noxious gases and particles. How effec-

A gleaming white acid plant occupies one end of the Kennecott copper smelter on the shores of Great Salt Lake. The plant's capacity is a million tons a year of sulfuric acid. The smelter itself is in the background, with the 1,200-foot stack looming over all.

tive are the pollution-control measures? Photographs made some decades ago show smelters surrounded by barren landscapes, apparently wiped clean of all living organisms. On recent visits to a few large smelters—the lead smelter in Herculaneum, Missouri, and the nickel smelter in Sudbury, Ontario—I saw no evidence of such devastation. On casual inspection, the vegetation in the neighborhood looks healthy. But biologists who study these environments report that subtler effects persist. And dangerous levels of metal residues in the soil will not dissipate anytime soon.

THE EMPIRE OF STEEL

A great crescent swath of the nation—from northern Minnesota, down along the southern flanks of the Great Lakes, and eastward through Ohio and Pennsylvania—has its industrial heritage in the making of iron and steel. From the blast furnaces of Bethlehem, Pittsburgh, Wheeling, and Gary came the metal to build railroads, automobiles, skyscrapers, bridges. For the better part of a century, steel was an emblem of national strength. Now the steelmaking crescent is often referred to as the Rust Belt. But there are still iron mines and steel mills to be seen, including a new generation of steel "minimills" that have sprouted up all across the country.

Iron is everywhere in the earth—indeed, the planet is mostly made of it—but the main North American sources are clustered around Lake Superior. Early mines there dug a soft red mineral called hematite, which was so rich in iron oxide that it could be dumped into blast furnaces with no purification. By the 1940s, however, the richest hematite ores were running out, and it looked like hard times for the Minnesota Iron Range. The region also had much larger deposits of an iron-bearing mineral called taconite, but it had never been mined. The iron content is only 25 or 30 percent—too lean for smelting without enrichment. Also, taconite is a hard, abrasive rock, which makes it a challenge for mining and milling. Finally, the iron in taconite is dispersed in tiny grains, too powdery to be put into a blast furnace.

A loading dock for iron ore at Ashland, Wisconsin, extends a quarter of a mile into Lake Superior. The dark gray chutes could be lowered into the hold of an ore boat; ore was dumped down the chutes from rail cars on the dock's upper deck. Docks of this kind are a familiar sight all along the shores of Lake Superior, although most of them have been idle for years.

Unloading of ore from the ships that ply the Great Lakes is done by specialized machinery that digs taconite pellets out of the hold. The dockside stockpiles have to last through the winter interruption of lake shipping. The unloading operation shown here is at a steel mill in Cleveland now operated by the International Steel Group.

Each of these problems was addressed and solved. The answer to the complaint that taconite is too hard to mine was the jet-piercing torch, a close relative of the channeling machine that cuts dimension stone in quarries. The problem of ore enrichment was solved by the lucky accident that most of the iron in taconite is in a magnetic form, and so it is easy to separate. As for the ore as fine as flour, the solution was to press it into pellets about the size of peas or marbles.

Taconite mining got started in the 1950s. At about the same time, new deposits of high-grade hematite were opened up in Labrador and Venezuela. Once again it looked like bad news for the Iron Range. Because of the processing needed, taconite cost more than the hematite ores. But the competition had a surprising outcome. Steel mills discovered that the spherical, uniform taconite pellets make an ideal blast-

Blast-furnace technology has changed less over the past 100 years than any other aspect of steelmaking. Here, at a steel mill in Trieste, Italy, the blast furnace is the tallest structure, toward the right side of the image, shrouded in scaffolding and topped by fume-collecting hoods and flues. The angled conveyor belts and the green elevator shaft carry raw materials—iron ore, coke, and limestone—to the air lock at the top of the furnace. The three upright tanklike structures in the foreground, each with its own smokestack, are stoves where air for the blast furnace is preheated.

furnace charge, allowing air to circulate more freely than the powdery natural ores. Despite the higher cost, taconite has become the predominant iron ore worldwide.

Still, the taconite mines and mills are not exactly booming. One reason is that Iron Range workers today compete not only against miners elsewhere but also against their own parents and grandparents. The main raw material for many modern steel mills is recycled steel, made from ore dug out of the ground by earlier generations.

Steel Mills. When the poet William Blake wrote of "dark Satanic mills," he couldn't have been looking at a steel mill because there were none in 1804. Nevertheless, whenever I visit a steel mill, Blake's phrase always comes to mind. What with the heat and the pounding noise, the dust and smoke, the red glow against the night sky and the pyrotechnic showers of sparks, it's hard not to see these places as infernal. And yet the process of making steel also produces some of the most hauntingly beautiful images found anywhere in the world of industry.

The big steel mills are called "integrated" plants because they handle the entire process: iron ore comes in one end, and finished steel goes out the other. At the input end, ships unload taconite pellets onto long stockpiles. Smaller heaps hold the other ingredients in the recipe for ironmaking: coke and limestone. (Coke is roasted coal—a fuel that's almost pure carbon, like a charcoal briquet except harder.)

Not far from the ore stockpiles rise the towers of the blast furnaces. Each furnace unit is a cluster of tall structures, including a smokestack and several silo-like "stoves" as well as the furnace itself. The furnace is a gigantic piece of equipment, 100 feet high or more, but nevertheless you may have a hard time seeing it because it is so thickly enshrouded in auxiliary equipment—flues and ducts for pollution control, scaffolding and stairways for workers, hoists or elevators for loading raw materials.

The basic idea of the blast furnace is not much different from that of the blacksmith's forge, where a bellows forces air through a bed of burning coal to produce a draft of superheated air. The blast furnace just does it on a larger scale and with greater intensity. Instead of a bellows, the blast furnace has powerful fans, and the air is heated to a searing temperature before it even reaches the furnace. The preheating is done in the tall stoves adjacent to the furnace tower. Each blast furnace has at least two stoves, and sometimes four or five. At any given moment some of the stoves are "on gas," or heating up, while others are "on blast," supplying hot air to the furnace.

The heated air is delivered through a fat duct that encircles the furnace about 20 feet above the base. This encircling duct is called the bustle, a name that has lost some of its descriptive power with changes in women's fashion. The pressurized hot air rises through the mix of ore and coke and limestone, igniting the coke and thereby raising the temperature even further.

Roughly speaking, the recipe for making iron is three cups of taconite to one cup of coke and half a cup of limestone. The chemistry that goes on when these ingredients are brought together is different from what happens in the smelting of copper. As described earlier, a copper smelter uses oxygen to lure sulfur away from the

Pugh cars transfer pig iron from the blast furnace to a basic-oxygen steelmaking furnace at the International Steel Group mill in Cleveland. The heat radiated by the cars is easily felt from the distance at which this photograph was made.

metal. But that can't work in the case of iron because the iron is already bound to oxygen. The oxygen is what needs to be removed, and it is carbon that acts as the seducer, carrying oxygen away in the form of gaseous carbon monoxide. Meanwhile, the limestone, the third ingredient, combines with other impurities to form a slag.

Once a blast furnace is started up, or "blown in," the mill will keep it running 24 hours a day, 7 days a week, for months at a time. Fresh ore, coke, and limestone have to be added at frequent intervals through a hatch at the top of the furnace. But there's a problem here: Because air is being blown in under pressure, opening the hatch would release a volcanic eruption of fiery debris. The solution is an air lock, just like the one in a space station. The top of the furnace is sealed by two conical plugs called bells. The upper bell is opened first, allowing the fresh ore to fall into the space between the bells. Then the upper bell is closed and the lower one is opened, dumping the charge into the body of the furnace.

An electric arc furnace is at the heart of steelmaking operations at the Nucor plate mill in Hertford County, North Carolina. The furnace itself is the giant cooking pot at right; the boxy apparatus at left preheats steel scrap and feeds it continuously into the furnace. In about 20 minutes the electric furnace can melt 250 tons of steel and heat it to 3,000 degrees Fahrenheit.

Flames erupt from the Nucor electric furnace when a load of scrap is added to the melt with the lid removed.

At the bottom of the furnace, which is called the hearth, molten iron collects in a puddle four or five feet deep, topped with a frothy head of slag. Every few hours the furnace is tapped to draw off these liquids by drilling out two plugs in the wall. The higher plug drains away the slag into a bucket called a slag pot or, more quaintly, a slag thimble. The lower plug is where the iron pours out—an orange stream so bright you have to shield your eyes to look at it.

The product of the blast furnace is pig iron. Supposedly, it got this name because it was cast into a series of molds lined up perpendicular to a central channel like a litter of piglets suckling at a sow. Whether or not this cute story is true, that is not the way it's done anymore. Instead of casting the iron into ingots (which would have to be remelted to make steel), it is poured into a strange railroad car called a Pugh car, with a heavy-duty tank shaped something like a football. The tank is pivoted at the pointy ends so that it can dump its load of molten metal. The Pugh car never travels far; it shuttles back and forth from the blast furnace to the steelmaking furnace.

From Iron to Steel. People sometimes think of steel as if it were iron with something added. The truth is just the opposite: steel is iron with something removed, namely, carbon. Pig iron from the blast furnace is about 4 percent carbon, which makes the metal weak and brittle. Steelmaking burns out the carbon and other impurities.

The first industrial-scale steelmaking machine was the Bessemer converter, where air is blown through liquid iron in a goblet-like vessel. When the batch is done, the vessel tilts and pours out the steel in a cataclysmic shower of sparks. But you'll only see this happen in old movies. The Bessemer converter was replaced early in the twentieth century by another steelmaking machine called the open-hearth furnace.

A carbon electrode glows brightly with residual heat as it is reinserted into the freshly charged furnace. The main furnace at the Nucor Hertford mill operates on direct current and so has just one electrode; many other furnaces use three-phase alternating current and have three electrodes.

Melted steel pours into a ladle as a worker observes *(above)*. After further adjustment of the steel's chemistry, the ladle is lifted by an overhead crane and transferred to the continuous casting machine *(below)*.

You won't find any of them in operation either. They've been replaced in turn by the basic oxygen furnace and the electric arc furnace.

The basic oxygen furnace harks back to Henry Bessemer's idea of blowing air through the iron. But by using pure oxygen instead of plain air, the process goes much faster; instead of several hours, a batch of steel cooks in half an hour or less. Oxygen is delivered through a lance, a tube lowered from overhead so that its mouth is a few inches above the surface of the liquid. It might seem surprising that just blowing oxygen over the surface of the iron would be enough to make anything interesting happen, but if your mental image is that of someone blowing gently over a bowl of hot soup, that's the wrong image. The blast of high-pressure oxygen comes out of the lance at supersonic speed, and it penetrates the liquid metal like a solid rod, thoroughly stirring it up.

When a batch of steel is ready, the whole furnace tilts over on hinge pins and pours the hot metal into a ladle. (Thimbles, bustles, ladles—yes, one begins to wonder about the burly image of steelworkers.) In years gone by, the ladle would pour out the metal into ingot molds, but now the molten steel goes directly into a continuous casting machine—a device so unlikely I can hardly believe it works, even though I've seen it in action. Think of it as a playground sliding board with a water-cooled copper funnel at the top. Liquid metal drops vertically into the funnel, and by the time it reaches the bottom of the slide, it is moving horizontally and has become a solid slab.

At a Nucor steel mill in eastern North Carolina I stood on the deck of a continuous casting machine, which throbbed gently beneath my feet. The casting supervisor explained that the funnel-like mold is kept in continuous motion to prevent the steel from sticking to the copper lining, in much the same way that one might jostle a frying pan to keep eggs from sticking. A glassy, lubricating powder is poured into the mold for the same reason. Making sure the steel flows through the mold smoothly and steadily is what the art of continuous casting is all about. A major blockage could destroy the mold and spill tons of hot steel. This was an interesting thought as I stood a few feet from the lip of the mold.

The tongue of glowing metal that emerges from the bottom of the casting machine is about 6 inches thick and 6 to 10 feet wide. At this point it has not yet fully solidified: it has a hardened crust but a taffy-like interior. A torch cuts the casting into slabs about 20 feet long, which move on to the rolling mill.

The rolling mill is the longest building on the site of a steel mill; it may well stretch out for half a mile. Why so long? Inside, there's a sort of pasta-making operation, where thick steel slabs are flattened into long, lasagna-like sheets or pulled into spaghetti-like bars and wires. When a slab is rolled down to a tenth of its original thickness, it grows to 10 times its original length. Further rolling to produce the thin sheets suitable for automobile body panels brings another tenfold elongation.

Minimills. Until the 1970s, integrated mills fed by blast furnaces were the only way to make steel. And then the minimill began to catch on. In a sense, minimills don't

make steel at all; they take old steel, melt it down, and reroll it into new forms. There is no iron ore, no coke, no blast furnace. The raw material comes from the junkyard rather than a mine; it consists of automobiles, old appliances, and factory wastes. A minimill is a recycling operation.

Mini is a relative term. The mills may be tiny compared with the behemoths of the industry, but they still cover enough land to build a major shopping mall or a foot-

Slabs of hot steel, still soft in the middle, roll away from the casting machine *(left)*; they will pass through a gas-fired oven to equalize their temperature before moving on to the rolling mill. Below is a spare mold for the continuous caster. The molten steel passes through the copper-lined slot in the middle; pipelines carry cooling water.

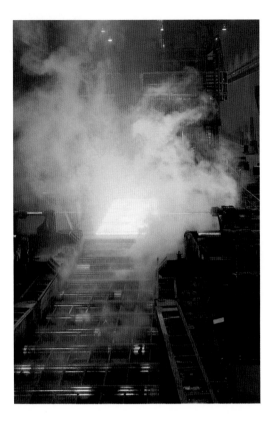

Entering the rolling mill (above), a slab goes through a high-pressure water spray to remove scale, producing clouds of steam. At the far end of the rolling mill (below), the plate is roughly 10 times its original length.

ball stadium. Instead of docks for ore boats and mountains of taconite pellets, the input end of a minimill features rail sidings and heaps of steel scrap. Cranes unload the scrap from gondola cars, sort it by type, and eventually load it into the furnace.

Most minimills use an electric arc furnace. This saucepot-shaped vessel is filled to the halfway mark with steel scrap and fitted with a heavy lid. Poking through the lid are carbon electrodes, much like pencil leads except they're at least a foot thick. The electrodes connect to a big electrical transformer out behind the furnace building, and the transformer connects to a high-voltage power line from a nearby utility plant. When the power is switched on, arcs leap from the electrodes to the load of scrap, producing enough heat to melt 100 tons of steel in 15 or 20 minutes.

Once the steel is melted, operations at a minimill are not much different from those at an integrated mill. A typical minimill has just two large buildings. The taller one houses the electric furnace and a continuous casting machine; the long and thin one is the rolling mill. Most minimills make only a few kinds of products, such as heavy steel plates, beams, and girders, as well as rebar—the steel reinforcement buried in the concrete of buildings and roadways. So far, minimills have not been able to make certain types of steel, such as stainless steel or impurity-free sheet steel, because there's not enough control over the composition of the scrap that serves as raw material.

The proliferation of minimills has transformed the economy and ecology of steel scrap. Integrated mills also use some scrap, but they could never take in as much steel as they put out, with the result that scrap was never worth much, and generations of worn-out cars and appliances were left to rust on rural hillsides. The minimills have brought a balance to supply and demand. There is a market for almost all steel scrap.

The minimills have also altered the geography of steel. No longer is steel made just where Appalachian coal meets Minnesota iron; scrap is everywhere, and so are the minimills. It's worth noting that Nucor Corporation, a minimill operator, is now America's largest steel producer.

ALUMINUM SMELTING

A century ago, aluminum was a rare and expensive metal. Drinking a beer out of it and then crumpling the can would have seemed a wildly extravagant gesture.

The ore for aluminum is bauxite, which is named for the French town of Les Baux de Provence. (But no one in the English-speaking world pronounces the word in the French way; over here it's "bawks-ite," not "bo-zite.") Very little bauxite is mined in the United States, and yet this country is the world's leading producer of metallic aluminum. The explanation is that the critical input to an aluminum smelter is not ore but electricity. The smelters go where the power is. A lot of bauxite comes out of sunny Jamaica, but it gets refined near the hydropower plants of the misty Pacific Northwest.

Before smelting can begin, the bauxite has to be processed and purified to yield alumina, a form of aluminum oxide. Alumina is a commodity in its own right. Small

grains of it form the grit on sandpaper; big crystals of it, laced with just the right impurities, are known by more glamorous names: rubies and sapphires.

To make the metal for a beer can, the smelter must decompose the alumina, breaking up the marriage of aluminum and oxygen. This turns out to be an even harder task than meddling in the romantic attachments of copper and iron. Heat won't do it; what it takes is electricity. In the same way that an electric current can split a molecule of water (hydrogen oxide) into hydrogen and oxygen atoms, electricity can separate aluminum oxide into aluminum and oxygen. But it's not an experiment you'd want to try on your kitchen table.

Before the process can even start, the alumina has to be dissolved in molten salts at a temperature of 1,800 degrees Fahrenheit. This is done in carbon-lined iron pots the size of a backyard swimming pool. Then a carbon electrode is lowered into the middle of the pot, and the power is switched on. Not much voltage is needed: roughly five volts, the output of three or four flashlight batteries. But the current drawn is considerable: between 50,000 and 150,000 amperes. The heat evolved by the passage of this current is all that's needed to keep the salt solution molten. Aluminum metal collects in a pool at the bottom of the pot.

In the aluminum smelting plant, dozens of these electrolytic cells are arranged in long rows called potlines, linked by heavy copper conductors that carry the electric current from one cell to the next. The potlines are housed in long shedlike buildings with a lot of ventilation. Fumes captured from the cells are carried by large ducts to a system of scrubbers. The fumes include carbon monoxide and carbon dioxide as

An aluminum smelter sprawls on the banks of the Columbia River at Goldendale, Washington. The long buildings house potlines—rows of electrolytic cells where metallic aluminum is extracted from the oxide by the action of electricity. Silos at the far end of the plant hold raw materials. Much of the other equipment between the potlines is for collecting and neutralizing hazardous fumes given off by the smelting process. In the foreground is the electrical substation that powers the plant, including banks of transformers and rectifiers to convert high-voltage AC into low-voltage DC. The source of the electricity is the John Day Dam, just to the right of the photograph. Not surprisingly, the buildings at the site are sheathed in aluminum.

well as hydrogen fluoride from the breakdown of the salts. The last of these gases is particularly worrisome; dissolved in water it forms hydrofluoric acid, the strongest of all acids. Air-pollution standards allow the release of very little hydrofluoric acid.

Another conspicuous part of an aluminum smelter is the power connection. A big plant going full blast consumes several hundred megawatts of electricity—the full output of a fairly large generating station. Hence, there will be at least one large power line entering the plant, and a substation that would be adequate for a medium-size city. Because of the extraordinary power demands, most aluminum smelters are built near a major hydroelectric power development. In the United States, the aluminum industry has grown up along the Columbia River and in the territory of the Tennessee Valley Authority. Canada's aluminum smelters are mainly near the hydroelectric resources of northern Quebec and along the St. Lawrence River.

MINING THE AIR

I suppose it's a sign of economic sophistication when even the air we breathe becomes an item of commerce—a resource to be exploited and marketed. But, of course, no one is really selling air. When the corner gas station annoyingly charges you a quarter to fill up your tires, what they're selling is air pressure. And as for those who mine the air, and who call themselves the air-products industry, what they sell is the separation of air into its component parts—nitrogen, oxygen, and small quantities of a few other gases.

The separation is done by distillation, the same process used to make whisky. In each case you bring a mixture of liquids to its boiling point. The more volatile fractions boil off first, thereby separating alcohol from water in the case of whisky, and oxygen from nitrogen in the case of air. The difference is that you have to light a fire to boil alcohol and water, but the boiling point of oxygen is almost 300 degrees below zero on the Fahrenheit scale, and the boiling point of nitrogen is even lower, by another 20 degrees. Thus, air separation requires not a fire but a refrigerator; it's a cryogenic process, done under supercold conditions.

Unlike mines that have to go where the ore is, air-separation plants can be built almost anywhere. They tend to follow their customers; if you find a steel mill, there's probably an oxygen plant nearby, to help run the basic oxygen furnaces. Every large city has a separation plant, if only to supply oxygen to hospitals and welders.

The heart of a separation plant is the coldbox, an insulated building that houses all the parts of the system that have to be kept chilled. Typically the coldbox is boot-shaped; the upright part of the boot is a tower eight or ten stories tall, with a much lower extension jutting off to one side. The tower is where the distillation is done; inside the sheathing and several thick layers of insulation is a distillation column much like those that define the skyline of an oil refinery. Liquefied air boils at the base of the column, and the vapors drift upward through a series of perforated metal

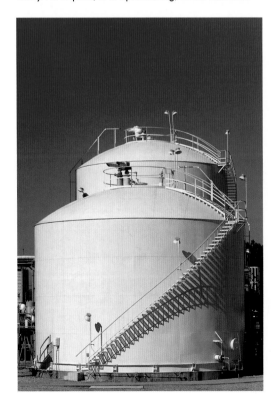

Insulated storage tanks for liquid nitrogen and liquid oxygen *(below)* are painted bright white to reflect solar heat and reduce evaporation—but the liquids boil continually nonetheless. Relief valves *(detail above)* prevent any buildup of pressure. The liquefaction plant, operated by Air Liquide, is in Spartanburg, South Carolina.

trays; at the same time, cooled vapor condensing into droplets on the trays drips back down into the pool at the bottom. In this way, gradually, the air molecules sort themselves out. Most of the nitrogen, which boils off a little more readily, winds up near the top of the tower, while most of the oxygen collects at the bottom.

The low wing of the coldbox houses all the equipment needed to liquefy the air in the first place and feed it into the distillation column. The basic scheme for doing this is the same one that runs a household refrigerator. When a gas is compressed, it heats up; when the same gas is allowed to expand, it cools off. The cooling effect is obviously what you want in a refrigerator, but you need a little ingenuity to make use of it. If you just compress some air and then immediately let it expand again, the air will merely heat up as it compresses and then cool down as it expands, and at best you'll be right back where you started. The trick is to compress the air and then draw off the heat while the air remains under pressure; then, when the already cool air expands, it cools still further, to a temperature well below the starting point. In the liquid-air plant, this compress-cool-expand cycle yields a temperature some 300 degrees Fahrenheit below zero.

The machinery needed to produce these temperatures includes large compressors and valves and sometimes turbines where the air expands, but the real heart of the operation is a device called a heat exchanger. Here, the warm incoming air is chilled by close contact with colder gases and liquids from further along in the process. The heat exchanger is a stainless-steel contraption the size of a bus, with a dozen or more flanged pipes connected to it. Inside, the various streams of liquids and gases flow through miles of labyrinthine passages. The fluids are separated from one another by thin membranes of metal, so that the substances don't mix but heat flows readily.

Apart from the coldbox, the biggest items at a separation plant are storage tanks. Big spherical tanks hold fluids under pressure. There may also be rows or racks of long cylindrical tanks with rounded ends ("bullet tanks"); these too are designed to withstand pressure. But the biggest tanks are insulated vessels for holding low-temperature liquids at ordinary atmospheric pressure. Inside these tanks, the oxygen or the nitrogen is always boiling off at a low rate, and it escapes through a vent in the roof—returning to the atmosphere it came from.

The coldbox and storage tanks of an air-separation plant are often painted bright white to reflect away as much heat as possible. Equipment that isn't painted is likely to glint with the cool sheen of stainless steel; the reason for choosing this expensive material is that ordinary steel gets brittle at the temperature of liquid air.

Ironically, after the plant expends so much trouble and energy sucking the heat out of air in order to make it a liquid, in the end the heat has to be put back in, to convert the liquid into a gas again. If the volume of liquid to be vaporized is not too large, it can be done with a big finned stainless-steel radiator, which draws heat from the surrounding air. You'll probably find such a vaporizer—partly encrusted with ice—next to the oxygen storage tank at your local hospital. Larger consumers (say, a steel mill) have to burn fuel just to reheat what was so expensively cooled.

The coldbox and distilling tower of the Spartanburg plant liquefy the air and separate the components. Nitrogen, which is a little harder to condense, collects at the top of the tower, oxygen below. The rectangular structure at the base is a complex heat exchanger.

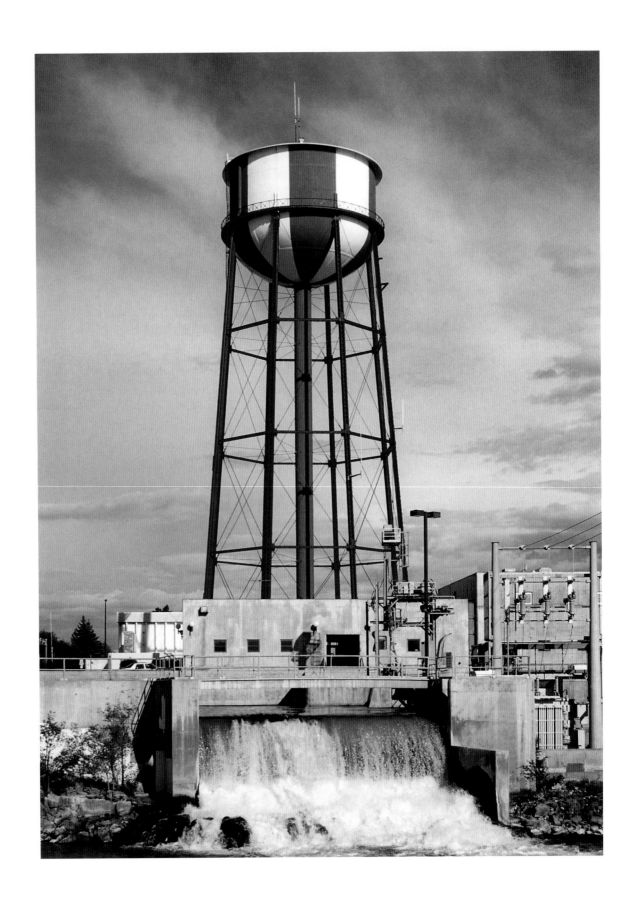

WATERWORKS

Y OU TWIST THE TAP, and a stream of pure water glints briefly in the light before twirling down into the dark hole of the drain. Most of us give little thought to where that water comes from or where it goes to, but there's no question it's one of the essentials of civilization. No human settlement can last long without a water supply—not a city or a village or a farm or even a campsite. Thus, the infrastructure for delivering fresh water and disposing of waste water is found wherever people live.

Almost all of the earth's water—some 97 percent—is in the oceans. The ice sheets of Antarctica and Greenland account for another 2 percent. That leaves only 1 percent for liquid fresh water, and most of that is deep underground, in the invisible reservoirs called aquifers. It seems hard to believe—especially when you're crossing the Mississippi or standing on the shores of Lake Superior—but all the rivers and lakes on earth hold only 0.1 percent of the world's water supply.

That small fraction of the water, however, is extraordinarily important because it circulates. Solar energy powers a planetary-scale distillery, in which water evaporates from the ocean surface and falls to earth again as rain or snow. The water is purified by this process, since evaporation leaves behind salts and other contaminants. Furthermore, when the water is raised above sea level, it acquires potential energy—a strong urge to flow back downhill again. That urge is exploited in canals, aqueducts, pipelines, and other structures that divert the water to where it's needed. Also, some of the gravitational energy can be converted into electricity when falling water is made to turn turbines and generators.

This chapter describes three broad categories of waterworks: first, dams, levees, and other structures for controlling natural watercourses; second, systems for collecting, storing, purifying, and distributing drinking water; and third, the corresponding systems for collecting, treating, and disposing of sewage. Water has such a central role in

A small dam and a spectacular red, white, and blue water tank *(opposite page)* are part of the municipal utility system of Idaho Falls, Idaho.

human life that waterworks also turn up in several other parts of this book. Irrigation is discussed in Chapter 3, on agriculture; hydroelectric plants have their place in Chapter 5, on power and energy; waterway navigation and all things nautical are taken up in Chapter 12, on shipping.

DAMS

At Hoover Dam, a stone plaque carved by sculptor Oskar Hansen portrays five uses of dams: controlling floods, improving navigation, providing water for irrigated agriculture, storing drinking water, and generating electric power. But whatever the blessings may be, dams have been a source of continual controversy in the United States. There has been squabbling between the two main dam-building agencies of the federal government, the Army Corps of Engineers and the Bureau of Reclamation (known as "Ace" and "BuRec"). There has been ideological conflict over government sales of hydroelectric power in competition with private companies. There have been bitter disputes over the environmental effects of dams. Even the recreational uses of dams have generated their share of strife, as white-water canoers do battle with water-skiers. In recent years the opponents of dams have not only opposed building new ones but also campaigned to tear down some of the existing ones.

Gravity Dams and Arch Dams. For the explorer of the industrial landscape, dams are hard to miss. They are among the largest of all man-made structures. They come in two main types: gravity dams and arch dams. A gravity dam relies on its mass to resist the force of the water penned up behind it; the structure is simply too heavy for the water to shove it aside or push it downstream. An arch dam, in contrast, resists

GETTING A LOOK

For many years, water projects were among the most welcoming of large industrial installations. Most of them are publicly owned. Reservoirs were open to fishing and boating, and many of the larger dams offered tours. Water-filtration plants and sewage-treatment plants got fewer visitors, but you could usually arrange to see your hometown facilities if you called ahead.

Since September 11, 2001, the situation has changed. Fearing sabotage, the authori-

ties have discontinued tours of some dams and other installations, and closed off public access

to many drinking-water reservoirs. A visit to a water-treatment plant is harder to arrange.

But the public is not totally shut out. One notable case is Hoover Dam, the most famous and photogenic of all American dams. Although officials worry that it might be a tempting target for terrorists, it remains open to visitors and receives about a million a year. The tour through the powerhouse has been slightly curtailed, and large trucks are forbidden to drive over the dam. Some of the facilities of the Tennessee Valley Authority also continue to welcome visitors.

the thrust of the water by bracing itself against the adjacent walls of the valley. An arch dam works like an arch bridge laid on its side, with the convex surface facing upriver. Arch dams can be built only in canyons or steep-sided valleys where strong rock faces anchor the ends of the arch.

An arch dam is easy to recognize by the graceful curve of the structure. The purest examples are remarkably light and spare, like a curved sheet of paper holding back a river. Most arch dams, however, are really hybrid designs that rely on mass as well as elegant geometry. Hoover Dam is the best-known example. It has an arch form and is carefully mortised into the sandstone walls of its site in Black Canyon, on the Colorado River 30 miles from Las Vegas. But the dam also gains stability from its sheer mass—some 7 million tons of concrete.

Gravity dams have a distinctive shape in cross section, being much thicker at the base than at the crest. Sometimes the upstream face is nearly vertical and the downstream face slopes away, so that the dam looks like a giant doorstop or bookend. More often both faces are inclined; the dam has the form of a berm or embankment. Sometimes the downstream face is supported by buttresses, like those of a Gothic cathedral. The broad foundations ensure that the water cannot topple the dam.

Shasta Dam, near Redding in the far north of California, is classified as a gravity dam: the great mass of the structure is what holds the water back. The mass is some 15 million tons of concrete. In cross section such dams have a characteristic wedge shape, much thicker at the base than at the crest. Shasta Dam was completed in 1945, mainly to impound water for the Central Valley Project, which supports irrigated agriculture in the central and southern parts of the state. The dam also has a hydroelectric generating station; the five large pipes, called penstocks, emerging from the face of the dam carry water to the power turbines.

Monticello Dam, west of Sacramento, California, is a concrete arch that relies on geometry rather than mass to resist the force of the impounded water. Seen in this photograph is the concave, downstream face of the dam (and a small hydroelectric installation at the base). The curved form allows the thin concrete membrane to transfer stresses to the adjacent walls of the canyon.

Concrete Dams. Dams are distinguished by their building materials as well as by their geometric form. Although there are lots of variations and hybrids, the two main groups are concrete dams and earth dams. All modern arch dams are built of concrete, but gravity dams can be made of concrete or earth or a combination.

A large concrete dam is not one big lump. A dam cast as a single piece would crack open as the concrete cured. To avoid such failures, engineers design contraction joints

Hoover Dam has the shape of an arch dam, but it is actually a hybrid structure, gathering strength from both mass and form. The dam is often ranked as one of the most exquisite of all engineered structures. It is fitted to its site so well that the gnarly canyon wall looks like an organic growth engulfing the mass of concrete.

into the structure every 50 feet or so. A contraction joint is a deliberately weakened section of the concrete where cracks can develop in a controlled way. Once the concrete is fully set, the contraction joints are grouted with a special cement pumped in under pressure through pipes embedded in the concrete for just this purpose. If you can get close to the face of a dam, you may be able to see the grouted joints.

What's inside a concrete dam? The smallest and thinnest ones are solid concrete all the way through, but major dams have a network of galleries and shafts rather like the secret passageways of an Egyptian pyramid. These interior rooms and corridors provide access for inspection and maintenance of the structure itself and of various instruments and mechanical systems.

Among the instruments installed in concrete dams are plumb lines hundreds of feet long. Each line is suspended in a narrow vertical well something like a laundry chute. The position of the weighted line with respect to the surrounding structure is monitored with a microscope near the base of the well. Since the plumb line always points straight down, any change in its position indicates movement of the dam, perhaps caused by settling of the foundation or flexing of the structure. Other instruments embedded in the dam include strain gauges and temperature sensors; seismometers are another common device, since earthquakes are a hazard to dams.

On the exterior surface of the dam, look for markers and monuments used in high-precision surveys that check for any movement of the structure. Sometimes red-and-white surveyor's targets are attached to the downstream face, where their positions can be monitored with instruments set up a few hundred yards farther down the valley. Today the sighting targets may be replaced by reflectors for laser distance-measuring equipment.

Earth Dams. An earth dam may seem unsophisticated—just a pile of dirt to block the river's flow—but in fact it requires very careful engineering. Internally, the dam probably has several layers arranged in wedgelike forms. Typically there is a core of hard-packed clay, whose main function is to prevent seepage through the dam. The core is flanked by shoulders of sand or rock, which offer mechanical support and added mass. The exposed surfaces of the dam need protection from waves (on the upstream side) and from rain (on the downstream side). The usual armor against wave action is a layer of broken stone, called riprap. The downstream face is often planted with grass.

An earth dam has to be even thicker in cross section than a concrete gravity dam. Resistance to overturning is not the only reason. The slopes must also be gradual enough to resist slumping and erosion. Thus, an earth dam may well be five times as wide as it is high, and it consumes much more material than a concrete dam of the same height. On the other hand, that material is much cheaper.

The greatest menace to an earth dam is water seeping through or under the embankment. In addition to building the dam with an impervious core, the foundation is sealed by grouting or by digging a cutoff trench under the dam and filling it

with watertight clay. The motive behind these measures is not a maniacal desire to hoard every last drop of water. The dam builders would be quite willing to let a little water go. The problem is that the escaping water may take part of the dam with it. Once a stream begins flowing through an earthen structure, it can wash out fine particles, thereby opening the channel for a stronger current, which in turn carries away more material, and so on. The effect is called piping, and if it is not controlled, it can quickly destroy the dam.

A WASHOUT

Teton Dam was built partly for flood control, which soon seemed a bitter irony. The towns and farms along the Teton River in southeastern Idaho had always been plagued by high water in the spring, often overflowing onto roads and fields. The new dam was going to end that annual inundation. Instead, it unleashed a roiling deluge that washed away whole neighborhoods and killed a dozen residents.

The Bureau of Reclamation built the dam in a steep-walled canyon a few miles upstream from the little town of Wilford. The dam was a wedge-shaped earth embankment 300 feet high and more than half a mile long at the crest, with a powerhouse nestled at the bottom on one side of the river and a concrete overflow spillway along the other bank. The reservoir of 260,000 acre-feet began filling in the winter and spring of 1976. The plan was to bring the water level up slowly, at about a foot a day, but the rains were heavy that year, and the level actually rose by three feet a day. By early June the water was approaching the inlet of the spillway. But it never quite got there.

Early on the morning of Saturday, June 5, workers noticed seepage on the downstream face of the dam, near the western abutment. Within an hour the flow was muddy—always a danger sign. The work crew brought in bulldozers to push more earth into the leak, but the machines were lost as gaping cavities opened up beneath them; the drivers barely escaped in time. A whirlpool formed in the

reservoir. Radio and TV reporters arrived, and alarms were sounded downstream. Then, just before noon, the whole western third of the dam gave way suddenly, disintegrating in a matter of seconds and releasing 80 billion gallons of water that scoured out the canyon below. The flood obliterated Wilford, then went on to destroy most of Sugar City and Rexburg, and caused substantial damage 30 miles downstream in Idaho Falls.

The ruin of the Teton Dam is still standing (but it's not easy to find—ask directions at the Flood Museum in Rexburg). During one of the investigations of the failure, part of the embankment was carved away to expose the inner structure. This is the only place in the world, as far as I know, where you can see a full-size dam in cross section *(see photo)*.

What caused the collapse? Blame has focused on two geological factors. First, the floor and walls of the canyon were full of cracks, which were not completely sealed off with grout. Second, the core of the dam, which was supposed to provide a watertight seal, was made of a silty soil called loess, which turns out to be great for growing potatoes but not so good for building dams. Teton was the largest dam ever to fail catastrophically. No dam that large has been built in the United States since the collapse.

Watauga Dam in eastern Tennessee is an embankment of earth and rock even broader in cross section than a concrete gravity dam. At Watauga the thickness of the dam at the base is four times the height. The view in the photograph is of the terraced downstream face.

If you see a small quantity of water leaking through a dam, don't panic. Even in a well-built and well-maintained earth embankment, a little seepage is common and nothing to be alarmed about. The damkeeper should be checking to make sure the water continues to run clear; a muddy flow suggests erosion.

Near the base of the downstream face you may find drains that carry off both seepage from the reservoir and rainwater that soaks into the dam itself. Without an exit, the seepage water would build up pressure in the soil, which might grow high enough to simply float the dam away. There may also be capped vertical pipes called piezometers; measuring the height of the water column in these tubes gives an indication of the pore pressure.

Spillways and Outflows. The focus of attention at many large dams is the spillway, where flood waters are channeled over the dam or away from it. Even at an earthen dam, the spillway is usually a concrete structure because water pouring over an earth embankment would quickly wash it away. Floodgates control the entrance to the spillway and thereby regulate the water level in the lake behind the dam.

Often the spillway looks like a giant water slide. The concrete is formed into a sensuous, smooth curve so water will flow over it in smooth sheets with little turbulence; this kind of flow minimizes damage during periods of heavy flood. Sometimes a "ski jump" at the bottom of the spillway helps to dissipate the energy of the falling water. The deep pool that the water plunges into is called a stilling basin.

One style of floodgate works like a rolltop desk. Each gate is a section of a cylinder, mounted horizontally and pivoted on the axis of the cylinder. Raising the gate allows water to flow underneath. This design, called a tainter gate, is much in favor today, but many other gate types are also in use. Perhaps the most common (especially in smaller dams) is a simple lift gate, a steel plate that is raised like a window

The spillway of Shasta Dam *(right)* is built into the middle of the dam's concrete face. The spillway provides a safe path for overflow in floodtime. The Shasta spillway is three times the height of Niagara Falls.

The spillway at Oroville Dam, a hundred miles south of Shasta, is a structure separate from the dam itself. It's a supersize waterslide or ski jump.

sash to open a channel into the spillway. Sometimes each gate has its own winch, which can be hand-cranked or electrically driven. Other dams have a traveling hoist that moves on a track from gate to gate, making adjustments as needed to regulate the level of the reservoir above the dam and the rate of flow in the stream below.

Every dam needs *some* provision for handling excess water, but it is not always a spillway on the face of the main dam. Sometimes an auxiliary dam elsewhere in the valley, called a saddle dam, handles overflow. Another solution is a drop-inlet spillway, also called a morning-glory spillway: a vertical tube that carries water into a conduit under or around the dam to a discharge below. The drop-inlet spillway works like the overflow protector of a bathroom sink.

A word of caution: The stream bed below a dam is a dangerous place. If the floodgates are opened, the water level can rise very quickly. Some dam operators sound a horn or siren before opening the gates, but you may not hear it through all the noise of rushing water.

Other Dam Sights. The turbines and generators of hydroelectric power plants are often near the base of a dam, or inside it, but they can also be miles away, with water conveyed to them by conduits called penstocks. All this is described in Chapter 5.

On a navigable river, every dam must have an associated lock where boats are raised and lowered. These interesting mechanisms are covered in Chapter 12.

Fish ladders and fish elevators do for fish what locks do for ships: provide a navigable route around the dam. In essence, a fish ladder is a series of many tiny dams that cover the same vertical drop as one big dam. Fish that swim upriver to breed, such as salmon and shad, are able to leap the small waterfalls; the young fingerlings

can likewise make their way down the ladder to the sea. The slope of fish ladders is usually gradual enough to make them look more like stairways than ladders. In North America the most elaborate system of fish ladders is on the Columbia River in the Pacific Northwest, supporting the salmon fishery. The Bonneville Dam, near Portland, has two ladders as well as a fish lock. Fish heading upstream enter the lock from the lower waterway; the lock then fills, and the fish swim out into the upper reservoir. On the Snake River fish are trapped and trucked around a series of dams instead of climbing ladders. Despite these measures, salmon runs continue to decline.

Booms, fences, cables, and nets stretched across a reservoir a few hundred yards above a dam are meant to keep large floating objects—not to mention boaters and swimmers—from passing through the floodgates and down the spillway or into the intakes of power plants and drinking-water systems.

Somewhere near a dam you will often find gauges to measure water depth in the reservoir or the elevation of the water surface. The gauge may be as simple as a pole marked in feet or meters, or it may be an elaborate instrument that communicates by radio with a central monitoring station. The more sophisticated gauges are housed in a shelter on the riverbank; the shelter is built over a well that is connected to the river through buried pipes. You might think that all lake and river levels would be measured with respect to sea level, but it isn't done that way. Each river system has its own benchmark level, so that comparing the official levels of two reservoirs, even nearby ones, will not necessarily tell you much about the true lay of the land.

The great enemy of the dam builder is silt, which fills up the reservoir and thereby renders the dam useless. On American rivers, the threat is most worrisome at the high dams on the Colorado, which is one of the world's dirtiest rivers. When you

The morning glory inlet is another kind of overflow device. When the water level in the reservoir rises to the lip of the flower-shaped inlet, the overflow is carried through a pipe to an outlet below the dam. The structure is at Watauga Dam.

Floodgates regulate the flow of water over a spillway and thus control the level of the reservoir. In the upper photograph at right are tainter gates at the Old River Control Structure on the Mississippi 80 miles upriver from New Orleans. Each gate is a section of a cylinder, which pivots around the axis of the cylinder. The lower photograph at right shows the floodgates of the Bonneville Dam near Portland, Oregon; they are flat panels that lift vertically. In the photograph below is an unusual small dam in Oklahoma City, where floodgates raised and lowered by hydraulic rams make up the entire structure of the dam.

look out over the Grand Canyon, ponder that everything you don't see there has been washed downriver. In its natural state, the Colorado carried 17 times as much silt as the muddy Mississippi. Now most of the Colorado's silt settles out in the still water behind dams. The reservoir behind the Imperial Dam on the lower Colorado is equipped with settling basins where silt is deliberately collected and then scraped out and piped back into the river below the dam. Other dams have scour sluices—gates that are briefly opened every few weeks to allow a surge of fast-moving water to carry away some of the accumulated solids.

Near the largest dams you may be able to find remnants of the construction process. A large earth-filled dam will probably have an equally large borrow pit nearby, where the material for the dam was excavated. Building a concrete dam requires a concrete plant, erected as close to the dam as possible. The plant itself may have been dismantled, but look for signs of hoists, tramways, or conveyor belts. One of the more surprising ancillaries to concrete-dam construction is a refrigeration plant. Concrete gives off heat as it cures, and a solid structure as large as Hoover Dam would take decades to reach a steady temperature if the builders relied on natural cooling alone. To speed the process, the body of the dam is laced with pipes through which chilled water or brine is pumped. At the Glen Canyon Dam the refrigeration plant produced more than a million pounds of ice per day. Like the concrete plant, the refrigeration plant is broken down and hauled away when it is no longer needed, but you may be able to spot some of the piping in the finished dam.

The relics of dam construction sometimes include whole towns. Boulder City, Nevada, near Hoover Dam, was built to house construction crews; so was Page, Arizona, near the Glen Canyon Dam, and Norris, Tennessee, near the Norris Dam (the first of the Tennessee Valley Authority projects). At Grand Coulee three towns were built—one for engineers, one for foremen, and one for laborers.

LEVEES AND FLOOD CONTROL

When the rains came in 1993, the whole country kept an eye on the levees of the Mississippi and Missouri Rivers. All through that summer we watched people battling the water with sandbags and boards. Most of them lost the contest. By the end of the summer, two-thirds of the levees along the upper Mississippi had failed.

A levee has the same basic job as a dam—to hold water back—but it labors under more difficult working conditions. When you build a dam, you can pick the best spot for it, where a strong foundation will support its weight; a levee has to be raised on loose and soggy riverbottom soils. Furthermore, a relatively small dam can plug up an entire river channel, but levees have to extend along hundreds of miles of riverbank.

The placement of the levee is a tricky decision. If you build it too close to the river itself, there will be nowhere for the floodwater to go. On the other hand, if you move the levee back, you are sacrificing some of the land it was meant to protect.

The traditional way of erecting a levee is to scoop soil out of shallow pits and pile it up into a berm, or embankment. The borrow pits are on the river side of the levee; after a few years they fill with silt from floods and may be hard to spot. Modern practice is to include as much sand in the soil as possible. Although the sand is more porous than other materials, it holds its form better when wet and is less inclined to slump. (Think of the difference between sand castles and mud pies.)

The completed embankment is usually planted with grass, which helps to prevent erosion. Trees are seldom allowed to grow because their roots would create danger-

A fish ladder at the Bonneville Dam is meant to provide a route around the dam for migratory salmon on the Columbia River. Each step in the cascade is about a foot high.

A river gauging station on a branch of the Potomac River measures the water level and automatically sends reports via satellite relay to a central monitoring service. The measuring instrument is in a well dug beneath the concrete enclosure. The station is near Franklin, West Virginia.

An earthen levee protects the New Orleans suburb of Algiers, Louisiana, from the lower Mississippi River. Above, a passing freighter looms over the embankment. Below, a water pipe crosses over the levee to avoid the risk of tunneling through it.

A long concrete barrier fitted with floodgates forms part of the Old River Control Structure at the point of closest approach between the Mississippi and the Atchafalaya Rivers. The structure regulates flow between the two river basins, not only for flood control but also to prevent the Atchafalaya from "capturing" the Mississippi.

ous passages through the bank. Even more hazardous are the burrows of rodents and rabbits. A single family of groundhogs could be the undoing of an entire floodplain. (In the legend of the little Dutch boy, the hole in the dike that he plugged with his thumb was a mole hole.)

Drainage. A levee is built to seal off the land from the river, but unfortunately it has the opposite effect as well. People living behind the embankment are cut off from the water's edge, and troublesome arrangements are needed wherever roads or pipelines cross the levee. Most important, the levee blocks the flow of floodwater *into* the river as well as *out* of it. Where a small stream needs to cross a levee, it can be carried through by a culvert pipe, equipped with a gate or check valve that can be closed in times of flood to prevent the river from backing up through the culvert. In other cases, pumps lift the streamflow over the levee. But culverts and pumps are practical only for the smallest streams. Where larger watercourses join the river, the levee must be extended back along the banks of the tributary, often for miles, until the land rises above the height of the main levee. In this way a system of levees along a river is broken into segments that span the intervals between tributaries. The segmentation has the advantage that if a levee fails, the flooding will be confined to the area protected by that segment, rather than spreading over the entire river valley. As a matter of fact, breaching one levee tends to relieve the pressure on others, since some of the floodwater is thereby drained off. Thus, one person's misfortune is another's good luck, a situation that leads to antisocial temptations.

Levee segments also define convenient administrative boundaries. In rural areas along many American rivers, levees are built and maintained by quasi-public commissions or associations. Along roadsides you might notice a small sign announcing that you are entering the Sny Island Drainage District or the territory of the Ferrier

A floodwall with a sliding gate protects Morgan City, Louisiana, from the rivers and bayous that surround it. During a flood, the gate would be closed, and the spot from which the photograph was made would be under water.

River Levee Commission. And at election time you might find people running for offices such as ditch warden and drainage commissioner.

When a levee is threatened by flood, overtopping of the embankment isn't the only hazard. Water also seeps through and under the structure, and crews must keep a lookout for sand boils—places where the seepage is so fast that it liquefies the soil. Ironically, levee failures often happen soon *after* the river reaches its highest crest, when the water starts to recede—in other words, just when residents are congratulating themselves for having withstood the onslaught. The reason has to do with the flow of water through the levee embankment. Rising water, seeping into the soil, plasters silt against the river-side face of the levee, which helps to seal it. But when the river falls, the direction of seepage reverses, and the silt is dislodged from the surface. This makes the levee more permeable and more susceptible to slumping.

Floodwalls. Where a river passes through a city, there is seldom room for an earthen levee, which might require a 60-foot-wide strip of land. The alternative is a floodwall, a concrete barrier as high as a levee but only about a foot thick. Even though the floodwall takes up less land, it still blocks access to the waterfront, and this is likely to be even more inconvenient in the city than in the countryside. Cables and pipelines can be routed through or under the floodwall, but dealing with tributary streams and storm runoff remains a challenge. There's not much sense in protecting the city from the river if neighborhoods are going to be inundated by storm drainage backed up on the landward side of the floodwall. Getting rid of the storm water often involves pumping it over the wall, which is expensive.

It's not only pipes and wires and drains that need to cross a floodwall, but also people and vehicles. Openings for streets and railroad tracks are equipped with massive steel doors that either swing closed on hinges or slide on rails, like a glass patio door.

THE MAN-MADE COUNTRY

In a book about man-made elements of the landscape, the Netherlands deserves a special place, for there the land itself is man-made. Almost the entire nation was reclaimed from the sea by human labor.

The site of the Netherlands is a great river delta, where the Rhine and several other rivers have piled up a bed of sediments in the North Sea. Early inhabitants of the marshes and sand banks built their homes on mounds—a few of which still exist—for refuge from high tides. But by about AD 1200 the building of dikes had begun, and it has remained a national obsession ever since.

The Dutch dikes not only protect existing land from storm surges but also, more remarkably, have created new land where once there was sea. The largest of the reclamation projects, the enclosure of the Zuider Zee, was begun in 1918. A 15-mile dike of clay and sand was laboriously constructed in open water between Holland and Friesland. The closing of the last passage through the dike in 1932 was a frantic effort to dump earth into the breach faster than the tide could wash it away. The finished structure dwarfs even the massive levees of the lower Mississippi. It is 300 feet wide, with a four-lane highway running along the top and space reserved for a rail line.

After the Zuider Zee was closed off, the inflow of river water gradually converted it to a freshwater lake (the IJsselmeer). Within the lake five giant tracts were diked off and pumped dry to become new land, or polders, with a total area of some 850 square miles. Much of the rest of the Netherlands had been created in the same way, albeit generally in smaller parcels. (Schiphol Airport, the international gateway to the Netherlands, lies on the Haarlemmermeer polder, reclaimed in 1852.)

Polders are already below sea level at the time they are created, and as water is pumped out of them, the soil collapses and they subside further. In some polders the land surface is now 30 or even 40 feet below the tide line. Keeping them dry requires constant pumping. Most of the Dutch windmills were built for this purpose, and 2,000 of them are still working at it, now supplemented by an even larger number of electrically driven pumping stations.

In the photograph below, made from the International Space Station in 2001, three large polders stand out in shades of brown and tan. The photograph was made in early spring, when much of the polderland had been plowed, whereas most other land nearby is pasture that was already green. The dike enclosing the IJsselmeer is at the upper right, the city of Amsterdam at the upper left. The three large polders, clockwise from the top, are the Wieringermeer, enclosed in 1930; the Noordoost, enclosed in 1942; and the Flevoland, enclosed in two stages in 1957 and 1968. The silty water above the Flevoland polders is the Markermeer, enclosed by the Markerwaard dike; eventually this area will be pumped dry to become another polder.

You don't have to go to the Netherlands to see polderlands. In California, where the Sacramento and San Joaquin Rivers empty into San Francisco Bay, 1,100 miles of levees enclose 260,000 acres of polder. The region has all the same problems as the Dutch polders, including continuing subsidence.

Most of the time, the doors remain open—indeed, they may be padlocked open to discourage mischief. If the river starts to rise, there's usually plenty of time to close the gates.

Like dams, floodwalls and levees have become a subject of controversy. The argument is that putting a river in a box may prevent minor flooding but makes the biggest floods worse. A river without artificial levees will gradually spread out over its floodplain; you get your feet wet, but the house isn't washed away. By confining a river, levees force it to climb upward instead of spreading out. When the levee ultimately gives way, the sudden onrush is catastrophic. According to this view, the right response to floods is not to raise the levees but to change land-use policies. People should not build houses on floodplains, and some of the land along rivers should be restored to natural marshes, which act as a buffer, slowly absorbing and releasing water.

On the other hand, a riverfront city that decides not to build a floodwall will draw criticism from another faction. In the spring of 2001, floods hit Davenport, Iowa, the last major town on the upper Mississippi without permanent riverfront barriers. The Federal Emergency Management Agency threatened to withhold aid because the city had refused a 1984 proposal to build a floodwall.

DRINKING WATER

We call it drinking water, but very little of it gets drunk. The rule of thumb for estimates of municipal water consumption is 150 gallons per person per day—more than enough to overflow the average bathtub. Obviously no one drinks that quota; most of it flows through the kitchen sink, the laundry tub, the dishwasher, the toilet, the shower, or the garden hose.

A typical municipal water system has four main components: a source of supply, a pipeline or aqueduct to carry the water from the source to the city, a facility for treatment and purification, and a distribution network. Often there are also reservoirs or tanks within the city capable of holding a few days' supply.

The Source. Some cities are fortunate to have a ready and reliable source of water they can simply dip into. Philadelphia draws its water from the Delaware River without any need for a dam and reservoir; Chicago taps Lake Michigan. Other cities need to impound water to ensure an adequate supply during dry spells. New York has reservoirs in the Catskills; Boston reaches out 60 miles west to the Quabbin Reservoir in central Massachusetts; and Los Angeles slurps up water wherever it can get it.

A drinking-water reservoir is much like any other artificial lake, but the smaller ones require special care and feeding. Or rather unfeeding. A major problem in reservoirs is the overgrowth of algae and other aquatic plants, which give the water an unpleasant taste. The root cause of such blooms is an excess of nutrients, usually washed in with the runoff from farms or lawns. The ideal solution is to control land

use on the watershed, but that is not always feasible. Hence, you will occasionally see workers puttering about on the reservoir in a small boat, towing a sack of copper sulfate, which kills the algae.

Another reservoir ailment is anoxia, in which the water below a certain depth becomes depleted of oxygen. Anoxia kills off fish and many other living things, while encouraging the growth of a different group of organisms, called anaerobic bacteria. The main trouble with anaerobic bacteria, from a human point of view, is that they stink. (For all I know, they may feel the same way about us.) They give the water a taint of sulfur. The solution is aeration, or in other words getting the oxygen back into the water. For a long time the favored method of aerating drinking water was to build a fountain, and so many reservoirs have impressive jets and ornamental cascades.

More recently, engineers have discovered that it's more efficient to bubble air directly into the reservoir. The equipment used to create the stream of bubbles is just like that in a home aquarium. A compressor on shore forces air through a pipe or hose to a "header" installed on the bottom of the reservoir; the header has many tiny orifices that break the airstream into bubbles, which drift to the surface. When the system is working, you'll see a spot out in the water that looks like it's boiling. Contrary to what you might think, the oxygenation does not come mainly from the bubbles themselves; instead, they create an upward current that "turns the lake over," bringing the anoxic bottom water to the surface.

Intake Structures. The point where water enters the municipal system can be as simple as an open-ended pipe or as elaborate as a multistory tower and gatehouse.

The intake is generally placed well away from shore to avoid collecting silt and other contaminants washed off the land. In a river, the best spot is usually in the main channel where the flow is greatest and the water is deepest. In lakes the intake structure may stand well offshore. Chicago's intake is an octagonal, fortress-like tower two miles out in Lake Michigan, reachable only by boat; it doubles as a lighthouse. Eight intake ports admit water into the center of the tower, where it is pumped to a purification plant onshore. At artificial reservoirs the intake structure is usually near the dam or incorporated into it.

Every intake has to be protected in some way from foreign matter that could be sucked into the port. The largest objects are stopped by a "trash rack," which is made of heavy iron bars or gratings. Finer screens are meant to exclude leaves, aquatic plants, and other small objects, not to mention fish. Fouling of these screens is a continual problem, which is solved mainly by limiting the velocity of the inflowing water so materials are neither plastered against the screen nor sucked through it. If the flow is kept to less than one foot per second, fish can escape it. To limit the velocity while still taking in a sufficient volume of water, the area of the screen has to be large. In cold regions, ice fouling is also troublesome. And the latest nuisance in some parts of the country is the zebra mussel, an imported mollusk that can clog an intake port with a single season's growth.

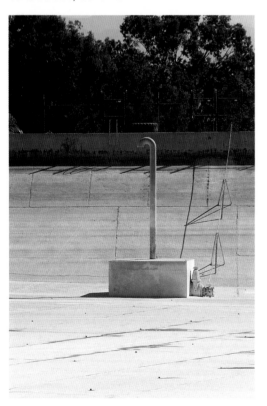

Machinery for aerating a reservoir, normally out of sight under the water, is exposed to view when the reservoir is drained. This concrete-lined basin is near Santa Barbara, California.

And what about the stories we've all heard of fish that come flopping out of the kitchen faucet into the ice cube trays, or that swim placidly around the toilet bowl? They are food for the alligators that make their home in city sewer systems.

Wells. Although lakes and rivers supply water to the majority of large cities, some municipal water systems draw part or all of their supply from wells, as do most homes in rural areas.

A completed well is not much to look at. With old-fashioned dug wells—the kind with a bucket on a rope—you could at least peer down into the subterranean gloom and time the descent of a dropped stone. A modern household well is nothing but a concrete slab covering a small-bore steel casing. Sometimes even the slab is omitted, so there is only a capped steel pipe.

Larger wells, suitable for municipal or community water supplies, sometimes have a shed erected over the wellhead; it houses the pump motor as well as electrical

Intake towers at Hoover Dam rise almost 400 feet from the floor of the reservoir. Note the bright white "bathtub ring" on the banks of the reservoir; it consist of mineral deposits left behind as the water level has fallen during recent dry years. (The photograph was made in 2004, when the water level was at its lowest since the reservoir first filled in the 1930s.) Because of the low water, the screening that keeps debris and fish out of the intakes is exposed on the, lower part of the towers.

A large well with a tank and pumping station is part of the municipal water supply system of Albuquerque, New Mexico. The elaborate electrical supply is needed to run pump motors that lift water from an aquifer that can be as much as 1,800 feet below the surface. In the background are the Sandia Mountains.

switchgear, valves, water-metering equipment, and access taps for water-quality testing. The building may look like little more than an outhouse, but on closer examination you'll find that the electric-power service is adequate for a medium-size factory.

The most interesting parts of a well are out of sight underground. People sometimes imagine that a well is like a straw sucking up water from an underground river or pond. Actually, except in caves, groundwater does not flow through open channels; it percolates through a layer of sand, gravel, or porous rock. The business end of the well has a screen that allows the water to pass into the well but holds back most of the sand or gravel.

In almost all modern water wells, the pump is at the bottom of the shaft. For small wells the electric motor that drives the pump is also submersible. Larger pump motors are installed at the surface and drive the pump through a long rotating shaft.

Smaller wells supply drinking water in Xenia, Ohio. The pumps themselves are underground; a bullet-shaped electric motor atop each wellhead drives the pump through a long shaft.

Aqueducts. Getting the water from the reservoir to the city is sometimes a bigger engineering project than building the dam and reservoir in the first place.

The Romans built aqueducts that worked like artificial rivers, following the "hydraulic grade line," the slope of naturally flowing water. This slope is typically about a foot per mile, which is so gradual that it can be hard to tell which direction is downhill. The Roman aqueducts had to follow the grade line because they were open troughs and therefore could not descend into valleys or climb over hills.

Grade-line aqueducts still have a place in modern water systems. When the terrain is favorable—for example, when the route is parallel to an existing river course, which necessarily follows the hydraulic grade line—they have the lowest construction cost. But when necessary, a modern aqueduct can cross hills and valleys with siphons and inverted siphons—sealed pipes that flow full under either negative or

positive pressure. A siphon climbs above the hydraulic grade line; an inverted siphon dips below it. Many aqueducts are composed entirely of pressure pipe from end to end so they can follow the terrain freely and ignore the hydraulic grade line.

Aqueducts (whether open canals or closed conduits) are usually designed to keep the water moving at about 5 feet per second, which is equivalent to roughly seven miles per hour—a fast jog. Knowing this velocity and also the cross-sectional area of the channel, a little back-of-the-envelope arithmetic will tell you how much water the aqueduct delivers. For example, a rectangular canal 10 feet wide and 5 feet deep has a cross-sectional area of 50 square feet. Multiply that by the flow rate of 5 feet per second and you get a water volume of 250 cubic feet per second. That's roughly 160 million gallons per day, enough for a city of a million and a half people.

Flow through an open canal is usually regulated by an adjustable weir, a ledge that the water must flow over. Raising or lowering the weir controls how much water enters the channel. Pressurized pipelines are controlled by valves, which are generally placed near high points along the route. These summits divide the conduit into sections that can be opened up for maintenance without emptying the entire length of the aqueduct. Also, pressure is at a minimum at the summit and so the valves are under the least stress there.

A large pipeline contains thousands of tons of water in each mile of its length. Even though the water is moving at only a few miles per hour, it carries the energy and momentum of a fast freight train. When the flow is suddenly stopped, all that energy must be dissipated somehow. If precautions are not taken, the pipeline can be torn apart by the resulting shock, which is called a water hammer. (Depending on

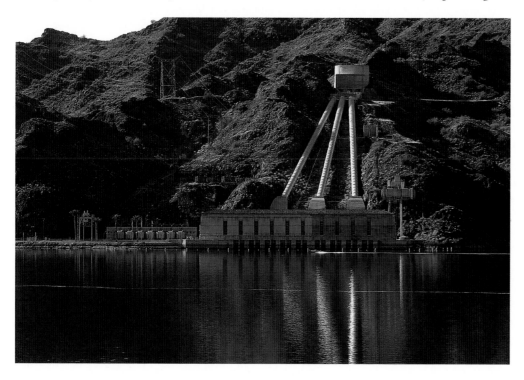

A pumping station on the California shore of Lake Havasu lifts water over the Whipple Mountains and into the Colorado River Aqueduct, which carries it more than 240 miles to the outskirts of Los Angeles. The pumping station consumes half the power output of Parker Dam (a few miles downstream) and has the capacity to lift a billion gallons a day. The pumps are in the large building at the water's edge; they push water upward through the three large steel conduits to a concrete structure where the aqueduct begins as a tunnel through a range of hills.

the state of your household plumbing, you might experience the same thing on a smaller scale when you suddenly shut the bathroom faucet.) The remedy for water hammer is a surge tank near each regulating valve. When a valve is closed, water is diverted into the tank, where it dissipates its kinetic energy; when the valve is opened again, water drains from the tank, helping to accelerate the flow through the pipeline.

Many aqueduct systems run entirely on gravity, which is certainly the motive force of choice: it's everywhere, it never needs lubrication, and it's free. The trouble is, it doesn't run uphill.

A TALE OF TWO CITIES

America's two largest cities both came close to dying of thirst.

New York survived on privately owned wells until the 1840s. (The best of them, according to contemporary accounts, was the Tea Water Pump, near what is now Foley Square in Lower Manhattan.) There was never enough water, and a series of fires finally prompted the city to seek a better supply, 40 miles to the north. The first Croton Aqueduct, tracing a path along the eastern bank of the Hudson River, delivered its water to a receiving reservoir in Central Park (which still exists), and from there it was piped to a grandiose distributing reservoir, decorated in an Egyptian motif, at Fifth Avenue and 42nd Street—later the site of the New York Public Library and Bryant Park.

By 1885 the city's population had quadrupled, and demand again outstripped supply. A larger dam was built at Croton, along with a new Croton Aqueduct. The right-of-way cleared for the buried aqueduct has created a pleasant path through suburban towns where joggers and dog-walkers have no idea that the city's drinking water flows a few feet below.

By the time the Croton system was complete, demand was already exceeding its capacity, and the Board of Water Supply began looking still farther north and on the other side of the Hudson River, in the Catskill Mountains. The water from there reaches the city through two deep tunnels, bored through

bedrock as much as 2,500 feet below the local terrain.

But that's not the end of the story. Since 1970 crews have been laboring to build a third tunnel, which won't be finished until 2020—even if all goes according to plan. The various branches of this new artery, 24 feet in diameter and 60 miles long, will allow some of the older tunnels to be closed off for repair for the first time in almost a century.

In Los Angeles, the construction of the water system is the stuff of myth and legend. They make movies about it. And why not? The story is full of drama, and not a little skullduggery.

The hero—or maybe the villain—was William Mulholland, who began as a zanjero, or ditch tender, and rose to become emperor of the Los Angeles Department of Water and Power. In 1904, when the 175,000 residents of the city were already running short of water, Mulholland conceived a plan to import it from the Owens Valley, 250 miles north and on the far side of the Sierra Nevada. By 1913 the Los Angeles Aqueduct was carrying more than 250 million gallons per day.

As one might guess, the ranchers and farmers of the Owens Valley were not happy with this arrangement. Over the next 10 years they attacked the aqueduct with lawyers and dynamite, but neither had any lasting effect. More devastating was the failure of the St. Francis Dam, in the hills north of Los Angeles. On the night of March 12, 1928, the dam washed

away and 12 billion gallons of Owens Valley water scoured out the valley of the Santa Clara River, killing at least 400. Mulholland suspected sabotage, but it seems the dam was built on geologically treacherous terrain.

By then the city's population was approaching two million, and draining the entire Owens River through the aqueduct would not have quenched the city's thirst. The new plan was to tap the Colorado River, with an aqueduct reaching 240 miles to a dam on the Arizona border. Again, folks elsewhere had other ideas about how best to use the water that Los Angeles wanted. In 1934 the "Arizona Navy"—actually a contingent of the state militia in a borrowed ferryboat—was dispatched to stop construction of Parker Dam. They failed, and so did a subsequent court challenge. Today, 4.4 million acre-feet of water from the Colorado River is pumped across the mountains.

Meanwhile the original Owens Valley aqueduct has been extended farther north to capture additional water, diverted from streams in Mono Crater, near the Nevada state line. The lower part of the aqueduct couldn't handle the additional volume, and so a second aqueduct was built alongside it. And the city has access to a third major supply: the canals of the State Water Project bringing water all the way from Oroville Dam, north of Sacramento, and Shasta Dam, up toward the Oregon border.

But the 15 million people of the Los Angeles metropolitan area are still thirsty.

A pumping station, when one is needed, tends to look remarkably like a hydro-electric installation. Like a hydro plant, the pumps are found somewhere in the neighborhood of a dam or reservoir, and they have enormous pipes coming in one side and equally impressive electrical connections on the other. But the pumping station is exactly the opposite of a hydroelectric plant. Instead of extracting energy from falling water to produce electricity, it consumes electricity to lift the water. (If you are unsure which one you are looking at, note that pipes from a pumping station rise *above* the surface level of the reservoir, whereas those for a generating station descend *below* reservoir level.)

THE FILTRATION PLANT

Some cities are lucky enough to have a source of water so pure that it can be piped into homes with no treatment beyond a small dose of chlorine for disinfection. New York City is in this happy category—although it has struggled to maintain water-quality standards in recent years. Elsewhere, the water has to be cleaned up to some extent before it's suitable for human consumption. This is the task of a filtration plant.

But a filtration plant does more than filter. In most cities it is the headquarters and nerve center of the water department. It is where the engineers and technicians hang out, where plans are made, where equipment is repaired, where the quantity of water is measured and the quality monitored. There's a laboratory on the site, and a control room. For all that, though, the filtration plant also does some filtering, as well as other kinds of water treatment.

A close look at the water in a reservoir, especially after a big storm, makes it obvious why filtration is often a good idea. At times the water carries a heavy load of silt and sediment you wouldn't want to brush your teeth with. And some of the invisible contaminants—viruses and bacteria—are even more unwelcome.

Let's follow a molecule of water through a filter plant. The aqueduct that supplies the plant discharges its water into a receiving reservoir that holds a few days' supply. From there the water passes through another trash rack and screen (let's keep those polliwogs out of the pipes!), and then into a mixing basin. This is a small tank, per-haps 10 feet deep, with a motor-driven agitator and equipment for adding various chemicals to the water. The agitator and its overhead motor look something like an outsize milkshake machine, and they have a similar effect on the water in the basin.

The milkshake being mixed is a dilute chemical concoction. In most cases the main ingredient—besides the water, of course—is alum, a form of aluminum sulfate, but there may also be iron compounds and tiny quantities of polymers. The chemi-cals are coagulants whose function is to bind together minute particles suspended in the water, creating much larger particles called flocs.

From the violently stirred mixing tank, the water passes into the quietest section of the plant, a settling basin where the flocs grow larger and heavier and very grad-

A rapid-mixing tank *(above)* begins the purification process in a water-treatment plant. The primary goal is to remove small solid particles suspended in the water—anything from silt to algae and other micro-organisms—by adding chemicals that cause the con-taminants to clump together in larger particles called flocs. After the chemicals are stirred in, the water flows through several more tanks with gentler agitation *(below)* and then into a larger basin, where the flocs settle to the bottom.

ually fall to the bottom. There are two styles of settling basins. In the circular ones, water flows in at the center and progresses toward the rim, slowing down as it goes, until finally it escapes over a weir that runs all around the perimeter. In rectangular basins the water enters at one end and flows at constant velocity until it is withdrawn at the other end. In both types of basins it is crucial that the flow be smooth and sedate and without turbulence, which would break up the flocs and prevent them from settling.

Walking along the deck beside a rectangular settling basin, you can watch coagulation, flocculation, and settling going on before your eyes. At first the water is somewhat murky and discolored, with its load of fine silt and organic matter. Then, ten steps farther on, the water begins to clarify, but it is suddenly full of tiny white or gray or brown flakes, the size of dandruff. Still farther along, the flakes become larger but fewer, as they coalesce. Then, very gradually, over the remaining length of the basin, the flocs begin to sink, like a slow-motion snowfall, until at the far end clear water flows out of the basin.

In some settling basins the flocs accumulate as a sludge on the bottom; every few weeks or months the basin has to be shut down, drained, and mucked out. More commonly today the sludge is continuously and automatically collected. Circular basins have slowly rotating rakes that gently push the sludge into a central drain. Rectangular basins have conveyor belts that carry the sludge to one end of the tank, where it is scraped off into a hopper. The sediment dries to a gray cake that's not hazardous or smelly, but it's not good for much of anything either. Generally it winds up in a landfill.

From the settling basin the clarified water passes into a bank of filters. These are nothing like the paper filters that you put in your coffeemaker. They are beds of sand and gravel and rock, six or eight feet thick. The water must find its way through the filter medium from top to bottom.

The earliest such filters, now known as slow sand filters, had a tightly packed bed of fine beach sand. Water forced through the filter would leave most of its larger particles at the surface or within the first few inches of sand. This deposit would clog the pores in the top of the sand bed, making it even more likely that additional solids would be trapped there. After a few days, algae and bacteria would begin growing on the rich, moist organic matter, creating a slimy film known by the wonderfully redolent German term *Schmutzdecke* (filth layer). Eventually the filter would become too clogged to continue in service. Then a few inches of fouled sand would be scraped off and the cycle would begin anew.

Slow sand filters are rare today. They have been replaced by the rapid sand filter, which is smaller and more efficient and has no *Schmutzdecke* to be raked up. The key idea in the rapid sand filter is to pump water at high velocity through carefully arranged layers of sand, gravel, pebbles, and rocks. Suspended particles of various sizes are trapped at different levels within the filter, instead of all accumulating at the surface. This makes efficient use of the entire depth of the filter bed.

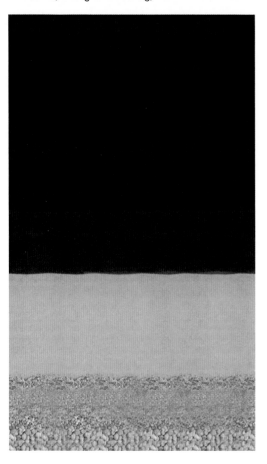

A rapid sand filter, shown here in a schematic diagram, consists of several layers of granular materials that have the remarkable property of spontaneously reassembling when the layered structure is disrupted. At the top is a thick stratum of powdered anthracite coal, then a layer of sand and finally a base of gravel and pebbles. The entire filter is about six feet thick. During normal operation, water flows in at the top and out at the bottom; during backwashing, the flow is reversed.

Backwashing a sand filter is a daily routine at many
water-treatment plants. The process begins with blow-
ing air into the bottom of the filter bed (upper photo) to
dislodge dirt deposited in the pores between sand
grains. Then water is pumped up through the bottom of
the filter to complete the washing process (lower photo).

A more efficient filter is one that gets dirty faster, and so it has to be cleaned more
often—typically once a day. The rapid sand filter also gets dirty through and through,
and accordingly it must be cleaned from top to bottom, not just on the surface. The
cleaning is done by an amazing process called backwashing. During normal operation,
a small amount of filtered water is set aside in a storage tank. When the filter needs to
be washed, this water is passed back through the bed in the reverse direction, coming
up from the bottom, at a carefully controlled rate. The reverse flow "fluidizes" the fil-
ter bed, suspending the sand, gravel, and other constituents in a roiling stew that shakes

loose the contaminants. At some plants, air is also bubbled up through the bed of granules, further disrupting the strata. The amazing part is that when the backwashing stops, the filter bed spontaneously reassembles itself into a tidy, layered structure.

At one filtration plant I've visited, a cross section of the filter bed had been put on display in a tall glass case, where its multicolored layers made it look like one of those Jello-chiffon dessert creations. At the top was a thick, jet-black bed of powdered anthracite coal, followed by a layer of reddish beach sand, and at the bottom a foundation of progressively coarser gravels and stones. The rationale for this specific assortment of layers is that the materials are arranged in order of increasing density, so the sequence of layers is stable even during the most violent backwashing. The lightweight powdered coal always rises to the top of the bed, and the heavy gravels remain at the bottom, with the fine sand in the middle.

After filtering, the last major step in water treatment is chlorination, which is meant to kill any microorganisms that might have survived passage through the earlier stages. At most large plants the chlorine is received in steel cylinders or delivered in bulk by tanker trucks and is stored as a liquid under pressure. This is the one dangerous aspect of the entire water-treatment operation. Chlorine is one of the poison gases of World War I, and a major release would be very unpleasant. Small water systems and those in densely populated neighborhoods can avoid the hazards of storing chlorine by making the gas as needed. The usual starting material is sodium hypochlorite, the active ingredient in chlorine laundry bleach. Whatever the source of the chlorine gas, it is dissolved in a small quantity of warm water, making a highly concentrated solution, which is then mixed with the output of the filters. Fluoride is added in a similar way.

The chlorinated water goes into a pure-water holding tank, which these days is likely to be quite large. Some storage capacity has always been needed to smooth out daily fluctuations in demand. (For the water department, the morning rush hour comes a little earlier and the evening rush hour a little later than for the traffic department.) Today water is being held for another reason: to give the chlorine more time to disinfect. A chlorine residence time of 24 hours is recommended, and so the tank must be large enough to retain a full day's supply.

Chlorination and other methods of disinfection are current areas of controversy. Regulatory agencies have been setting stricter standards for bacterial and viral contamination of drinking water, which seem to demand higher levels of chlorination. But some of the excess chlorine combines with organic molecules in the water, creating chlorinated compounds that are also subject to stringent limits. Thus, another disinfectant may be needed. The leading candidate is ozone, a highly reactive form of oxygen created in electrical sparks and high-voltage discharges—and, ironically, a major component of air pollution.

Two general aspects of the operation of a filtration plant are worth mentioning. First, the operators tend to make cautious and gradual adjustments to the process, rather than dramatic interventions. If the acidity of the water begins increasing, they

The outflow of wash water gives a dramatic indication of the quantity of material trapped by a sand filter. The filter bank on the right is near the end of the wash cycle, and the water is flowing clean, in sharp contrast to the brown water from the filter bank on the left, where backwashing is still in process.

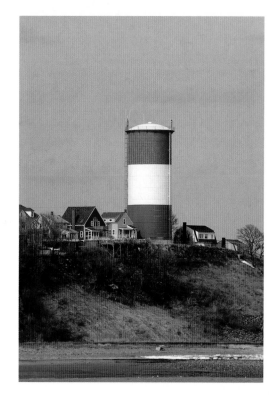

Rooftop tanks made of wood staves bound by iron barrel hoops are the most traditional choice for water storage, still found in many large cities. These specimens are atop a building in Midtown Manhattan.

The standpipe, a cylinder set on end, is another type of water tank with a long history. This tricolored one is on a hill in Winthrop, Massachusetts.

might add a very small dose of alkali to correct the imbalance, then wait half an hour to gauge the effect of this action before making any further changes. The idea is to keep the system at a point of stable equilibrium rather than constantly fiddling to maintain an unstable balance. The second distinctive trait is an emphasis on redundancy and reserves. There is more than one of everything at the plant, so that no single failure can shut down the entire operation. One large pump or mixer might be cheaper, but two small ones are more reliable.

WATER TANKS AND TOWERS

On the flatlands they are a familiar sight and almost a civic institution. If you want to know where you are, you look for the name of the town blazoned on the water tower. There's a fair chance you'll also find the spray-painted nickname of the high school football team and the name of somebody's sweetheart.

Elevated water tanks and towers serve two purposes (apart from their role as billboards). They store water that is ready for consumption, and they pressurize the distribution system. If the water were kept at ground level, it would have to be pumped to your faucet. Furthermore, the pumps would have to be big enough to keep up with peak demand, and every interruption in their operation would leave you without water. By using the pumps to lift water 100 feet in the air, a city can get by with smaller pumps that run all night. Moreover, brief outages go unnoticed.

The past century has seen a fascinating stylistic evolution in water tanks. The oldest ones are cylindrical wooden barrels, with vertical staves held together by steel

In a gallery of multilegged, elevated water tanks (opposite page), older forms with conical roofs are in the upper row and more rounded shapes in the lower row. From top to bottom and from left to right the tanks were photographed in Red Rock, Arizona; Gulfport, Mississippi; Percival, Iowa; Platte City, Missouri; Greensboro, North Carolina, and Keokuk, Iowa.

The water tank as monumental architecture: The imposing brick and masonry structure in the photo below holds the water supply for the town of Manistique, on the Upper Peninsula of Michigan.

hoops. Notice that the hoops are more closely spaced toward the bottom of the tank, where the pressure is greatest. The best place to look for these tanks is on the roofs of older city buildings. Manhattan has a splendid collection of them.

The wooden tanks were later displaced by riveted or bolted steel. Typically a steel cylinder is mounted on spindly Erector Set–legs and topped with a conical hat like the one the Tin Woodman wears. You'll find examples of this style at many an abandoned factory.

In the 1930s a less angular design became popular. The tank itself has smooth curves where the sides meet the roof and the floor, and the curvaceous motif is continued in supporting legs, which are tubular columns rather than angular girders. The steel panels are welded rather than riveted. Tanks in this basic style are still being built. They come in sizes from about 50,000 gallons to three million gallons. The biggest ones have a squashed look, since the tank is made wider but not much higher.

The most dramatic innovation in the evolution of water tanks was the sudden appearance of the monopod—the spherical or ellipsoidal tank supported by a single flaring stalk. The first of these was built in 1939 (in Longmont, Colorado), but they didn't begin to catch on until the 1950s. Since the 1970s they've been sprouting like mushrooms. (They even look a little like mushrooms.)

The Watersphere and Waterspheroid, as they are officially called, are creations of Chicago Bridge and Iron (CBI), a company that dominates the world market for water tanks. CBI is also the inventor of the fluted column, in which a flat cylindrical tank with a funnel-shaped bottom is perched atop a broad column that looks like it's made of corrugated cardboard. The corrugations, or fluting, are functional: they improve the rigidity of the thin-shelled tube. Waterspheres range in size from 25,000 to 150,000 gallons; Waterspheroids go up to two million gallons, and fluted columns can hold up to three million.

What's inside a water tank? "Water" is not a sufficient answer; there's more to it than that. In all the CBI monopod designs, the column, or pedestal, is dry. Pumped water reaches the tank at the top of the structure through a riser pipe, perhaps a foot or two in diameter, inside the stalk. There is also an overflow pipe, just in case. A ladder inside the pedestal ascends to the base of the water tank and then continues through a tube that passes right through the middle of the tank to a hatch in the roof. (Topologists will recognize that because of the tube through the middle, the Watersphere is really a Watertorus.)

The fact is, most of the volume inside the base of most water towers is totally vacant. Some towns have installed pumps and valves there; some use the space for storage; at least one fluted column tower has a fire station in its pedestal.

A selling point of the monopod towers is security. Because the ladder is inside the pedestal, behind a steel door, the tanks are less vulnerable to unauthorized decoration. Some of the *authorized* decoration is bad enough. There is the inevitable golf ball on a tee, as well as at least one baseball, an eight ball, and a yellow smiley face. There are two nearly identical Peachoids, one in Clanton, Alabama, and the other in

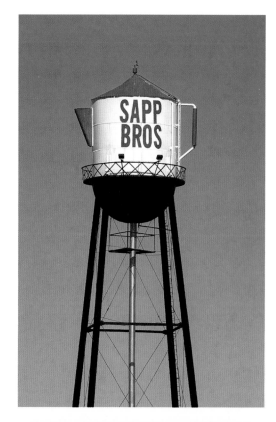

The largest water tanks tend to be fat and flat rather than tall and thin. For a tank of any given storage capacity, the horizontal configuration yields a more nearly uniform pressure in the water supply line.

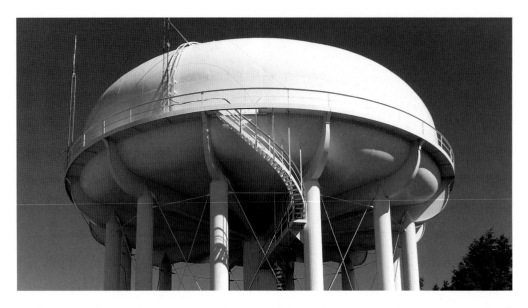

Gaffney, South Carolina. But I prefer any of these fantasy motifs to towers painted insipid swimming-pool blue. The towers are interesting structures, and they should stand out against the sky, not hide in its glare.

WATER DISTRIBUTION

Suppose your house is on the other side of town from the waterworks, at the end of a long pipeline that serves everyone else before it gets to you. In the middle of the night, when no water is flowing, the pressure at your faucet is the same as it is everywhere else in town. This is the static pressure, determined entirely by the "head": the difference in elevation between the storage reservoir and the point of measurement. But when your neighbors get up in the morning and flush the toilet and turn on the shower, you may get only a feeble trickle from your tap. As water flows through the pipe, pressure is used up in overcoming the friction that opposes the flow. The greater the flow rate and the greater the friction, the greater the loss of pressure.

Avoiding problems like this one is what the design of water-distribution systems is all about. One solution is to build a water tower or reservoir at each end of the line. This layout is very common for small municipal water systems: the main pumping station and reservoir are on one side of town, and an elevated tank is on the opposite side. During the night, water is pumped across town into the tank, then during periods of high demand, water flows into town from both ends of the pipeline.

The pressure in a residential water system is about the same as it is in an automobile tire—somewhere between 30 and 40 pounds per square inch. The minimum allowable pressure is usually considered 20 pounds per square inch, which corresponds to a head of 46 feet. The maximum pressure is about 60 pounds; beyond that, plumbing fixtures begin leaking.

Monopod water tanks *(opposite page)* seem to invite whimsical decoration. From top to bottom and left to right the tanks shown are in Rosemount, Illinois; Rend Lake, Illinois; Gaffney, South Carolina; Germantown, Maryland; Monrovia, Maryland, and Deer Island, Massachusetts.

Shutoff valves are an essential component of every water distribution system, but in most places they are hidden underground; these exposed valves are in San Diego, where freezing is not a concern. The valves are the tall red devices with threaded shafts at the top. The blue objects between the valves are backflow preventers, which ensure that water flows through the pipe in one direction only.

Cryptic metal placards fastened to a wall in Budapest, Hungary, are markers indicating the location of valves and other utility equipment under the adjacent street.

Keeping pressures within the allowable range gets tricky in a hilly area. The pressure needed to get water to the hilltop houses could burst pipes in the valley. Lots of water systems are divided into pressure zones, with connections between them through pressure-regulating valves.

Almost all the components of the water-distribution system are out of sight underground. They are protected from many hazards there, but the main reason for burying them is to avoid freezing. The depth varies with latitude, from 3 or 4 feet in southern U.S. states to 6 or 8 feet in northern states and 12 to 15 feet in Alaska (where the pipes freeze anyway). Textbooks on water-system engineering state that supply mains are generally installed on the north side of the street in the Northern Hemisphere and on the south side in the Southern Hemisphere, so the sun will warm them. In both hemispheres they are supposed to be on the east side of north-south streets, on the premise that the afternoon sun is warmer than the morning sun. I have not been able to confirm that these principles are actually followed in practice. My sample is small and confined to the Northern Hemisphere, but I have found roughly as many cities that violate the rules as follow them, along with a great many middle-of-the-roaders.

Water mains are out of sight and mind until they break—at which point the evening news has film of geysers in the street and caverns that swallow up parked cars. How does it happen? Does a big pipe just get tired and suddenly blow open like a bursting balloon? That can happen, but it's not the usual way. More often, a small leak develops and remains undetected for weeks or months. The escaping water washes away the soil supporting a section of the main. Eventually the unsupported span of pipe is so long that it breaks. That's when the slow leak becomes a flood.

At each intersection in a water-supply grid, valves are installed to isolate the segments in case there is a break or it needs to be closed for repair. But valves on large

pipelines are expensive and troublesome (they stick and they leak, and they need periodic maintenance), so the water company doesn't install any more than are absolutely needed. There is a simple rule of thumb for solving this problem. At any "+"-shaped intersection, three of the four branches should be protected by valves; and at a T-shaped intersection, two of the three branches have valves.

Occasionally you may see water-company workers driving long steel stakes into the ground and then pulling them up. They are looking for leaks by testing for soft, moist ground. They may also simply listen for the sound of water running underground with a sensitive microphone and amplifier. Other signs of leaks are high flows of water in the middle of the night, excessive flows in sewers, and areas of unusually green vegetation.

Fire Fighting. To a larger extent than you might guess, the design of water systems is influenced by the needs of fire fighting rather than household supply. Often, what provoked a city to build its first large-scale water system was a fire. And fire fighting probably dictates the maximum performance of the system.

Fighting a fire requires both a large volume of water and high pressure. Installing big enough pipes is all that is necessary to deliver the large volume. Getting the pressure right is more difficult. If the pressure at the hydrant is high enough for effective fire fighting, household fixtures spring leaks. One solution is booster pumps that are turned on only when needed, thus minimizing the damage to residential plumbing. For a long time New York City had a completely separate high-pressure distribution system used only for fire fighting. There were high-pressure zones in Manhattan and Brooklyn, with their own heavy-duty fire plugs. The separate high-pressure system was shut down in 1957, because pumper trucks had made it obsolete.

The kind of fire hydrant seen at the curbline in most North American cities is called a dry-barrel hydrant. The valve that shuts off the flow of water is not in the exposed part of the hydrant but is several feet underground, where the barrel of the hydrant connects to the water main. This design has two advantages. First, because the barrel is dry, there is no problem with freezing. Second, the hydrant can be

Fire hydrants differ considerably in shape and even more so in paint scheme, but most of them have the same functional parts. There are usually three connection points: a large hose fitting for supplying water to a pumper truck and two smaller outlets for fire hoses. The hydrants below, from left to right, were found on the streets of San Diego; Tryon, North Carolina; New Orleans; Ljubljana, Slovenia; and Chicago, Illinois. The brass hardware in the photo above, mounted on the facade of a New York City office building, is a connection point where fire hoses can pump water into the building's sprinkler system and the standpipe that supplies hose connections on each floor.

Early sanitation works: above, a drain discovered in the ruins of the Greek and Roman town of Velia in southern Italy; below, a privy at Eckley Miners Village, near Hazleton, Pennsylvania. (The untrammeled snow suggests the outhouse is strictly for show.)

designed so that even if it breaks off at the ground line, the valve will remain closed. In the movies, a car running into a fire hydrant never fails to produce a geyser, but in real life this is rare.

Most hydrants today have three outlets—two for a two-and-a-half-inch fire hose and one for a four-inch "suction hose." The larger hose gets connected to a pressure-boosting pumper truck.

In many areas hydrants are color-coded to indicate their flow capacity. In one scheme the barrel of all hydrants is painted chrome yellow and the bonnet color gives the capacity: green for hydrants capable of delivering 1,000 gallons per minute or more, orange for 500 to 1,000 gallons per minute, and red for less than 500 gallons per minute. There are also various ways of marking the locations of hydrants, such as colored bands high on utility poles and reflectors embedded in the centerline of the street. Years ago on a visit to upstate New York, I was puzzled by the pennants on tall metal poles planted next to all the hydrants. That was in October; I understood their function when I returned in snowy February.

WASTE WATERS

The amazing thing about modern sewers is how modern they are. Although large-scale projects for bringing drinking water into the city go back to antiquity, pipes to carry it out again are a comparatively recent idea. In the 1750s the latest high-tech fad—and the epitome of fashion for middle-class New Yorkers—was the backyard privy. It eliminated the need to empty chamber pots and "ordure tubs." The first sanitary sewers were not built until a hundred years later. (Ancient Rome had sewers, and so did medieval Paris, but they were primarily for storm runoff.) Sewage-treatment plants became common only after World War I. Thus, in many places sewer systems were an afterthought that lagged behind such other modern conveniences as electricity and the telephone.

Sewer Lines. In many ways, a sewer is harder to design and build than a water-supply system. The water-distribution pipes that come into your home operate under pressure, which means they can go uphill and downhill as necessary. But sewers, with rare exceptions, flow only partially full, which means the sewer engineer is entirely at the mercy of gravity. The starting point for a sewer is typically seven or eight feet underground—where an iron pipe exits through the foundation of a building—and it's all downhill from there. In a system that works strictly by gravity flow, the pipes must slope continuously downward all the way to the sewage-treatment plant.

The exact slope of the pipes is also more important than it would be in a drinking-water conduit. Sewers transport not just water but also materials that are politely known in the trade as "solids." To keep the solids moving, the water has to flow at a speed of two or three feet per second. To maintain that speed, a large sewer needs

to dip by about 50 feet per mile of length, and small sewers are even steeper. The profile of the whole system has to be carefully planned before construction begins; if a certain sewer turns out to be a couple of feet higher than one that's supposed to dump into it, that's big trouble. And details count. For example, where the stream of sewage is forced to make a turn, the curvature causes a little extra friction that tends to slow the fluid down, and so the slope needs to be increased slightly where the pipeline turns a corner.

Because sewer lines start at foundation level and descend from there, you shouldn't expect to see much of them except during construction and repair. They are generally the most deeply buried of all utilities. It also follows that if you go looking for a sewage-treatment plant, you're not going to find it on a hilltop. Often it's on the very lowest land in town.

The small tributaries of a sewer system are called laterals. They dump into submains, which in turn lead to mains and trunks and interceptors. The entire system is arranged in a branching pattern, like a tree. Not coincidentally, it is the same kind of pattern formed by a natural drainage system made up of streams and rivers.

Small sewer pipes are generally made of vitrified clay, which is a glazed, bricklike material. The smooth surface helps to speed the flow and prevent clogging, and it also resists erosion by all the abrasive grit swept along. Larger sewers are made of cast concrete, which lacks these special properties but is cheaper and stronger. Both materials are brittle, so they need to be firmly supported underneath and protected overhead. If you watch a crew laying a sewer line, you'll see that they treat the sections of pipe very gently as they lay them in a bed of crushed stone.

A sewage pumping station in Superior, Wisconsin, is helpfully labeled for easy identification. (Many others are deliberately disguised.)

Sewer lines and access manholes, generally subterranean structures, are exposed here where a sewer crosses a streambed near Butner, North Carolina.

Pumps and Lift Stations. In hydraulic heaven, every city is built on a hillside, with the drinking-water reservoir above and the sewage-treatment plant below. Water flows straight through by gravity alone. The geography of cities in this world is not always so convenient, and sewage sometimes has to be lifted from a low-lying neighborhood to a sewer at a higher elevation.

Sewage pumps face some challenges that don't arise for water-supply pumps. It's those "solids" again. Sand and grit erode the metal of the pump, and the occasional chunk of wood can get jammed in the impeller. But the biggest problem is rags, which get wrapped around drive shafts, jammed into crevices, and snagged on guide vanes. How do rags get into sewers? Do people flush old clothes down the toilet? Do socks get sucked into the drain of the washing machine? Who knows.

STREET ART

Mimi Melnick and her late husband, Robert A. Melnick, were Los Angeles art collectors who wanted a manhole cover to hang above their sofa. After some preliminary investigation, they

decided that sitting beneath a 300-pound slab of cast iron wasn't such a good idea after all, and they settled for a photograph instead. Over the years they collected hundreds of photographs, publishing many of them in their book *Manhole Covers* (MIT Press, 1994). A worldwide community of "drainspotters" carries on the tradition.

The raised patterns on manhole covers are not merely decorative. They are there to improve traction, originally for horses but now for automobiles and pedestrians. They also help to identify the manhole to utility crews. For

example, a beehive pattern of hexagons *(photo at left)* has been adopted by telephone companies since the 1920s. Other commonly seen markings include waffle-iron, basketweave, and diamond-shaped patterns. Most of the designs are standardized, but the Melnicks found some real collectors' items, and several cities have commissioned artists to produce special decorative covers.

Many covers identify not only the utility

company that owns the manhole but also the foundry that made the lid. The Neenah Foundry in Wisconsin has its name on manholes in hundreds of U.S. cities. In recent years there has also been a blooming of manhole covers labeled "Made in India" *(photo above)*. You might think that any advantage in labor

costs would be wiped out by shipping charges, but evidently not.

Manhole covers are manufactured as sand castings; that is, molten iron is poured into a mold made of moist sand. The cover and the seating ring in which it rests are then machined individually so that they mate without rocking. (A loose cover clatters annoyingly every time a car runs over it.)

The vast majority of manhole covers are circular, supposedly because lids of other shapes could fall into their own hole. (Also, a round cover can be rolled out of the way instead of carried.) Nevertheless, nonround lids do exist. A good place to look for equilateral triangles

is Nashua, New Hampshire *(photo above),* although Nashua has recently caved in to the round-lid standard.

The typical sewage lift station doesn't look like much from the outside. It's a small brick or cinderblock building, usually unattended. Immediately inside the door are the motors that drive the pumps, and a control panel; this electrical equipment is kept above ground level to minimize damage when the station is flooded. And the hazard of flooding is almost unavoidable; lift stations have to be built in low-lying areas, because that's where the sewage needs to be pumped out.

What you can't see from outside the little brick building is that it has a basement going down two or three stories. Beneath the motor room, the structure is divided into a wet well (which receives the incoming sewage) and a dry well (which houses the pumps, driven by long shafts from the motors above). With the two-well arrangement, no one has to dive into a pit full of sewage when a pump needs maintenance.

Another piece of equipment to be found at most sewage pumping stations is a diesel generator. The pumps cannot be allowed to stop for more than a few hours at a time, lest sewage back up into somebody's basement.

In residential neighborhoods some sewage-pumping stations are trompe l'oeil fantasies, dressed up to look like the split-level or colonial houses that surround them. If you look closely, it's not hard to spot these disguised pumphouses: the heavy-duty power connections, the big ventilating fans, and the diesel generator in the backyard are all tip-offs. Furthermore, the windows are often fakes, with curtains and shutters adorning a blank wall.

Manholes. At the surface, the only visible sign of most sewers is a line of metal manhole covers every hundred yards or so. But not all manholes give access to sewers; some open into vaults for water pipes, gas mains, or electrical and telephone cables. The lids of sanitary-sewer manholes tend to have more vent holes than other kinds of manhole covers. There is a reason for this: decomposing sewage gives off methane gas, which is flammable. So it's not a smart idea to drop lighted matches into a manhole to see what's down there.

The main function of sewer manholes is to provide access points for cleaning out the deposited solids. In earlier years this was a routine chore. Workers would climb down one manhole and push brushes through the sewer pipe to the next manhole, using long iron rods. Paris still has a cadre of *égoutiers* to flush out its famous sewers, but elsewhere in the world manual cleaning is uncommon. Nowadays, robotic vehicles roll through the pipes carrying TV cameras. And the emphasis is on designing sewers with enough slope that the solids keep moving and the pipes seldom need cleaning. Nevertheless, the placement of manholes is still determined by the maximum reach of the cleaning rods. As a rule, there is a manhole wherever the sewer pipe changes direction or slope, and every 300 or 400 feet on straight runs.

What's life like down a manhole? I've only visited two, and perhaps they've been showplace manholes, specially selected to make a good impression, but for what it's worth I found them not unpleasant. Temperatures were moderate, although the humidity was high. I climbed down a ladder of aluminum rungs set into the mason-

Fancy manhole covers are from Ljubljana, Slovenia (upper photo), and Budapest, Hungary (lower photo).

ry wall like giant staples. Sounds were deadened. The fragrance was strong but not overpowering. I saw no creepy-crawlies—no alligators, no rats, not even a spider.

The Sewage-Treatment Plant. If sewers are modern, then sewage-treatment plants must be postmodern. Although the basic concepts have been known for decades, the plants have been changing dramatically in the past 25 years. Even the name of the place has been evolving too fast for the sign painters to keep up. Although I persist in talking about *sewage treatment*, the preferred term long ago became *waste-water treatment*, and now it's *water reclamation*. At one town in New Jersey the site of the sewage plant and the garbage dump is officially an *environmental park*. By any name it smells as sweet.

A sewage plant has much in common with a filtration plant for drinking water. They are both in the business of purifying, and they use much of the same equip-

A sewage-treatment plant operated by the Metropolitan St. Louis Sewer District is seen in an aerial photograph made as part of the Urban Areas series of the U.S. Geological Survey. The circular vats at the bottom of the image and the rectangular tanks at the far right are clarifiers, or settling basins, where solids fall out of suspension. Just above the circular clarifiers are reactors for the activated-sludge process; half of the reactors are full of roiling, aerated fluids and half have been drained for maintenance. The tanks at the top of the image, with glinting metal roofs, are sludge digesters.

ment to do it. But whereas the treatment of drinking water is mainly a chemical and physical process, sewage treatment is very biological.

Before considering the details of treatment, we should pause to examine the nature of the material being treated. Sanitary engineers rate sewage as strong, medium, or weak, like coffee, according to the concentration of solids. They also classify sewage according to age and color. By all accounts, fresh sewage is the best sewage; it's gray. When sewage goes stale, it tends to turn black because of the action of anaerobic bacteria. Despite what's on everybody's mind when first looking into a 50,000-gallon vat of raw sewage, the largest contributions come not from the toilet but from the kitchen sink and the laundry.

The overall aim of treatment is to separate the solids from the liquid. Methods vary, but the devices mentioned here can be seen in any region of the United States.

The trash rack. Separation begins by intercepting the gross debris: two-by-fours, old galoshes, soggy tennis balls, lost spoons, goldfish. The trash rack is a grill or coarse mesh of iron bars installed where the sewage stream first enters the plant. Instead of a rack, some plants have a comminutor—a garbage grinder tough enough to chew up and spit out anything that comes down the pipe.

The grit chamber. You know all that sand you washed off in the shower after you came in from the beach? Here it is. The grit chamber is a tank or channel where the incoming sewage pauses just long enough to deposit the densest of the particles it has carried through the sewer system. The trick is to segregate out the sand and gravel—also eggshells, coffee grounds, and watermelon seeds—without collecting too much fine organic matter. One way to accomplish this is to gently agitate the sewage with a stream of air bubbles injected into the bottom of a tank. Another method is to run the sewage through a long sloping flume at just the right speed to settle out the grit but keep the finer matter in suspension.

These front-end devices—the trash rack and the grit chamber—are very simple, but they cause a disproportionate share of trouble for plant operators. The grit can't be allowed to flow through the rest of the plant because it could damage pumps and other equipment, and so it has to be disposed of elsewhere. For the grit itself, that's not a problem; it's mostly sand and could go to an ordinary landfill. Unfortunately the grit is contaminated with what the sanitary engineer describes as "putrescible matter," meaning that if it doesn't stink now, it will soon.

Settling tanks. From the grit chamber the sewage flows into a series of settling basins, much like those of a drinking-water purification plant. Here the pace slows. There is no agitation to disturb the peaceful, reflective waters. At the surface of each tank, a skimmer collects buoyant, greasy scum. Meanwhile, a rake or conveyor belt inching along the bottom collects the heavier sludge that settles out. (Scum and sludge: no sweet-smelling euphemisms here.) When the liquid leaves the settling basin, after a few hours in residence, it has lost most of the suspended particles.

The settling basins are usually the largest structures in a sewage plant. There are always at least two, and a large facility may have dozens. (Lots of little ones are pre-

Primary clarifiers, at the input end of a treatment plant, receive raw sewage and perform preliminary separation of solids and liquids. Heavy solids are raked off the bottom of the tank, and floating scum is skimmed from the surface. The two clarifiers below are identical in structure; one of them has been drained for repair. The photographs were made at the North Durham Water Reclamation Facility in North Carolina.

A trickling filter is a bed of crushed rock, providing a substrate for a community of living organisms that thrive on the nutrients in sewage. The filters shown are at the water reclamation facility in Henderson, North Carolina.

ferred to one big one so that a tank can be shut down for maintenance without crippling the whole operation.) Many older plants have round settling basins, anywhere from 20 to 200 feet in diameter. Water wells up in the center and gently overflows at the edges; the sediment falls to the bottom, where a rotating rake scrapes it into the central pit. Today's fashion in settling basins is more straight-edged. Rectangular tanks save space since they can be nestled together more efficiently, and they save concrete since adjacent tanks can share a wall. Some of the basins are as long as a football field.

The settling process is known in the sewage trade as primary treatment. Some plants do nothing more, though that's becoming rare. Secondary treatment can take either of two forms: trickling filters and the activated-sludge process.

Trickling filters. A trickling filter looks like a giant's lawn sprinkler. Four long arms extend radially from a central mast, turning slowly and spraying sewage from nozzles all along their length. The liquid falls onto a bed of coarse broken stone, contained

in a circular concrete vessel 8 or 10 feet deep. The effluent trickles through the stones to drains underneath.

From an engineering point of view, it doesn't look like there's much going on in a trickling filter. All the action is at the microscopic level, in the film of slime (that's the technical term) that builds up on the stone. This layer is, in fact, a whole ecological community that thrives on the nutrient broth raining down from above. Bacteria are the main sustainers of this ecosystem, but there are also many protozoa and a few algae, fungi, and insects, as well as various worms and even snails. One insect likes the environment so much it is known as the filter fly. It becomes a pest in the summer.

Activated sludge. The biology of the activated-sludge process is similar to that of the trickling filter, but it takes place in a different setting. Instead of raising a crop of bacteria on a solid surface, the activated-sludge process allows the microorganisms to swim free in a deep vat, called a reactor. To keep them growing, it's necessary to supply lots of air, and in some cases pure oxygen.

The vigorous aeration makes it easy to pick out an activated-sludge reactor among the many other water-filled vessels in a sewage plant. There are two kinds of aerators. Mechanical units work like a washing-machine agitator, whipping the sewage up into roiling, frothy mounds, like the Colorado River in flood, only browner. The alternative is to pump large volumes of air or oxygen to diffuser heads at the bottom of the tank so that streams of bubbles rise up through the liquid, as in an aquarium. Either process has a heavy energy demand for the large motors that drive the agitator or run the compressors.

Concrete partitions divide an activated-sludge reactor into several cells, which the sewage passes through in sequence. Each cell hosts a somewhat different communi-

The activated sludge process has replaced the trickling filter at most sewage-treatment plants. The biological community consuming the sewage is kept in suspension rather than forming a film on a solid substrate. Vigorous aeration, as seen in the nearest of the tanks here, helps to maintain an appropriate environment for the microorganisms. At this installation sewage flows slowly from the far end to the near end through a series of concrete cells. Each cell has a somewhat different complement of organisms, which digest different components of the sewage. In some cells a heavy froth collects on the surface—which can become solid enough to support the growth of grass. The system shown is at the North Durham facility in North Carolina.

Final stages of sewage processing: a secondary clarifier; sand filtration; ultraviolet disinfection; the outfall, where reclaimed water returns to the stream.

ty of organisms, which consume different nutrients in the sewage. For example, one group of bacteria breaks down proteins and other nitrogen-containing molecules into ammonia; another group converts the ammonia into nitrate compounds; and a third group decomposes nitrates into elemental nitrogen and oxygen. An outward sign of the biochemistry going on in the reactor is the development of froth on the surface. The froth begins as a light foam, like the head on a mug of beer; as it ages, it grows darker and stiffer, becoming more like a meringue. Sometimes the froth becomes solid enough to support the growth of grass and weeds.

Clarifiers. After the turmoil and turbulence of the activated-sludge reactor, the sewage returns to meditative tranquility. The clarifier is another settling basin where sludge slowly settles to the bottom and is gathered into a hopper. Compared with the gray soup in the first settling basin, the water here is marvelously transformed. You can see clear to the bottom, 10 feet below, where green algae wave serenely as if in a meadowland stream. In some cases, the clarifier is followed by a sand filter, much like the one at a drinking-water filtration plant. The filter's job is to remove any solids that remain in suspension.

In the activated-sludge process, some of the sludge collected from the clarifier or the sand filter is recycled back to the reactor. That's what makes the sludge "activated." As in making yogurt or bread, you have to save a bit from each batch to inoculate the next one.

Disinfectants. The water that comes out of the clarifier *looks* clean, but it still has a high titer of bacteria and other organisms, some of which you wouldn't want to go swimming with. Most sewage plants add a heavy dose of chlorine or chlorine-based bleach as the final step in water reclamation. Another option for disinfection is ultraviolet light. The treated effluent runs through a chamber fitted with hundreds of high-intensity ultraviolet lamps—a tanning parlor for sewage. The radiation is strong enough to kill most of the microorganisms.

Sludge digestion. The watery part of the sewage is now on its way downstream, but we are not yet finished with the solids. The sludge collected from the settling basins, clarifiers, and filters needs further treatment. The heart of the process is digestion. Here, the cast of microbial good guys and bad guys exchange hats: Digestion is an anaerobic process; it can take place only in the absence of oxygen. It also requires elevated temperatures. It is therefore done in tanks with a sealed roof to keep out air and with some means of temperature control, such as coils heated by warm water.

The main products of digestion are heavier solids, which settle to the bottom of the vessel, and gases such as carbon dioxide and methane, which rise to the surface and are carried off through a pipeline. The production of flammable methane is the telltale trademark of a sludge-digestion operation. You will generally see a flare near the digester for burning excess gas. Larger plants collect the gas and use it as fuel—if only for warming the digester—but the flare is usually present anyway, as a safety measure.

The fate of solids. Even digestion can't make sludge go away altogether. At this point, the material is dark brown or tarry, bubbling with gas and still saturated with

water. The gas soon dissipates, but the water has to be removed. In some places, the sludge is spread out on drying beds, where it gradually comes to look like thick cracked mud. This drying process is slow, especially if the weather doesn't cooperate. A quicker method squeezes the water out in a filter or a centrifuge.

Finally, at the very end of this long chain of concentration and separation steps, something must be done with the residue from the sludge dryer. It makes a fairly good fertilizer and soil conditioner, but not for fields where edible crops will be grown. The other choices are to bury it in a landfill or burn it in an incinerator.

STORM-WATER DRAINAGE

Early sewer systems were meant to carry away both household waste waters and rainwater. And why not? It all wound up in the same place anyway—dumped directly into the nearest lake or river. But now that sewage is being treated, we need to separate the two streams.

Sludge digesters at Deer Island, the treatment plant that handles sewage for the city of Boston and many surrounding communities, are immense egglike vessels. Sludge digestion is an anaerobic process, and so the vessels must seal out air.

The solution adopted in many cities is a hybrid arrangement. A single network of drainage pipes carries both sewage and rainwater runoff, but large storm flows are diverted away from the treatment plant and dumped directly into whatever body of water gets the outflow. The diversion isn't hard to accomplish and doesn't require someone to go out and turn a valve every time it starts raining. Think of an underground chamber with a small outlet pipe at the bottom and a much larger outlet several feet higher up the wall. As long as the rate of inflow is not too great, all the water in the chamber drains through the small lower outlet, which goes to the treatment plant. In a storm, the chamber fills, and the overflow goes into the larger, higher, diversion channel. Thus, whenever it rains hard enough, some untreated sewage is washed into a lake or stream. This situation is not ideal, but there is a mitigating factor: it can happen only when there is a large volume of rainwater to dilute the sewage.

More than 900 cities in the United States have combination sewers and storm drains. Mostly they are the older cities in the Northeast and the Great Lakes region, but there is also a concentration in the Pacific Northwest. The combined systems were built years ago, when pollution standards were looser, and it's now considered too expensive and disruptive in downtown neighborhoods to disentangle the two networks. But no one is building new sewers this way. All recent drainage plans provide separate conduits for sewage and storm water.

The design challenges for sanitary sewers and for storm drains are rather different. The flow rate in a sanitary sewer varies within a fairly narrow range, but a storm sewer has to be equipped to carry a trickle or a torrent. A small city might produce 10 million gallons of sewage a day, but a single thunderstorm could dump a billion gallons in an afternoon. Thus, storm drains have to be larger than sanitary sewers. On the other hand, they don't have to be buried as deeply because they don't need to be connected to basement building drains.

The most visible parts of a storm-sewer network are the street-level inlets, usually installed in the gutter or the curb. For obvious reasons they are placed at low points along the street. Often the hardest places to drain are underpasses beneath highways and railroad tracks.

In older storm-drain systems every inlet has a catch basin, a pit below the inlet grating that is three or four feet deeper than the level of the drain pipe. The idea is to trap leaves and dirt in the basin, so that they don't clog up the drain line. A work crew cleans the basin once a year or so. New storm drains adopt another approach. There is no basin to trap debris; instead, the inlet has smoothly curved and steeply sloped surfaces, designed so that anything falling into the drain will be flushed all the way through the system. (If you lose your car keys down a storm-sewer grate, pick an old one—you will have a better chance of getting them back.)

In recent years the philosophy of storm-drain design has changed in another way as well. For a long time the primary aim of drainage engineers was to conduct away as much water as possible as quickly as possible. But this policy amounts to sweeping the problem downstream: the puddles are drained from local parking lots and streets,

A curbside storm-water inlet in Toledo, Ohio, has an aptly fluid decorative motif.

but the receiving lake or river has to deal with a massive surge of dirty water, all arriving at once. The new approach emphasizes slowing the runoff rather than speeding it on its way, and wherever possible absorbing it at the source. Instead of smooth concrete pipes and canals, rainwater is collected in grass-lined ditches and swales, and the flow is slowed further by detention ponds. Many zoning and building codes now limit the "impervious area" of a lot—the fraction of the space covered by buildings and pavement—so that more rainwater will soak into the soil where it falls, and less will flow into the storm drains.

LIVING UNDER THE RIVER

No city takes flood control and storm-water management more seriously than New Orleans. Most of the city lies below sea level. And the meaning of "below sea level" is different here from what it is in a place like Death Valley, where the sea itself is nowhere in evidence and

its level is an abstraction, a line painted on a pole. In New Orleans—wedged between the Mississippi River and Lake Pontchartrain—there is nothing abstract about sea level. In some neighborhoods you have to climb a stairway to get from the street to the waterfront.

New Orleans differs from Death Valley in another way as well: it rains there, lots. Thus, it's not enough to build a wall around the city to keep the water out; the water that falls out of the sky has to be removed too. Since all the surrounding waters are at a higher elevation, there's no choice but to pump it.

The storm drains of New Orleans are cavernous canals that lie under the city's major streets—on the same scale as New York's sub-

way tunnels. Away from downtown, the tunnels deliver their water into even larger open canals, most of which flow toward Lake Pontchartrain. Then, at the levee, pumping stations lift the water up and over. There are 22 of these stations, with a combined capacity to move 30 billion gallons a day. Someone has calculated that the pumps could fill the local football stadium, the Superdome, in 35 minutes.

The biggest of all the pumping stations is No. 6, which straddles the 17th Street drainage canal on the northwestern boundary of the city. The entire front of the building is an enormous trash rack (photo above), with rakes that scoop debris directly into dump trucks. Inside the long brick building, 15 pumps and the motors that drive them are lined up in a row (photo right). They are axial-flow pumps—essentially a ship's propeller spinning inside a big pipe. Some are 12 feet in diameter and some are 14. What makes all these machines most remarkable is that they are antiques: most

of them have been running since before 1920.

Ironically, the greatest long-term threat to the well-being of New Orleans is not that it will drown, but that its lifeblood will dry up. Eighty miles upriver, a major portion of the annual Mississippi flood escapes westward into the Atchafalaya River, which takes a shorter path to the Gulf of Mexico. In the long run, the Atchafalaya channel will "capture" the flow of the Mississippi, creating a new outlet near Morgan City, Louisiana. On a geological time scale, this event is nothing remarkable: The mouth of the river has been wandering for millennia. But people who live along the river would strongly prefer that it stay put, and so the Army Corps of Engineers has erected a massive complex of concrete spillways and control gates to regulate the division of waters between the two channels. For now the situa-

tion is under control, but eventually the Atchafalaya channel will win, and the port of New Orleans will be struggling to keep the river flowing rather than to keep the river out.

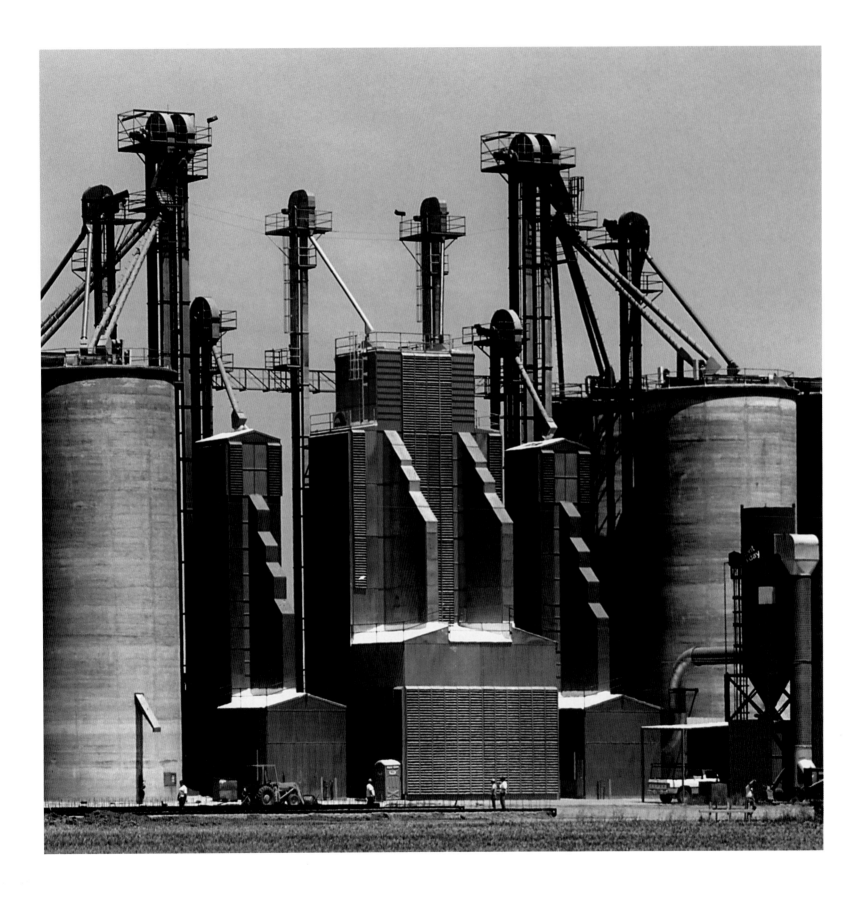

FOOD AND FARMING

WHAT IS A CHAPTER ON FARMING doing in a book about the *industrial* landscape? Most of us put agriculture in a different mental category from, say, mining or manufacturing. But farming is also an industry. Indeed, it was the first industry, and it remains in many ways the most essential. If we can't eat, we can't do much of anything else. And, as industries go, agriculture is a highly developed and mechanized one—at least in the highly developed and mechanized parts of the world.

Agriculture can also claim a place in this book because it is the industry that has the largest-scale impact on the landscape. Farming takes up more space than anything else people do on the planet, by a wide margin. Farmers and ranchers own fully half the land area of the United States, and they also use and shape large tracts of publicly owned land. When you look down on the countryside from an airplane window, agriculture almost always marks the clearest and most extensive sign of human presence.

WHERE HAVE ALL THE FARMERS GONE?

When America was young, we were a nation of farmers. In the years before 1800, at least 90 percent of the people made their living from the soil. A century later, 40 percent of the population was still farming. But the 2000 census found only 1 percent of Americans living on farms. This exodus from agriculture has to be counted as one of the most stupendous turning points in all of human history. For several thousand years, just about everybody was a farmer; now, all of a sudden, almost nobody is. The change overtook us within just a few generations.

What allowed so many people to leave the farm—or *compelled* them to leave—was a phenomenal spurt in agricultural productivity. In the old days, it was either grow your own or starve. On average, it took the labor of nine farmers to feed and clothe

The monumental grain elevator on the opposite page has a prosaic function—it is a warehouse for farm produce—but it also represents an imposing landmark, with stylistic elements that suggest both Art Deco skyscrapers and the flying buttresses of a Gothic cathedral. The tallest, slender towers are mechanical conveyors that lift the grain; the large concrete cylinders are storage bunkers; the red and metallic structures in the middle are grain driers. The elevator is in Williams, California, in the Central Valley.

10 people—a ratio that imposed a pretty stringent limit on how many of us could run away to be Young Urban Professionals. Today, a tiny fraction of the population grows enough for all of us. More than enough: in most years the United States has huge surpluses of food to ship overseas.

This transformation of modern agriculture sounds like a rip-roaring success story—and it is. Feast is better than famine; no one would trade the present bounty for the melancholy vision of Thomas Malthus, who predicted two centuries ago that population would inevitably outrun the food supply. And yet the miracle of agricultural technology has had costs too. Farmers themselves have paid the steepest price: their own efficiency has put many of them right out of business. And the dislocations extend far beyond the midwestern corn belt. Because of those exported surpluses, farmers in Bangladesh and Zimbabwe are thrown into direct competition with well-capitalized—and sometimes subsidized—American agribusiness.

Viewed at the scale of continents and centuries, the depopulation of rural America seems to be a consequence of implacable economic forces. Seen at closer range, however, vast demographic trends always seem to resolve themselves into a family gathered at the kitchen table to deal with a crisis. As times get hard, the farm squeaks by on grit and goodwill for a few years. Then some last reversal—bad weather, falling prices, rising interest rates—brings the gavel down. We long to comfort them with the thought that they'll find better lives elsewhere, or their children will. I've heard it said that both the cause and the effect of civilization is getting people off the farm and into the city. But I wouldn't say it to a family whose land has just been sold at auction.

Ironically, there are also places in the United States where agriculture faces exactly the opposite challenge. As cities grow—swelled in part by fleeing farmers!—the

GETTING A LOOK

Farming is done out in the open air. Drive down a country road every week for a year and you'll see all the stages of the agricultural cycle, from plowing and planting to harvesting. If you want a closer look, many farmers are happy to show visitors around. A few have even turned hospitality into a business sideline: "agrotourism" is gaining popularity, especially in Europe.

Sadly, though, some kinds of farms are not as welcoming as they once were. Following an outbreak of hoof and mouth disease in Britain in 2001, American dairy farmers and cattle ranchers became wary of visitors who might inadvertently carry the infection from overseas.

Chickens and turkeys are also susceptible to various diseases that people might transmit.

A number of other factors have strained

relations between farmers and their neighbors in recent years. Controversies over genetically modified crops, over the humane treatment of farm animals and over dozens of environmental issues have left some farmers feeling besieged. Farther up the food chain, in the large companies that manufacture and sell agricultural products, secrecy and suspicion of curious strangers are commonplace. Trying to get inside a meat-packing plant is like trying to penetrate the CIA.

All the same, much of the farmer's work remains on exhibit for all to see. The photo at left shows a "trait plot" near Hastings, Nebraska, where a farmer has planted test rows of more than a dozen varieties of corn. Neighboring farmers often gather by the roadside to check the results of such tests.

suburbs encroach on surrounding rural land. Rising real estate values and taxes drive the farmers out. (But it's pleasanter to leave with a fat check from a developer than with a bankruptcy settlement.)

In recent years the population exchanges between city and country have gotten even more complicated. Former farmers don't necessarily have to move to the city to find a job; the job may come to them. It's not unusual nowadays for large factories and other kinds of industrial establishments to be built in small towns or out in open rural terrain. Even in places where farms continue to thrive, most people earn their living in other ways. In Iowa the landscape is dominated by farming, and on the radio you'll hear corn and hog prices, but 95 percent of the people of Iowa are not farmers.

THE LAY OF THE LAND

It's no surprise that land has a special role in agriculture. After all, a plot of land isn't just a place to put a farm; it *is* the farm. The nature of the land—slope, drainage, soil conditions—helps determine what kinds of plants or animals can be raised there. In turn, farming practices have a major effect on landforms.

The power of agriculture to transform landscape is nowhere more obvious than in the checkerboard country of the American Midwest and Great Plains. Whether you see those arrays of square fields from the air or by driving the relentlessly straight rural roads, which always intersect at right angles, it is abundantly clear that this swath of the earth's surface has been carved up along boundaries defined by human whim, not by natural geographic features. The grid of fields in Kansas is even more regular than the grid of streets and avenues in Midtown Manhattan.

The creator of the great midwestern checkerboard was the most passionately agrarian of all American presidents—Thomas Jefferson. Long before he was elected president, he drafted the Land Ordinance Act of 1785, which spelled out how new territories west of the Appalachians would be surveyed and subdivided. The law mandated uniform, square townships measuring six miles on a side. Each township was divided into 36 square sections, each with an area of one mile, or equivalently 640 acres. According to Jefferson's original plan, sections were to be sold whole and at a minimum price of a dollar an acre, but $640 was more than many farmers could afford, and 640 acres was more land than they could cultivate. Revisions of the law allowed subdivisions into quarter sections (160 acres) and then quarter-quarter sections (40 acres). A quarter-quarter section remains the basic unit of land measure in the Midwest today. Most of the individual square fields are 40 acres. However, the size of a farm (which typically consists of several fields) has grown over the years to almost reach Jefferson's ideal; the national average is now a little less than 500 acres.

Trying to impose a rectilinear grid on the curved surface of the earth is bound to fail at some point; squares and spheres just don't get along that well. If you drive north along one of the roads that separate townships, you will probably have to make

Agricultural land in Minnesota and North Dakota *(above)* forms a geometric array of uniform squares, overlain but undisturbed by the squiggles of the Red River and its tributaries. The photograph was made from the Space Shuttle Atlantis 125 miles overhead. The squares outlined by roadways are a mile on a side. Note the discontinuity just above the middle of the image, where some of the squares are shifted left or right to accommodate the curvature of the earth. Along the lower Mississippi River *(below)*, land is divided not in uniform squares but in narrow strips and wedges aligned perpendicular to the river. In this photo, made from the Space Shuttle Columbia, New Orleans is at the extreme right. Both images are reproduced courtesy of Earth Sciences and Image Analysis, National Aeronautics and Space Administration.

Farmhouses widely scattered over the landscape are the norm in most areas of rural America, putting a considerable distance between neighbors. The quilt of fields shown here, with farmhouses and other buildings casting long shadows in the evening light, is in northern Indiana.

a jog to the left or the right every 24 miles. The adjustments are needed because roads running north and south come closer together as they approach the earth's poles.

Not all farmers' fields are square. In most of Europe and along the eastern seaboard of North America, farms tend to be much less regular in shape, with boundaries following streams or ridges. French surveyors who divided the land of the lower Mississippi valley adopted a "long lots" system, with skinny tracts of land oriented

European farm families tend to live together in small villages surrounded by their fields. The half-dozen villages visible here are in northern France.

perpendicular to the riverbank. This scheme gives everyone a little frontage on the river and a share of both floodplain and higher ground. Complicated adjustments are needed at every bend in the river.

The family farm, a revered institution in American life, is both an economic unit and a geographic one. It is the prevailing economic unit—98 percent of American farms are family-owned businesses. It is a geographic unit in the simple sense that American farm families tend to live on the land they work, often at a considerable distance from their neighbors. This practice is worth noting because it is not nearly as common elsewhere in the world. In Europe and Asia there are not so many isolated farmhouses; multiple families tend to live together in a village, which is surrounded by the fields the families own and tend. Housing patterns make a major difference in the way of life of farm families, but they also alter the look of the rural landscape even for those just passing through. The single-family enclave, with house and barn huddled together in one corner of a giant cornfield, is a peculiarly American sight.

FENCES

The boundary lines that surveyors staked out generations ago are traced out on the land today by walls and fences. Their function is generally the same everywhere, and yet they are some of the most regionally distinctive features of the agricultural landscape.

Every fence has an essential ambiguity: is it meant to keep something in or to keep something out? Early European practice was to let animals roam freely, and to surround cropland, orchards, and vineyards with walls or fences to protect them from livestock. This was also the early practice in much of the American West, where cattle grazed the open range, and settlers who wanted to farm had to enclose their crops. But later the situation was reversed. These days, pastures and grasslands for cattle are fenced to keep the animals in. Fields planted in crops such as corn and soybeans often have no fence at all.

Traditionally, walls and fences were made from local materials. In New England, where farmers continually unearth boulders as they plow their fields, it makes sense to haul the rocks to the edge of the field and pile them up in walls. As Robert Frost observed, "Something there is that doesn't love a wall"—those heaps of stone need frequent mending.

In Virginia and some other areas of the Southeast, the "worm" fence was an early favorite. Made of rough-hewn rails piled up in a zigzag pattern, it used half again as much timber as a fence laid in a straight line, which seems an extravagant waste. But the worm design had a compensating advantage: it saved the labor of digging holes and setting posts in them. Evidently labor was worth more than timber.

Parts of Normandy and the southern counties of England have neither stones nor trees; the traditional "fences" there are living hedges. For those familiar with suburban

A worm fence *(above)* seems wasteful of both land and timber, but it requires no nails or other fasteners, nor much labor. This specimen is a reconstruction at Cades Cove in Tennessee. The double-stake rail fence *(below)* can also be built without nails. The design is attributed to colonial farmers, although the fences now seem to be found only along the Blue Ridge Parkway, built in the twentieth century.

Barbed-wire fencing has been the dominant technology for enclosing animals for more than a century. The individual barb *(above)* is locked in place between two twisted strands of steel wire. Often the wire is stapled to wood posts *(right)*. In areas of Kansas *(below)*, wood was in such short supply that fence posts were carved of limestone. Quarrying, cutting and hauling the stone was a costly process; on the other hand, the stone posts have survived where wood or steel ones would have decayed long ago.

ornamental hedges, it might seem surprising that mere bushes would be enough to keep cattle and sheep confined, but the Normandy hedges are formidable thickets. In World War II some of them were nontrivial obstacles for Sherman tanks.

The American prairies are also short of both stones and lumber, and hedges were tried there as well. (The most popular shrub was the thorny osage orange.) It's amusing to imagine what the West would look like today if this fad had caught on. As it turned out, the prairie hedges had barely become established when they were superseded by one of the most important inventions in the history of the West: barbed wire. John W. Gates, a salesman who made and lost several fortunes in barbed wire, explained its key virtues in his understated style: "It's the best fence in the world. As light as air. Stronger than whisky. Cheaper than dust." The last point wasn't just marketing hyperbole. By 1897 the price of barbed wire was so low that you could fence an acre of ground for $2.

There is controversy over just who deserves the credit for inventing barbed wire, but there's general agreement that it happened in the small town of De Kalb, Illinois, in the 1870s. The crucial innovation was not the idea of attaching a sharp barb to a strand of wire; that had been tried before. The key idea of the De Kalb inventors was to twist together *two* strands of wire, in order to keep all the barbs from bunching up in one place. And this design had another benefit: the two twisted strands help to control thermal expansion and contraction. Whereas a single strand would either sag in summer or snap in winter, paired strands kept a more nearly constant tension.

Barbed wire brought the farmer and the rancher into the industrial economy. Unlike fences built of local materials, barbed wire came from a distant factory, it had to be shipped by rail or wagon, and it was bought with cash. It also became a cultural arti-

fact as much as a technological one, an emblem of the "range wars" between cattlemen and farmers. In the twentieth century, barbed wire took on more sinister associations: it protected the trenches of World War I and then defined the perimeter of the Nazi death camps in World War II. Today you can find it in abundance at any prison. It's worth noting that barbed wire manufactured for enclosing or excluding people—especially the type called razor wire—is far more vicious than any fencing used for animals.

A more recent replacement for barbed wire is the electrified fence, which you can recognize most easily by looking for the bright yellow or red plastic insulators that hold the wire a couple inches away from the fence posts. Like barbed wire, electric fencing is meant to sting, not to kill. The animals are supposed to learn to avoid getting shocked. The voltage is quite high—5,000 volts, give or take—but it is parceled out in brief pulses and limited to low currents. Reportedly, touching the wire is painful but not dangerous; I've never been brave or careless enough to find out.

Many other fence styles are in agricultural use—chicken wire, split rails, the white-board fences of horse country. But it's interesting to note one type of fence that is not much used on the farm: the chain-link or woven-wire fence. Extremely common elsewhere, it surrounds everything from playgrounds to prisons. Apparently it's too expensive for most agricultural uses.

BARNS

The barn is the unmistakable icon of American agriculture and rural life. And on many farms—especially those with dairy herds—the barn tends to be the center of operations, so the symbolic role is entirely appropriate. But it's also a distinctively American symbol. Farms in much of the world have no building that quite corre-

White plank fences, too costly for most kinds of farms, are the favored choice of horse breeders. At this farm near Hereford, Maryland, adjacent enclosures do not share a fence line but have a wide alley between them to separate horses that don't get along. This arrangement nearly doubles the amount of fencing needed.

An immense red barn *(right)* has an overhang providing shelter for the stable doors—a characteristic feature of barns in the Pennsylvania style. This one is near Bedford, Pennsylvania, south of Altoona.

Two barns of historical note *(below)*: The reconstructed upland crib barn at Cades Cove in Tennessee (according to one account, the most-photographed barn in America), and the circular stone barn at Hancock Shaker Village in Massachusetts.

sponds to the American barn. British farms, for example, usually have separate buildings for housing cattle, for stabling horses, and for storing hay and grain. American practice is to combine all these functions (and others) in one big barn.

The classic American barn has two levels. In the nineteenth century, the floors had clearly distinguished functions. The lower level was the stable floor, where animals were kept, including both livestock (cattle, sheep, pigs) and the draft horses that did the work of the farm. The second story was the threshing floor, where grain was separated from chaff at harvest time. This floor also stored baled hay as winter feed for the animals; sometimes a third level, a hayloft, provided additional storage.

These functions have changed over the years, yet the architecture of the barn remains. The space once occupied by workhorses and their harnesses may now be given over to machinery. Threshing is done in the field, but upper levels of the barn may still hold hay bales.

A challenge of barn design is providing vehicle access to both levels. In hilly terrain, the barn can be set into a slope, with upper and lower entrances on opposite sides. On flatter ground an earthen ramp may lead to the threshing-floor level.

Regional Variations. In an age when houses and offices look pretty much alike from coast to coast, barns are some of the most geographically distinctive buildings on the American landscape.

The upland crib barn, found in the Appalachian states, is the smallest and most rustic of barns. It is thought that it evolved from the corncrib—a ventilated bin for storing and drying feed corn. The ventilation remains: the walls are made of logs, without any chinking to seal the spaces between them. The original simple crib sprouted additions in various directions: a tool shed on one side, a pair of cribs with a drive-through shed between them, a hayloft overhead.

New England, from Massachusetts north into Maine, is the territory of the connected barn. Entire farmsteads, including barns, carriage houses, woodsheds, and various other buildings, are joined to the main house and to one another by covered passages. Presumably this style of architecture was invented when some farmer got tired of trudging through the snow to milk the cows. The house is always at one end of the series of buildings, never in the middle, so the farmer doesn't have to track mud through the parlor to get from the barn to the chicken house.

The Pennsylvania barn is a style distinctive enough to support scholarly investigation of its distribution, which includes not just its namesake state but also much of Maryland and Virginia. The diagnostic feature is an upper story that overhangs the lower floor on one long side of the barn, creating a sheltered area at the entrance to the stables. The design has been traced back to Swiss and German models.

The model for the three-bay barn might be the cathedral at Chartres, where the great nave is divided by aisles into a center section and two side sections. The three-bay barn has a long central open area (parallel to the ridgepole) with rows of stalls on either side, usually under a lower roof.

The Dutch barn is the most classic American barn style—the one in the calendar photo, in children's books, in toys. It has a gambrel roof (steeper at the bottom, flatter at the top), and the doors are at the ends of the building, not on the broad sides. According to architectural historians, this most barnlike of all barns is a fairly recent innovation; the gambrel roof was never seen until after the Civil War. There are also examples with curved roofs that came along only in the 1920s. Although the style is called Dutch, it is unknown in the Netherlands.

The three-bay barn, a style common throughout the Appalachians, has a tall central nave and two lower wings, like many Gothic cathedrals. This barn is near Boone, North Carolina.

The hay hood is found on the gable end of barns in many regions; the hood on this Tennessee barn is larger and more elaborate than most. The hood supports and shelters a pulley used for hoisting hay into the loft.

This variation on the theme of the Dutch barn has an ogive-shaped curved metal roof. The barn is near Middleburg, New York.

A large circular barn built by the Shakers in the nineteenth century has been preserved at Hancock Village in western Massachusetts. A few other circular barns have also survived. One reason they have been preserved is that they were always unusual, and perhaps a little pretentious.

In Arizona and southern California barns are much less common than they are in the East and Midwest. Even dairy farms in the Southwest are likely to have an open hayrack of some kind instead of a real barn, and the only shelter offered to cows is a metal deck that looks like a carport.

More Features of Barns. A pointed extension of the roof at one end of the barn is called the hay hood. It supports and shelters a pulley used to load hay into the loft. (If you happen to spot a stout hook lower down, under the eaves of the barn, it has a different purpose. It's where you hang the carcass when you're butchering an animal.)

Ventilation of a hayloft is crucial. The cupolas mounted along the ridgelines of many barns serve this purpose. There may also be vents in the gable ends, as well as owl holes. (The owls are invited in to eat mice.) Older barns seldom have windows; newer dairy barns may have lots.

Not all barns are red—not even a majority of them—and yet red is so much more common on barns than on other buildings that the stereotype must have something behind it. Another point to note is that many barns are sided with boards running vertically, whereas this is rare on houses. Then there's the matter of advertising: very few houses are painted with billboards for Mail Pouch Tobacco.

Joseph Glass, who has done a statistical study of the orientation of barns, reports that more often than not the roof beam is parallel to the main road, and the farmhouse is located along the axis of the roof beam. The front of the barn typically faces south.

The new look in farm storage buildings is the metal-sheathed shed. This one, near Worcester, New York, retains the two-level layout of the traditional barn—not to mention the red paint.

When farmers today go looking for storage space, they seldom find it in the barn. The building that makes sense now is a prefabricated metal shed, ordered from a catalogue or a Web site and erected over a concrete pad. The metal building is cheaper; it goes up quicker; it is easier to clean; it has larger clear spans to make room for jumbo-size equipment. All it lacks is charm. Then again, maybe charm is something that comes with age to any building that survives long enough. In a few decades we may be hearing pleas to preserve the Sears and Roebuck toolshed.

MAKING HAY

One of the basic challenges of farming is that crops grow only in warm weather, but animals expect to eat all year round. Even assuming you can harvest enough food to last through the winter, how can you keep it from spoiling? In the case of grass for feeding cattle, there are two strategies. Hay is preserved by drying: You cut the grass and let it lie in the sun until the moisture drops below the level that will support the growth of mold. (Thus, the saying, "Make hay while the sun shines.") The alternative to hay is silage, which is essentially pickled grass. It is moist fodder preserved by excluding oxygen and encouraging a certain kind of fermentation.

The haymaking process is ancient. In the traditional method (as described by Thomas Hardy or Tolstoy), tall grass was cut with sickles or scythes, left to dry in the field, and then raked up into soft mounds and carted to the loft for storage. Over the colder months, it would be dished out to the cattle with a pitchfork. By the way, this quintessentially countrified activity was once a specialty of the borough of Queens, New York. The fields and meadows of Queens were harvested every year to feed the herds of dray horses across the river in Manhattan. The carts and barges that carried the hay into the city also returned the manure to Queens.

The rural haymaking routine was transformed by reaping machines, which took the place of the sickle and scythe, and then by the mechanical baler, a tractor-driven contraption that rakes up the cut grass, compresses it into densely packed bricks tied with twine or wire, and spits the bales out into the field. The bales are so solid and durable that they've been used as building material, staked together and covered with mud to form the walls of houses.

The traditional hay bale is rectangular and measures about two feet by two feet by four feet. There are also jumbo bales, the same shape but more than twice as big in every dimension. They weigh half a ton. And beginning in the 1970s, giant rolls of shredded wheat began appearing mysteriously in farmers' fields, like something deposited by extraterrestrials on a high-fiber diet. These big round bales are made by an ingenious machine that gathers up the carpet of cut grass on a moving belt and twirls it into a sort of jelly roll. Sometimes the round bales are wrapped in plastic.

Silage is made from the same raw material as hay, but it is treated differently. The difference is something your nose can tell you. Hay is sweet and grassy; silage is pun-

A towering wall of hay bales, 8 feet thick and stacked almost 20 feet high (above), awaits loading and shipment along a roadside in the Imperial Valley of southern California. In that area hay is a cash crop; elsewhere most farmers grow hay for their own livestock. Round bales (below) litter a North Carolina field at the end of summer. Often the round bales are left in the field rather than stacked in the barn. The outermost layer may deteriorate in the weather, but because the bales are so large, the percentage of hay lost is not great, and the saving of labor is considerable.

Another haymaking technology: the wooden drying rack, called a *kozolec*, is a national emblem of Slovenia.

gent. It's like the green goo you scrape out from under the lawn mower after a few wet weekends.

The standard way of making silage is—no surprise here—with a silo. The tall cylinder with a hemispherical dome on top, standing next to the dairy barn, is another icon of the farm. But the tower silo is not part of some time-tested agricultural

Tower silos, which preserve grass and other forage crops for use as animal feed through the winter, are an American invention and one of the most familiar sights on American farms. Early examples were built of wood, stone and brick; these are made of interlocking staves of cast concrete, held together by steel hoops. The cluster of six silos stands on a hill at an agriculture experiment station in North Carolina.

tradition. The technology is only a little more than a hundred years old. The basic idea is to pack the moist forage into an airtight space. The metabolism of the still-living plant cells quickly uses up the remaining oxygen, making conditions favorable for a group of bacteria known as *Lactobacillus* (a group that also, incidentally, brings us yogurt). The bacteria produce large quantities of lactic acid; if all goes well, the acidity suppresses the growth of other, less welcome microorganisms.

The first silos were loaded by hauling bucketfuls of feed to the top of the tower with a rope and pulley. Now it's done with a clever little machine that chops the grass into small bits and blows it up a long chute. At the very top of the silo the chute makes a graceful arc and drops the fresh material on top of what's already stored inside. For unloading, another machine is lowered on cables from a tripod under the roof of the silo. The unloading machine scrapes up the top layer of silage and heaves it through one of a series of doors in the side wall. As the silo empties, someone has to climb a ladder to move the unloading gear down to the next lower door.

This scheme of operation has one potential problem: it is a first-in, last-out system. The fodder harvested at the end of the growing season gets fed to the herd right away, and the oldest silage remains squished at the bottom until the very end of the winter. It would be better to use a first-in, first-out protocol. In many newer and fancier silos this is achieved with a bottom-unloader, a device like a giant chainsaw that's installed in the foundation of the silo. It always serves up the oldest material first.

Early silos were built much like wine barrels or water tanks, with vertical wood staves held together by circumferential steel hoops. For the most part, only ruins of those wood silos remain; they've all rotted away. But the same stave-and-hoop style of construction has been adapted to silos made of cast concrete staves, and these are apparently quite durable. The other common silo, especially in the Midwest, is the Harvestore, which is made of glass-coated steel colored a gorgeous cobalt blue. The glass coating is needed because the acidic silage would quickly eat through bare metal.

Tower silos are a standard fixture on dairy farms in many areas, but it turns out that there are other ways to make silage that don't require such elaborate equipment. The very first silos, as a matter of fact, were merely trenches or pits. Such horizontal silos have been making a comeback in recent years, although they are so inconspicuous you may not know they are there. The horizontal silo is a trench or an above-ground bunker where mown grass is compacted to preserve it. To exclude air, it is covered with a sheet of plastic, often held in place by old tires.

FARM MACHINERY

Go to any midwestern county seat and follow the main road out beyond the new high school and the Wal-Mart, and somewhere along the way you'll find a row of farm implement dealers. Lined up along the curb will be the big harvester combines and the eight-wheel articulated tractors, with dozens of accessories and attach-

Deep-blue Harvestore silos are made of glass-coated steel plates. The glass protects the metal from the corrosive effects of acidic silage. The technology evolved from the glass-lined domestic water heater. The silos shown here are on a dairy farm in Colwich, Kansas.

A tractor on crawler treads cultivates an Oklahoma wheat field in swaths 50 feet wide. The use of rubber treads rather than ordinary wheels helps to distribute the tractor's weight over a broader footprint, alleviating problems of soil compaction. The implements pulled by the tractor are harrows and chains, preparing the soil for winter wheat.

ments—grain wagons, plows, seed drills, mowers, rakes, balers, sprayers, augers. But what is most eye-catching is the array of colors. One dealer's lot is all deep green, another's is red, and across the street everything is painted yellow.

Why does farm machinery look like this? (Cars and trucks, you'll note, tend to draw on a rather different palette.) There may be practical reasons for combines and tractors to be painted the way they are, but I can't help thinking there's an aesthetic principle at work as well—that farmers really prefer these pure, intense, primary colors.

On a dealer's lot in Kingfisher, Oklahoma, a big red eight-wheel farm tractor awaits a buyer. This one is an articulated machine: a hinge allows it to bend in the middle, reducing the turning radius.

Tractors. The Latin word *tractor* means "puller," and the early mechanical tractors were looked on mainly as something to pull the plow through the soil, replacing the horse or the ox in this role. Tractors still pull, but they do a lot else besides. Like the Veg-O-Matic, the tractor slices and dices, rakes and bales, plants and sprays. It's at the center of an entire system of mechanized agriculture.

The red Ford tractor I rode at my uncle's farm when I was a boy—it would be a collector's item today—had large-diameter drive wheels at the rear to maximize traction and much smaller front wheels to make steering easier. Tractors today look quite different. They have four big wheels, or even eight. In most cases all the wheels are driven by the engine. Some tractors also have all-wheel steering, which not only lets the machine turn around in a smaller space but also keeps the front and rear wheels in the same ruts during a turn. The engine is a big diesel. The transmission has a dozen or more forward speeds, and several choices in reverse as well.

What makes the tractor more than just a puller is the PTO—the power take-off shaft. This is where you connect the hay baler or the mower or the rotary cultivator or any of dozens of other attachments.

When my uncle drove his little red tractor, he perched on a metal-pan seat, protected from the sun by a straw hat. Times have changed. The modern farmer sits in an air-conditioned cab with an audio system, an air suspension seat, and a beverage cooler. The latest in high tech is a scheme called precision farming. The tractor is equipped with a GPS (Global Positioning System) receiver, which continually monitors its position in longitude and latitude. An on-board computer uses the geographic information to dispense pesticides or fertilizers according to the needs of each small section of the field. In the most sophisticated versions of this system the computer actually drives the tractor, although the farmer is supposed to be there in the cab—and not napping.

Harvesters. Most of the machines that work a farmer's field—for plowing, tilling, planting, fertilizing, and so on—are towed behind the tractor. But harvesting is different. Wheat, corn, soybeans, rice, cotton, and several other major crops are gathered from the fields by specialized, self-propelled combine harvesters. Through most of the year these mammoth and very expensive machines sit idle in the barn. Then, about the time baseball season winds up, they roar to life all over the Midwest for a few weeks of frantic activity.

The machine is called a combine because it does both reaping (cutting the ripe grain) and threshing (separating the wheat from the chaff, or the equivalent process for other crops). Mechanizing these operations was the goal of an intense century-long competition, beginning with the spindly horse-drawn reaper of Cyrus McCormick, first demonstrated in 1831. Threshing machines were invented soon

A combine harvests corn plants for silage and loads the chopped fodder directly into a truck in a field near Plainville, Kansas. The drivers of the two vehicles have to dance a careful duet to get all the crop into the truck without risking a collision.

A combine can be equipped with various kinds of cutter heads for different crops. The rotating flail mechanism seen here, which harks back to McCormick's earliest reapers, works well with small grains or, as in this Ohio field, soybeans.

after, and there were some early attempts to link reapers and threshers, but the combine didn't really begin to catch on until well into the twentieth century.

Harvesting grain has always been a bottleneck in the farm economy. Nature takes its sweet time growing the crop, all summer long, but once it ripens, it has to be cut and dried in a hurry or it will rot in the field. The ages-old tradition in wheat farming is to cut the stalks with a scythe, bind them into sheaves, and stand the sheaves up in shocks, the picturesque conical mounds that adorn the months of October and November on a million calendars. Threshing was done at the barn, in two stages. First, the sheaves were dumped on the threshing floor and beaten to separate the grain from the ear. Then, the grain was winnowed—tossed in the air to let the wind carry away the lighter chaff.

In the belly of a modern combine, all this happens in seconds. A cutter head mows down a swath as much as 50 feet wide and gathers the stalks onto a moving belt. Inside the machine, a rotating drum fitted with "rasp bars" beats the grain off the ears. Most of the grain falls through a perforated concave housing under the drum, while the straw is carried on to a series of vibrating belts or chains called straw walkers, which are meant to shake loose any grain still mixed with the straw. At the back of the machine, the straw either is chopped and spread over the field as mulch or is left in rows for gathering. Meanwhile, the collected grain goes through a cleaning process that's much like the winnowing of old. A blast of air from a fan carries away chaff. It also creates a high plume of dust that generally makes it easy to find a combine at work, even when it's a mile or two away in another field. If you get close, you'll be covered with a fine coat of dandruff-like fluff. The cabs of some combines are pressurized to keep the dust out.

The big combines hold more than five tons of grain, but they cover the ground so fast that they need to be emptied out every half hour or so. Trucks or wagons pull

up alongside, and the grain is pumped out through a long chute. Often it's done with both the truck and the combine on the move.

It's not unusual to see combines working the fields long into the night, with banks of floodlights mounted above the cab. Even with these gargantuan machines, getting the harvest in before the weather turns is still an urgent matter.

Specialized harvesting machinery has been developed for other crops as well—cotton, tomatoes, sugar cane. There are even machines that grip the trunk of a cherry tree and shake all the fruit loose.

GRANARIES

Although a grain crop is harvested in a matter of weeks, it is consumed throughout the year. Hence, we need someplace to store it—a buffer between the variable supply and the steady demand.

Some of the grain is stored right on the farm. The standard container is a galvanized steel bin, cylindrical with a conical top, bigger than a silo but usually not as tall. The bins you are most likely to see on a farm hold anywhere from 10,000 to 100,000 bushels. (A bushel is eight gallons; it works out to 50 or 60 pounds of grain. In an average year, a 500-acre farm might produce 60,000 bushels of corn.) Some of the bins have a dryer attached: a fan with an oil- or gas-fired burner that blows warm air up through the stored grain.

On-the-farm storage is essential for farmers who grow crops as feed for their own livestock, but grain destined for the market usually bypasses the farmer's own storage bins and goes straight from the field to a grain elevator. The local elevator is both a landmark and a social institution in thousands of small towns. Often it is run as a

A steel tank with a conical roof provides on-the-farm storage for harvested grain. The long tubular device is an auger, powered by the tractor; inside the tube, a helical screw lifts the grain to the top of the bin. The farm is in Palmyra, Nebraska.

A large country elevator near Dixon, California, has both concrete silos and metal tanks for grain storage.

cooperative by an association of nearby farmers, increasing their bargaining power as both buyers and sellers. The typical "country" elevator holds a few million bushels of grain. Eventually, the grain will move from the country elevator to a larger "terminal" elevator and then to buyers both at home and abroad.

Like many city kids, I was once flummoxed by the term *grain elevator*. An elevator was something I rode in an apartment building, and I couldn't imagine what it could have to do with grain. The original meaning of the term simply referred to a means

A bumper crop is a boon to the farmer, but it may leave the local elevator short of storage capacity. Once all the bins are filled, there is no choice but to store grain in heaps on the ground. Here workers in Dixon cover a heap with plastic sheeting, held down by old tires.

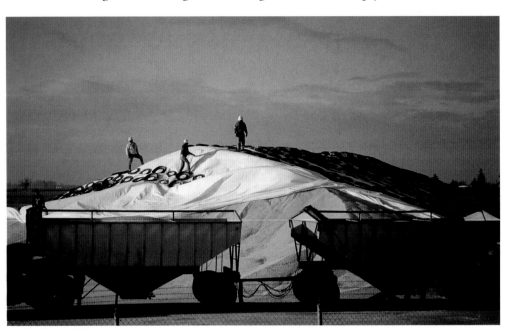

for getting the stuff up off the ground, to keep it dry and out of easy reach of rats and mice. But the modern grain elevator *does* have something very much like the elevator of a high-rise apartment building. The lifting mechanism is a train of buckets attached to a chain that continually circulates around pulleys at the top and bottom. For some reason this device is called a leg.

One familiar type of grain elevator is a row of tall cylindrical bins made of reinforced concrete. Typically there is a leg at each end of the row, with a conveyor belt that runs horizontally across the top. Grain is lifted by one of the legs, and then carried by the overhead conveyor to a selected bin. Out of sight in the foundation of the structure is another conveyor belt that carries grain unloaded from the bottom of the bins.

Although thousands of these concrete-silo grain elevators punctuate the midwestern skyline, not a lot of new ones are being built. Instead, new elevators are using the same kind of corrugated-steel cylindrical tanks seen on the farm, but in larger sizes. At one of these "tank farms," there's a central bucket elevator and an "octopus" of tubes or chutes through which grain is carried to the various bins. A rotary selector allows the grain to be directed into a particular tube.

At a country elevator at harvest time, trucks from the surrounding farms line up to wait their turn to unload. Each truck pulls onto a scale, and while the load is being weighed, a sample is retrieved by a hydraulically driven probe called a trier, which looks like an oversize hypodermic syringe. The samples are tested for moisture and the presence of foreign matter—factors that determine quality and price.

Once these formalities are over, the truck drives to the unloading shed and dumps its load through a metal grate into a below-grade "boot." From the boot, the grain is hoisted by one of the bucket elevators to the overhead conveyor or distributor. Before going into a storage bin, the grain may have to pass through a drier. Grain that is too moist—in the case of corn, anything greater than about 14 percent water—is at risk of going moldy. Ideally, the farmer does not harvest the crop until it reaches the ideal moisture range, but that depends on the cooperation of the weather. If the grain has to be brought in while it's still moist, the farmer will be charged for the cost of drying it. Most elevators use a tower drier—a tall metal structure where the grain tumbles slowly downward through rising heated air.

A country elevator is only the first stop on the long journey of grain that farmers dispatch into international commerce. Much of the produce of the American Midwest goes by barge down the Mississippi or by rail to ports on the Great Lakes and is then shipped overseas. The grain elevators at the international ports are built on a much larger scale than the small-town country elevators—some of them hold billions of bushels—but in other respects their operations are not much different.

A serious problem at all grain elevators is dust. It's not just messy or inconvenient; fine dust rubbed off the grain is flammable and an explosion hazard. The remedies are to suppress the dust (for example, by coating the grain with oil), to collect it (by putting a big fabric "sock" over vents where dust would escape), and to avoid sparks

At an elevator in Minden, Nebraska, a hydraulic probe called a trier sucks up a sample of soybeans. The sample will be tested for moisture and other properties before the truck is allowed to unload.

by putting all electrical equipment in explosion-proof housings. The most frequent explosion source is the bucket elevator, probably because the moving buckets tend to accumulate static electricity. For this reason the elevator legs are usually placed outside the building, and their housings have explosion vents—outward-opening doors meant to direct any flames in a safe direction.

The modernist architect Le Corbusier was an admirer of American grain elevators, suggesting that their regularity and modularity could serve as a model for other kinds of buildings. At least one later architect took the suggestion seriously. The Hilton Inn at Quaker Square in Akron, Ohio, occupies the shell of a former elevator. If you're in town for the night, you can rent a round room in one of the silos.

MILLING GRAIN

In Amarillo, Texas, a terminal elevator with almost 200 concrete silos occupies a substantial tract of land along the downtown railroad right of way. Grain from many smaller elevators is consolidated here and shipped to customers. The silvery tower in front of the bank of silos at right is a grain drier.

More than any other technology, agriculture brings the ancient and the modern face-to-face. The operations of reaping, threshing, and winnowing come from a time-honored tradition, but we still practice them, even if the tools look very differ-

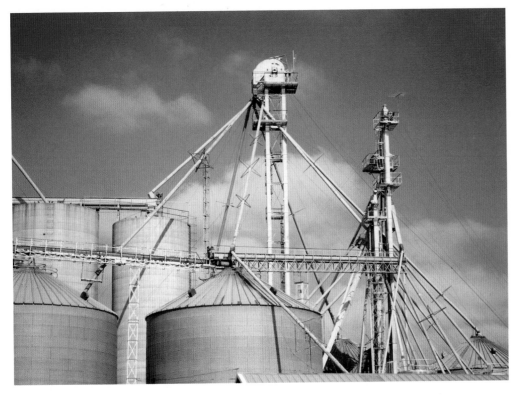

The "octopus" of an elevator *(left)* distributes grain to various storage bins. The tallest structure in the middle of the image is the elevator itself, called a leg; the two parallel tubes house an endless chain of buckets that lift the grain to the pinnacle and then pour it into one of the sloping chutes. The elevator is in Garden Prairie, Illinois. Another octopus design *(below)*, atop an elevator in Phillipsburg, Kansas, has a rotary distributor.

ent. It's the same with grinding grain, or milling. The modern flour mill is a huge industrial enterprise, but what goes on inside is not so different from the work of an eighteenth- or nineteenth-century gristmill built on the banks of a swift stream.

Many of those water-powered mills have been preserved or restored, and some of them, operated as museums, are still grinding grain. At the heart of the operation are two disk-shaped stones, one fixed in place and the other rotating above it. The bearing surfaces where the two stones meet are dressed—carved in a pattern of grooves and ridges, perhaps radial spokes or arcs like flower petals. The stones do not grind directly against each other; they would be quickly destroyed if they did. They are kept minutely separated, so that only the grain between them is ground, not the stones.

Today, the disk-shaped millstones have been replaced by cylindrical steel rolls, but they too are dressed with grooves or ridges in spiral patterns, and they work much the same way. Grain passes between two rolls turning at somewhat different speeds, so as to apply a combination of crushing and shearing forces. The process is a delicate one; the rolls don't just pound the grain to powder. On the first pass, the miller tries to gently "break" the grain without grinding it too heavily. The object is to isolate the layer known to botanists as endosperm but called middlings by millers. The term *middlings* is a bit confusing because it suggests mediocre, but in fact it is the best part of the grain, with the purest and most valuable starch. The middlings need to be separated from the germ in the middle of the kernel and from the bran on the outside.

Grinding is only half of the milling process; the other half is sieving or sifting, which is how the various components are actually separated. The mixture is spread

Grain-hauling trucks line up at a terminal in Toledo, Ohio, where agricultural products are loaded onto ships for export via the Great Lakes and the St. Lawrence Seaway.

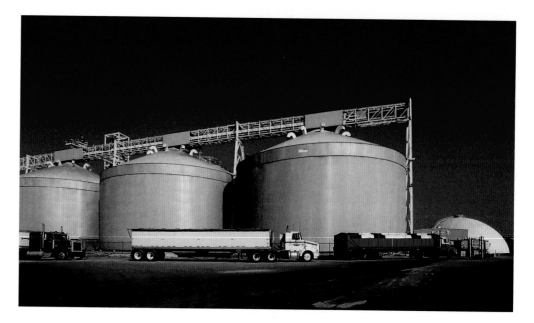

out and shaken on a screen or cloth; the parts that fall through are unders while the bigger pieces left behind are overs. The sifting is done in stages from coarse to fine.

Turning millstones takes a lot of energy, which is why flour milling has a strong association with the development of wind and water power. The historical center of the flour milling industry in the United States is Minneapolis, where the mills could exploit the drop in the Mississippi River at the Falls of St. Anthony. Some of those mills are still running, although they have long since been converted to electric power.

A traditional millstone—this one is a modern reproduction at Cades Cove, Tennessee—has a pattern of grooves on the grinding surface. Modern mills use steel cylinders rather than stone disks, but the pattern of incised lines remains important.

Milling Corn. Wheat makes our daily bread, and its products are known to all. The ultimate fate of all those billions of bushels of corn is a little less familiar. A little of it is dry-milled just like wheat flour; that's where we get cornmeal muffins. Dry-milled corn is also the source of hominy grits, which go straight to the breakfast table for southerners but have to be turned into cornflakes for the rest of the country. Grits are also the main raw material for distilling whiskey, so you can both start and end your day with them.

But if those were the only markets for corn, Iowa would go out of business. The bulk of the corn crop goes through a quite different process, called wet milling, and comes out the other end in forms you would never recognize. It's called wet milling because the grain is steeped in warm water for a couple of days before the milling begins. The soaking is partly to soften the kernels, but there is also a biological aspect to the treatment, involving the same lactic-acid bacteria that help preserve silage and make yogurt. The softened and fermented corn is ground up, and then the starch is separated from the germ and the bran in a series of washing steps (analogous to the sifting steps of flour milling). The starch is the most valuable product, but most of it goes through yet another transformation. It is converted by treatment with acids or enzymes into corn syrup and fructose, the main industrial sweeteners; these sugary

liquids are hauled away in 65,000-gallon rail cars to wherever it is they make Twinkies and Tootsie Rolls. Meanwhile, the germ of the corn kernels is pressed to extract corn oil. The residue left behind by this process, along with the bran and the steep liquor, becomes a high-protein component of animal feed. The overall wet-milling process is remarkably efficient: 99.5 percent of the original feed-stock winds up in products.

Biofuel. Corn has another destination as well: the fuel tank of your car. Alcohol made from corn has become a major gasoline additive.

The alcohol in question is ethanol, which is the drinking kind (although it's deliberately made undrinkable in the biofuel plants). Production of alcohol for fuel is an adjunct to the wet-milling process. The raw material is a starch slurry, which is treated with enzymes to break down some of the starch molecules into simple sugars, then pumped into fermentation vats with a dose of brewer's yeast. The fermentation takes a day or so, then the "beer" goes to a distillery tower.

Growing fuel instead of sucking it out of the ground seems like an attractive proposition. It's a home-grown energy supply, and crops are a renewable resource, unlike petroleum. However, there's a catch. Critics argue that American agricultural practices are so energy-intensive that growing the corn and converting it to alcohol actually consumes more energy than the ethanol provides. In other words: the more agrofuel we make, the more oil we have to import in order to grow it. Gasohol producers dis-

Wet-milling of corn is a large-scale industrial enterprise, where multiple streams of raw materials, intermediates, and finished products all have to move simultaneously through a series of process steps. Seen here is a small part of a wet-milling plant operated by Roquette America in Keokuk, Iowa.

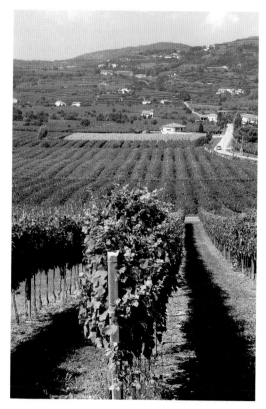

Grape vines are trained to waist-high trellises on hill-sides near the town of Soave in northern Italy. The grapes are pressed to make white wine.

pute this conclusion, but even in the most optimistic analysis, the energy balance is a near thing. The industry could not exist without tax incentives. Gasohol with 10 percent ethanol gets a tax break of about 5 cents per gallon, which means the cost of the ethanol itself is effectively reduced by 50 cents per gallon.

OTHER CROPS

More than other industries, farming tends to have a distinctive local personality. Why do they grow corn in Iowa and tobacco in the Carolinas and cotton in Texas? Climate and soil conditions have something to do with these choices, but infrastructure also matters. To raise a crop, you need a local supplier of seed, machinery, fertilizer, and so on. And you need a market for the finished product; if the nearby coop buys only corn and soybeans, it won't be much help when you drive in with a truckload of pineapples. All these factors have led to regional patterns of specialization.

Soybeans. Although tofu and tempeh are no longer obscure health-food items in the United States, their popularity is hardly enough to explain the millions of acres of soybeans planted throughout the Midwest. The beans are grown for their oil, which goes into everything from margarine to cookies, and for animal feed.

Soybeans were almost unknown outside Asia until 1900, but in the past 50 years they've become a mainstay of American agriculture. One reason is that they neatly complement the growing of corn. Much of the same equipment can be used for planting, harvesting, and storing both crops. And soybeans are legumes (like peas), members of the family of plants that help put nitrogen back in the soil, reducing the need for fertilizer. As part of a crop-rotation pattern, they can also reduce the need for

Peach orchards covered with shade cloth create an eerily shrouded landscape near the small coastal village of Velia in southern Italy. The canopy is mainly meant to prevent "sunburn" on the fruit, although it also affords some protection from birds.

Trees grown for papermaking stand in dense and very orderly ranks on a plantation near Venice. In North America most trees raised for pulp and paper are pines or other softwood species, which have longer cellulose fibers and therefore produce stronger paper. European mills favor hardwoods that produce whiter paper with less need for bleaching. The trees shown here are nearly large enough for harvesting.

pesticides. Each field is planted in corn some years and soybeans other years; because pests that attack corn don't like soybeans, and vice versa, the rotation helps both crops.

Orchards. Growing apples, oranges, or walnuts requires a longer time scale than field crops. The interval from planting a tree to the first harvest is at least a few years. In the case of olive groves along the Mediterranean coasts, the trees outlive the farmers—and maybe the farmers' children and grandchildren too.

Tending plants that are large and long-lived allows the farmer to lavish individual attention on them. Orchard trees are carefully pruned and sometimes even bandaged. In the citrus groves of Florida, the worry that makes the nightly news once every few years is the threat of frost. For a long time the defense against frost was building bonfires. Now the weapon of choice is a powerful fan, which doesn't try to heat the air but just keeps it circulating so that the chill doesn't settle onto the ground.

As you drive past an orchard, you can't help noticing the geometric arrangement of the trees, as straight rows come into focus first in one direction and then another. Why are the trees planted in such rigid patterns? Columns and rows make it easier to drive equipment through the orchard. They also help keep the trees properly spaced—not so close that they impede each other's growth, not so far apart that they waste land. But in this respect the planters have missed a bet. They could make more efficient use of land by planting in a honeycomb pattern, with trees at the centers of hexagons instead of at the centers of squares.

Vineyards. If orchard trees are carefully tended, grapevines are positively pampered, especially in domains of origin that have name recognition when pasted on a wine bottle. The vines are tied to waist-high trellises, spaced just far enough apart for workers to walk between them. Many of the vines in both Europe and America are "chimeri-

A century-old pulp-and-paper mill in Canton, North Carolina, was long owned by Champion International but now is operated by an employee-owned company called Blue Ridge Paper Products. The two tallest structures, in the foreground, are boilers where bark and other wastes are burned to generate steam and electricity for the operation of the rest of the plant. The other large building, to the right of center, is where wood fiber is converted to pulp.

At a timber-processing yard in Mobile, Alabama, wood chips are unloaded by the brute-force method: simply upending the entire truck.

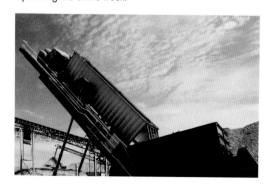

cal"—the roots, which may be hundreds of years old, are of one grape variety, and the rest of the plant is a different type, grafted to the root stock in a surgical procedure.

There are machines for harvesting grapes: Think of a tractor on stilts that straddles each row of vines and drubs the bunches of grapes into plastic wrap. But in Bordeaux you will not see these machines picking the grapes for next year's *premier cru*. That's done by highly skilled hand labor.

Paper. The book you hold in your hands is an agricultural product, made mostly from trees grown on pulpwood plantations. A widely cited statistic says it takes 17 trees to make a ton of paper. This number was first publicized in the 1970s by a company called Conservatree, which was then a distributor of recycled paper products. Conservatree, now a nonprofit organization, has since revised its calculations and suggests a range of 12 to 24 trees per ton. The trees in question are rather spindly—6 to 8 inches in diameter.

However many trees it takes, converting them into paper is a dramatic transformation; the wood fiber has to be taken apart and put back together again. Logs are stripped of their bark (which is burned as fuel), the wood is chopped up into small chips, and the chips are reduced to a mush called pulp. Then, in a machine the length of a football field, the pulp becomes paper in a matter of seconds. The wet pulp is spread on a porous screen, where much of the water drains away; then more water is blotted up by felt cloths and cooked out by steam-heated rollers. All of this happens while the paper-to-be is moving at 60 miles an hour.

Sugar. In Florida and Louisiana, stands of sugar cane 15 or 20 feet high are cut and trucked to local mills, where the cane is pressed between rollers to extract the sweet

juice. Then water is driven off in a succession of heated vacuum vessels, concentrating the juice in stages until eventually the sugar crystallizes. The process requires a great deal of energy, most of which comes from burning the bagasse, the leftover husks of the canes. The evaporation also releases great clouds of water vapor, which makes the mill a distinctive sight over the green cane fields.

The product of all this work is not yet ready for the sugar bowl on the breakfast table. It is "raw" sugar, with enough impurities to turn it tawny brown. The final purification is done in a refinery, where the sugar is dissolved in water again, filtered, and then put through another cycle of concentrating and crystallizing.

Vegetables. Peas, carrots, cabbage, beans, tomatoes, squash, cucumbers, lettuce: they're all so different, and yet there is at least one element they have in common—an urgency about the harvest. Even if vegetables are destined to be canned or frozen, they have to be picked within a narrow interval between unripe and rotten. There's seldom more than a day or two of leeway.

The rush to market presents a logistical challenge. Does a farmer buy the highly specialized machine that strips peas from the vine and pops them out of the pod, only to use it one day a year? Finding labor for picking by hand is no easier. And if everyone's peas ripen on the same day, the packing plant can't possibly accommodate them all. Because of problems like these, most vegetables are grown under prearranged contracts. Buyers for supermarket chains or packing companies supply the necessary harvesting equipment and coordinate its deployment.

The products of this system are not held in high esteem by consumers. The supermarket tomato is a particular target of derision and opprobrium. No one has any-

Clouds of vapor erupt from a sugar mill in southern Louisiana, where cane grown locally is pressed to extract the sugary sap. Most of the plumes from the plant's many stacks are merely water vapor evaporated from the cane juice, but there is also some smoke from the burning of bagasse, the crop residue.

A cotton combine combs through a North Carolina field, stripping away the white fiber and the seeds, leaving behind bare stalks. In preparation for the harvest, the plants have been chemically defoliated.

thing good to say about it, except that it's always there, even in February. The tomato you buy in Boston has been on the road a week or more, coming from California or Mexico. It was picked hard and green, then artificially ripened (or at least reddened) by exposure to ethylene gas.

But modern agricultural technology has its little triumphs too. One of them is black plastic sheeting, the same material used to make garbage bags. Laid down over strawberry or carrot plants in spring, it warms the soil during the day, which speeds plant growth, and retains warmth at night, as a guard against frost. The result is locally grown vegetables earlier in the season and farther north than would otherwise be possible.

Cotton. Everyone has heard of cotton's role in the economy of the American South during the slavery era, along with stories about the invention of the cotton gin, the depredations of the boll weevil, and the back-breaking labor of "chopping" cotton (which is not harvesting the crop but weeding it). Today, cotton still has the reputation of a finicky crop, vulnerable to weeds and pests, but much of the labor has been replaced by machinery and chemicals. Herbicides and pesticides can be applied more liberally to a nonedible fiber than to crops destined for the dinner table. At the end of the season a final spraying of defoliant turns the leaves crinkly brown or strips them off the plant entirely, easing the job of the harvesting machine.

Cotton fields at harvest time are a bizarre sight. One side of the field is festooned with the bright white tufts of burst-open bolls; on the other side, where the machine has passed, stalks are stripped bare. Along the roadside are dirty-white bales of uncleaned cotton, the shape of a loaf of bread but the size of a truck trailer. The bales are hauled to the local cotton gin, which has the same role in this agroeconomic system as the grain coop farther north. The gin separates the lint (the white fibers) from the seeds, both of which are valuable commodities. The seeds are pressed for oil.

At a cotton gin near Brawley, California, giant loaves await ginning, the separation of fiber from seeds.

Bees. It's a scandal: migrant agricultural workers are trucked from farm to farm, made to labor from dawn till dark, housed by the thousands in cramped wood boxes. And at the end of the growing season, all the sweet fruit of their labor is taken from them.

The services of bees are essential to dozens of crops; there is no other practical way to pollinate the flowers. Beekeepers rent out their hives for a week or two at a time as various crops come into bloom. You'll see the box hives set up in a shady corner of the field. There's a very clever idea in the design of these boxes that goes back more than a century, to experiments done by the Reverend Lorenzo Langstroth, an amateur apiarist in Miami, Ohio. Langstroth observed that when bees build honeycomb in natural cavities, they always leave the same amount of space—about three-eighths of an inch—between disconnected parts of the structure. This is just about the width of a bee, the amount of room it needs to pass comfortably through the spaces. So by building the hive box with all the pieces three-eighths of an inch apart, Langstroth induced the bees to build honeycombs on removable frames, which could be pulled out without damaging the structure or angering the bees.

A few years ago, there was much gloom and dread over the arrival in the United States of "Africanized" bees, thought to be so aggressive and unmanageable that they would be the end of beekeeping, if not the end of the world as we know it. The Africanized bees are here, at least in the Southwest. Beekeepers aren't happy about it, but so far it hasn't been anything like the movie. More worrisome, perhaps, is the spread of certain parasites and diseases that have decimated honeybee populations.

Socializing honeybees form a "beard" at the entrance to a box hive. The hive entrance is the slit at the bottom of the wood panel, just wide enough to admit bees. These particular bees are not just social insects but also Socialist insects: the hive is in the National Botanical Garden of Havana, Cuba.

DAIRY FARMS

The social life of dairy cows is endlessly fascinating. I used to watch them queuing up to be milked, late in the afternoon, at a small farm in southern Minnesota. They all knew their place in line. If one of the herd happened to be a little late—and the question of what errand or distraction might make a cow late is in itself a matter worth pondering—the others would save her place. I'm told that the order of the lineup at the door of the milking parlor does not necessarily correspond to the herd's dominance hierarchy. There *is* such a hierarchy, to be sure; every cow knows who's bossy. But for some reason the dominant cows don't choose to go first at milking time, nor do they wait until last; they tend to be in the middle of the pack.

As for the social life of dairy farmers—there's precious little to be said about that. More than any other type of farming, dairying enforces a relentless routine. The cows must be milked twice or even three times a day, every day, all year round, whatever the weather, even if you have the flu, even if it's your wedding day, even if the Green Bay Packers are playing in the Super Bowl. In between milkings, the cows have to be fed.

The cows you are most likely to see on an American dairy farm are Holsteins, also known as Friesians—the tall, lanky ones with black-and-white splotches and prominent hip bones. Among all breeds of cows, Holsteins are the most efficient convert-

ers of feed into milk. In round numbers, a healthy Holstein cow takes in 300 pounds of food and water every day and puts out 60 pounds of milk.

Most dairy cows wear at least one earring—a bright plastic tag with an identifying number. On larger farms each milking cow is also equipped with an RFID tag; "RFID" stands for *radio-frequency identification*, and the tag works much like the E-ZPass transmitter that gets you through tollbooths on the turnpike. As the cow steps up to a feeding station or enters the milking parlor, a sensor records the RFID number, and a computer notes the animal's weight and how much it eats or how much milk it gives. Dairy farmers keep scrupulous records of these things.

The Life and Times of a Dairy Cow. In the bucolic vision of life on a dairy farm, the cows are turned out into the pasture every morning, where they munch the grass, chew the cud, swat flies, and socialize. Such picnic days are not unknown on modern farms, but they are mainly for younger animals—heifers—that have not yet started giving milk. Millions of cows never set hoof in a pasture. Some of them live in stanchion barns, where each cow is held in an assigned stall by a steel head restraint. But stanchion barns seem to be on the way out, at least on larger farms. The new fashion is the free-stall barn, which is like an open-space office: there are stalls, but they're not assigned, and the cows are free to wander among them. Have free-stall barns become popular because they're nicer for the cows? Maybe, but it's also worth noting that the new arrangement eliminates a lot of bother for the farmer getting cows into and out of stanchions.

Whatever the rooming arrangement in the barn, the cows certainly have all their material needs catered to. Their meals are brought in by room service. Under one plan, there's an all-you-can-eat salad bar of hay, silage, and other roughage, plus special supplements of grain and vitamins doled out to individual cows according to their potential for producing milk. More often today, each cow gets a TMR—a total mixed ration. What this means, a farmer explained to me, is that it's all blended into a uniform mash so the cow can't have her dessert without eating her vegetables.

The life cycle of a dairy cow goes something like this. At the age of 14 or 15 months she is bred for the first time and nine months later gives birth to her first calf. She'll give milk for another 10 or 12 months. Then, after a brief respite, she'll be bred again, or "freshened," and the cycle starts over. The typical cow has a career of five years; the best producers may be kept on another year or two. When milk production drops, the cow is culled from the herd. What happens to her then? About half the beef in fast-food hamburgers comes from culled dairy cattle.

The Milking Parlor. The term milking parlor seems so quaintly Victorian, it summons up an image of cows lounging on divans and drinking sherry. But what the place itself reminds me of is Jiffy Lube. The standard arrangement for modern milking parlors is called a herringbone: the cows line up on both sides of the parlor facing outward, with their business ends angled toward the center. Between the two rows of rumps is a con-

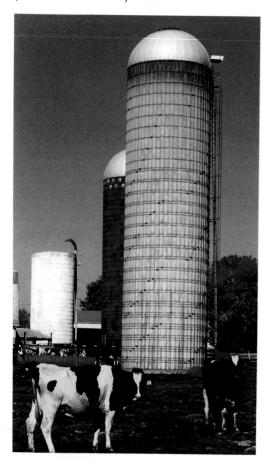

Holstein dairy cows hang out in the barnyard on a farm in New California, Ohio. Holsteins originated in the Netherlands, apparently from the interbreeding of black herds and white herds. They now make up 90 percent of American dairy cattle.

crete pit two or three feet deep, where the milker works eye-to-eye with the udders. Long gone are the three-legged milking stool and the tin pail; they have been replaced by vacuum-operated milking machines of gleaming stainless steel and glass. As each cow settles down in a stall, the milker washes the udder and swabs the teats with disinfectant, then attaches the four suction cups of the milking machine. This part of the machine—the part that hangs onto the cow—is called a claw. When the machine is running, there's a soft, rapid ticking, like a metronome. Typically, milking takes just five minutes or so; when it's done, the claw drops off automatically.

European dairies have been leading the way toward even greater automation in the milking parlor. Robotic milking machines take over the last hands-on operations, cleaning the udder and attaching the claw, so cows can mosey in and out on their own and be milked whenever they feel like it. The robotic milkers are just beginning to appear in North America.

From the cow, the milk is piped through a filter and into a refrigerated bulk tank. Usually the bulk tank and the various vacuum pumps and controllers for the milking machine are in a separate building or shed. Every day or two a stainless-steel tanker truck hauls the raw milk to the dairy for pasteurizing and packaging.

Not all dairy cows are black and white. These Brown Swiss belong to a breed that is generally considered less productive than Holsteins, but their milk is favored by many cheesemakers. The cows were photographed at a small North Carolina dairy. (What is the sharp-edged nose ring for? I asked the farmer. She explained that it controls a behavioral problem: some cows attempt to suckle at the teats of other cows.)

BEEF CATTLE

Dairy farmers and cattle ranchers are both in the business of raising cows. But if you drive west from the dairy country of Wisconsin and Minnesota into the ranch-and-range land of the Dakotas, you'll note a dramatic change in the nature of the farmsteads. A dairy farm is a capital-intensive business, with barns, silos, and big tractors, plus all the gleaming stainless steel of the milking parlor. A cattle ranch is seldom such a showplace of ag-tech. Sometimes there's nothing to see but a fence, a pickup truck, and a hay baler.

Beef cattle outnumber dairy cows 10 to 1 in the United States. Although it's hard to count because of large fluctuations from year to year, in round numbers the nation keeps 10 million dairy cows and 100 million head of beef cattle.

The original idea of raising beef cattle—some 8,000 or 10,000 years ago—was to feed the animals something that they can digest and we cannot, namely, grass. And cattle ranching is still concentrated in grassland regions too dry or otherwise unsuitable for more intensive styles of farming. But beef cattle no longer eat only grass. The red meat in the butcher's display case comes from corn-fed cattle.

Cattle Ranches. Most beef cattle start their lives in the grassland environment. They are born into the rancher's cow-and-calf herd and spend roughly a year there. The usual practice is to keep calves on the ranch until they reach a weight of about 600 pounds; then some are selected as breeding stock, and the rest are sent to a feedlot for "finishing."

Black Angus steers graze a central Kansas pasture in late summer. They are likely candidates for the feedlot within a few weeks or months.

On another Kansas farm, a portable corral is set up at the edge of a field for the annual ritual of separating cows from their calves. The herd is enticed into the enclosure, and then cows and calves are sorted to opposite sides of a partition. Finally the calves are moved through a narrow chute and loaded onto a trailer.

Impressions of ranch life are inevitably influenced by the lore and legacy of the old West—rodeos, roundups, and range wars. From watching the movies you might suppose that the way to gather cattle is to chase them on horseback with a lariat; a rancher in Kansas explained a less-strenuous technique: "I just drag a round bale behind the pickup, and they come right along." One article of western lore that remains important is the corral; this is where cattle are branded, vaccinated, dehorned, castrated, dipped in pesticides, and ultimately loaded onto trucks when their time comes. But the corral is no longer a permanent, wood-fenced enclosure; it is a portable device, made of prefabricated steel-pipe fence panels, which can be set up in the field where it's needed.

Feedlots. Steers go to slaughter when they reach a weight of 1,200 pounds or so, a little less for heifers. If they were kept on grassland, it would take them three or four years to reach this weight. Sent to a feedlot and fattened on a diet rich in grain and other starch, a steer can reach market weight before its second birthday. Furthermore, the marbled meat from such corn-fed cattle commands a higher price.

"Corn-fed" does not, of course, mean corn on the cob. It is hard, dried corn that has to be cracked open in a grinder to make it digestible. Small feedlots buy rations already prepared and mixed, but a large lot will have a mill of its own to prepare the feed. The corn is mixed with a little roughage (hay or silage) and various nutrient supplements. An Oklahoma feedlot I visited steams and flakes the corn rather than merely cracking it. The manager explained that although the steaming process is costly, the animals prefer the cooked grain and gain weight about 5 percent faster. He offered me a taste. It's not the best popcorn I've ever had, but it's not the worst either.

Generally speaking, feedlot operators do not own the cattle in their lots; they take them on consignment and charge a fee for their care and upkeep. The lot is divided into many small pens so that each owner's livestock can be fed and tended separately.

Some 20,000 head of beef cattle occupy hundreds of pens at the Wheeler Brothers Feedyard in Watonga, Oklahoma. Each pen holds a consignment of animals from a single owner. They will spend anywhere from 120 to 220 days at the feedlot reaching a market weight of roughly 1,200 pounds. Feeding and tending the cattle is not the only concern of the feedlot managers; they also pay close attention to prices and markets, since small adjustments in the timing of transactions can make the difference between profit and loss.

The feed is delivered to troughs, called bunks, which are often placed just outside the pen so the cattle have to poke their heads through the fence to reach their food. This setup avoids the problem of animals stepping in or defecating into the feed bunks, and it allows the bunks to be refilled by a truck or tractor driving along the fence line. Often the water trough is on the opposite side of the pen, which forces the animals to move around a little, within the admittedly narrow scope available to them.

Feedlots that accommodate 10,000 or 20,000 head are common, and a few of the biggest yards can handle 100,000. With so many animals confined in a small area, you might think that sanitation would be a major problem, but I am told that the pens need to be raked out only every year or two. Mud, on the other hand, can be a misery for both the animals and their keepers. One reason the feedlot industry is concentrated in a band that runs from the Texas panhandle north through western Kansas and eastern Colorado is that the area is very dry.

At the Wheeler Brothers feedlot, rations are dispensed in troughs, called bunks, along the fence line. The main component of the feed is steam-flaked corn.

PIGS AND POULTRY

Even more than the beef business, the raising of swine and poultry has been transformed into an industrial-scale, assembly-line operation. Hogs once grew from "farrow-to-finish" on a single farm. Now one farm raises piglets until weaning, then passes them along to a nursery for 8 or 10 weeks; finally, they go to a finishing operation, where they are fattened to market weight. The hogs are owned by none of these farmers but rather by a meat-processing company. Chickens are raised under the same kind of contract arrangement.

The modern hog spends its entire life indoors. The swine barn is a long, low-roofed shed with a slotted floor over a waste pit. The interior is divided into pens that hold 8 or 10 animals each. Social interactions limit the size of the pens; if too

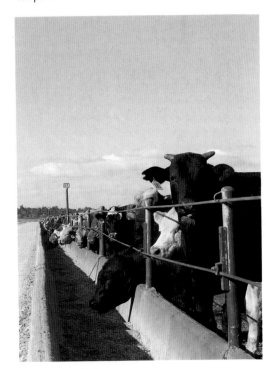

many animals are put together, the less assertive ones will never get a chance to eat. Even with small pens, feeding time is an occasion for squealing and grunting. The feed is dispensed from overhead bins, and ideally all the pens receive their ration at the same time. If the feeding is done sequentially, those who have to wait tend to get restive, like passengers in the back of an airplane waiting for the meal cart to come down the aisle. Also, once the overhead bins have discharged their contents into the

THE CATTLE GUARD

A barbed-wire fence that can stop a 1,500-pound cow is also a pretty serious impediment to people. Gates allow pedestrians through, but they're a nuisance to drivers. Every time you cross the property line, you have to stop the car, get out and open the gate, get back in the car and drive through, get out again to close the gate—and then maybe round up the animals that sneaked through while you were busy. The gadget that eliminates all this rigmarole is the cattle guard, an item that's as common on the back roads of cattle country as manhole covers are in the city.

The idea of the cattle guard is to leave an unfenced opening where a road or driveway crosses a fence line but put something on the ground that livestock won't cross. The classic choice is a grill made of steel pipes or rails spaced a few inches apart and laid across a shallow pit. Cars and trucks can bump their way across this grating, but cows and other hoofed animals are deterred by the precarious footing.

There is some disagreement over just why livestock shy away from the array of bars in a cattle guard. Cows do seem to be very careful about where they put their feet, and they get skittish on slippery surfaces. But that's not quite the end of the story. Someone discovered that most cows will be stopped by the mere optical

illusion of a cattle guard—a pattern of dark and light lines painted on an asphalt roadway. Do these trompe l'oeil barricades really work, and if so why? Is a cow's eyesight so bad that it's fooled by such crude fakery? Or is the repulsion just a matter of bovine fashion sense—a horror of stripes? One might even speculate that the effectiveness of real cattle guards has nothing to do with the risk of falling between the rails, that the cows are just reacting to the pattern of parallel lines. A crucial question is whether a naive cow—one that has never seen a real cattle guard—will turn away from the painted kind.

The inventor of the cattle guard has not been identified. The devices were installed first on railroad rights-of-way, and then were adapted to roads almost as soon as the automobile reached the prairies. James F. Hoy tells all that's known in *The Cattle Guard: Its History and Lore.* The guard photographed at left lies athwart a road in Topock, California. Above, Kansas cows gaze, perhaps resentfully, across a barrier.

Poultry farming has become an industrial-scale enterprise and has also moved indoors in recent years. This long, windowless shed is one of several housing turkeys at a farm in Oak Flat, West Virginia. The funnel-like canisters in front are feed bins.

feeding troughs below, the bins are refilled immediately, while the hogs are busy eating. If the refilling were done just before the next feeding, the noise would provoke more anticipatory commotion.

The swine industry has run into controversy lately over the question of what to do with the waste flushed out of the barn. Current practice is to pump it into open pits called lagoons—although the fancy, romantic-sounding name has done nothing to improve their reputation. Neighbors complain about odors. And some lagoons have failed or overflowed during storms, draining into the nearest watercourse.

Chickens are raised in long sheds that look much like hog barns in their exterior architecture, but the living arrangements inside are quite different. As a rule, cages are suspended from the ceiling so there are no obstructions to hosing out the floor underneath. Often there are two or three tiers of cages, offset in a stairstep arrangement so that droppings from the upper levels don't land on the heads of birds below.

Chickens are susceptible to heat stress, and so a poultry barn will usually have large ventilating fans. Inside, there may also be fogging nozzles, like the ones that spray the produce at the supermarket. What there won't be, in many cases, are windows. In nature, hens lay eggs only in the spring and summer, but they can be fooled into laying year round if the duration of daylight is at least 14 hours. Many farmers find it easier to maintain this schedule in a closed barn with artificial illumination.

IRRIGATION

The cradles of civilization on the Nile, Indus, and Tigris-Euphrates Rivers all depended on irrigated agriculture. Some historians suggest a double connection between irrigation and these early experiments in urban life. Irrigation was necessary to boost productivity, so that some of the people could leave the land and live in the city. At

the same time, large-scale social organization was needed to build and manage the irrigation projects. The role of irrigation is similar today. It's a way of boosting the intensity of agriculture—getting more out of the land than natural rainfall would support. And it still takes a lot of organization and cooperation.

Canal-and-Ditch Irrigation. From ancient Mesopotamia to the Central Valley of California, canal-and-ditch irrigation is the time-tested way to get water to the land. It's essentially a river system in reverse: water flows from a main canal into a smaller branch canal and then into still smaller "laterals" and eventually into a particular field. Building such a system requires no technology the Sumerians didn't have, but it does demand some very sophisticated hydraulic engineering. The canals have to maintain the right slope to keep water flowing over long distances, and fields have to be carefully leveled or else the water will puddle in some areas and leave others dry. Farmers in the paddy-rice regions of Asia have had the surveying skills and the earthmoving technology to solve these problems for thousands of years. American farmers have lately come to rely on laser leveling instruments.

Irrigation projects also require sophisticated social engineering. The farmers served by a canal have to agree on how to allocate the water. In many places, the decisions are made by district irrigation boards, which occupy an interesting middle ground between private enterprise, public utility, and elected governmental body.

A farmer's water allocation is expressed in acre-feet—one of the more vivid units of measure in common use. A farmer entitled to three acre-feet could in principle build a dike around the fields and flood them to a depth of three feet. In practice, the water is parceled out over the growing season.

The flow of water through the canals and laterals is controlled by gates and weirs. Some of these devices serve to balance the flow of water throughout the system. Others—called delivery gates or headgates—control the flow out of a canal into the fields. Usually there's one headgate for each 40-acre parcel. The gate is opened for some fixed period of time each week, scheduled in advance.

Sprinkler Irrigation. The main alternative to canal-and-ditch irrigation is to water the crops the way a suburbanite waters the lawn and garden—by spraying water over the tops of the plants. Doing this on an agricultural scale requires a lot of expensive hardware: pipes, sprinkler heads, a pump to supply pressure. It's also less efficient than the ditch system, since a lot of the water evaporates before it ever reaches the roots of the plant. Nevertheless, sprinkler systems are the predominant form of irrigation in much of North America. One reason is that their maintenance requires less labor. Even more important, they can be made to work on sloping ground.

The most remarkable of the sprinkler systems are the center-pivot irrigators, if only because they create one of the most intriguing sights you'll ever see from an airplane window. Flying over the Great Plains in summer, when you look down on the checkerboard pattern of square fields, many of the squares have checkers on them!

Irrigation canals divert water from rivers or other natural bodies and spread it over farmland. In the upper photo, water pumped from the Mississippi River wells up into a canal in Avoyelles Parish, Louisiana. This is an area that gets plenty of rainfall, but additional water is needed to flood rice paddies. In the lower photo, headgates regulate the flow of water into farmers' fields near Heber, on the Mexican border in southern California. This area would be a desert without the All American Canal, which captures much of the flow of the Colorado River.

The inscribed green circles, which just reach the boundaries of a 160–acre quarter section, are the areas watered by center-pivot irrigators.

At close range a center-pivot irrigation system is an unlikely-looking plumbing job. An aluminum pipe a quarter of a mile long and six inches in diameter is held high over the cornstalks on spindly, two-wheeled A-frame carriages. Mounted every few yards along the top of the pipe, or hanging below it on flexible hoses, are rotating sprinkler heads much like those used for watering lawns. The contraption looks too flimsy to support its own weight, much less to move under its own power.

Center-pivot irrigation was invented in the early 1950s by Frank Zybach, a farmer on the flatlands of eastern Colorado. His key idea was a scheme for propelling the sprinkler system. Your first thought might be to put a big motor in the middle and swing the whole apparatus like a baseball bat, but that would never work. A quarter mile of water-filled pipe is much too heavy and not nearly stiff enough. The only way to move the pipe is to give each supporting carriage its own means of propulsion. But then there's the problem of coordinating all those separate motors, which have to go at different speeds because they are at different distances from the center. If one carriage moves even slightly faster or slower than it should, the pipe will twist and eventually break.

A center-pivot rig *(above)* begins its slow circular tour of a square field near Plymouth, Washington. The two outrigger sprinklers at the end of the boom provide a little extra coverage in the corners of the square. The pivot end of a center-pivot system *(below)* near Alma, Nebraska, is erected over the well that supplies the water A seal near the top of the pyramidal frame connects the rotating pipe to the stationary segment.

The polka dot landscape in the photograph at the top of the page is a densely packed array of center-pivot systems near Garden City in western Kansas. Each small dot covers a 40-acre field; there are also a few larger dots in 160-acre fields. The photograph was made by the Landsat 7 satellite in September 2000. Because the satellite's camera is sensitive to wavelengths outside the visible range, colors are unnatural: red indicates growing vegetation. The lower photograph above, made in October 2004 by an astronaut aboard the International Space Station, shows center-pivot irrigators at an oasis called Al Khufrah in southern Libya. Because Libya's surveyors were not bound by Thomas Jefferson's preference for square lots, the circular irrigators are packed more efficiently on a hexagonal lattice. (Photographs reproduced courtesy of the U.S. Geological Survey and the National Aeronautics and Space Administration.) At right, another solution to the square-field problem: a side-roll irrigator, which moves across the field in a straight line, sprays newly planted crops in Tulelake, California.

Zybach's ingenious solution was to let the deflection of the pipe itself regulate the speed of each carriage. The farmer adjusts a control to set the speed of the outermost carriage, which begins a slow march around the perimeter of the field. At the start, none of the other carriages has yet begun to move. The movement of the outermost carriage therefore tends to bend the pipe. Before the bending can cause any damage, however, it is detected by a sensor at the next carriage, which switches on its motor and moves just enough to bring the pipe back into alignment. The bending and the compensating movement propagate inward from carriage to carriage all the way to the center, in a snakelike wave of rotation.

The placement of the sprinkler nozzles along the pipe calls for some delicate calculations. Suppose they were installed at uniform intervals. Because the pipe moves faster at the perimeter than near the pivot, the central parts of the field would get soaked while the peripheral areas would remain parched. Compounding this problem is the loss of pressure as water flows outward through the pipe, which again favors the inner over the outer regions. The solution is to mount fewer and smaller sprinkler heads near the center.

At maximum speed, a center-pivot system on a quarter-section plot can make one revolution in about 12 hours; at this rate it moves at the same speed as the hour hand of a clock. The usual speed is slower—perhaps three or four days per revolution.

What about the awkward problem of fitting a round peg in a square hole? A circle inscribed in a square covers less than 80 percent of the area; the unwatered corners go to waste. A number of inventors have struggled to overcome this handicap with various articulated appendages and other devices that reach out into the corners, but they have not caught on with the majority of farmers. The economic reality is that center-pivot irrigation is employed mainly on marginal land, where the loss of production from the corners is not crippling. If the surveyors of the Northwest

Territory had taken as their model not a checkerboard but a honeycomb, center-pivot irrigation would now be more efficient. A circle inscribed in a hexagon covers more than 90 percent of the available area.

Other Sprinkler Systems. A more radical solution to the corner problem is to build a sprinkler system that crawls along the ground in a straight line. One approach adapts the same technology of motorized A-frame carriages. The long pipe stretches out across the width of the field, and all the carriages advance in unison. This scheme not only fills in the corners but also eliminates all the problems of having different sprinkler heads moving at different speeds. Unfortunately, it also loses the advantage of having one end of the pipe stationary over the wellhead; now the water connection has to be made through a flexible hose or by pumping water out of an open ditch dug parallel to the direction of travel.

Another style of moving sprinkler, known as a side-roll irrigator, looks superficially like the Zybach-inspired systems but works a little differently. It also has a long pipe fitted with sprinkler heads, but instead of being held aloft by A-frame carriages, the pipe passes through the centers of large spoked wheels. The water supply, again, has to come through a flexible hose. When I first saw one of these, it seemed obvious at a glance how the device works: the wheels roll along in unison, carrying the pipe with them. But then, after a second glance, I wasn't so sure. The sprinkler heads are attached to the pipe, so how do they stay upright as the pipe rolls? The answer is that the side-roll irrigator does not move continuously, as the center-pivot one does. The unit is set in place, sprinkles for a while, then is drained and rolled along 50 feet or so to the next section of field. After each move, the sprinklers are set upright.

The simplest sprinkler system is just a big water cannon, much like the ones that firefighters use. With enough pressure, it can launch a stream of water a quarter of a mile. Sometimes it is set up on a platform that's slowly pulled across the field by a cable and a winch, with a flexible supply hose trailing behind.

Trickle Irrigation. In places where water is especially scarce or crops are especially valuable, there's a new style of irrigation that's very precise and efficient, though also expensive. Small plastic tubes and miniature nozzles deliver water to individual plants, almost drop by drop. No water is wasted in the spaces between plants. But threading the tubes through the field takes a lot of work, and the nozzles tend to clog.

Drainage and Salinity. With farming in dry terrain, the obvious problem is how to get enough water to the crops. How to drain water away from them hardly seems like an issue worth worrying about, but in fact it's crucially important. Irrigation turns out to be like the body's circulatory system: You need both arteries and veins.

Drainage is key because the water applied to crops is never pure; it always includes some dissolved salts. When the water evaporates, the salts stay behind, gradually poisoning the soil. To combat salinity, some of the water must be allowed to drain away.

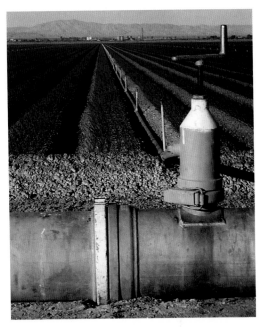

Irrigation in California's Imperial Valley has made the arid region one of the nation's most productive, but it has also brought the intractable problem of salt accumulation in the soil. Much of the valley is below sea level, so water cannot drain away; as it evaporates, salts are left behind. The very deep furrows between rows in this field help keep salt below the plants' root zone.

On-farm drainage ponds look like swimming pools in these southern California fields. The ponds are impoundments for saline runoff.

A farmer injects anhydrous ammonia into the soil after the fall harvest near Atlas, Illinois. By the time the next crop comes up in the spring, soil bacteria will have converted the ammonia into nitrogen compounds the plants can absorb. The ammonia flows from the tank through a rubber hose to nozzles that release the gas a few inches below the soil surface.

Salinity problems get particularly nasty in arid regions (where there's not a lot of water to spare) and at low spots in the landscape (where water flows in but there's nowhere to send the drainage). A case in point is the region west of Bakersfield and Fresno, California. When canals began delivering irrigation water in the late 1960s, salinity troubles began almost immediately. The answer offered by the U.S. Bureau of Reclamation was the San Luis Drain, meant to carry waste water 200 miles north to the San Joaquin River. But the drain never quite made it as far as the river; construction halted at Kesterson Reservoir in Merced County. The reservoir began filling with saline runoff, and soon there were complaints of waterfowl dying. A 1985 court ruling closed the entire system, and the reservoir was filled in. But the story is not over yet. Parts of the drain have been reopened; more lawsuits are pending.

The alternative to drainage canals is to keep the salt on the farm. When you fly over farmland in southern California today, you might get the impression that every field has a bright blue swimming pool. Those are evaporation ponds, built at the low point of each field and fed by underground drain pipes. Farmers are unhappy with them because they sacrifice 10 or 15 percent of the land area. The ponds also leave behind concentrated residues that are going to have to be dealt with somehow, someday.

AGRICULTURAL CHEMICALS

It takes more than seed and sunshine to raise a crop of corn these days; farmers rely on a variety of fertilizers, pesticides, and herbicides to maintain yield. Increasingly, they also rely on protective traits bred or engineered into the genes of the plant itself.

"Nurse tank" for anhydrous ammonia holds 1,000 gallons as a liquid under pressure.

Fertilizers. For many centuries the most important fertilizer was the manure of farm animals, which was readily available and had to be disposed of anyway. In the nineteenth century there was a vogue for seafood as a supplementary fertilizer. On Long Island, oysters that you would now pay $20 a dozen to slurp down on the half shell were shoveled onto potato fields by the ton.

The fertilizer business was changed forever in the 1840s with the commercial exploitation of guano from islands off the coast of Peru. Guano is a natural product and a common substance; it's what birds leave behind. They leave it in places other than equatorial Pacific islands, but the deposits there were particularly deep and rich because the islands get almost no rainfall. Like oil today, the Peruvian guano was a resource that governments sent gunboats to defend or acquire. President Millard Fillmore nearly took us to war over bird droppings.

Not too long after the excitement of the Fillmore years, chemists came to understand what it is about guano that makes it a useful fertilizer. The key component is the element nitrogen, which is an essential building block of proteins and other molecules of life. Most plants can absorb nitrogen only in certain chemical forms, known as fixed nitrogen, which guano contains in abundance.

Supplying fixed nitrogen to crop plants is an even bigger business now than it was in Fillmore's time, but guano is no longer the source. Nitrogenous fertilizer has become a product of the petrochemicals industry. The nitrogen itself comes from the air and is free for the taking, but converting it to fixed form means combining it with hydrogen to make ammonia, and the hydrogen comes from natural gas.

Ammonia seems an unlikely substance to promote plant growth. It's an acrid gas—the smell is known to anyone who has ever changed a baby diaper—and in concentrated form it's dangerous enough to require rubber gloves, goggles, and a special respirator. If you were to spray it on your lawn, it would kill the grass. Nevertheless, this is the fertilizer that powers much of American agriculture.

In late fall or early spring you'll see farmers towing lozenge-shaped white tanks behind the pickup truck. The tanks hold a thousand gallons of anhydrous ammonia. (*Anhydrous* just means "without water"; it distinguishes this stuff from the solution of ammonia in water that you might use to wash windows.) Inside the tank, the ammonia is a liquid under pressure, but it boils away to a gas as soon as the pressure is released.

In the field, you'll see the ammonia tank pulled along behind a tractor, with a plowlike device that releases the gas under the soil. You might well think that the gaseous ammonia would simply blow away, but when it is injected a few inches below the surface, it quickly dissolves in soil moisture. Losses are small. There may be wisps of white clouds drifting away behind the plow; these visible traces are not actually ammonia (which is colorless) but water clouds formed because the ammonia is very cold when released from liquid form.

Ammonia is always applied after the harvest or before planting. The ammonia itself is of no direct use to crop plants; it has to be taken up by bacteria in the soil, which convert it to other nitrogen compounds, such as nitrates. The conversion to nitrates

An insect trap, installed at the edge of a cotton field in North Carolina, is used for taking a census of pest species and deciding whether or not to apply insecticides.

A crop-spraying aircraft makes a low-altitude pass over a field near Calipatria, California, in the Imperial Valley.

can also be done as an industrial rather than a biological process. Apart from anhydrous ammonia, the other main fertilizer is ammonium nitrate. This white powder has an unusual combination of properties: miners use it to blast open the countryside, and farmers use it to improve the soil. Go figure.

Pesticides and Herbicides. Crops have been attracting pests at least as far back as the biblical plagues of locusts, but it seems the idea of trying to poison or repel insects with chemical sprays didn't come along until the nineteenth century. One of the first of those insecticides was pyrethrum, derived from chrysanthemums; interestingly, related products are still popular today—partly, no doubt, because something made from chrysanthemums seems so innocent it couldn't possibly be dangerous. Other early pesticides would never win approval today: they were lead and arsenic compounds, as well as forms of cyanide.

The most famous—or infamous—insecticide, DDT, was developed during World War II. Its first uses were not for agriculture but for the control of disease vectors, such as the mosquitoes that spread malaria and yellow fever. Later, however, enormous quantities of DDT were sprayed on cotton crops. Then came Rachel Carson's book *Silent Spring*, which blamed DDT for a decline in songbird populations and warned that extinction threatened the bald eagle. A decade later, DDT was banned in the United States. Whether it was guilty of the crimes charged seems less sure now than it did then, but a more general conclusion is beyond dispute: the use of pesticides can have unintended and unpleasant consequences.

Insecticides are still applied to cotton as well as many orchard and vegetable crops, but farmers claim to be more sensible about their use these days. For economic as

well as environmental reasons, they are urged to spray only in response to an actual infestation, not to prevent one that might or might not arise. One sign of this strategy is that you may see colorful insect traps, looking like some sort of decorative party lantern, planted on the margins of farmers' fields. They are there to take a census of bugs. When insecticides are called for, they might be applied with a sprayer towed behind a tractor, or from the air.

Herbicides fall in a different category from insecticides. A herbicide is a substance that kills plants—which seems like the last thing a farmer would want to do in a field where crops are growing. But in most fields, crops aren't the *only* plants growing. The trick is to make the herbicide selective so that it kills the weeds while sparing the crop.

In some cases selectivity is not hard to achieve. Crops such as corn and wheat are grasses, known to botanists as monocotyledons, whereas most weeds are broadleaf plants, or dicotyledons. The two groups are about as far apart as they can get and still be flowering plants, so it's not surprising that there are chemical compounds that inhibit the growth of broadleaves while having little effect on grasses. The first of these broadleaf herbicides, developed in the 1940s, was 2,4-dichlorophenoxyacetic acid, better known as 2,4-D. In later years it attained notoriety as an ingredient of Agent Orange, the defoliant sprayed by the U.S. military in Vietnam.

Herbicide strategies get trickier when the crop plant is also a broadleaf, as with soybeans or cotton. One approach is a "preemergence" herbicide, which is sprayed on the field after the seeds are in the ground but before the shoots come up. If the timing is just right, the herbicide kills the fast-growing weeds, but breaks down or washes away by the time the crop plants have appeared.

The new trick is to make the crop plant resistant to the herbicide. This is the basis of "Roundup Ready" cotton and soybean varieties being marketed by the Monsanto company, which also (no coincidence) makes the herbicide Roundup.

THE BACKLASH AGAINST INDUSTRIAL AGRICULTURE

Advances in agricultural productivity have kept the world's plates overflowing for the past 50 years. It's hard to argue with that kind of success, and yet it often seems that no one is happy with current farm policies and practices, or with the state of the food supply. American farmers (as well as European and Japanese ones) complain they are being squeezed economically and their way of life is in jeopardy. Farmers elsewhere say they have it even harder, especially when rich countries export their crop surpluses and call it foreign aid. Consumers worry about pesticide residues on crops, hormones in meat, and the threat of mad-cow disease, not to mention tasteless tomatoes. Environmentalists warn about the erosion of topsoil, the exhaustion of aquifers, and a system of high-intensity agriculture that may not be sustainable. Some biologists are concerned about the loss of genetic diversity in crop plants. Nutritionists say the American diet stinks and we're all getting fat.

One response to these grievances has been a resurgence of small-scale, low-tech, family farming. On the outskirts of every city now there are boutique farms growing organic "heirloom" tomatoes, raising "free-range" chickens, making a bit of goat cheese, and selling all their produce at the local farmer's market on Saturday morning. From the point of view of global agriculture, the output of these suburban backyard farms hardly even registers as economic activity. In the production of commodity calories, they'll never compete with the Iowa corn farmer or the Texas feedlot. But who wants to eat commodity calories?

OIL AND GAS

IN THE FALL OF 1973 one piece of the industrial infrastructure emerged from the unnoticed background of American life and became a public obsession. The gas pumps had run dry. Any filling station with fuel to sell had a long queue of cars snaking around the block. Prices soared. Getting the tank filled—a chore that had been so routine it slipped beneath the level of conscious attention—was suddenly a challenge that called for strategy and guile, not to mention getting up before dawn.

The gasoline shortage lasted only a few months, but it made a strong impression on those who lived through it. For a while, American automobiles became smaller, lighter, and less thirsty. A pipeline from the North Slope oil fields across Alaska was quickly approved and built. A 55-mile-per-hour national speed limit was enacted as a fuel-saving measure and ruled the roads for more than 20 years.

The causes of the 1973 oil crisis were more political and economic than technological, but the event nonetheless prompted much sober thought about life in a world of finite resources. The oil was not running out in 1973, and 30 years later we are still pumping it out of the ground at a furious pace, but the idea that it will not last forever is now taken seriously—even by some of the oil companies. (Though evidently not by the owners of sport utility vehicles.) This chapter looks at the infrastructure of the oil industry, from the well through the refinery to the filling station. Finally, there is a section on the natural gas industry, which has a rather different culture.

THE NATURAL HISTORY OF AN OIL WELL

The drilling rig is the universal emblem of the oil industry: a tapered steel derrick, usually depicted with a gusher shooting up through the middle of it. Your chances of ever seeing a gusher are nil. Even to see a drilling rig, you need to be in the right

A spherical tank in Port Arthur, Texas *(opposite page)*, forms part of a network of refineries, tank farms, and pipelines that delivers the petroleum products to keep the United States fueled and lubricated. Spherical tanks are used to hold fluids under pressure; commonly they contain liquefied petroleum gas (LPG), which is mainly propane. The stacks atop the tank are vents for emergency pressure-relief valves. The bright red pipes carry water or foam for firefighting.

place at the right time. In the life cycle of an oil well, drilling is a relatively brief phase. It lasts a few months, and then the derrick is broken down and hauled away. You're much more likely to happen upon a well in the production stage, which can last for decades. Nevertheless, drilling is where it all begins, and it's certainly the adventurous side of the oil business, the world of wildcatters and roughnecks.

The Drilling Rig. When drilling is under way—when the rig is "turning right and making hole"—several things have to be going on at once. The bit, or tool, which does the actual drilling, has to be simultaneously turned and pressed downward so it will cut into the rock. At the same time, a lubricant called drilling mud has to be pumped down to the bit to carry away the cuttings.

The bit is connected to the surface by the drill string, which is not a string at all but a rigid, hollow steel pipe assembled from sections called joints, which are typically 30 feet long. A well 15,000 feet deep would need 500 such joints. The drill string transmits both vertical forces (pressing the bit down into the rock and later hauling it out again) and the twisting force that turns the bit. It also carries the mud through its hollow bore.

The twisting force comes from the rotary table, a motor-driven, spinning platter set into the main deck of the drilling rig. The rotary table grips a special length of drill pipe called the kelly, which has a square or hexagonal cross section matching a bushing in the center of the rotary table. The table turns the kelly in the same way that a wrench turns the head of a bolt; the kelly then turns the rest of the drill string. As the kelly is being spun by the rotary table, it is also free to slide vertically through the table so downward force can be maintained.

The drill string is not pressed down from above; pushing on a pipe three miles long would accomplish nothing but buckling the tubing. Instead, the downward force is provided by a series of extra-heavy joints of drill pipe, called drill collars, installed immediately above the bit; it is the weight of the collars that drives the bit

GETTING A LOOK

Much depends on where you live. If you come from the "oil patch" in Texas and Oklahoma, then the sight of a drilling rig is not a novelty, and a sucker-rod pump nodding over an oil well probably seems as commonplace as a traffic light. In other parts of the country, including most of the populous Northeast, oil wells are exotic rarities. The distribution of refineries is also patchy. There are clumps of them along the Gulf Coast and near some major ports, but fewer inland.

Virtually all elements of the oil-and-gas infrastructure are privately owned and receive visi-

tors only by special arrangement. On the other hand, petroleum is largely an "outdoor" industry: the machinery is not hidden away behind closed doors, and there is much to see from outside the fence.

Something to keep in mind, however: Refineries have often been cited by the FBI and the Department of Homeland Security as potential terrorist targets. If you stop by the roadside to take pictures, you may attract the attention of plant security forces or the local police.

A gleaming white rig in Marlow, Oklahoma, drills for natural gas on behalf of Chesapeake Energy. The rig is a portable, temporary structure. The derrick folds up and breaks into pieces for transport, and the rest of the rig consists of modules the size of truck trailers, which can be hauled to a new site and reassembled in a matter of days. The two stairways lead to the main drilling deck. The house-trailer–size shelter on the near side of the deck is the "doghouse," where instruments and controls are kept out of the weather. The ramp on the right side is used to lift 30-foot "joints" of drill pipe up to the main deck. The rig is capable of reaching depths of about 15,000 feet; the well in Marlow was expected to tap a natural-gas reservoir at roughly 7,000 feet.

into the bottom of the hole. Ordinary joints of drill pipe weigh a few hundred pounds; drill collars weigh 2,500 pounds or more. An entire 15,000-foot drill string could weigh well over 200,000 pounds. The weight is supported by a block and tackle suspended from the top of the derrick, so that the string remains under tension at all times. Every few minutes a brake on the main hoist emits a loud squawk or honk as it automatically adjusts the tension—one of the characteristic rhythmic noises of the drilling site.

The routine of drilling is to keep "making hole" until you have "drilled down the kelly." When the top of the kelly shaft has almost reached the rotary table, the crew stops the rotary, hoists the drill string far enough to expose the top of the uppermost joint of drill pipe, and inserts a clamping device called the slips, so that the drill string cannot fall down the hole. Then the kelly is disconnected from the string, another joint of pipe is inserted, and the kelly is reattached to the top of this joint. The joints fit together by means of threaded couplings, female at the upper end and male at the lower end. The threads are tightened with "tongs" that work much like a plumber's pipe wrench but weigh several hundred pounds.

At intervals, the drill bit needs to be changed. This requires hauling the entire drill string back up to the surface, a process known as tripping out. The converse process, naturally, is tripping in. As a rule, the drill pipe is not broken down into individual joints during such a round trip; instead, sections two or three or four joints long (doubles, thribbles, or fourbles) are stood on end in a rack inside the derrick.

Top Drives and Downhole Motors. The rotary-table-and-kelly mechanism for spinning a drill bit was a marvel of ingenuity—and it takes great ingenuity to keep it running. In recent years simpler schemes have come into use.

The drilling crew on the rig in Marlow adds another 30-foot joint of pipe to the drill string. Above, the threaded end of the new joint is wrestled into position; at right, the threads are tightened with enormous wrenches known as tongs. During this operation the entire length of drill string is supported by a two-handled device called the slips clamped to the drill pipe where it goes through the main deck.

A drilling rig with top drive has an electric motor mounted on vertical rails. The motor's drive shaft connects directly to the drill string, eliminating the need for both the rotary table and the kelly. Furthermore, the motor's torque can be used to tighten the threaded connections between joints, thereby dispensing with tongs as well. With top drive, smaller crews can drill faster and with less fuss.

So why don't all rigs have a top-drive motor? It turns out that mounting the main drive motor aloft requires a much stronger and heavier derrick, which takes longer to set up, break down, and transport. Most rigs move to a new hole every few months, so the loss of portability is a serious drawback.

Another drilling innovation puts the motive force at the opposite end of the drill string. Instead of turning the entire string from the surface, a motor is installed just above the bit, near the bottom of the hole. Thus, only the bit turns; all the rest of the drill string is locked in place. This down-hole motor is not an electric one. It runs on mud! It is a turbine that extracts energy from the stream of drilling mud pumped through it. (A dentist's drill works the same way, except that the fluid driving the turbine is air—thankfully—instead of drilling mud.)

The big advantage of a down-hole motor is that it facilitates steering the drill. Various geological strata can deflect the drill, and a correction is needed to bring it back to vertical alignment. Moreover, the driller may *want* to deviate from the vertical. Sometimes multiple wells fan out like spider legs from a single drill site to reach widespread areas of a petroleum reservoir. (This practice is particularly common in offshore drilling.) Sometimes a well is drilled vertically into an oil-bearing stratum and then deflected horizontally, so that it can collect oil from a larger region. Such wells can be drilled even with a conventional rig, but the technique is easier and more precise with a down-hole motor.

Mud. Drilling mud is the least obvious and most ingenious aspect of rotary drilling technology. It is also a major expense—a considerable fraction of the capital investment needed to drill a well goes into buying mud.

Drilling mud cools and lubricates the cutting edges of the drill bit, but that could be done as well by an ordinary fluid such as water or oil. Mud has other important properties. First, because it is a thick, viscous fluid, it holds the cuttings in suspension; rock chips carried away from the bit do not settle back to the bottom of the hole even when circulation stops. Second, because it is dense (sometimes twice the weight of water), it resists the pressure of fluids deep underground. And it leaves behind a coating of clay that helps to seal the bore of the well.

Legend has it that the first drilling mud was made by driving cattle through a bog, but the modern stuff is a high-tech product. The main ingredient is bentonite, a kind of clay. The density can be increased by adding barite, a mineral rich in the heavy metal barium.

A mud pump draws the mud from an open pit or tank and forces it down the bore of the drill string to emerge through nozzles in the bit. It returns to the surface

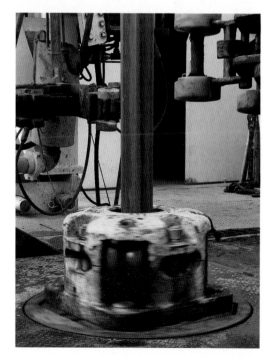

The main hoist on the Marlow rig *(above)* supports the drill string and regulates downward force. The black hose supplies drilling mud, which is pumped down the bore of the pipe. The rotary table *(below)* turns the drill string, transmitting twisting force to a special length of drill pipe called the kelly.

A pair of mud pumps *(right)* drive the rig's circulatory system. Each pump has three pistons that force the mud down the bore of the drill string and back up the annular space surrounding the pipe. The white sheds in the background house the prime movers—the diesel-driven generators that power all the machinery on site.

The blowout preventer *(below)* is the most important safety apparatus on the drilling site. The preventer, installed at ground level under the drilling deck, is a heavy casing with several hydraulic rams that can seal off the well if high-pressure fluids underground threaten to escape. The main controls for the preventer are on the drilling deck, but auxiliary controls *(bottom)* are at ground level for use in emergencies. The red steel cylinders hold hydraulic fluid under pressure in case the hydraulic pump should fail.

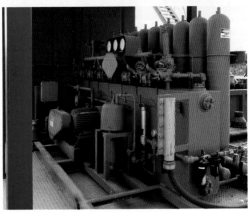

through the annular space between the drill pipe and the wall of the hole or the casing. The crew keeps a close watch on the returning mud; it's an important source of information on what's happening at the bottom of the hole.

Layout of the Drilling Site. The rig and derrick are at the center of the drilling operation, but there is much else on the site as well. On the ground near one side of the rig structure are the prime movers—the engines that supply power for all the machinery. The prime movers are usually diesel engines, three or four of them, with a total power output of several thousand horsepower. The mud pumps, as major consumers of power, are placed nearby.

The mud pits take up most of the space on another side the rig. Yet another large area is occupied by the racks where hundreds of joints of drilling pipe are laid out. Other racks hold the larger and heavier pipe called casing. From the racks, joints of pipe or casing are dragged up a ramp to the drilling floor.

Elsewhere on the site is a row of trucks and trailers that provide office space (and sometimes living space) for the numerous consultants and contractors who are an essential part of the drilling operation. For example, one specialist provides the drill bits (known in the trade as tools). Another offers instruments and expertise in well logging—recording the geological strata the drill passes through.

There is more office space up on the main deck of the rig, in a structure amiably named the doghouse. It serves as lunchroom and gathering place for the crew, as well as housing a full set of controls for the rig.

Gushers and Blowouts. In the Hollywood version of the oil business, every well is a gusher. As crude thunders out of the ground, the crew celebrates in a rain of black gold, like boys playing in a mud puddle. A real gusher would be a disaster: the rig might be destroyed, fire is likely, people are injured or killed, and at the very least there is a monumental mess to be cleaned up.

The main line of defense against such events is the blowout preventer, a series of valves mounted at the wellhead, under the deck of the drilling rig. The valves close off the well as soon as it starts to "kick," or show evidence of back pressure. A common sign of kicking is mud that continues to flow out of the well even when the pumps are shut off. The kick can usually be stopped by closing an annular preventer, which seals off the annular space around the drill pipe. If the annular preventer fails, there are ram preventers that crush the drill pipe or, if necessary, shear it off. Once the kick is controlled, the crew will mix a heavier mud and pump it into the hole.

The blowout preventers are operated by hydraulic pressure. The main control levers are on the drilling deck, but a second set of controls is placed some distance away for emergency use. In case the hydraulic pump fails, a battery of steel cylinders stores hydraulic fluid under high pressure.

Casing and Completion. After a well is drilled but before it goes into production, several more steps need to be taken.

First, the well must be cased and cemented. Casing is a steel pipe that lines the hole, keeping the oil out of overlying formations. In a deep well there will be several strings of casing, nested one inside the other, though only the innermost goes all the way to the bottom.

If you are in a petroleum-producing area and you see pipe being unloaded at a rail siding or hauled by truck, it is more likely to be well casing than drill pipe. The reason is that much more casing is needed, since each string of casing remains permanently in the hole, whereas drill pipe is reused many times. For the same reason, casing is a major item of expense in completing a well.

Casing is bonded to the well with cement. It must have taken a fair amount of courage the first time this procedure was tried. Having gone to the trouble and expense of drilling a well and setting casing, the crew then fills the casing with cement. If this cement hardened in place, it would totally plug the well. The trick is to pump in another fluid (such as mud) under enough pressure to force the cement out the bottom of the casing and up the annular space to the surface.

Even when the final casing is set and cemented, the well is not yet ready to produce. There is work yet to be done deep underground. Diagrams of wells often make it appear that the drill string pokes through the roof of a hollow cavity filled with liquid oil, but there are no such cavities deep in the earth. Oil and gas are actually dispersed in microscopic pores within a matrix of solid rock. Fluids move so slowly through this material that a newly drilled well could recover oil and gas only from within a few feet of the well bore. For drilling to make economic sense, this range has to be extended, allowing a larger volume of oil and gas to flow into the well. The process is known formally as well stimulation or well development, but workers in the oil field call it fracking (short for *fracturing*). Passages through the rock are opened up by explosions, by acids, or by surges of high-pressure water; then the cracks are held open by pumping in sand or a synthetic "proppant" of crush-resistant spherules.

Another Chesapeake Energy well near Marlow had been drilled but was not yet in production when the photograph above was made. The wellhead at left still has a blowout preventer in case any high-pressure fluids are encountered. The apparatus at right is a sand separator, used to remove excess sand or "proppant" from the product stream. The proppant *(below)* consists of crush-resistant synthetic spherules that are forced down the well to prop open cracks in the oil- or gas-bearing rock layers.

After fracking, a well might tap the hydrocarbon resources within a few hundred feet of the well bore. This is still a small area—perhaps 10 acres. It is sobering to reflect that the oil and gas recovered from under such a small patch of ground can be

The offshore drilling rig Rowan Gorilla III is seen in an outfitting yard along the Sabine River between Texas and Louisiana, where many such rigs are built and maintained. It is a jack-up rig: the platform moves up and down the tall, girder-like legs. When this photograph was made, in 2001, new legs were being constructed. At one time, the Rowan Gorilla III was the world's largest jack-up rig. For a sense of scale, note that the basket dangling from a crane on the left side of the rig holds two workers, just visible in silhouette.

worth enough to repay the multimillion-dollar cost of drilling the well. But there is another side to the economics of oil. Most existing wells (as opposed to newly drilled ones) have a much smaller output. The national average is just 11 barrels per day, and thousands of wells produce less than a barrel. Even when the price of oil is up around $50 a barrel, owning an oil well is not guaranteed to make you a millionaire.

Offshore Drilling. Boring a deep hole in the ground is hard enough on land; it's even harder when the ground is hundreds of feet beneath a hostile sea. Offshore drilling is therefore worth the cost and the risk only when the potential payoff is very large.

In the Gulf of Mexico and in several areas along the California coast, drilling operations begin not very far offshore: you can see drilling platforms from land or reach them with a small boat. For that matter, you can see a lot of the equipment onshore, where rigs are brought into port for maintenance. The machinery is so big you don't have to get very close to get a clear look.

Exploratory offshore drilling is done with a mobile drilling unit. These barges and ships and specially built semisubmersible rigs remain floating over the hole as drilling proceeds. The vessels are either anchored in place or held on station by a dynamic positioning system, which uses computer-controlled thrusters to correct any drift. In shallow water another kind of mobile unit is popular: the jack-up rig. It is a barge with three or four tall girder-like legs that slide vertically through openings in the hull. The rig is floated to the drill site with the legs raised; then they are lowered to the sea floor and the hull is jacked up above water level.

In proven oil and gas fields, drilling is done from permanent platforms, which also serve as production platforms once all the wells have been completed. These platforms are much larger structures than mobile drilling units. Some have a frame of hollow steel tubes, which is floated to the site on its side and then turned upright and sunk by flooding the tubes. The tallest structure in the world is a platform of this kind, called Bullwinkle, 150 miles southwest of New Orleans in the Gulf of Mexico. It towers 1,600 feet over the sea floor and almost 300 feet over the sea surface.

Wells in Production. After all the commotion of drilling is over, a producing well is a lonely place. A few pipes and valves poke out of the ground, and there may be a pump, but there is seldom anyone around to tend this equipment. An employee known as a lease pumper is supposed to stop by once a day to check the machinery and note the amount of oil produced, but automated instruments for remote monitoring are making even these brief visits unnecessary.

A free-flowing well—one with enough pressure underground to push the oil or gas to the surface—is fitted with a tall stack of valves and gauges called a Christmas tree. Why so many valves? They serve various functions—taking samples, regulating flow and pressure, providing an emergency shutoff—but there's also deliberate redundancy. If a valve fails and needs to be replaced, the job is a lot easier if there is another valve below it that can close off the flow while repairs are made.

A producing well near Marlow, Oklahoma, is fitted with a Christmas tree: a stack of valves for controlling the flow of oil and gas that come out of the earth under their own pressure. On this well the upper valves regulate flow through the well's central tubing; some of the lower valves give access to the annular space between the tubing and the well casing.

The pressure that drives oil to the surface usually dissipates long before the oil itself is exhausted. Thereafter, the oil must be lifted artificially. The standard device for doing this is a marvelous contraption with a marvelous name: the sucker-rod pump. The business end of a sucker-rod pump is far below the ground, installed in the tubing at the level of the oil-bearing formations. The basic components are a cylindrical chamber and a pair of one-way valves, arranged so that the chamber fills on the downstroke and lifts the oil on the upstroke. The pump chamber is driven by the sucker rods—a string of rigid steel bars that run inside the production tubing. Like drill pipe, the rods come in joints about 30 feet long, but they are solid rather than tubular and only a half inch to an inch in diameter.

The topside part of the sucker-rod pump is a distinctive sight in any oil-producing region. The sucker rods are connected by a short length of wire rope to one end of a walking beam, which is held 10 or 12 feet off the ground and pivoted in the middle. The other end of the beam is driven by a crank arm, so the beam rocks about its pivot point, periodically raising and lowering the sucker rods. A large, curved guide at the sucker-rod end of the beam is called the horse's head, and the name couldn't be more

A sucker-rod pump extracts oil from a well when subterranean pressure is insufficient to bring it to the surface. The pump's raised beam (called a walking beam) rocks back and forth, driven by the motor and linkage at left. This nodding motion is transmitted to the vertical string of steel sucker rods, which extend all the way to the bottom of the well. A pump chamber at the bottom lifts oil on the upstroke and refills on the downstroke. The pump in this photograph is on farmland near Plainville, Kansas.

A battery of tanks near the well site holds oil pending shipment to the refinery. Not all the tanks are for oil. Storage is also needed for the salt water that comes out of the well.

apt, for it is horselike not only in form but also in its nodding motion. The face of the horse's head is an arc of a circle, and so it converts the rotary motion of the walking beam into linear motion of the sucker rods. It is important that the rods move straight up and down into the well, without any bending stresses.

The power source for a sucker-rod pump is usually an electric motor, but there are other possibilities, particularly for wells in remote areas. If a well produces significant amounts of natural gas, some of it can be used to fuel a small engine.

Sucker-rod pumps run at a stately, unhurried pace—typically 10 to 20 strokes per minute. Often you'll see a pump standing idle, but that doesn't necessarily mean the well has been shut down. Most pumps run intermittently—a few hours on, a few hours off—to avoid distorting the distribution of oil, water, and gas in the underground formations. Overpumping can permanently impair a well's production, and can even affect neighboring wells.

Up close, a sucker-rod pump makes a moaning or sighing noise with a slightly syncopated rhythm, since the upstroke puts more load on the machinery than the downstroke. In the evening the sound has a lullaby quality to it. The whole apparatus looks like a holdover from the age of steam. In fact it's not quite that old, but the design has changed little since it was introduced in the 1920s. Some individual pumps from that era are still running.

Field Processing of Oil and Gas. The fluids that come out of an oil well are not ready to be poured into your gas tank. They are not even ready to be pumped to the refinery. The stuff that comes out of the well doesn't qualify as crude oil until some preliminary field processing is done. The equipment for this processing is sometimes at the wellhead; more often a single processing station serves a cluster of nearby wells.

A heater-treater at one of the Marlow wells has the job of separating gas, water, and oil, which come out of the well mixed in a frothy emulsion. Under heat and pressure, the gas bubbles out of the mixture, while the denser water settles to the bottom.

Something to notice about a field-processing station is the bed of gravel that surrounds each piece of equipment. In the event of a small leak, the absorbent gravel reduces the fire hazard. It also makes leaks easier to spot.

The main task in field processing is to separate oil, gas, and the salt water that inevitably comes up out of the well along with the more valuable fluids. The simplest kind of separator is a settling tank with one inlet pipe and three outlets—at the top for gas, in the middle for oil, and at the bottom for water. A characteristic feature of these gravity separators is a sight glass used to gauge the water and oil levels. Gravity

SWEET-AND-SOUR HYDROCARBON SOUP

Crude oil comes in many varieties, some of which sound like the names of exotic coffees: Louisiana Sweet, Nigerian Bonny Light, Jobo Crude, Suez Blend, Escalante, Oriente, West Texas Sour. Every oil field produces something a little different.

The classification of crudes depends on two factors: heavy versus light and sweet versus sour. Heavy crudes are viscous, tarry, and dense; they have a higher concentration of the "bottoms" that give refiners fits. The sweet and sour crudes have nothing to do with soups served in Chinese restaurants. The difference between them is that sour crudes have more sulfur. I'm told that the terms came about because people really did taste the oil to detect sulfur.

Petroleum is not a single chemical substance but a complicated broth with many ingredients. Most of the molecules are hydrocarbons: little assemblies of carbon and hydrogen atoms (shown here as black balls and

methane ethane propane

white balls respectively). The carbon atoms are linked to one another in chains and necklaces, adorned with hydrogens. The rule for forming the molecules is that every carbon atom has four bonds, or places where it needs to be

attached to something else—either another carbon atom or a hydrogen atom. A hydrogen atom has just one bond.

The simplest hydrocarbon is methane, in which a solitary carbon is festooned with four hydrogens. Ethane has two carbons holding hands, with six hydrogens filling up the rest of the bonds. Propane and butane are chains of three and four carbon atoms, again with hydrogens stuck everyplace there's a free bond.

Organic chemistry has a wonderful vocabulary with the unusual property that once you've

normal butane **isobutane**

learned the full name of a molecule, you've also learned all about its structure. (What if people's names were similarly informative?) This is not the place for a discourse on hydrocarbon nomenclature, but one basic principle is worth knowing. Beyond the four smallest molecules (methane, ethane, propane, and butane), the names are easy to remember if you know how to count in Greek: pentane has five carbon atoms; hexane, six; heptane, seven; octane, eight; nonane, nine; and decane, ten.

Not all the hydrocarbons are linear chains, with carbons lined up like beads on a string.

Among the four-carbon molecules, the straight-chain version is called normal butane, but there is also a branched version, identical in composition, called isobutane. Larger molecules have many possible isomers, or rearrangements. A

isooctane

particularly important one in the petroleum industry is isooctane, which has eight carbon atoms in a lopsided cruciform shape. Isooctane is the gold standard for gasoline. The octane number posted on the gas pump is defined as 100 for pure isooctane, and other fuel constituents are ranked in comparison with it.

In addition to forming straight and branched chains, hydrocarbons can also fold up on themselves to form ring-shaped molecules. Cyclopentane has five carbons in a closed ring. Ring molecules in one important family are called aromatics (because many of them have strong aromas). Benzene is the prototype of this group; it is usually represented as a six-carbon ring with alternating single and double bonds. Both the cyclic and the aromatic rings are present in gasoline, and benzene is a major constituent of some heavier products, such as jet fuel and diesel fuel.

A fully developed oil field seen from far overhead is a dense netting of wells, roads, and pipelines embroidered on the landscape. Each of the small white squares in this photograph is the gravel pad of a well and its associated equipment. As a rule, there can be no more than one well for each 40-acre parcel of land, which accounts for the regular spacing. The bright white lines threading through the array of wells are access roads; there is also a fainter and less orderly network of pipelines that gather the oil. The area shown is the Wasson oil field in west Texas, where drilling began in the 1930s. The gray patch near the bottom of the frame is Denver City, Texas. The photograph was made by an astronaut aboard the International Space Station from an altitude of about 200 miles.

separators are very slow, and thus, if you see one, it is probably safe to assume that the wells it serves are producing at a low rate.

Higher-capacity separators operate under pressure. Like other pressurized vessels, they can be recognized by their beefier pipe fittings, by their rounded form, and by the presence of pressure gauges and relief valves. One type is a long pressure vessel filled with hundreds of baffle plates where droplets of liquid collect and then drain to the bottom. The well stream enters at one end of the separator, and outlets for gas and liquids are near the opposite end. A variant has two horizontal barrels stacked one above the other; the lower barrel accumulates the liquids.

Separators work quite well for dividing gas from liquid, but they are less efficient in removing water from oil. Often, further treatment is needed to dry the hydrocarbons. Crude oil is not considered saleable if it has more than 1 percent water, and dehydration requirements for gas are even more stringent.

Crude oil and water are difficult to separate because they form a frothy emulsion, with the water dispersed in microscopic droplets. One way of breaking up the emulsion is to heat it, which causes the droplets to coalesce in the same way they do in an overheated béarnaise sauce. The usual device for this purpose is called a heater-treater. It can be either a vertical or a horizontal vessel with a burner underneath, typically fired by natural gas from the well. The stack for flue gases is a distinctive feature. As the heated emulsion breaks down, oil is drawn off from the upper part of the chamber and water from below.

The drying of gas is more of a safety issue than a commercial one. At high pressure, water vapor combines with some of the hydrocarbons in natural gas to form solids called hydrates, which can plug up valves and pipelines. Because hydrates are never seen in everyday life, they have an air of the spooky about them, and yet they

Pipeline bridges *(above)* are one of the few places where petroleum pipelines emerge from underground. Both of these bridges carry pipelines across the Colorado River near Needles, California. The one at right was built as a highway bridge (for U.S. Route 66); the much lighter suspension structure above was designed from the start for pipeline duty. Even where pipelines run underground, surface markers reveal their route. The marker below is along Black Bayou in southern Louisiana.

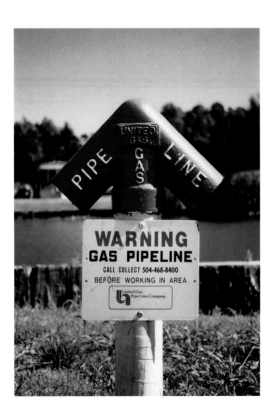

are more than a phantom menace to the gas industry, which expends much effort to suppress their formation.

The salt water separated from the oil and gas, called produced water, is quite a nuisance. It is corrosive, which makes it hard to store, and it is contaminated not only with salts but also with petroleum residues, which means you can't just dump it in the nearest ditch. A common solution is to put it back where you got it, by pumping it into another deep well. If you can arrange to pump it into the bottom of an oil-bearing formation, it will help to drive the oil up and out.

PIPELINES

Oil is sold by the barrel, and in the beginning that's how it was shipped too. Horse-drawn wagons hauled barrels to the railhead, where they were stacked on flatcars. Today the barrel remains a unit of measure (in the petroleum industry it's equal to 42 gallons), but oil, like cheap wine, never sees the inside of an actual barrel. Bulk shipments move over water by tanker and over land by pipeline.

The pipeline system is organized like a tree. Small collector pipelines in the oil field, called flow lines, are the fine roots of the system. They gather crude oil from many wells and bring it to the field processing station. Somewhat larger pipes carry the oil to the terminus of a main-line pipeline, which supplies refineries hundreds of miles away; this is the trunk of the tree. The products of the refinery are then distributed through another system of main-line pipes, which divide into smaller and smaller branches until they reach distribution depots—the leaves of the tree. The natural-gas pipeline system has a similar architecture, except there is no refinery, and the final twiglike branches go all the way to individual homes.

The pipe used for petroleum and natural-gas transport is made of high-strength steel. Joints 40 feet long are welded together in the field, and then the welds are x-rayed to check for defects. High-quality construction is critical because the pipeline

operates under high pressure, and a leak can be explosive. Because pipelines are buried, you won't see the steel pipe along most of its length, but the pathway is not hard to spot. There should be an identifying marker or warning sign wherever the pipeline crosses a road. One place where the pipeline may emerge from the concealment of its burrow is at a river crossing. Sometimes the pipeline is carried on a highway or railway bridge, but there are also bridges specially built for pipelines alone. Some of these structures are elegant and spare suspension bridges.

Corrosion is the great enemy of petroleum pipelines. It's not simply that the steel pipe rusts in the ground; chemical interactions with the soil set up electric currents that actively eat holes in the steel. To slow corrosion, the pipe is coated with various

HOT OIL FROM THE ARCTIC

The most spectacular pipeline—and the most controversial—is the one that carries crude oil from Prudhoe Bay, on the shores of the Arctic Ocean, 800 miles across Alaska to Valdez, on the southern coast. What's most unusual about the Trans-Alaska Pipeline is that you can see it. Over about half its length, the pipeline is not buried but instead is raised on stilts.

The reason for the elevated style of construction is that the arctic soil is permafrost: below a thin surface layer, the earth remains frozen year round. The oil running through the pipeline has the consistency of hot fudge, and the temperature of it too—up to 180 degrees Fahrenheit. If this scalding fluid were pumped through the permafrost, the result would be a boggy mess.

Even with the pipeline propped up off the ground, the builders had to take extraordinary measures to keep the foundations frozen. Ordinary steel or concrete posts would have conducted heat down into the soil. To keep the permafrost permanently frosty, the posts are fitted with "heat pipes" that actively transport heat upward. At the bottom of each post is a reservoir of liquid ammonia. If the base warms, the ammonia boils, extracting heat from the soil and carrying it upward; at the top of the post the heat is radiated away by fins; then the ammonia condenses again and dribbles back into the reservoir. There are 76,000 of these

supporting posts along the route, and most of them extend 50 feet into the subsoil. (More drilling was needed to set the posts than for all the oil wells at Prudhoe Bay.)

The pipeline is four feet in diameter, and the insulation wrapped around it adds another foot. At any given moment during normal operations there are nine million barrels (or 400

million gallons) of petroleum in transit through the pipeline—enough to fill several tankers.

Over most of its route the pipeline is in wilderness terrain, and a big question was how herds of migrating animals would cross the pipeline route. The first proposal was to lift the pipe high enough to let animals pass under. This plan was fine for moose, but caribou are too wary to go under such a bridge. So the

pipeline had to go under instead: at 24 "sag bends," a section of the pipe is buried in a specially prepared bed that's able to withstand cycles of freezing and melting. In a few cases a refrigeration plant pumps chilled fluid through the soil to keep it frozen.

In the first few weeks after the pipeline was filled, there were several leaks as well as a major fire. The years since have brought further incidents but no major catastrophes. As is well known, the worst accident with Alaskan oil happened not on the pipeline but in Prince William Sound, where the *Exxon Valdez* was sailing through a deep, wide, and straight passage but nonetheless managed to run aground. (Photograph reproduced courtesy Bureau of Land Management.)

An anticorrosion device along the route of a petroleum pipeline in southern Arizona generates a small voltage opposing natural electric currents that would eat away the metal of the pipe. The circuitry for the protective device is inside the metal box; the black-and-white disk above the box spins when the current is flowing, as an indicator that the system is working.

A pumping station near Deming, New Mexico, boosts the pressure in a long-distance gas pipeline. The pumps (or compressors) are inside the metal-clad building; the tall cylindrical structure is a heater-treater for moisture removal.

kinds of gunk and wrapped with tape. A further measure is to plant "sacrificial anodes"; these are metal plates that protect the pipeline by allowing their own substance to be etched away. (You have one in your water heater, for the same purpose.) If electric power is available along the right-of-way, small transformers and rectifiers can feed direct current into the ground as another way of combating corrosion.

If you have ever turned on a faucet and been annoyed waiting for the water to run hot, consider how long it can take for a parcel of oil to make its way through a transcontinental pipeline. I was surprised to learn that oil does not gush through a pipeline in a high-velocity torrent, like water from a fire hose. The oil moves at a comfortable walking pace—between three and five miles an hour. Thus, a batch of oil pumped into one end of a 1,000-mile pipeline takes a week or two before it comes out at the far end. The quantity of oil in transit in the largest pipelines is greater than the storage capacity of the largest tank farms. The Trans-Alaska Pipeline, which is four feet in diameter and more than 800 miles long, holds nine million barrels.

Pipelines for refined petroleum products tend to be smaller than crude-oil lines, but their operations are more complicated. The products pipeline may run a few thousand barrels of gasoline, then some kerosene and some diesel fuel before more gasoline. Cutoff valves have to be turned at just the right moment to divert each batch to the proper destination; it's like sorting railroad cars in a moving train, except that you can't see the train.

Don't the various products get all mixed up as they flow through the pipeline? In most cases the mixing is confined to a small portion of each batch, and it doesn't cause much trouble. If two products are seriously incompatible, an inflatable sphere can be popped into the line to keep them separated.

Because their operations are strung out over such long distances, pipeline companies have always been on the leading edge of communications technology. Even the very first crude-oil pipeline, built in 1865 in western Pennsylvania and running a

A pig launcher stands out above other equipment in the pipe yard of a pumping station near Houma, Louisiana. A pig is a device sent through the pipeline to clean or inspect it. The spherical bulge in the elevated section of pipe is a valve that isolates the launcher tube. With the isolation valve closed, a cap is removed from the end of the launcher tube, and the pig is inserted; then the cap is replaced and the isolation valve is opened. Fluid diverted through the launcher tube drives the pig out into the pipeline through the downward-sloping tube.

total distance of six miles, had telegraph wires strung parallel to the pipeline itself. The telegraph was replaced successively by the telephone, the teletype, and microwave radio links. Now many pipelines have fiber-optic cables buried along the right-of-way. One pipeline company abandoned petroleum transport altogether and went into the telephone business—then came back again.

Pumping Stations. Pipeline transport is more energy efficient than shipping the oil by either rail or road, but it still takes hundreds of thousands of horsepower to push the oil through the pipe. The pushing can't all be done from the starting point; booster stations are needed every 60 to 100 miles.

The pumps themselves are usually roofed over to protect them from the weather, but sometimes the end walls of the pump building are left open to prevent explosive vapors from accumulating. Outside the pump building is a neatly arranged but still complicated-looking array of brightly painted pipes, mounted on stanchions that hold them a foot or two off the ground. There are valves everywhere to control the flow of oil through the system.

Another item to look for in the pipe yard is a pig launcher. Animal lovers can rest easy: as far as I can tell, no one has ever tried to run a live pig through an oil pipeline. A pig, in the pipeline trade, is a big brush or scraper or squeegee that gets pushed through the line to clean the inner surface. Most of them are shaped more like bullets than pigs; some look like oversize toilet brushes. I'm told the name comes from the squealing sound that a pig with metal scrapers makes as it passes through the pipe.

The pig launcher is a short section of pipe arranged as a branch from the main line, like the on-ramp of a highway. The several valves controlling the launcher are initial-

Four pumps driven by 6,000-horsepower motors occupy an open-ended shelter in Port Fourchon, Louisiana. The station pumps oil received at the Louisiana Offshore Oil Port (LOOP), a docking facility 20 miles out in the Gulf where large tankers unload. A submarine pipeline delivers the oil to the Port Fourchon pumping station, which then moves it farther inland.

A tank farm associated with a refinery in Philadelphia dwarfs the refinery itself (the small area at upper left with plumes of water vapor rising on a winter day). The farm includes both cone-roof tanks and floating-roof tanks. (Most of the latter appear to be empty.) The white tanks are designed for volatile fluids; by reflecting solar energy, the white paint helps reduce evaporation. The black tanks are meant for more viscous materials, which warmth from solar energy helps to keep flowing.

ly set to seal and drain the launcher barrel, so the pig can be loaded through a door at the end of the barrel. Then the valve settings are changed so the flow of oil drives the pig into the pipeline and carries it along to the next pumping station. There it is retrieved in a pig trap, which looks the same but points in the opposite direction.

The rubber spheres that separate product batches are launched and retrieved with the same kind of equipment. And these days there are also "smart pigs" that carry instruments through the pipeline to check for corrosion and other flaws.

TANK FARMS

Rather than *tank farms,* it might be better to call them *tank orchards*, for the tanks are arranged in neat rows and columns like fruit trees. Tank farms are found at the terminals of pipelines, at refineries, and at tanker ports. There are smaller collections of tanks—perhaps we should call them tank gardens—at the depots where petroleum products are stored for retail distribution.

Seen up close, a tank farm seems immense; from another perspective, however, the tanks are amazingly small. Given the world's thirst for oil, they hold only a few weeks' supply. A typical refinery has enough tank capacity to store two weeks' worth of crude stocks and four weeks' worth of refined products. It seems petroleum molecules are like those insects that emerge from the earth, live and breed for a day, and then perish. Oil is burned up no more than a few weeks after it comes out of the ground.

Floating-roof tanks are the usual storage vessels for crude oil and for gasoline, both of which have volatile constituents that would evaporate into the empty space above the liquid surface in a conventional tank. In these tanks, at a marine terminal in Trieste, Italy, spiral stairs climb the outside of the tank, then telescoping stairs descend to the floating surface of the roof.

Berms surrounding oil tanks create a gently rolling landscape at a pipeline terminal in Greensboro, North Carolina. The berms offer both fire protection and pollution protection. They are sized to confine any possible leak from a tank.

Most petroleum tanks are cylindrical. Their proportions tell an interesting story. The smallest tanks are taller than they are wide—perhaps 20 feet in diameter and 30 feet high. Going to larger sizes, the diameter increases much more than the height, so the tanks become fatter and flatter. The very largest are 400 feet in diameter and 60 feet high (big enough to house a respectable professional sports arena). The reason for making big tanks shallower is that the internal pressure at the base of the tank depends on the height of the fluid inside, not the total volume. Higher pressures call for stronger and more expensive walls.

The walls are fabricated from curved steel panels welded together in a bricklike pattern. The most common size for the plates is 8 feet by 30 feet. (Knowing this, you may be able to estimate the size of a tank by counting panels.) Because the pressure is greatest at the bottom, the lower plates are thicker—an inch to an inch and a half—tapering to about half an inch thick at the top.

Cone Roofs and Floating Roofs. The simplest cylindrical tanks have a conical roof, usually supported internally by a post in the center, like the pole of a circus tent. But tanks of this kind are suitable for only a few petroleum products, such as diesel fuel and home heating oil. These are liquids with a low vapor pressure: they have little tendency to evaporate. In contrast, crude oil and gasoline include volatile components whose fumes would fill the vacant space in a partially filled cone-roof tank. The vapors would then be driven out through the roof vent whenever the tank was filled. The loss of the volatiles would be costly, and would also be a source of air pollution.

The secret to storing gasoline and crude oil is the floating-roof tank, in which the roof is a pan or a buoyant platform floating on the surface of the stored liquid. Because there is no vacant space, very little of the liquid can evaporate. As the tank is filled and emptied, the floating roof slides up and down inside the shell. At the perimeter of the roof are seals or gaskets that wipe against the inner surface of the shell.

Firefighting provisions at the Trieste tank farm include permanently installed water cannons as well as berms. The red panels on the walkway around the perimeter of the tank are also firefighting stations.

Among all geometric forms, a sphere is the best adapted to resisting pressure, since outward force is exerted equally on all parts of the vessel. In the petroleum industry spherical tanks typically hold propane and similar light molecules that remain in liquid form only if they are kept under pressure of a few hundred pounds per square inch. Berms are not needed around such tanks, because any leaks would be released in gaseous form. A wind sock is more useful; it indicates which way to run in an emergency. These tanks are at the ChevronTexaco refinery in Pascagoula, Mississippi.

From above, floating-roof tanks are easy to recognize, since the roof platform is likely to be well down inside the hollow shell. Even from ground level, you can identify a floating-roof tank if you know what to look for. There is usually a flange, called a wind girder, that circles the tank near the top; on the largest tanks it is a walkway with a handrail. The wind girder adds stiffness to the shell; it is not needed in a tank with a conventional roof, since the roof itself stiffens the structure.

A problem with the floating-roof design is that it not only confines liquids under the roof but also allows rainwater to accumulate on top of the roof. How do you drain a roof that's lower than the surrounding walls? The usual solution is a flexible drainpipe that gathers rainwater from the middle of the roof, carries it through the petroleum compartment, and exits through the wall of the tank near its base. Another solution is to put a floating roof inside a tank that also has a conventional cone roof.

Pressure Spheres. The most distinctive tanks are the spherical ones, which hold liquids that have to be put under pressure to keep them from boiling away. (Propane and butane are the main petroleum products in this category.) The spherical form is chosen because it offers the best resistance to internal pressure, which can reach 250 pounds per square inch.

Spherical tanks are supported on piers that reach up to near the equator of the tank, like a baseball resting on fingertips. They cannot sit on the ground the way a flat-bottomed tank does, not only because they might roll away but also because all the weight would bear down on a single point. It so happens that a sphere is the most efficient of all shapes for a tank, in the sense that the largest volume of fluid is enclosed by the smallest quantity of steel. Nevertheless, spherical tanks are more expensive than cylindrical ones because the fabrication is so tricky. Many spherical tanks have a spiral staircase that follows a particularly elegant compound curve to the top.

Fire Berms. If a tank farm isn't an orchard, maybe it's a rice paddy. Notice that each tank sits in the middle of a large square plot, surrounded by an earthen berm three or four feet high. Pipelines, roads, or walkways that cross the berm go up and over rather than under or through. The area enclosed by the berm is calculated to hold the entire contents of the tank in the event of a major leak.

THE REFINERY

A petroleum refinery is the *ne plus ultra* of industrial landscapes. The maze of pipes, towers, vessels, stacks, and fuming vents defies comprehension. Looking at all that intricate plumbing, you cannot possibly trace the path of any given molecule. Still, it's possible to make some sense of the chemistry happening in a refinery. There are two important aids to understanding. First, some kinds of equipment are used over

An oil refinery suggests the image of a metropolis for hydrocarbons, the pipe manifolds like expressways, the distillation towers like skyscrapers. The refinery depicted here, in Rodeo, California, northeast of San Francisco, was one of the first on the West Coast of the United States. It has changed ownership several times; at last report it was operated by ConocoPhillips.

and over throughout the plant (although no two structures are exactly alike). Thus, if you recognize a feed heater or a fractionating tower in one place, you'll know it wherever you see it. Second, the refinery is organized into distinct units, each with a specific function. Although you may never be able to deduce the purpose of every pipe and pressure vessel, you might be able to identify some of the major units.

Petroleum refining is the prototypical process industry—a style of manufacturing where things are made in continuous streams rather than discrete batches. Oil could be refined one vat at a time, the way you make soup in a pot, but refinery engineers always try to avoid that mode of operation. Whenever possible, they design a process so that raw materials flow in steadily at one end and products come out the other.

Feed Heaters. Most of the processing steps in a refinery require high temperature—often a few hundred degrees Fahrenheit (enough to bake bread), sometimes a thousand degrees or more (enough to peel paint). For this reason, just about every unit has a furnace, or feed heater, where the incoming fluid is brought up to temperature.

The heaters have a distinctive shape, with sloping shoulders leading up to a tall metal smokestack, somewhat similar in form to an old stone fireplace with its chimney—but the hearth is the size of an entire house. At the bottom of this structure is a burner. The walls are lined with steel pipes, through which the feedstock is pumped.

A refinery might have a dozen of these feed heaters. The fuel burned in them is a mixture similar to natural gas, drawn from the refinery stream itself.

Fractionating Columns. Crude oil is a mixed-up stew of hundreds of chemical compounds. The refinery's first task is to sort them into groups according to molecular size. A fractionating column, or tower, is where the sorting gets done.

The tower, together with its feed heater, is a fancy kind of distillery. It separates substances according to differences in their boiling point. Inside the column are perforated baffles, called trays, stacked up one above the next at intervals of a few inches or a few feet. When the hot feed liquid is pumped into the bottom of the tower, much of it boils away. The vapor begins rising through the perforated trays, but meanwhile other fractions are condensing into a liquid and trickling back down through the same trays. As these counterflows continue, the most volatile molecules accumulate near the top of the tower while the heaviest sink to the bottom. Each intermediate component finds its own natural level.

Fractionating towers come in a range of proportions, from squat and tubby with a pronounced midriff bulge to extremely tall and svelte. A refinery could have dozens of columns. They are the objects that stand out most clearly against the horizon and give the refinery its characteristic skyline.

Reactor Vessels. Not every tall, metal, cylindrical structure with pipes around it is a fractionating column. There are also large vessels with other purposes, such as housing chemical reactions. Some of these reactors are upright and oblong, like a distilling

A feed heater at the ChevronTexaco refinery in Pascagoula, Mississippi, heats crude oil to about 700 degrees Fahrenheit. There are gas burners near the bottom of the unit; tubes lining the walls carry the feedstock to the heater. This furnace and another just like it at Pascagoula process 325,000 barrels of oil per day.

tower, but they are not as tall. Other reaction vessels are laid on their side. Many of them are built to withstand high pressure (up to 5,000 pounds per square inch, which is roughly the pressure at the bottom of the ocean). The pressure vessels have steel walls as much as six inches thick—not that you can tell from the outside. But there is a tell-tale sign of high pressure: hemispherical end caps give the vessel the shape of a medicine capsule. It's called a bullet tank. Vessels that don't have to hold high pressures have blunter ends, closer to the shape of a soup can.

Heat Exchangers. At many places in a refinery, one stream of fluid needs to be heated and another needs to be cooled. A process engineer will never miss a chance to swap some heat in a situation like this. The two fluids are run through a heat exchanger. From the outside, a heat exchanger is just a long cylindrical tank, usually laid on its side. Inside is a big bundle of tubes. One fluid is pumped through the tubes; the other passes through the surrounding space, called the shell. Heat flows through the walls of the tubes from the warmer fluid to the cooler one.

Pipes, Pumps, and Valves. Pipes knit together the fabric of the refinery, carrying the various raw materials, products, and intermediate stocks, as well as steam, water, pressurized air, and exotic fluids such as hydrogen sulfide and ammonia. The racks of piping may look chaotic, but there is nothing ad hoc about their design; there are thick volumes of standards and specifications, and the piping engineer must have a lawyerly mastery of all this literature.

Many pipes are insulated. Some are heated with steam. A feature of these pipelines is an expansion loop, which looks like the hump in an inchworm. Expansion loops absorb the strain when the pipe expands or shrinks in response to temperature changes.

For every pipe, there's a valve—or a bunch of them. One engineering manual estimates that 8 percent of the equipment budget for a refinery goes into valves. There are two broad categories. Stop valves are meant to be either fully open or fully closed, like light switches. Throttling valves offer continuous control over the rate of flow. These days, most throttling valves are remotely operated from a central control room.

Pumps in a refinery have a prominent place in the minds of the operators, but they are hard to spot from a distance. They are usually mounted at ground level or below because they need to be lower than the vessel they draw fluid from. Pumps consume most of the refinery's electric power.

The Crude Unit. The crude unit is where the refining process begins. In the early years of the petroleum industry, it was also where the process ended. That is, the crude unit was the entire refinery.

The crude unit is a distillery with three main parts: a feed heater, an atmospheric-pressure distillation column, and a vacuum distillation column. This last item is also called a flasher. The atmospheric column is taller and slimmer; the flasher has a distinctive shape, with a bulge in the middle.

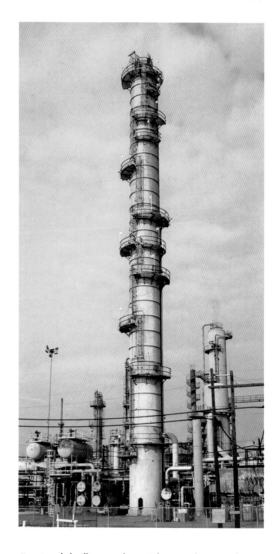

Fractional distillation column *(above)* is the most characteristic feature of a refinery. The catwalks at various levels provide access for maintenance. The column is at the Premcor refinery in Port Arthur, Texas. Insulated, bullet-tank reactors *(below)* are at the ExxonMobil refinery in Chalmette, Louisiana.

In the feed heater, the oil is brought to a temperature of 700 degrees Fahrenheit, then it is pumped to the atmospheric tower. In the molecular sorting process that goes on inside this column, the four lightest hydrocarbons—methane, ethane, propane, and butane—float to the top, where they are drawn off as overheads, or light ends. The fraction withdrawn at the next lower level is called straight-run gasoline. A century ago, straight-run gasoline was all the gasoline there was. Now quite a lot else gets blended into the product, and straight-run gasoline is a minority constituent.

Below the straight-run gasoline tap, several more fractions are drawn off: naphtha, kerosene, and something called gas oil. Then, at the bottom, there's the even heavier

One of the crude units at Pascagoula has two main components (apart from the furnace). The atmospheric distillation tower, in the background at right, operates at normal air pressure and separates the lighter components of the oil. The vacuum distillation tower, or flasher, in the foreground, boils heavier oil fractions at reduced air pressure.

stuff that just won't boil under these conditions; it's called resid (short for residual oil), or simply bottoms, and it is the problem child of the refinery. If you try to boil it away by raising the temperature further, the oil breaks down chemically, like sugar turning to caramel. But some of the resid will boil if you keep the temperature constant while lowering the air pressure (just as water boils a little easier in Denver than in Dallas). That's what the vacuum tower is for. The air pressure inside is about a third of normal.

The fractions boiled off in the vacuum tower are known as flasher tops. They are a raw material for the manufacture of motor oils and other lubricants. But if there are flasher tops, there must also be flasher bottoms! This is the residue of the residue—the gunk that won't boil even in a partial vacuum. Some of it can be made into asphalt for road paving and roofing.

The Gas Plant. The refining of the lightest fractions is the mirror image of what happens to the heavy resid. Where lower pressure helps the heavy stuff boil, higher pressure helps the light stuff condense. The gas plant has several tall and very slender towers. Height is needed because the light gases are hard to separate from each other, and so the columns must have a large number of trays. The towers can be narrow in cross section because the volume of material being handled is much smaller than it is in the crude unit. The pressure is about 200 pounds per square inch.

Four products come out of the gas plant. Methane (which is also the main ingredient of natural gas) fuels most of the refinery operations. Ethane can also serve as fuel, but it's worth more as a raw material for making other chemicals, including plastics such as polyethylene. Propane is marketed as liquefied petroleum gas. Butane is the clear liquid in all those transparent plastic cigarette lighters, but that's not where most of it goes. It's also a vital component of modern gasolines.

The Catalytic Cracking Unit. Cracking is just what the name suggests: Taking big molecules and snapping them apart into smaller pieces. For example, a chain of 16 carbon atoms might be cracked into a 10-carbon unit and a 6-carbon unit. The aim of this process is to convert some of the heavier petroleum fractions, for which there is little market, into more gasoline. The molecular bone-breaking could be done by heat alone, but a catalyst—a chemical facilitator—makes it happen faster and with better control over the outcome. Catalysts for this process are mainly zeolites, which are lacy molecular structures with many voids where hydrocarbons can lodge.

For some reason, "cat cracking" is better known to the public than other refinery operations. But I had a wrong impression of how it's done until a visit to a refinery clued me in. I had imagined feedstock trickling through a big vessel packed with beads of catalyst. That type of cat cracker, I learned, has been obsolete for decades.

The cracking process favored today is called fluid catalytic cracking. The catalyst isn't actually a fluid; it's a solid, but it's ground into a powder so fine that it flows. The catalytic powder and the feedstock are mixed at a temperature of about 900 degrees Fahrenheit and pumped upward into a reaction chamber. The chemical cracking

The fluid catalytic cracking unit at Pascagoula includes a reactor, at left, where all the hydrocarbon chemistry takes place, and a much larger regenerator, at right, where the catalyst is restored to its active state by a blast of hot air that burns away carbon deposits. (Photograph courtesy Stephen Renfroe, ChevronTexaco.)

Hydrocracking *(above)* is done in tall pressure vessels, where the feedstock and added hydrogen can percolate through a bed of granulated catalyst. A reformer *(below)* has pressure vessels filled with precious-metal catalysts. Both units are at Pascagoula.

takes place almost instantly. Then the catalyst and the hydrocarbons have to be separated so the catalyst can be reused. The separation is accomplished with a cyclone, a device that spins the mixture so the denser catalyst migrates to the periphery while the hydrocarbons remain near the center.

The fluid cracking unit has two more major parts. The fractionating column is similar to the atmospheric tower of the crude unit; it separates the products of the cracking reaction. The regenerator, which is usually the biggest part of the entire cracking unit, receives the spent catalyst and gets it ready for reuse. With a blast of hot air, it burns away carbon that plugs up the pores of the zeolite.

Where does the carbon come from? Answering this question calls for a detour into the chemistry of cracking. The feedstock molecules are chains of carbon atoms surrounded by hydrogens. You might think of such a molecule as a long banquet table. The carbon backbone is the table itself; the attached hydrogen atoms are the chairs lined up along both sides of the table and also (this is important) at the head and the foot. When the table is broken somewhere in the middle, two more chairs are needed to fill the newly created positions at the head and foot of the two fragments. The chairs cannot be created out of nothing—and neither can hydrogen atoms. Thus, the cracking of molecules leads to a hydrogen deficiency. The needed hydrogens are stripped away from other feedstock molecules, leaving behind a residue of carbon.

Hydrocracking. What if you ran a catalytic cracking operation but added hydrogen to make up for the deficiency? Roughly speaking, that's what happens in hydrocracking. It is another way of converting heavy oils into gasoline components. As it happens, hydrocracking is done in the kind of fixed-bed catalytic reactor that would be so old-fashioned in cat cracking. This arrangement works because no carbon clogs the catalyst bed, and there is no need for continual regeneration.

The main hardware components of a hydrocracker are hefty pressure vessels, usually of the bullet-tank variety. They are mounted vertically so the feedstock can trickle through the catalyst bed. Also part of the hydrocracking unit are feed heaters, separators to recover unreacted hydrogen gas, and—as everywhere—a fractionating column.

The Reformer. The reformer is not a crusading politician, but it *is* a do-gooder of sorts. It rearranges various hydrocarbon molecules, making them better citizens in the world of motor fuels. The basic idea is to take abundant straight-chain molecules and either give them branches or twist them into closed rings. These geometrically intricate molecules burn more smoothly in automobile engines.

The reformer is another unit based on high-pressure reaction vessels. They are filled with an unusual catalyst, rich in rare metals such as platinum and rhenium. The catalyst is worth a few million dollars, and it has to be replaced every few years.

Alkylation. The cracking units break big molecules into small ones; the "alky" plant does the opposite, gluing together small molecules to make bigger ones. The catalyst

for this process is different from the various powders and granules used elsewhere in the refinery: it is liquid sulfuric acid. Another difference is that the alkylation unit doesn't have a feed heater. Instead, it has a big chiller, where the feed is cooled to 40 degrees Fahrenheit. This is the temperature where the acid does its best work.

The Bottom of the Barrel. The heaviest part of the crude oil has always been the unloved, good-for-nothing, poor relation of the petroleum industry. The stuff burns fine, but it is gooey, stinky, dirty, and almost worthless. Since the 1980s many refineries have built new facilities for dealing with the bottom of the barrel; they can't make the noxious elements go away, but they can concentrate them. The process is called coking, and it cooks the oil until there's nothing left but carbon. A coking unit is a bizarre—and unmistakable—sight. It has drilling derricks, as if the refinery has had the good fortune to strike oil on its own property! But a closer look reveals something odd about these drilling rigs. They are mounted a hundred feet in the air, atop tall steel vessels. The vessels are the coking drums, where the overheated residual oil breaks down, releasing all its volatile constituents but leaving behind a hard mass of solid carbon. The drilling rigs are needed to break the carbon loose. A coking drum is cooked for about

An alkylation plant at Pascagoula *(above)* also relies on bullet-tank pressure vessels, but in this case they are laid down horizontally; one large tank is visible here behind the smaller drums of a heat exchanger. The unit also includes several tall distillation towers.

Coking units at Pascagoula are an unmistakable feature of the skyline. The coke is formed in the tall drums in the base of the tower. Oil heated to more than 1,000 degrees Fahrenheit breaks down there, releasing all volatile components and leaving a residue of solid carbon. The drilling derricks are needed to remove this carbon after a drum has filled up. The drill creates a pilot hole through the carbon deposit, and then a high-pressure jet of water washes out the rest.

Flares flicker against the evening sky in Pascagoula. Most of the time, the flame atop each stack is just a pilot light. In emergencies, however, the stacks are required to dispose of very large quantities of flammable gases. One reason the stacks are so high is that the heat of an emergency flare could injure anyone closer to the flame than about 100 feet.

24 hours, until it fills up with carbon. Then the drum is allowed to cool, and the rig drills a pilot hole through the coke. A jet of water flushes out the rest of the carbon.

Petroleum coke is soft, spongy, and intensely black—like concentrated grime. It's too full of sulfur to be burned anywhere subject to air-quality standards. What becomes of it? Most of it is exported to less fastidious countries.

Product Blending. Refiners convert as much crude as possible into transportation fuels—diesel, jet fuel, and especially gasoline. Most of the molecules that go into gasoline have five to eight carbon atoms, but when I learned how gasoline is blended, one ingredient surprised me. It is butane, lighter than the rest, with just four carbons. I was surprised because butane boils at about 30 degrees Fahrenheit, and so I thought it would vaporize and escape.

Butane is added to gasoline for two reasons. First, it has an excellent octane number (the main rating of gasoline quality), and it's much cheaper than other high-octane additives. The second reason is the low boiling point, which turns out to be an advantage as well as a drawback. Vaporizing is just what gasoline needs to do in order to burn in an automobile engine. Having some highly volatile components is particularly important for starting the engine.

For economic reasons, refiners would put as much butane as possible in a gasoline blend, but they can't overdo it or much of the butane would evaporate. The evaporation rate varies with the weather. Hence, motor fuel is blended differently according to the season; gas for a cold climate has more butane. When you fill your tank, you may notice wavy fumes emerging from the filler pipe. That's butane escaping.

The Flare Stack. People who give refinery tours tell me the flare—that pulsating flame held aloft on a tall mast—is what everyone asks about first. At one time, refineries burned off large volumes of gas that was not worth selling, but that's not done anymore. In many refineries today the flare does not burn at all during routine operations. It is there for emergencies. If the gas plant has to be shut down suddenly because of a leak or malfunction, all the gases coming out of the other refinery units have to go somewhere. Burning is the safest and cleanest way to get rid of them.

The burner at the top of the flare is a lot like the one in a gas grill or broiler: it has many rows of small nozzles that spread the flame over a wide area and let air reach all parts of it. Bars called flameholders, mounted just above the nozzles, create turbulence to mix the gas and air. Some flares have additional nozzles to inject steam, which, paradoxically, makes the flame burn cleaner.

Sulfur Recovery. Much of the sulfur in crude oil is released in the refinery as hydrogen sulfide, the rotten-egg gas. Until about 1970 the common way of getting rid of the hydrogen sulfide was to add it to the refinery fuel gas. There were two problems with this practice. First, it made the refinery a very stinky place. Second, it added a heavy load of sulfur oxides to the atmosphere. The burning of hydrogen sulfide has been stopped now; instead, a recovery plant converts it to elemental sulfur or sulfuric acid. People still hold their nose when driving by a refinery, but it's just a reflex. The ones I have visited lately don't smell at all.

THE GAS STATION

If the refinery is an exotic locale that most people never visit, the gas station is all too familiar. Still, even at the retail end of the oil industry, a few technological elements often go unnoticed.

The gas station has undergone a curious evolution. Early in the twentieth century, gasoline was sold as a sideline by general stores, with a pump at the curb. Then came the classic filling station, which grew into an emporium for all things automotive. The gas station of the 1950s, clad in gleaming white porcelain enamel, as if it were some giant kitchen appliance, would not only fill your tank but also change your oil, fix your flat tire, and replace your leaky muffler. Since the 1980s another

The metropolis for hydrocarbons also has a skyline like that of a large city. This is the ExxonMobil refinery in Chalmette, Louisiana, observed from a ferry crossing the Mississippi River.

The American gas station began as an adjunct to the general store and has returned to that role. A modern station *(right)* in Altoona, Pennsylvania, sells food, beer and cigarettes as much as fuel. An antique gas pump *(below)* is on display as a novelty item at a shop in Greensboro, North Carolina, that was once a gas station but has long since been converted to other uses. The glass vessel atop the pump allowed the buyer to confirm the quantity of gasoline before pouring it into the fuel tank. (The price on the placard is 17 cents per gallon.)

transformation has come over the gas station: it has fused with the convenience store. Now you can buy gas and pick up a six-pack of beer and a bag of chips all in one place, but you need to go elsewhere—to an array of more specialized franchises—for an oil change, a tune-up, a muffler, or tires.

In other ways too the gas station has come full cycle. The early curbside pump was often a self-service device. Later stations were staffed by squadrons of uniformed attendants who not only operated the pump but also swabbed the windshield, checked the oil, and issued Green Stamps. Now (except in a few states) we have returned to the tradition of do-it-yourself. Pumps equipped with a credit-card reader allow the entire gas-buying experience to be completed without having to speak with a human being.

Meanwhile, the gas pump itself—the heart of the gasoline retailing business—has also been evolving. Early pumps were cranked by hand, and the gasoline flowed first into an elevated glass vessel graduated in gallons (or liters). That way the purchaser could verify the quantity of fuel before it was allowed to flow by gravity into the car's tank. This principle continued to be honored in vestigial form for decades thereafter; as late as the 1950s some pumps still had a glass globe where you could watch the gasoline flowing through and turning a plastic turbine.

Early retailers stored gasoline in drums or barrels. Then, as larger volumes were sold, the storage tanks were moved underground, both to save space and for safety. But burying the tank didn't necessarily bury the problem. Steel tanks would eventually corrode and begin leaking. For a long time, minor leaks were not considered a serious issue. The only cost of a leak was the lost gasoline. Now, an undetected underground leak is the gas station owner's worst nightmare. Hundreds of tons of contaminated soil have to be dug out and carted off to be treated as hazardous waste.

To deal with the leak hazard, steel tanks have been replaced by corrosion-free fiberglass ones. Monitoring wells are drilled all around the tanks for early detection of any hydrocarbons entering the soil or groundwater. Sometimes electronic leak-

detection systems are installed. And in some parts of the United States, above-ground storage tanks have come back into favor, especially for kerosene and diesel fuel, which don't present quite as much risk of fire and explosion as gasoline does.

A typical tank at a large gas station holds 12,000 gallons, and there are three or four such tanks to accommodate the various grades of fuel. The tanks are filled through fittings recessed into the concrete apron that surrounds the pumps. A standard color code is developing for the various hatches and access caps. White marks regular unleaded gasoline, blue is the "plus" grade, and red is premium gas. Yellow and green are for diesel fuel—the yellow for low sulfur and the green for high sulfur. Finally, brown is for kerosene. Monitoring wells are danger orange: you wouldn't want the delivery truck to pump 10,000 gallons of high test into the well.

In heavy-smog areas, gas stations have had to make another change in the past few years: installation of vapor-recovery systems. As noted earlier, gasoline is rich in butane, a very restless molecule that takes any opportunity to escape confinement. The gas-pump nozzle is fitted with a plastic hood, and an interlock switch allows the pump to run only when the hood is seated to the filler pipe. Suction draws the vapor through the hood and into a separate hose, eventually to be returned to the storage tank.

NATURAL GAS

The petroleum and natural gas industries are twins separated at birth. They have a great deal in common; indeed, the raw materials for both industries often come out of the same hole in the ground. And yet they have grown apart, developing different cultures, practices, and technologies.

The word *gas* is a variation on *chaos*. The name was suggested by Jan Baptista van Helmont, a seventeenth-century Flemish chemist, who thought of gases as unruly spirits given off when solids or liquids are heated. (Or maybe he was just being whimsical.) The stuff we burn today is called natural gas to distinguish it from manufactured gas, which came first historically. All through the nineteenth century and up until about 1950, gas was made by roasting coal or heavy oil residue to drive off a mixture of combustible gases—mostly carbon monoxide but also some hydrogen and methane and traces of much else. The gas was a poor fuel, not to mention a deadly poison. The residue left behind at the gas works, called coal tar, was even nastier. Efforts to find a use for the stuff or somehow get rid of it were so assiduous and enduring that they pretty much created the discipline of organic chemistry, and then the dyestuff and pharmaceutical industries.

Manufactured gas hasn't been manufactured for decades, but many of the companies that once made it live on as distributors of natural gas. Companies such as Brooklyn Union Gas and Public Service of New Jersey converted their old gas works to storage facilities for the new fuel flowing in from Texas and Oklahoma. The most difficult part of the conversion was readjusting millions of stove burners and furnaces.

The storage of gasoline and other fuels at retail outlets has also undergone significant change. Originally stored in elevated tanks, gasoline was later put underground for safety reasons. But in many jurisdictions, tanks are now being installed above ground again because of the risk of underground leaks. The tanks below are at a gas station in western Missouri.

Natural gas is a simpler fuel than manufactured gas. It is mostly methane, the smallest of the hydrocarbon molecules, with few impurities except water. The only major products of combustion are carbon dioxide and water. One impurity is deliberately added to the gas before it enters the distribution pipelines: methyl mercaptan, the mercury compound that gives gas its characteristic smell. (The methane itself is odorless.)

Natural gas became a practical fuel only with the construction of long-distance pipelines. Other forms of transport are simply too expensive, which means that gas fields not served by a pipeline can't be developed. The gas in the Alaskan North Slope, for example, has no way of getting to market, and so it is reinjected into the wells, where it helps push oil to the surface. A few oceangoing tankers carry natural gas as a cryogenic liquid (temperature −259 degrees Fahrenheit), but commerce in liquefied natural gas (LNG) has not taken off the way the world petroleum market has.

Gas received through the pipeline system is stored in the equivalent of tank farms. The traditional storage facility is a gigantic, collapsible canister, or holder, that rises

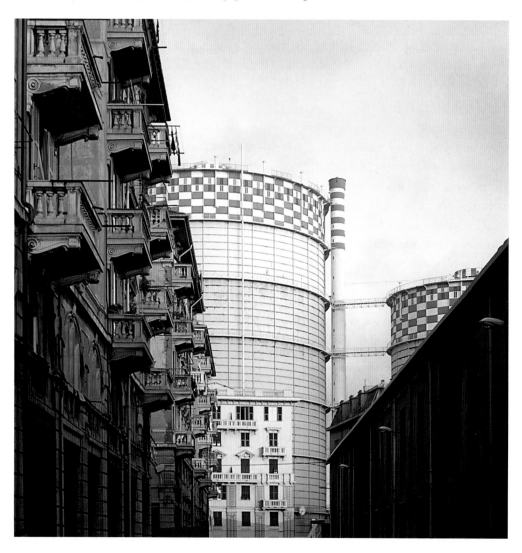

Rigid-walled gasholders loom over a residential neighborhood in Genoa, Italy. Most tanks of this kind were built in the years before 1950 and originally held manufactured gas. Often they are the tallest structures in the area and therefore serve as important landmarks. Inside the tanks, a floating piston separates a layer of natural gas below from air above.

Telescoping gasholders in Elizabeth, New Jersey, rise and fall within their skeletal framework as gas is stored and withdrawn. In this photograph the gasholders are fully deflated. They had probably been emptied in preparation for demolition; since the photograph was made in 1998, the gasholders have disappeared, like many others of their kind.

and falls with changes in the supply and demand for gas. In New York City, generations of commuters came to know two such gas tanks near the Long Island Expressway in Elmhurst, Queens. The Elmhurst tanks have been demolished, like many others around the country.

The holders are essentially inflatable structures, but made of telescoping sections instead of elastic fabric. The walls are concentric rings called cups, linked together by flanges. As the empty tank starts to fill with gas, the innermost cup (which supports the slightly domed roof) starts to rise. A flange at the base of the innermost cup engages a matching flange at the top of the next cup, which is therefore pulled up as the filling continues. The joint between sections is sealed with water, which is steam-heated in winter to keep it from freezing. At full extension a typical holder stands more than 200 feet high and holds roughly 10 million cubic feet of gas. The sections are held in alignment and braced against wind loads by a cylindrical exoskeleton of steel trusses.

Because of the immense volume, the pressure needed to inflate the holder and lift the cups is very slight—only about one-half pound per square inch above atmospheric pressure. You could blow up the holder with your breath, like inflating a giant beach toy, but it would take a while—perhaps 300 million lungsful.

There are also rigid gas holders, which look from the outside like plain steel tanks without moving parts, but inside they have a floating piston that separates gas in the lower part of the holder from air above. (Keeping air and gas apart is the point of all the storage arrangements. Gas is reasonably safe as long as there's no oxygen present.)

The big low-pressure gas holders are being replaced by high-pressure tanks and by insulated tanks for liquefied gas. The typical facility for high-pressure storage is a bank

High-pressure cylinders store natural gas in a smaller volume than the low-pressure rigid and telescoping gasholders. The cylinders are in Ljubljana, Slovenia.

Liquefied natural gas (LNG) is stored at even greater density than the high-pressure gas. The LNG depot at right, shrouded in foggy vapors on a winter afternoon, is in Everette, Massachusetts.

A vent stack on a city sidewalk serves as a pressure-relief valve for the gas distribution system. Ordinarily the valve is closed and nothing escapes through the vent; if a malfunction causes higher-than-normal pressure in the mains, it is better to release the gas through a vent like this one than through leaks in underground pipes or inside buildings.

of cylindrical tanks with hemispherical end caps. Each tank is about the size of a house trailer. A dozen of these tanks have as much capacity as one of the giant old telescoping holders. That's about a day's supply for a major city.

Larger stocks are stored in liquid form. Six hundred cubic feet of natural gas condenses into a single cubic foot of LNG. Tanks for LNG don't have to withstand pressure, but they have to be insulated to keep the fuel cold and be built of materials that retain their strength in the deep freeze. In addition to the storage tanks, an LNG depot will have a refrigeration plant, and an evaporator, which accomplishes the opposite task—making the gas a gas again.

Although storage technology has changed, the gas-distribution system remains much as it has been for the past century. The gas moves through underground pipes at very low pressures—roughly 0.2 pound per square inch over atmospheric pressure. This is less than the change in barometric pressure when the weather turns from fair to stormy. Thus, the gas is very gently wafted through the pipes. The low pressure entails much larger pipes than a high-pressure system would need. Some of the gas mains in New York are six feet in diameter. Low pressure is a safety feature; it reduces the leakage rate when a pipe fails or a pilot light goes out.

The traditional material for gas mains is iron pipe, but distribution lines laid in the past few years are mostly plastic, colored bright yellow for easy identification.

FUTURE FUELS

Petroleum is routinely called the life's blood of the industrial economy. The metaphor is not far-fetched. Societies rely on oil and gas to meet a share of almost all energy demands, and transportation in particular would go nowhere without petroleum. The

least hiccup in supply sets off international alarms. Nations do not hesitate to go to war over oil.

But given the central and essential place of petroleum, it's startling to reflect that oil is a newcomer to the world economy. The industry sprouted up only a little more than a century ago. What's more, in another century or so it will have disappeared. The petroleum age is a brief episode in human history.

No one knows how much oil remains in the ground, or how much of it can be recovered at acceptable cost. There is surely more to be found and more to be pumped than the oil companies know about—or tell about. Nevertheless, it is beyond dispute that the resource is a finite one. We are burning petroleum much, much faster than the earth is making it. And there is reason to suspect that we are near the high-tide mark in oil production and consumption. In the United States, petroleum production peaked in 1970. Production elsewhere has continued increasing since then, but Kenneth S. Deffeyes, a geologist with close ties to the oil industry, has argued that the upward trend cannot last more than a few years; indeed, he has gone on record predicting that the peak will come on Thanksgiving Day in 2005. After that, he says, it's all downhill—although the decline might take as long as the build-up did.

The prospect of running out of oil a few decades from now should not be cause for panic or despair. Given that the whole infrastructure of the petroleum industry was built in less than a hundred years, there should be plenty of time to create its replacement. It will be interesting to see what that replacement is, and what new features it adds to the technological landscape. And if a world without gasoline seems unimaginable, look back to the 1850s, when a world without whale oil and a whaling industry must have seemed equally unlikely and forbidding.

THE HYDROGEN ECONOMY?

In one popular vision of the post-petroleum future, automobiles powered by fuel cells will convert hydrogen and oxygen into electricity, producing only water as exhaust. President George W. Bush endorsed this idea in his 2003 State of the Union address, and proposed a $1.2 billion program of research.

The two main elements of this plan—the use of fuel cells and the use of hydrogen—are almost independent. If fuel cells turn out to be the best way of powering automobiles, they can probably be made to work on fuels other than hydrogen. At the same time, if hydrogen is readily available as a fuel, it could be put to work in many ways, including burning it in engines much like those of cars today.

The big question is where the hydrogen comes from. Some envision splitting water molecules either electrically or by applying intense heat. Thus, the hydrogen economy would be a closed cycle: water would be broken down into its constituent hydrogen and oxygen molecules, which would later recombine to make water again. It's an attractive scheme, but an energy source is needed to make the cycle run. The electricity or the heat for prying apart water molecules would have to come from a power plant of some kind. In other words, hydrogen is not an energy source in this plan; it's merely an intermediate or carrier, a way of turning coal or nuclear power into automobile fuel.

There is another source of hydrogen. Almost all hydrogen used in industry today is made from methane, by stripping the four hydrogen atoms away from the single carbon atom in a methane molecule. There's a well-developed technology for doing this, and it could be scaled up to supply larger quantities of hydrogen as fuel. Nevertheless, this process will not solve the problem of what to do after the oil and the gas run out. The methane from which hydrogen is made comes from natural gas.

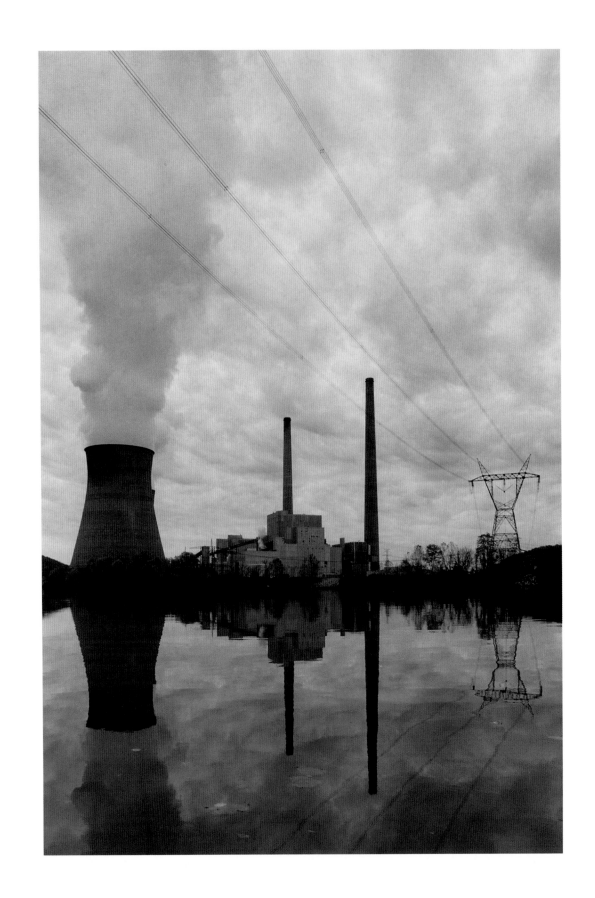

POWER PLANTS

W HEN HENRY ADAMS, the one-of-a-kind American historian, wanted to contrast the medieval with the modern, the emblems he chose for those two eras were the Virgin and the Dynamo. The earlier age, he said, expressed its highest aspirations in building cathedrals consecrated to a spiritual ideal; in our time we exalt the generation of electricity—another invisible but powerful essence. The votaries of the electric cathedral certainly agree with this assessment. A textbook for power-plant operators says of the generating station, "It is like a shrine or source of unfailing light which must be given ceaseless attendance once it is brought into being."

Adams encountered the dynamo at the Universal Exposition in Paris in 1900. A century later the electric power plant is no longer a novel piece of machinery that can attract a crowd at a fair, but it is more central than ever to daily life. And the dynamo (or generator, or alternator—they are all terms for the same machine) still seems an apt symbol of both the hopes and fears invested in industrial progress.

Electricity has become the standard currency of the energy economy. It is not in itself a natural resource that you can dig out of the ground or pump from a well, but other forms of energy are converted into electricity for convenience of distribution and use, just as the body converts a variety of foods into a few simple sugars that circulate to all the tissues. Thus, the power plant doesn't create energy; it merely transforms it. The chemical energy locked up in coal, for example, is captured in the heat and pressure of steam, then passed on to the kinetic energy of a spinning turbine, and finally converted into electric current in the generator.

Three kinds of power plants are scattered around the American landscape. Fossil-fuel plants, which burn coal, oil, or natural gas, make up almost two-thirds of the nation's generating capacity. Nuclear plants tap energy from the disintegration of uranium atoms. Hydroelectric plants are found only where the water is—or more specifically where the water runs downhill.

The John E. Amos power plant *(opposite page)* presides over a moody moment on the banks of the Kanawha River, a few miles from Charleston, West Virginia. From left to right the major structures are cooling towers; the blue-clad buildings that house boilers, turbines, generators, and other machinery; two smokestacks; and a tower for the transmission line that delivers electric power. The coal-fired plant, operated by American Electric Power Company, has three units, each with its own boiler and generator. The total capacity is 2,900 megawatts, which makes the plant one of the 10 largest in the United States.

A few other energy sources are also squeezed to produce the juice of electricity. Their contributions are smaller, but the machinery is no less interesting to look at. Wind power, in particular, creates haunting landscapes. This chapter discusses only generating plants; the transmission and distribution lines that carry electricity to the consumer are the subject of Chapter 6.

FOSSIL-FUEL POWER PLANTS

A big coal-fired power plant going full blast burns a thousand tons of coal an hour, or 30,000 pounds of coal a minute. It generates a billion watts of electricity—in round numbers, enough for a million households. These inputs and outputs are conspicuous features of the plant. On one side you'll see long trains of hopper cars unloading coal, which is heaped up in mountainous black stockpiles. On the other side is a high-voltage switchyard, with electrical transmission lines disappearing over the horizon. Another "output" of the plant is also obvious: a tall smokestack, surrounded at its base by pollution-control equipment.

What's not so easy to see—at least from the outside—is everything that happens between the burning of the coal and the generation of the electricity. The paragraphs that follow describe a power plant in terms of three major flows. First, we follow the fuel from the coal pile through the furnace to the exhaust plume and the ash pit. Then, we trace the circulation of water and steam. Finally, we consider the flow of electricity from the generator into the transmission network.

Coal Handling. When you need to move coal at a rate of 30,000 pounds per minute, it's not done with shovels, or even with bulldozers and trucks. Large power plants all have conveyor systems; one of the most characteristic sights at a coal-fired plant is a conveyor slanting across the skyline. Closely related to conveyors are the stackers that

GETTING A LOOK

Few power plants today are ready to welcome the casual visitor, but some will accommodate you if you call ahead. For a long time, nuclear power stations were more readily accessible than fossil-fuel plants. Companies in the nuclear industry were eager to promote and explain their technology, and so they offered tours on a regular schedule; some plants had educational exhibits and even souvenir shops. All that changed in the aftermath of September 11, 2001. Security is now very tight at nuclear power plants, and visitor programs have been shut down, at least for the time being.

Hydroelectric power stations are more likely to be open to visitors, especially plants owned by public agencies. Similarly, the operators of wind power and solar power installations are accustomed to getting queries from the public and requests for tours.

In areas where the climate isn't too severe, many coal-fired power plants are built on an open, unsheathed framework, so that the furnace and boiler are visible. Even when the boiler is enclosed, the coal-handling equipment is exposed, and so are smokestacks, cooling towers and some pollution-control devices.

The Homer City power plant, 40 miles east of Pittsburgh, is a startling apparition when it suddenly comes into view as you come over the crest of a hill. The buildings housing the boilers and generators are dwarfed and partly hidden by ancillary equipment: three smokestacks in the center, three cooling towers at right, and a series of storage bunkers and conveyors for coal, receding into the distance at the left. The plant has its own coal-cleaning facilities. It was designed to burn coal mined in the immediate vicinity but now gets some of its fuel from more distant sources. Owned by a subsidiary of Southern California Edison, the plant's three units can produce more than 1,800 megawatts of electricity.

pile coal into storage heaps and the unstackers that eventually retrieve it from those heaps. These machines have immense pivoting booms that nod up and down and sway from side to side like long-necked dinosaurs ponderously grazing on the coal.

The coal can't just be piled up and forgotten. It needs watching because of a process called weathering, which is just oxidation, or, in other words, slow-motion burning. If the process gets out of hand, the burning will no longer be slow. The largest piles are inspected daily for hot spots. A worker probes the coal with a long steel rod, inserted like a cake tester. If a section of the rod comes out too hot to hold in the bare hand, the pile needs attention.

From the storage pile the coal moves by conveyor to a bunker or silo, and then another conveyor carries it into the plant. You might imagine lumps of coal being shoveled into a furnace by sweating, shirtless men, but that's not how it's done. All the big plants burn pulverized coal, which is blown rather than shoveled into the furnace. Big lumps are broken into smaller lumps by tumbling in a crusher, which looks like the drum of a giant clothes drier. Then the smaller lumps are pulverized in another rotating drum, this one with steel balls inside or hammers on hinges. The result is a powder as fine as beach sand.

Oil-Fired and Gas-Fired Plants. Coal is the fuel of choice at most U.S. power plants, but there are exceptions, especially in the New England states and some areas of the West Coast. The obvious difference at an oil-burning plant is that you'll see a tank farm instead of a coal pile.

The oil burned by utilities is nothing like the light fuel oils used for home heating. You'll have a more accurate impression if you think of roofing tar. Thick, black, and corrosive, it is the "bottom" product of the refinery, in both the literal sense (it

Lights glimmer within the "exoskeleton" of the Mayo plant, operated by Progress Energy near Roxboro, North Carolina. The plant has two coal-fired boilers supplying steam to a single turbine and generator. Most of the photographs on the next 10 pages were made at the Mayo plant.

The coal pile at the Mayo plant could keep the boilers fired for a few weeks if supplies were interrupted. The two conveyors in the background carry coal to and from the stockpile; it is piled up by a device called a stacker, with a swiveling arm. The coal is retrieved from the pile by an underground auger and then is brought into the plant by the upward-sloping conveyor line in the foreground.

comes out of the lowest tap in the distillation tower) and in the sense that it commands the lowest price (see Chapter 4).

The major problem in handling heavy oil is that it won't flow in cold weather; it needs to be melted. To run through the fuel pipeline, the oil has to be above 100 degrees Fahrenheit; to be sprayed as tiny droplets from a burner nozzle, it needs to be about 200 degrees. Thus, steam heating coils are installed in the storage tank. The fuel pipeline may also have steam tracer lines to keep the oil fluid.

Natural gas is a less troublesome fuel and much cleaner, but also more expensive. It is burned mainly in urban power plants where air pollution levels allow no alternative. A utility-scale plant takes its gas not from the municipal gas mains, which operate at low pressure, but directly from high-pressure transmission lines.

A pulverizer on one of the lower levels of the Mayo plant grinds coal to a powder fine enough that it can be transported by blowing it through a conduit. Almost all of the coal-handling equipment inside the plant is tightly sealed, so you can walk through galleries of machinery that process thousands of tons of coal each day and never see any sign of the coal itself.

The Firebox. When you think of a coal-burning furnace, it's natural to imagine a bed of coals glowing on a hearth, but that's the wrong image for an industrial-scale burner. The powdered coal is treated like a fluid, not a solid. It is sprayed into the firebox.

The ideal is to burn it all, every last particle. Complete combustion not only gets full value out of the fuel; equally important, it minimizes waste-disposal and pollution problems. Anything that doesn't burn will eventually have to be hauled away.

The key to full combustion is maintaining the right ratio of fuel to air, and making sure they are mixed thoroughly. In most large furnaces there are two air streams. A big fan blows the powdered coal into the firebox and starts the combustion process. Then an even bigger fan adds secondary air, with an effect somewhat like that of the afterburner on a jet engine. The fans are usually of the centrifugal type, with a snail shape, like the blower in a hand-held hair drier. But the power-plant fans are built on a totally different scale. They are as big as a two-story house, and the duct work that carries their output is big enough to drive a truck through.

Plants of this kind run 24 hours a day, not so much because there's always demand for electricity but because shutting them down and starting them up again takes hours.

The startup process—called lighting off—is not just a matter of striking a match. Burning a special ignition fuel, usually kerosene, warms up the firebox enough to establish a stable flame pattern before the primary fuel is switched on. It's a difficult process to manage. If the flame goes out, the furnace has to be purged with air to get rid of unburned fuel, which otherwise might explode on reignition. Once the furnace is lit, another 10 or 12 hours may pass before the plant comes up to full power.

Flue Gases. Inside the furnace a pillar of fire rises 100 feet or more. Even where the flame zone ends, the gases remain extremely hot—up to 3,000 degrees Fahrenheit. The idea guiding the design of the plant is to let none of this heat go to waste.

The pathway of the combustion gases is arch-shaped: up through the boiler, then horizontally across the top of the furnace building, then partway back down again. All along this route the gases pass through devices that extract heat in various ways. First is the boiler, where—obviously enough—the heat is used to boil water and make steam. Then, near the top of the arch is a superheater, which raises the temperature of the steam far above the boiling point. Farther along, on the downward arc, are reheaters, which pour more heat into steam that has already passed through the first stages of the turbine. Then comes the economizer, where water is preheated on its way to the boiler. And even after all this superheating and reheating and preheating, the flue gases are still not quite done with their day's work. Their last task is to preheat the air that blows the fuel into the firebox.

Air Pollution Control. Once upon a time, the spent flue gases would have gone straight up the smokestack, carrying a substantial load of unpleasantness with them. The problem is not smoke, as it would be from a malfunctioning fireplace. Smoke

Coal burners are lined up along one wall of the firebox. The coal-and-air mixture ascends through silver-colored conduits from the pulverizers three floors below; it passes through a scroll-shaped fan housing and is injected into the furnace. The green pipe carries liquid fuel used only in starting up the boiler.

Air-handling equipment at the Mayo plant includes a maze of immense ducts. The green machine at lower left is a fan with intake filters that reach a height of about 20 feet. This "forced-draft" fan pushes air into the furnace. An "induced-draft" fan—the smaller green device at ground level right of center—sucks air and flue gases out of the furnace. The effects of the two fans are balanced in such a way that the pressure in the firebox is slightly negative; thus, any small leaks draw air inward instead of spewing fumes outward. The path through the maze of ducts proceeds upward from the forced-draft fan, through a regenerator, then to the left and downward into the boiler building. Flue gases emerge from the furnace in the uppermost duct. They can make a circuit through a selective catalytic reactor at the upper right, or they can bypass this device and descend directly through the regenerator to the induced-draft fan. From there they flow on to the stack, just visible in the background. The regenerator, which sits at the intersection of the two streams of gases, is an air-to-air heat exchanger. The housing, just above the forced-draft fan, is an octagon; inside, a honeycomb disk some 30 feet in diameter rotates slowly. The disk absorbs heat from the exhaust gases and gives it up to the combustion air.

comes mainly from incomplete burning of carbon, which isn't tolerated in utility-scale power plants. But coal is not pure carbon. For one thing, it includes a mineral residue that just won't burn. The heavier part of this residue, called bottom ash, winds up in a pan at the bottom of the firebox. The lighter part is fly ash, and it is carried along by the flue gases as a fine gray dust. Coal also includes at least a little sulfur, which burns to produce sulfur dioxide, the precursor of sulfuric acid and acid rain.

Power plants today are required to capture nearly all the fly ash before it escapes up the stack. The two main technologies are baghouses and electrostatic precipitators. The baghouse is just a filter. The bags are long and thin—maybe six inches in diameter and 20 feet long. Each bag is closed at the top but open at the bottom, where the dirty gas flows in to keep the bags inflated. Gases pass through the fabric, leaving the ash behind as a dust cake on the inner surface. As with a vacuum cleaner, the bags need to be emptied from time to time. In most plants this is done by briefly reversing the flow of air, driving the dust out of the bag and down into a hopper; some units also have a shaker that thrashes the bag back and forth.

Electrostatic precipitators rely on subtler physics than a vacuum-cleaner bag: they work on static cling, the force that makes a toy balloon stick to the wall after you rub it on your clothes, and that sometimes makes your clothes stick to you. Inside the precipitator are many parallel rows of vertical metal plates, with flue gas flowing horizontally through the lanes between them. Hanging down into the spaces between plates are fine wires energized with several thousand volts of electricity. Electrons are repelled from the wires and flee to the metal plates; along the way they attach themselves to passing particles of fly ash, which then stick to the plates. A "rapper" mechanism shakes the collected dust loose, and it falls into a hopper below.

The precipitator may sound like an exotic piece of machinery, but electrostatic air cleaners for the home work the same way. Laser printers and photocopiers are also based on the same principle of lending an electric charge to fine particles.

Neither baghouses nor precipitators can capture the sulfur dioxide in the flue gases. That's a job for a scrubber, which relies on chemistry rather than physics. The scrubber sprays the flue gases with a slurry of lime; sulfur dioxide combines with the lime to form calcium sulfate, or gypsum. The scrubber is a set of large cylindrical vessels (typically 60 feet high and 20 feet in diameter) where the flue gases enter at the bottom and travel upward through the descending mist of lime slurry. The scrubber comes last in line in the processing of the flue gas, after the precipitators or baghouses. A telltale sign of a scrubber in operation is pure white steam pouring out of the stack in the summer. (In cold weather the flue gases may form a visible vapor trail even without the moisture added by scrubbing.)

Not all plants have scrubbers. Some utilities have been able to meet air-quality standards without a scrubber by burning low-sulfur coal (much of it from the Powder River Basin in Wyoming). More controversially, a number of older plants are "grandfathered," or exempt from regulations enacted after the plants were built.

In recent years yet another pollution-abatement technology has begun to appear at coal-fired power plants. Selective catalytic reduction treats the flue gases with ammonia to deal with oxides of nitrogen (usually denoted NO_x). The NO_x is created when nitrogen and oxygen in the air combine in the high-temperature flame of the furnace. Ammonia reacts with the NO_x to form nitrogen and water.

The Stack. Even from miles away, the stack of a power plant is an impressive structure. Up close, what seems most remarkable is not the height but the girth and the bulk.

"Rappers" on the roof of the electrostatic precipitator knock the accumulated dust free, letting it fall into the storage hopper. Each rapper is the size and shape of a baseball bat. Inside is an electromagnet that pulls a steel plunger upward, then allows it to fall again, producing a sharp knock. The rappers are energized at seemingly random intervals, producing a haunting, syncopated music. (The rhythm seemed more modern jazz than rap.)

Electrostatic precipitators at Mayo remove fine particles of ash from the flue gases. The gases enter through the vertical duct at right, pass through the precipitators from right to left and exit downward through the vertical duct at left. Inside the four-story-tall precipitators are electrically charged plates and wires, which trap the particles and deposit them in hoppers below.

The function of the stack has changed over the years. Early industrial smokestacks worked much like a fireplace chimney: buoyant, warm air rose through the stack, creating a natural draft to draw fresh air into the furnace. The taller the stack, the more powerful the natural draft. Today, with a long twisting pathway from the firebox to the stack, natural draft is not nearly enough to keep the air moving, and large fans force the flue gases into the base of the stack. The height of the stack is determined not by the need for natural draft but by the requirement to release effluents high in the air, where they will be diluted and dispersed.

A plant that has three or four stacks has a corresponding number of firebox-boiler-turbine units. But a single stack doesn't necessarily mean that a plant has only one unit. Sometimes the stack is divided internally into multiple flues. In addition to the gigantic main stacks, most plants have a few smaller, stubbier stacks as well. They serve small auxiliary boilers. What looks like a small smokestack may also be a steam vent.

The Boiler. The boiler of a toy steam engine is a little steel tank with a fire under it. But outside of toyland, a boiler is all about tubes, not tanks.

The power-plant boiler is one of those artifacts whose evolution is so complicated that you can't really understand how it works without also knowing where it came from. The distant ancestor of the modern boiler had a firebox of brick, with a bundle of water-filled steel tubes running through the middle of the combustion zone. Flames swirled among the tubes, boiling some of the water. As boilers got bigger and hotter, the firebrick lining became a factor limiting performance. No material could withstand the heat of the furnace for very long. As a strategy for cooling the brick, some of the boiler tubes were run vertically down the inner face of the furnace walls. Today, boiler tubes have been removed entirely from the interior volume of the combustion chamber; all the tubes are installed in "waterwalls" that line the furnace. The tubes are closely spaced and welded together with a webbing of steel to form solid, airtight panels. The brick that the waterwalls were once protecting has now been eliminated altogether.

Considering the inferno inside the firebox, the environment surrounding a big furnace is surprisingly benign. The waterwalls are swaddled in insulation. You can get up close to them, standing inches away from a 2,000-degree torch roaring 10 or 15 stories tall. The space is warm, but not uncomfortable.

The tubes in the waterwall are called risers, because heated water and steam rise through them. Another set of tubes, called downcomers, carry water flowing in the opposite direction. Risers and downcomers are joined at both the bottom and the top of the boiler to form a continuous loop. The entire assembly of boiler tubes, weighing hundreds of tons, hangs from the roof of the building, with no rigid supports underneath. This arrangement allows for expansion and contraction—the length may change by a foot or more—as the boiler heats up and cools down.

At the very bottom of the boiler, where the downcomers and risers meet, some plants have a mud drum, a cylindrical vessel that takes its name from what you find

A scrubber in action at the Homer City power plant produces a characteristic white plume, which consists of water vapor from the scrubber condensing as it cools. Only one of the three units at Homer City is equipped with a scrubber, and so only one of the stacks emits a plume. Utility companies occasionally get complaints about these conspicuous emissions, but it is the stacks *without* a visible plume that are probably releasing greater quantities of sulfur.

A boiler hangs from the roof, allowing it to expand and contract with changes in temperature. The full weight of the boiler is supported by a forest of steel rods that hang from the uppermost girders of the frame and connect to the roof of the boiler structure. The cuplike objects at the very top of the frame are steam vents used for either routine or emergency releases. The cups are filled with baffles meant to reduce the noise level of high-pressure releases.

inside when the boiler is shut down for an overhaul. All the rust and scale accumulate there. The corresponding structure at the top of the boiler is the steam drum, which in fact is more than a drum; it's a complicated piece of apparatus, with lots of machinery inside. The main business of the steam drum is to separate steam from water so the vapor can be drawn off and piped to the turbine, while the liquid is recirculated through the downcomers. To those of us whose experience of steam comes mainly from teakettles, separating steam from water seems easy: the vapor just wafts off the top. But under the conditions inside a power-plant boiler—temperature 675 degrees Fahrenheit, pressure 2,600 pounds per square inch—steam and water are hard to tell apart. It takes a kind of centrifuge, or cyclone, to separate them.

A feature of all boilers, required by law, engineering codes, and insurance regulations, is a pressure-relief valve. Early in the age of steam, boiler explosions were a notorious technological hazard. Both railroad locomotives and the stationary boilers of factory steam plants were blowing up with enough regularity to inspire public dread comparable to modern worries about nuclear-power accidents. Those at greatest risk were the engineers who tended the boilers, and yet they resisted regulation of their work. Nevertheless, a safety measure was imposed and remains universal today: every boiler has a valve that automatically vents off steam at some preset pressure not too far above the normal working pressure. The valve is installed on the steam drum. It relies on the simplest kind of spring-loaded mechanism, which pops open if the internal pressure ever exceeds the strength of the spring. Modern boilers are equipped with other valves, tied into the central computer control system, that allow finer regulation of pressure. But the mechanical safety valve is there in case the computer ever crashes or someone falls asleep at the switch.

Downcomers form part of a loop of piping that allows water to circulate through the boiler. The photograph looks upward along one side of the boiler. Behind the metal sheathing and a layer of insulation is the waterwall, where pipes called risers are heated by radiant energy from the furnace. Some of the water in the risers turns to steam, but the rest is recirculated through the downcomers.

The steam drum *(below)* is where risers and downcomers join at the top of the boiler. Visible here is just one end of a long drum, filled with machinery for separating water and steam. Pressure-relief valves *(above)* are on the roof just above the steam drum. They are simple mechanical valves, which discharge steam whenever the pressure in the boiler is great enough to compress a spring that holds the valve closed.

The Turbine. The steam-driven turbine is a close cousin of a jet engine. In both machines, a hot, high-pressure gas spins a series of fanlike turbine wheels. In the process, the gas expands and cools. It's a simple idea, but power-plant turbines rated at a billion watts of mechanical power are not simple to build or operate.

The turbines and the generators they drive are the most expensive hardware in a power plant. Often, they are built atop their own special concrete-and-steel foundation, separate from the rest of the plant. This is done to control vibrations in the rotating machinery and to maintain precise alignment in the bearings that support the long, spinning steel shaft that runs through both the turbine and the generator.

A single stage of the turbine consists of a stator wheel (which is rigidly fixed to the frame of the turbine) and a rotor wheel (attached to the rotating shaft). Steam is steered through vanes in the stator and then passes through the blades of the rotor, turning it by the same principles that run a windmill or a waterwheel. The blades and vanes have graceful airfoil shapes, and they are carefully machined from fancy steel alloys that can withstand extremes of temperature, pressure, and mechanical stress, as well as a corrosive environment. A single blade breaking off would destroy the entire machine as the debris crunched through the downstream rotors and stators.

Typically a turbine has three units, all mounted on the same shaft. Steam straight from the boiler and superheater is fed into the high-pressure turbine, where it expands and cools somewhat. The steam then goes back to a reheater unit, where its temperature comes back up to about 1,000 degrees Fahrenheit, although the pressure is not restored to its original level. This warmed-over steam then goes through the intermediate-pressure turbine, where again it expands and cools. Finally, the steam passes into the low-pressure turbine. Note that the same quantity of steam—the same mass of water molecules—goes through all three turbines, but because the pressure drops in each unit, the volume of steam increases. As a result, the intermediate-pressure turbine has to be bigger than the high-pressure unit, and the low-pressure turbine is the largest of all. Judging from their relative sizes, you might guess that the big low-pressure turbine is doing most of the work, but the truth is just the opposite. The little high-pressure turbine puts out 60 percent of the total horsepower, and the massive low-pressure unit supplies only about 15 percent.

Turbines are so large and complex that their most mundane auxiliary equipment is more imposing than any of the machines most of us meet in everyday life. Pumps and motors larger than an automobile engine are needed just to keep the turbine supplied with lubricating oil. The bearings and seals along the main shaft also require large accessory pumps.

Another vital auxiliary is the governor that regulates turbine speed. The classic speed-control mechanism is the flyball governor, which became an icon of the industrial age and a textbook example of the concept of feedback control. The governor has two weights (the flyballs) attached to hinged arms that spin around a vertical shaft at the same speed as the turbine. As the shaft turns faster, the balls are flung outward, and the hinged arms are lifted up. A linkage attached to the flyball arms then closes the

The high-pressure turbine is the smallest but most powerful of three turbine units at the Mayo plant. The turbine itself is hidden under a thick blanket of insulation.

steam valve a little, slowing the turbine and allowing the flyballs to sink back toward their resting position. In this way the turbine is slowed every time it tries to speed up and is sped up every time it tries to slow down, so a steady speed is maintained. The flyball governor was the world's first version of cruise control. The feedback principle is still at the heart of turbine control, although now it's all done by computer.

The Condenser and Feedwater System. To make a turbine spin, it's not enough to push steam into the inlet port; you also have to let it out at the exhaust port. Lowering the pressure and temperature at the outlet is the job of the condenser, where the steam gives up the last of its heat. As the steam condenses, its volume is greatly diminished, and so the pressure falls too. Indeed, the pressure in the condenser is less than atmospheric; there's a partial vacuum, which actually sucks steam out of the turbine.

The water that collects in the bottom of the condenser is distilled water, which is usually considered the ultimate standard of purity. But the water needs further treatment, called polishing, before it can be returned to the boiler. Any minerals deposited inside the tubes of the boiler would clog up the arteries and could cause a dramatic kind of heart attack: the deposits would act as an insulating blanket, allowing the metal wall of the tube to overheat. If a tube splits open, everyone within a few miles of the plant hears it.

To remove suspended solids, the water is filtered through sand or charcoal, and magnetic separators extract particles of rust. An ion-exchange column works just like a residential water softener to eliminate troublesome magnesium and calcium compounds. Other chemical treatments adjust the pH—the acidity or alkalinity—and remove dissolved oxygen, which can attack metals.

Governors and throttle valves mounted on the high-pressure turbine control the speed and the power output of the entire turbine-generator unit.

The condensers are among the largest devices in the plant, because they receive steam at its lowest pressure, when it has fully expanded. Here only the end caps of two condenser units are visible; they are the gray metal structures with flanges that look somewhat like gear teeth along their side walls. The giant blue ducts carry spent steam.

The treated condensate, as well as fresh "makeup water," goes back around the loop to be boiled into steam again. Pushing the feedwater into the boiler is not easy, however; the feedwater pump has to overcome the entire head of steam pressure. The feedwater pump is the biggest of the many pumps in a power plant. It is usually mounted on the floor below the turbine and generators and is driven either by a very large electric motor or by a steam turbine of its own. In either case the pump consumes 2 or 3 percent of the raw power output of the plant.

Generators. The ultimate purpose of everything in the power plant, from the coal pile through the furnace and the boiler to the turbine, is to spin the shaft of the generator and create an electric current. Generators rely on an effect called electromagnetic induction, discovered 150 years ago by Michael Faraday and Joseph Henry. Induction creates a voltage in a loop of wire whenever a magnetic field moves through the loop. In a generator, the magnet that creates the field is on the spinning rotor; the loop of wire is wound on the unmoving stator that surrounds the rotor.

In the type of generator used with a steam-driven turbine, each turn of the magnetic rotor produces one cycle of alternating current (AC) in the stator coils. The current flows first one way through the coil and then the other, like a tide sloshing in and out. A rotor turning at 60 revolutions per second generates alternating current at a frequency of 60 cycles per second, or 60 hertz. This speed—usually stated as 3,600 revolutions per minute, which amounts to the same thing—is the standard throughout North America. Every generator connected to the U.S. power grid is adjusted to this rate of rotation. In Europe similar generators turn at 3,000 revolutions per minute, producing power at a 50-hertz frequency.

The feedwater pump is mounted at the low point of the plant in terms of elevation, but it is the point of highest pressure in the entire boiler system. The pump has to attain this pressure in order to force water into the boiler against the head of steam.

Here's something to puzzle over: to generate an electric current, you need a strong magnetic field, but to create a strong magnetic field, you need an electric current (because the magnet on the rotor is an *electro*magnet, a big coil of wire with a current flowing through it). Where does the current for the rotor magnet come from? The answer is that it comes from another, smaller generator called the exciter, mounted on the same shaft as the main generator. But the exciter also has a rotor magnet that needs a current—where does *that* come from? It comes from an even smaller generator, the pilot exciter. At this point it sounds like we're going to have an infinite regress of smaller and smaller generators, but in fact there's a stop to it. The pilot exciter is small enough to operate with a permanent-magnet rotor. In this way the many megawatts of the main generator are bootstrapped from the feeble stirrings of permanent magnets like the ones that hold notes on your refrigerator.

Big generators are remarkably efficient. Out of all the mechanical energy cranked into turning the shaft, the generator converts between 98 and 99 percent into electricity. But if a generator's output is a billion watts, internal losses of just 1 percent add up to 10 million watts of heat—the equivalent of 10,000 toaster ovens running at the same time. Getting rid of this heat is a major challenge.

The stator windings, where the heaviest currents flow, are water-cooled. The conductors in these coils are hollow copper tubes, and water is pumped through them at high speed. If running water through a high-voltage machine seems contrary to common sense, the cooling medium for the rest of the generator will strike you as even more unlikely. It is hydrogen gas. Hydrogen is chosen because among all gases it is the best possible coolant; the lightweight molecules carry off heat more effectively than those of heavier elements. The lightness also reduces "windage" losses, the energy spent moving the rotor through the atmosphere. But hydrogen has had a reputation for danger ever since the Hindenburg accident in 1937; infusing the flammable gas into a generator full of hot metal and high voltages seems to invite disaster. But hydro-

Pumps and valves below the turbine deck circulate gaseous hydrogen through the generator, cooling the rotor windings.

The single generator for the Mayo plant is rated at more than 750 megawatts. The generator is inside the cylindrical shroud in the foreground; the similarly shaped housing behind it is the low-pressure turbine.

The switchyard of the Gordon Evans Energy Center in Colwich, Kansas, is a thicket of transformers, switches, circuit breakers, lightning arresters, and other high-voltage devices. The main function of the switchyard is to raise the voltage to a level that can be transmitted long distances.

gen burns only in the presence of oxygen; the key to using it safely is to exclude all air. Before the generator is filled with hydrogen, it is purged with carbon dioxide.

Electricity is carried away from the generator on bus bars, heavy copper or aluminum conductors rated to carry as much as 40,000 amperes of current. (The heaviest wires you'll find in your home are limited to 100 or 200 amperes.) The bus bars are as thick as tree limbs, and they may be encased in protective tubes that make them look even thicker. They lead to a switchyard outside the plant.

The Switchyard. Although the generator puts out prodigious currents, the voltage level is only moderate by power-company standards—usually between 10,000 and 30,000 volts. Right outside the wall, a transformer boosts the voltage to a much higher level—often 230,000 or 345,000 volts, and in a few cases as high as 765,000 volts. The high voltage allows the power to be transmitted long distances with relatively little loss along the way.

The transformers and their related switches and circuit breakers are set up in a fenced-off area called the switchyard, which can be as large as the rest of the plant. The devices here are essentially the same as those in the substations at the other end of the transmission line, where the power is brought back down to lower voltages for distribution to neighborhoods. This machinery is discussed in the next chapter.

The switchyard brings power *into* the plant as well as providing a way out. A typical generating station absorbs 4 to 7 percent of its own electrical output for running machinery such as fans and pumps. When the plant is starting up, much of that equipment has to be running before the main turbines and generators are cut in. The start-up power is supplied by other stations on the power grid, and brought in over the same transmission lines that normally export the plant's own output.

What happens if all the power plants in a system are shut down at the same time? Until 1965 plant operators thought they would never have to answer this question

because such an event seemed so unlikely. But on November 9, 1965, a blackout in the northeastern United States left some cities dark for more than 12 hours. One reason it took so long to restore power was that generating stations didn't have enough emergency power to restart without help from their neighbors—who were, of course, in the same predicament. The utilities companies promise it won't happen again.

COMBUSTION TURBINES

The demand for electricity fluctuates by the millisecond. When you turn on the coffee pot and the toaster in the morning, a power plant somewhere has to respond by opening the throttle a little. Demand also fluctuates on longer time scales. People use more electricity during the day than at night, and in most places they use more during the summer than the winter. It would be a great convenience to have generating units that could be run only at times of peak demand. Big coal-fired stations are

A power plant in a different architectural style, with all the machinery fully enclosed, the Ravenswood Generating Station is a well-known landmark for New Yorkers; it occupies a conspicuous site in Queens, just across the East River from Manhattan. It is also a landmark for power engineers, the home of a generator known as Big Allis (built by the Allis-Chalmers Corporation), the first generator capable of producing 1,000 megawatts. The plant burns natural gas, with oil as a backup fuel.

A combustion turbine at the Gordon Evans Energy Center supplements the capacity of the plant's main steam units. The tan duct extending horizontally to the right is the air intake. It arches over the generating unit. The turbine itself is in the square tan enclosure; it is followed by a flared horizontal exhaust duct, and then the large gray exhaust stack, which discharges vertically to reduce noise.

not well suited to this duty, because they take hours to start up and shut down. Hydroelectric plants are much more flexible in this respect, but, on the other hand, you can't build Hoover Dam just anywhere. The solution adopted by many utilities is a machine known as the combustion turbine or gas turbine. Power company employees call them jets, and for good reason: they evolved directly from the engines that power jet aircraft.

A combustion turbine relies on the same physical principle as the steam turbine in a coal-fired plant: hot, high-pressure gases expand against the vanes of a turbine wheel, exerting a force that causes the wheel to spin. But instead of steam, the hot gases are the products of combustion. Fuel and air are mixed, compressed, and ignited inside the turbine, where they expand and thereby turn the rotor vanes. The fuel is usually natural gas. A single combustion turbine has a power output of 10 to 100 megawatts, but it's easy to build clusters of them with larger aggregate power.

Combustion turbines are less efficient than the best steam turbines, but they have compensating advantages. First, of course, they can be started and shut down in a matter of minutes, sometimes just by pushing a button in a distant control room. Furthermore, because they don't require as much land as a full-scale power plant and because they burn cleaner fuel, they can be put closer to cities, which relieves congestion on electric transmission lines. And jets can supply start-up power for larger conventional plants. For this last reason, many steam plants have a few combustion turbines on the site. The standard mode of operation is to keep the big, efficient "base load" plant running all the time, and start the jets only at times of peak demand.

Combustion turbines vary in appearance, but a common feature is that the turbine itself is overshadowed by air intakes and exhaust stacks. One reason for the large

intake and exhaust structures is that free flow through the system improves efficiency. But there's another reason: the structures are engineered to suppress noise. If the turbine plant has neighbors, noise is likely to be a major issue. Jet engines are no quieter on the ground than they are on airplanes.

NUCLEAR POWER PLANTS

It began as the technology of megadeath. Then in the 1950s and 1960s "the peaceful atom" promised a life of ease through boundless energy—electricity so cheap no one would bother metering it. By the 1970s the tide had turned again, and nuclear plants were regarded as a menace, at least in the United States. Who can say where this roller-coaster history will end. At the moment nuclear power looks like a zero-growth industry, but even if no one ever builds another nuclear plant, the existing ones will remain a major part of the energy infrastructure for decades. Almost 100 commercial power reactors supply about 14 percent of the nation's electricity.

A nuclear generating station has much in common with a coal-fired power plant. They are both closed-loop steam cycles. The generators and electrical switchgear are almost interchangeable, and the turbines are very similar. The only important difference lies in how the steam is produced.

The energy source in a nuclear plant is the disintegration of uranium, the heaviest of the naturally occurring chemical elements. In certain uranium atoms, the nucleus—the dense core of protons and neutrons—can spontaneously split in two. The splitting, or fission, is especially likely to happen after a nucleus absorbs an extra neu-

Four new combustion turbines in Kearney, New Jersey, are seen under construction in 2002. The air intakes are atop the buildings that house the turbines.

Eight combustion turbines are lined up along the East River north of the Ravenswood plant; there are eight more elsewhere on the site, as well as a new "combined cycle" plant that uses the hot exhaust from a combustion turbine to produce steam. In the units seen here, the brick-red stacks are air intakes, and the taller corrugated metal ones are exhausts. Visible in the background is the Queensborough Bridge.

tron. As the nucleus breaks apart, it gives off a small jolt of energy, which is what ultimately gets turned into electricity, and it also emits a few spare neutrons, which can go on to induce the splitting of other uranium nuclei. In this way a chain reaction gets started. It's just like one of those pyramid marketing schemes, except it works.

To keep the chain reaction going, all that's needed is a sufficient number of susceptible nuclei in a small enough space. It's also essential to control the reaction. Other substances come into play here. Water tends to enhance the reaction because it slows down neutrons, and slow neutrons are more likely to be absorbed by uranium nuclei. Carbon and boron tend to damp out the chain reaction by absorbing neutrons and making them unavailable. All three of these substances have roles in power reactors.

For use as reactor fuel, uranium oxide is molded into cylindrical pellets a third of an inch in diameter. The pellets are very heavy, and always warm with a glow from within. Or so I'm told. I've never held them in my hand. Outside of a few fuel-processing installations, the bare pellets are never seen. They are stacked up inside tubes made of a high-temperature zirconium alloy, and then the tubes are welded shut.

Nuclear power has certainly not made electricity too cheap to meter. The uranium fuel is not free, and the capital costs of building a plant have turned out to be

Two pressurized water reactors stand side by side at the Arkansas Nuclear One site on Lake Dardanelle near Russellville, Arkansas. The domed cylindrical structures are the containment buildings; behind them are the fuel-handling facilities At left in the foreground is part of a large cooling tower. The two units are siblings but not twins; they were built a few years apart from slightly different designs and components.

daunting. All nuclear construction is governed by special engineering codes, with elaborate schedules of inspection and maintenance. Every pipe and valve must bear a "Code N" stamp, which raises the price more than gold plating would. Someday, dismantling the plants may wind up costing even more than building them did. And there's also the cost of dealing with radioactive wastes and spent fuel.

Worldwide, there is considerable diversity in the design of nuclear plants, which may signify that engineers have not yet built enough of them to reach consensus on how best to do it. Just two designs dominate in the United States: the pressurized water reactor (PWR) and the boiling water reactor (BWR). Only those types are described here.

The Pressurized Water Reactor. The distant ancestor of the PWR is the U.S. Navy's program to develop nuclear power for ship propulsion. The defining feature is a reactor core fully immersed in liquid water, which is kept under so much pressure that it cannot boil even though the temperature reaches 600 degrees Fahrenheit. About two-thirds of the operating American reactors are PWR types.

The PWR relies on an indirect, two-stage process to drive the turbine and generator. Water heated in the reactor core is pumped to a steam generator, where it heats and boils water in an entirely separate circuit; it is the fluid in this secondary loop that drives the turbine. There is no exchange of fluids between the two loops; this is the safe-sex version of nuclear power. Because the steam that drives the turbine never enters the reactor, the chance of radioactive contamination should be slight.

A PWR has a distinctive profile. The containment building, which houses the reactor, is a tall cylinder with a domed lid. Deep inside is a massive steel pressure vessel, and inside that is the reactor itself. Also in the containment building are the steam generators and pumps to drive the circulation through the primary loop. The pumps stand three stories tall and are powered by electric motors of 4,000 to 7,000 horsepower. Each pump has a flywheel that will keep it running for a few seconds after a power failure—long enough for other emergency cooling systems to kick in.

The reactor vessel is shaped like a medicine capsule standing on end, 40 feet high with steel walls nine inches thick. It weighs close to a million pounds, which means it can only be shipped by barge or rail. (There are no 500-ton highway trucks.) The inner surface is clad with half an inch of stainless steel as a defense against corrosion. And corrosion is a serious worry. In 2002 a work crew at the Davis-Besse nuclear plant near Toledo, Ohio, discovered a spot on the lid of the reactor vessel where acid had eaten away the entire thickness of the wall except for the stainless-steel cladding.

Within the reactor itself, several thousand fuel rods are packed into a volume about the size of a high-ceilinged bathroom. There are also control rods made of boron carbide—a compound of two neutron "poisons," or absorbers. With all the control rods in place, neutrons are blotted up quickly enough that a chain reaction can't sustain itself. The control rods are lifted out through the top of the pressure vessel to start the nuclear reaction. In the event of a power failure or some other malfunction, the rods fall back into place by gravity.

Adjacent to the containment building is the fuel-handling building—often larger in volume though less distinctive in shape. Here fuel-rod assemblies are stored, both fresh ones awaiting installation and depleted ones removed during refueling. The fuel rods can't simply be stacked on a shelf. They are kept immersed in water both as a radiation shield and as a coolant. The deep vat of water, known as the swimming

THREE MILE ISLAND

Sometimes an industrial accident seems to have the fatal momentum of a Greek tragedy. Terrible things keep happening, but nobody understands why until it's too late.

March 28, 1979, was a bad day on Three Mile Island, in the Susquehanna River south of Harrisburg, Pennsylvania. In the small hours of the morning, a shift foreman and two other workers were doing routine maintenance in one of the two nuclear power plants built side by side on the island. Both plants are pressurized water reactors. The maintenance work was in what would seem to be a noncritical section of the plant—the polishers that remove minerals from feedwater in the secondary cooling loop. But events in that obscure corner of the plant had consequences the whole country soon heard about.

The work crew was blowing compressed

air into one of the polishers, and apparently the pressure drove water into an instrument air line, one of many small pneumatic tubes used for sensing and controlling conditions in the plant. The clogging of this particular air line had the effect of closing valves that controlled the flow of feedwater through the polishers. With the supply of water cut off, the main feedwater pumps shut down automatically (or, in power-plant argot, "tripped"). Three emergency feedwater pumps immediately started up, but they were unable to deliver any water because another pair of valves had mistakenly been left closed. The improper position of these valves was discovered and corrected eight minutes later, but by then a great deal else had happened.

Less than a second after the main feedwater pumps tripped, the turbine and generator

tripped in turn. In the next three seconds, the pressure within the reactor and the primary coolant system rose to 2,255 pounds per square inch, at which point a relief valve opened up, draining steam and water from the reactor vessel into a tank at the bottom of the containment building. After another five seconds, the reactor itself tripped, and control rods were automatically inserted to halt the nuclear reaction.

Although this fast-paced cascade of emergencies sounds quite dire, there was as yet no reason for alarm. Turbine and reactor shutdowns are not routine events, but operators are trained to deal with them. In this case the two operators on duty in the control room immediately set out to perform what seemed to be the most urgent tasks—double-checking the status of the turbine and generator to be cer-

pool, is a beautiful sight. The water is still and clear; technicians test its purity and clarity by reading fine print at the bottom through binoculars. The fuel assemblies hang on racks 25 feet below the surface. The most active elements are enveloped in the haunting blue glow of Cerenkov light, which is emitted when electrons streak through the water faster than the speed of light in water.

tain that these expensive pieces of machinery would not be damaged. The all-knowing chorus in a Greek tragedy might have warned them that bigger worries were looming, but the operators at Three Mile Island did not have the benefit of such a warning.

Over the next two hours the scene in the control room grew more hectic; at one point 60 operators, supervisors, engineers, and others struggled to stabilize the system. The main focus of their attention was maintaining the right water level in the primary cooling system. Most of the time, the level seemed to be too high. A vessel called the pressurizer is supposed to be kept half full of water and half full of steam; the operators thought it was filling with liquid water, which would make it hard to control pressure in the system. Hence, they throttled back emergency systems that were pumping water into the reactor. Actually, the water level in the pressurizer was never too high; it was dangerously low. The operators had been misled by their instruments. The underlying source of the problem was yet another valve malfunction: the pressure-relief valve that had popped open three seconds after the start of the accident should have closed just 10 seconds later, but it remained open, allowing a massive leak. The stuck valve was not discovered until more than two hours later, by which time most of the primary coolant had boiled away.

By now it was too late to avoid serious damage to the reactor. Although inserting the control rods had halted the nuclear chain reaction, radioactive decay was still producing

about 30 megawatts of heat, which could not be removed fast enough. Parts of the reactor core crumbled and melted. Also, the overheated zirconium-alloy cladding on the fuel rods reacted with steam to produce hydrogen gas, raising fears that a hydrogen explosion might rupture the containment building. The explosion never came; it turned out there was too little oxygen present to create an explosive mixture. Throughout the accident there were only small releases of radiation.

It took a month to coax the reactor into a safe state, and it took more than 10 years to

clean up the mess. Several commissions investigated the accident—Greek choruses chanting of catastrophe after the fact. Factors cited as contributing causes included management policies that allowed the plant to run with emergency feedwater valves closed, operator train-

ing that put too much emphasis on one kind of accident and neglected other possible failures, and a reactor design that may have been too skittish for reliable control. Most of all, the investigators criticized the man-machine interface. The operators could easily have averted the damage if only they had known what was happening inside the containment building, but the hundreds of meters and gauges in the control room failed to communicate the information they needed. The indicator for the crucial relief valve showed that it had been ordered to close but did not register its true position.

What lessons should be learned from Three Mile Island? Opinions vary widely. Opponents of nuclear power interpret the accident as a demonstration of just how dangerous and uncontrollable the technology is. Proponents look at the same evidence and argue that the accident shows the inherent safety of nuclear reactors, since just about everything that could have gone wrong did go wrong, and yet there was no serious harm done to public health. Both sides would rather not see any further demonstrations of this kind.

Today the empty shell of the failed reactor still stands on Three Mile Island, next to its older sibling reactor, which was shut down after the accident but was restarted in 1986. (In the photograph on the opposite page, the active unit is on the right, the corpse on the left.) General Public Utilities, the operating company, built a visitor center and souvenir shop, where you could buy Three Mile Island tee-shirts and cookbooks. But at last report the visitor center was closed.

Every year or two, the reactor needs to be refueled. This is not like gassing up the car; it's more like a major engine overhaul, with the added complication that it's done underwater. The first step is to pop open the top of the reactor vessel. The containment building has a crane built in for lifting off this 150,000-pound item. Then a section of the building above the open reactor vessel is flooded to a depth of 15 feet; this pool of water connects via a tunnel with the swimming pool in the fuel-handling building. Old fuel assemblies are pulled out of the reactor and transferred through the water-filled tunnel to the fuel-handling building. New fuel elements come back in through the same tunnel, which you might think of as something like a pass-through between kitchen and dining room.

Another building houses the control room. It's airtight and pressurized (so that any leakage is outward). The ventilation system can close off all air intakes in seconds. The reason for these features is not hard to guess: in the aftermath of an accidental release of radiation, it's helpful if the operators can stay on the job, and survive.

One peculiarity of nuclear power can make the operator's job especially tense. When something goes wrong in a fossil-fuel plant, shutting off the fuel and air puts

A nuclear generating station built around boiling water reactors also has two independent units, but the entire plant fits under one roof. The reactors are in the wing of the building at the left; the fuel-handling facilities are in the lower wing at right. The multicolored forms in the foreground, which look like picnic tents, are actually the roofs of tanks for holding various water-treatment chemicals. The power plant, operated by the Tennessee Valley Authority, is in Browns Ferry, Alabama.

out the fire. With a nuclear reactor, dropping the control rods quenches the nuclear chain reaction, but that's not the end of the story. The fuel elements continue to produce megawatts of heat for hours afterward because of the ongoing radioactive decay of fission products. There's no switch or valve that turns this process off. As a result, the reactor needs a continuous supply of water for cooling even after a shutdown.

The standard operating nightmare for nuclear plants is the dreaded LOCA—the loss-of-coolant accident. With a major leak in the primary loop, the water in the reactor vessel will quickly boil away; if the coolant is not replaced within seconds, the fuel rods will overheat and melt.

The Boiling Water Reactor. As the name suggests, the BWR system allows the primary coolant to boil on contact with the hot fuel in the core, so that liquid water and steam coexist in the reactor vessel. The steam is piped directly to the turbine and then condensed and pumped back to the reactor. Thus, the steam circuit has only a single loop, rather than the two-stage process of the PWR. The BWR design has the virtue of simplicity. On the other hand, steam running throughout the plant will routinely pick up low levels of radioactivity from its passage through the reactor. And if something goes wrong—such as the sudden failure of a fuel rod—large quantities of radioactive material would enter the steam circuit.

A BWR power plant looks very different from a PWR. Gone is the domed containment structure and the separate fuel-handling building. The reactor, the swimming pool, and all the rest of the nuclear machinery are in a single large building. Many BWR plants do have one visually distinctive feature: a very tall stack—as tall as one you might see at a fossil-fuel plant, though not as big around. The function of this unusual stack is explained later.

The BWR does have a containment structure; it's just smaller than the kind used in a PWR, so it fits inside a conventional building. The BWR containment is either cone-shaped (with the pointy end up) or lightbulb-shaped (with the screw-in end up). The reactor vessel is suspended near the top of this structure. Below it is a pool of water meant to absorb and condense steam released in the event of an accident.

The presence of steam in the reactor vessel requires some changes in the way the reactor core is designed. Because steam lines must connect to the top of the reactor vessel, the control rods can't enter the core from overhead; they have to come up from the bottom. But if the control rods are under the reactor, you can't count on gravity to insert them in an emergency. The emergency shutdown mechanism uses hydraulic pressure to drive in the rods, then latches them in place mechanically. The reliability of the hydraulic system is critical.

Because the water driving the turbine in a BWR plant passes through the intense radiation of the reactor, it becomes mildly radioactive even when nothing is leaking. Water quality is critical. Any minerals present will be transmuted to radioactive elements; if they are thereafter deposited in the turbine or elsewhere in the system, they can make the whole plant "hot." But even pure water is susceptible to irradiation; both

The elegant spire at Browns Ferry is designed to launch trace releases of radioactive gases high into the atmosphere where they should disperse. The form of the chimney—a slender hyperboloid—is calculated to give the gases maximum upward velocity.

Fan-driven cooling towers at the Browns Ferry nuclear station have a trapezoidal form, wider at the top, so that warm water cascading down the side will wash away any ice buildup. Each of the 16 shrouds atop the cooling unit houses a large fan blade that draws air in through the sides of the structure and discharges it upward.

the hydrogen and the oxygen atoms of the water molecule can be converted to radioactive forms. Worse still, radiation can break apart a water molecule, so that the hydrogen and oxygen are in gaseous form. The gases separate from the steam in the condenser, and a whole subsystem of the plant is needed to capture and dispose of them. That's what the tall stack is for. It is designed to launch the emissions well up into the atmosphere, where the gases disperse. The release of radioactivity is quite small. The Nuclear Regulatory Commission sets the maximum allowable amount at a level believed to be completely safe, and plants routinely stay below 1 percent of that standard. But that doesn't always set the neighbors' minds at ease.

COOLING TOWERS

Ever since the accident at Three Mile Island, the cooling tower has been the sinister symbol of nuclear power. Television reports on nuclear issues set the mood with a haunting image of the towers, often with a cloud of white vapor drifting above them, hinting at some toxic release. This choice of icon could not be less appropriate. In the first place, not all nuclear power stations have cooling towers, and not all cooling towers are installed at nuclear plants. Second, most cooling towers look nothing like the tall, tapered chimneys that have acquired such menacing associations. Finally, the cooling tower is not where the stinger is in nuclear technology. Nothing radioactive passes through it, and any release of radiation would have to come from elsewhere.

There is another irony in the evil reputation of the cooling tower. The reason for building the towers is not that utility companies earn money from them. On the

contrary, they are a concession to environmental preservation. Their main function is the protection of aquatic life.

A perfect power plant would convert all the heat liberated by burning fuel or by a nuclear reaction into electricity. Real power plants fall short of that goal. For a coal-fired plant, only about 40 percent of the heat energy is captured in electric power; nuclear plants do even worse, with an efficiency of only about one-third. All the rest is waste. A nuclear plant with an electrical output of 1,000 megawatts must get rid of 2,000 megawatts of waste heat. The flow of water needed to carry off that heat can amount to 500 million gallons per day. This is more than enough warm water to provide a luxurious daily bath for the population of New York City. It's also enough to parboil the fish in a small river or lake. The cooling tower dissipates some of that heat to the atmosphere.

Cooling towers come in two basic types: the fan-driven tower, which is more common but less conspicuous, and the natural-draft tower, which is the one that has entered the public imagination. The choice between them is one of balancing operating costs against capital costs.

Fan-Driven Towers. The typical fan-driven cooling tower is a long, boxy structure, roughly 50 feet wide and 50 feet tall and as much as several hundred feet long. The end walls are of solid construction, but the long side walls consist of louvers to allow for the inflow of air. The warm water is pumped to the top of the tower and falls

A variation on the design of the forced-draft cooling tower wraps the structure into a circle and puts the fans in the center, drawing air through from the perimeter. This tower is at the Coal Creek power station in Underwood, North Dakota.

through a labyrinth of wood or plastic slats called fill. Meanwhile a fan pulls air through the fill into a central void, and then exhausts it upward. Thus, in the fill there is a cross-flow: water trickles downward while air is drawn inward.

Seen from the end, the cooling tower has the form of an upside-down trapezoid. It is wider at the top than at the base, and so the louvered walls slope inward. This shape is chosen to control icing in the winter. Because of the inward slope, the louvers are continually washed by warm water.

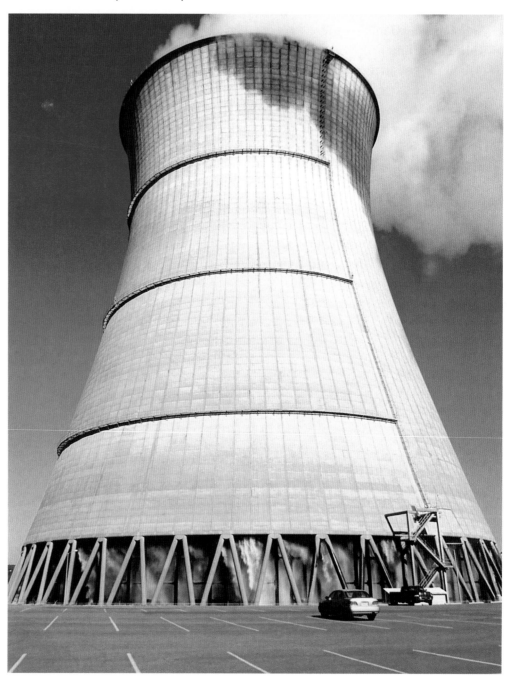

The natural-draft cooling tower at Arkansas Nuclear One has the classic hyperbolic form, tapering to a narrow throat and then opening to a slightly wider diameter. The shape is designed to produce optimum air flow for a given temperature difference between the water and the surrounding atmosphere. At the base of the tower (detail on opposite page), the water is broken up into fine droplets to maximize evaporative cooling.

The fan in a power plant–size cooling tower can be 30 feet in diameter, driven by a motor of 200 horsepower or more. A flared shroud surrounding the blades increases the velocity of the air and reduces noise. Along the inner surface of the fill, lining the central void, are special louvers called drift eliminators. "Drift" is the mist of tiny droplets that get caught in the airflow. The drift eliminators force the airstream to make a sharp turn as it accelerates toward the fan, shedding entrained droplets.

Natural-Draft Towers. The 200-horsepower motor in a fan-driven cooling tower consumes about 150,000 watts, which is a lot of power even for the electric company. On a muggy day the fans can claim 3 percent of a generating station's electrical output. A cooling tower that needs no fan has an obvious advantage in daily operating cost; the disadvantage is that it costs more to build such a tower in the first place.

A natural-draft tower works just like the chimney above a fireplace. In both cases there is a source of heat at the base—burning logs in the fireplace, water from the condenser in the cooling tower. Air is warmed by the heat source and expands, thereby becoming less dense than the surrounding air. The buoyant air rises inside the chimney or the tower, and new air is drawn in at the bottom to replace it. The new air is heated in turn, so that a sustained draft is established.

The optimum size and shape for a chimney depend on the temperature of the heat source and the outside air. With a very hot fire (as in a modern home furnace) even a narrow, straight chimney can draw effectively. The lower temperature of a log fire demands a wider flue. In a cooling tower the temperature difference is only about 10 degrees, which means the chimney must be wide, tall, and carefully shaped. The largest towers are 300 feet in diameter at the base and 500 feet high.

The characteristic shape of a natural-draft tower, which maximizes the airflow, is a hyperboloid (based on the mathematical curve called a hyperbola). Rising air tends

A small hydroelectric installation at the base of the Monticello Dam west of Sacramento, California, exploits a high head (about 400 feet) to extract about 11 megawatts of energy from a modest flow of water. The turbine is a Francis wheel, a hybrid type that draws on both the impulse and the reaction principles. The discharge is above the level of water in the tailrace.

to cool because of the lower atmospheric pressure at increasing altitude; as a result the air loses its buoyancy and its ascent slows. The hyperbolic taper of the cooling tower compensates for this tendency by maintaining a nearly constant pressure up to the throat, the narrowest section of the tower. The slight outward flare above this point allows the air to expand and accelerate upward. The exhaust nozzles of rockets and jet engines have the same shape for the same reasons.

What is inside a natural-draft tower? Mainly nothing. The shell of the tower is raised up on concrete pillars or on a triangulated trusswork of steel, leaving channels on all sides for air to flow in. Just inside this perimeter is an assembly of louvers, fill, and drift eliminators little different from the one in a forced-draft tower. All this apparatus occupies a narrow ring at the bottom of the tower; the rest is a cathedral-like empty space, open to the sky.

HYDROELECTRIC POWER

Waterpower has a history going back to antiquity; it was a thriving and sophisticated technology long before electricity entered the scene. Waterwheels driving elaborate systems of belts and shafts ran the textile factories of New England, sawed timber in the West, and ground grain into flour everywhere. Some of these early mills have been preserved or restored, and a few of the waterwheels are still turning. Nevertheless, apart from sites of antiquarian interest, waterpower now means hydroelectric generation. The waterwheel has evolved into the turbine, much as the paddle wheel of early steamships has evolved into the propeller. The belts and shafts for power transmission have been replaced by electrical lines.

In the 1930s waterpower supplied about 40 percent of the electricity in the United States. Hydroelectric capacity has increased since then, and yet the proportion of all power coming from hydroelectric plants has fallen to only about 15 percent. The reason is that other power sources have grown much faster. The trend is likely to continue, simply because the best spots for hydropower are already occupied.

Two factors determine the power available from falling water: the height of the fall (called the head) and the quantity of water. A little water plunging off a high cliff can produce the same amount of power as a large mass of water falling over a low ledge.

The head and the quantity of water determine what kind of turbine a hydroelectric plant is likely to use. With a high head but only a modest volume of water, the turbine of choice is a Pelton wheel. It works on the impulse principle: nozzles direct high-speed streams of water against curved buckets on the rim of the turbine wheel. The wheel—or runner, as hydraulic engineers prefer to call it—is not immersed in water but turns in air. Pelton wheels spin very fast, and so they are used with the same kind of generator employed in steam power plants. The Pelton wheel and the generator are mounted on a horizontal shaft, which turns at 1,800 or 3,600 revolutions per minute in order to produce the North American standard 60-hertz power.

For lower heads and higher flows, turbines of another type work better. The runner, which has curved vanes rather than buckets, is immersed in a stream of water that flows through it. The runner is mounted on a vertical shaft and enclosed in a spiral scroll case, shaped like a snail shell. Water enters horizontally and flows inward to the runner, then makes a 90-degree turn as it is deflected by the vanes, and exits

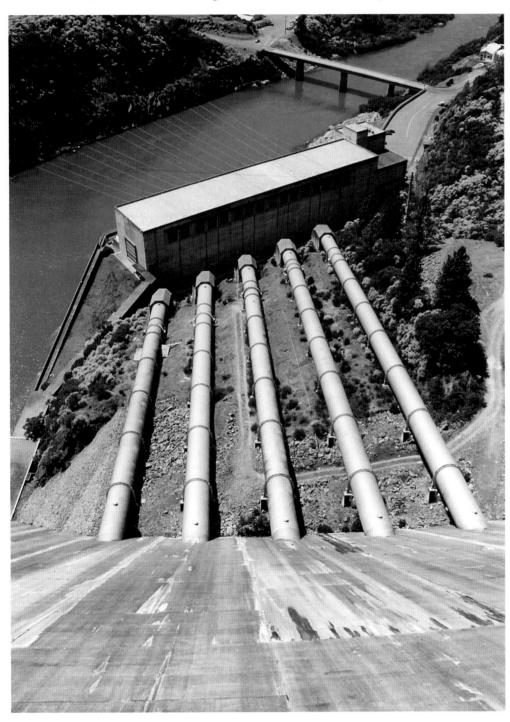

At the Shasta Dam in northern California, five large penstocks emerge from the face of the dam to drive turbines in the powerhouse below.

A surge tank at the crest of a hill helps to smooth the flow of water through penstocks feeding a small hydro-electric plant below the Fontana Dam in North Carolina.

downward, parallel to the axis of the turbine shaft. Turbines of this type—called a reaction-wheel turbine—turn much slower than the Pelton wheel.

The low rotation speed of the reaction-wheel turbine calls for a different kind of generator, one that can produce 60-hertz alternating current when turning at only a few hundred revolutions per minute. In the high-speed generators used with steam turbines, the rotor has a single pair of magnetic poles, much like an ordinary bar magnet. The generator produces one cycle of alternating current for each revolution of the rotor; thus, 60 cycles per second requires 60 revolutions per second, or 3,600 revolutions per minute. To generate the same output frequency with a machine that turns more slowly, you need a rotor with more pairs of poles. If the rotor is a cluster of 12 pairs of north and south poles, spaced equally around the perimeter, then on each revolution the output current will go through 12 alternating cycles. The generator produces 60-hertz power when turning at only 5 revolutions per second, or 300 revolutions per minute.

The generators employed in slow-turning hydroelectric plants have as many as 60 pairs of poles, yielding 60 hertz at a rotational speed of just 60 revolutions per minute. These generators are larger in diameter than the high-speed machines, in order to make room for the many rotor windings, but they can be shorter in the other dimension. Because the generator is mounted on a vertical shaft, it takes the form of a squat cylinder on the powerhouse floor, with a small turret at the center that houses the main thrust bearing supporting the shaft of both generator and turbine.

Often the powerhouse of a hydroelectric project is built into the structure of a concrete dam, usually at the foot. Water drops down through passages within the body of the dam, turns turbines installed near the baseline, and then rushes out into the tailrace. In other cases the powerhouse is a structure separate from the dam, possibly miles away. Water is conveyed from the reservoir to the powerhouse through a penstock, which is typically a welded steel pipeline 10 or 15 feet in diameter. Look for a surge tank above the penstock somewhere along the run. It is needed to smooth changes in the rate of flow as the load on the turbine varies. The tank is designed to be about half full during normal, steady-state operations. If the gates suddenly open wider, calling for more water, the surge tank is drawn down momentarily to help meet the demand. When the gates close suddenly, the surge tank is even more important: it gives the moving water somewhere to go as it decelerates, preventing the hard knock called water hammer.

The environment in the generator gallery of a hydroelectric plant is calmer than the turbine hall of a fossil-fuel plant. Gone is the shriek of steam. The noises are all low notes—hums, buzzes, groanings, rhythmic vibrations that you feel rather than hear. Workers—if there are any—can converse as quietly as in an office. The control of a hydroelectric plant is also less hair-raising than that of either a fossil-fuel or a nuclear plant. Power output is regulated by gates that control the flow of water through the penstock and into the turbine. An automatic governor system adjusts the gates to track variations in load and keep the generator turning at a constant speed.

One of the features of hydroelectric plants most welcome to power dispatchers is that they can be started up and shut down at a moment's notice. It takes as little as two minutes to get a unit up to speed and synchronized with the power grid. This makes hydroelectric power attractive as a means of satisfying short-term peak loads. When you come home in the evening and switch on the lights and the TV, somewhere a gate in a penstock has opened very slightly and sent a few gallons more down the penstock.

OTHER ENERGY SOURCES

Fossil-fuel plants, nuclear reactors, and hydroelectric plants account for 99 percent of the electric power generated by utility companies in the United States. Everything else—all the "alternative" energy technologies—amount to just 1 percent, and so they are pretty marginal in economic terms. But the alternative energy sources have a conspicuous place in the landscape and in public consciousness, even if they don't yet make much of a dent in the energy budget. And their contributions are growing. Three of these technologies are described here: wind power, solar power, and geothermal power.

Blowin' in the Wind. Wind power, like waterpower, has a long history. The Old World windmill, with its broad cloth-covered blades, or sails, goes back at least 800 years. But wind technology has been evolving rapidly in recent decades, and modern windmills look nothing like their ancient prototypes. They are tall and spindly, with narrow blades like those of an airplane propeller but on a vastly larger scale.

Generators at Hoover Dam are mounted with the shaft vertical. Each generator has 60 pairs of magnetic poles and turns at 60 revolutions per minute to produce 60-hertz alternating current. The eight generators seen here are on the California side of the dam; there are nine more on the Nevada side.

They aren't even called windmills anymore; the preferred term is *wind turbine*. Moreover, they are usually brought together in large wind farms, with hundreds of turbines lined up along ridges or scattered across a broad plain.

In the United States, wind farming got its start as a California thing; for a time, that one state produced half of the world's wind energy. The three biggest California wind-energy areas are at Tehachapi Pass, 100 miles northeast of Los Angeles, where a range of hills separates the Central Valley from the Mojave Desert; Altamont Pass, near the town of Livermore east of San Francisco Bay, where another range of low hills divides the coastal plain from the Central Valley; and San Gorgonio Pass, in the southern California desert near Palm Springs, where once again the wind has to rise over hillsides to reach an interior valley. All three areas have major highways running through them, so you can easily get a look at the machinery. (Some European wind-energy developments are even more tourist-friendly, with visitor centers and picnic areas out among the fields of turbines.)

The great boom in California wind power was launched by tax laws in the 1970s that encouraged experiments with alternatives to fossil-fuel and nuclear power plants. But the first generation of wind turbines proved to be expensive and unreliable, and they had a hard time competing against the more mature smokestack technologies. The result was a slump in the wind-power industry during the 1980s. Today the wind is rising again, however, even though some of the tax incentives have expired. The new generation of wind turbines—many of them built in Europe or inspired by European designs—are more efficient and cheaper to maintain, and they also work in a wider variety of wind conditions. One result is that wind farms are no longer just a California crop; you'll find them in Texas and Minnesota and Iowa and Vermont, and in years to come they may well sprout on hillsides everywhere. The state with the richest potential for wind power is North Dakota; if the farmers of North Dakota farmed wind instead of wheat, in principle they could supply a third of the electricity consumed in the United States.

As of 2003, the total capacity of wind turbines in the United States was about 6,000 megawatts—the equivalent of five or six nuclear plants. Europe has far more wind-energy capacity: well over 14,000 megawatts in Germany alone, and another 6,000 in Spain.

When you look at the spindly propeller-like rotor of a modern wind turbine, and compare it with an old multiblade farm windmill, you might conclude that the new technology is letting most of the wind slip by without getting any benefit from it. But that's an illusion; the new wind turbines rely on different physical principles. The older windmills are drag devices: they arrange the blades so that the wind pushes against a broad surface. The modern ones are lift devices: the air passes over an airfoil, like an airplane's wing, and pulls the blade through the air. Drag devices produce higher torque (turning force), but in most conditions they extract less energy from the air.

Just as engineers have varied the thickness of the blades over the years, so too have they disagreed over the ideal number of blades for a wind turbine. The blades are

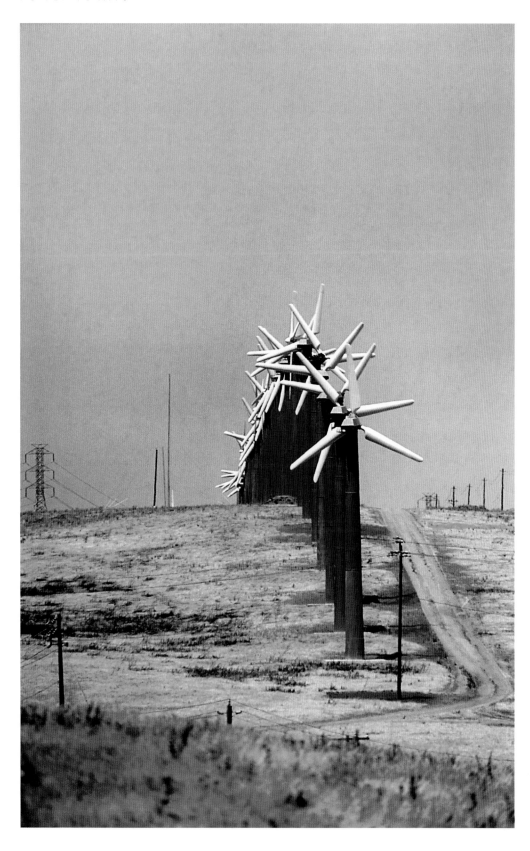

Wind turbines are planted in strict military rows along a northern California hillside, but their twirling blades are not to be disciplined. The wind farm is near the town of Tracy.

expensive, and so a design with fewer of them might be expected to reduce the cost of the machine. The minimum number, obviously, is one, and one-bladed rotors have actually been tried. They look funny, to say the least; even though a small counter-weight keeps the machinery in balance, the visual impression is of something dramatically out of kilter. But that's not the big problem with one-bladed designs; more serious is that the one-bladed rotor has to turn faster to produce the same energy output as a turbine with more blades, and higher speed brings more strain and noise.

Two-bladed rotors have a subtler problem. The balance of the blades is perfect, but trouble comes whenever the wind shifts direction and the turbine has to swivel—or yaw—to stay pointed into the wind. When both blades are vertical, there's no resistance to yawing, but as the blades turn toward the horizontal, the inertia increases. This cyclic change in resistance to yawing—going from maximum to minimum twice in every revolution—creates vibration and stress, shortening the life of the blades.

The vast majority of modern wind turbines have exactly three blades. Apparently three is just enough to solve the problems of speed, balance, and vibration; any more than three would be a needless expense.

A wind farm near Tehachapi Pass in California had more than 600 turbines spinning when this photograph was made in 1999. The machines are planted like orchard trees in a rectangular array; many other wind farms line the machines up along ridges.

Still another contentious issue in wind-turbine design is whether to mount the blades on the upwind or the downwind side of the machine. Letting the blades trail behind the rest of the turbine has one big advantage: the blades can act like a weather vane, automatically turning the machine to face into the wind. When the blades are mounted in front, some kind of steering mechanism is needed to sense the wind and forcefully pivot the turbine whenever the direction shifts. Nevertheless, the prevailing design has the rotor in the front, with a complex power-steering unit to keep it properly pointed. The reason is one I never would have guessed: with a rear-mounted rotor, the turbine blades pass through the "wind shadow" of the tower structure on every revolution. The result is a cyclic variation in wind force that can set the blades vibrating, thus creating yet another source of fatigue and premature failure.

All this leads to a portrait of the typical wind turbine. It has three blades, each about 50 feet long, made of carbon fiber or some other lightweight ultrastrong material. The blades are attached to a hub, which in turn pokes out the front of a streamlined housing called a nacelle. Inside the nacelle, which is the size of a moving van, are the generator, a gearbox, and other machinery needed to control the turbine. The nacelle is mounted atop a hollow steel pylon, 100 feet high, 20 feet in diameter at the base, and tapering gradually toward the pinnacle.

On top of the nacelle you might notice a small airplane-shaped weather vane—just like the ones you see on suburban lawns. This is the sensor for the mechanism that keeps the turbine facing into the wind. Elsewhere on the wind farm, scattered among the massive turbines, are tiny, spinning cups of anemometers on tall masts. These instruments are there to keep records of wind speed for use in analyzing turbine performance and also to shut the turbines down if winds approach dangerous levels.

Every wind turbine is designed for a limited range of wind speeds. Too little wind, and it's not worth starting up. Too much, and the machine could destroy itself. On some turbines the blades can be "feathered," or twisted so that the wind won't spin the rotor, when speeds get into the danger range. Others have aerodynamic "spoilers" with the same purpose. The final line of defense is a mechanical brake that binds the main shaft—but the operators at a Tehachapi wind farm told me they're not eager to use that one. Climbing the tower in a gale to tighten down the brake is more excitement than they're looking for.

In normal operation most wind turbines spin at a fixed rate. You might think they would speed up and slow down as the wind varies, but instead they are designed to adjust the pitch of the blades so that the speed stays constant even as the energy output changes. Running at a constant speed makes it easier to maintain the steady frequency of the alternating current that the turbine supplies to the power grid. The speeds are slow enough that you can count the revolutions. At one big wind farm in northern California I found that the turbines were making 40 turns per minute; at another farm down the road the speed was 72 revolutions per minute.

Most wind turbines turn clockwise, as seen from the hub side of the rotor. But there's no fundamental reason for this, and a few machines spin the other way.

The three-bladed propeller of the wind turbine drives a generator inside the nacelle, atop the mounting mast. Within the range of operating wind velocities, the turbine rotates at constant speed, adjusting the pitch of the propeller blades to regulate power output.

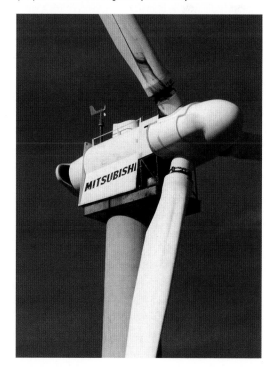

The most unusual of all wind-turbine designs is the vertical-axis machine developed by D. G. M. Darrieus. Instead of an airplane propeller, it's an eggbeater: a vertical shaft with two thin blades bent into bow shapes so that they can be attached at the top and the bottom. The big advantage of the Darrieus design is that it responds equally well to wind from any direction, with no need to pivot when the wind shifts. Also, the generator can be mounted at ground level, which makes it more convenient for maintenance and allows a lighter structure. Nevertheless, the design seems to have gone out of fashion. In the northern California wind farms, the few Darrieus machines still running were looking pretty tired and careworn when I last saw them.

A large wind farm, with hundreds of turbines, makes a powerful visual impression. From a great distance, they look like cheerful daisies or sunflowers planted in neat

LONG BEFORE THE WIND FARM, THE FARM WINDMILL

The multibladed, pinwheel-like farm windmill was an American invention in the middle of the nineteenth century that became an icon of American rural life. By the 1890s the windmills were a standard item in the Sears catalogue, and traveling salesmen peddled dozens of brands to farmers throughout the Midwest and Southwest. An estimated 100,000 of them are still at work in the United States, mostly pumping water on ranches in the western states. In aggregate they may put out 250 megawatts.

The most famous brand of farm windmill is the Aermotor, designed by Thomas O. Perry. At the peak of production in the 1890s some 20,000 per year were being made. The company is still in business, in San Angelo, Texas, where they manufacture about 500 windmills annually.

Most farm windmills are erected directly over a well shaft. A crank arm connected to the fan wheel operates a piston down in the well tube. Some later models have gearing to reduce the speed of the pump and increase the force available. The windmills come in many sizes, but the most common ones have a rotor eight feet in diameter and can pump up to 10 gallons per minute.

As with other styles of wind machines, the big challenge in building a farm windmill is making sure it doesn't fly to pieces when the

wind blows too strongly. Over the years, designs were equipped with spring-loaded vanes or centrifugal weights or other contraptions to furl the blades or turn the fan wheel parallel to the wind when the speed reaches dangerous levels. Another engineering issue is the need to grease the bearings of a windmill mounted atop a tower 20 or 30 feet tall. Nobody ever wanted to climb up there in the

middle of the winter. A number of tricks were tried (including hinged towers that fold in half to bring the works down to ground level), but nevertheless there are a lot of squeaky old windmills out there.

The Fairbanks-Morse New Eclipse model pictured below was still twirling cheerfully, despite a peppering of bullet holes, when I photographed it in northern California in 1999.

Darrieus wind machines take a different approach to wind energy: by mounting the rotor on a vertical axis, they eliminate the need for a turbine to swivel to follow changes in wind direction. These four Darrieus turbines were photographed in the Altamont Pass of northern California.

rows. As you get closer, their gargantuan scale becomes apparent. And when you finally stroll among them on an afternoon with a fresh breeze, there is something almost comical about their appearance: they seem to be waving their arms frantically, signaling to unseen friends on the next hill, or else they are turning cartwheels along the ridgelines like overexcited children. The sounds of the turbines are equally extraordinary: the swish of the blades slicing through the air, the whir of the gearbox, the hum of the generator, an occasional groaning or booming as turbines yaw with the shifting wind. Some wind machines produce a deep-bass whomp-whomp-whomp as the blades pass through the wind shadow of the support pylon. And I've even heard a few squeaky wheels.

As it happens, these very sights and sounds have become an impediment to further development of wind energy: people don't want to see wind turbines on the skyline, or hear them. There's an irony in this. Modern wind power began as a "clean" alternative to nuclear and fossil-fuel plants, promoted by environmental activists and resisted by utility companies that were skeptical of the economics. Today the opposition to wind power comes from environmental groups that see the turbines as despoiling the landscape. There is also concern about the turbines as a hazard to birds, which sometimes wander into the blades at night or in fog. Meanwhile, the utilities have begun to warm up to wind energy, as costs have come down.

Wind has proved itself as a supplement to conventional power sources. But if we want to rely on it for a large fraction of the base load, there's a problem: you can't tell the wind when to blow. In the argot of the power engineer, wind is not a "dispatchable" energy source. This puts a limit on wind's total contribution to the energy budget, but we are still far from reaching that limit.

Mirrored-trough solar collectors cover a square kilometer of the desert floor at Kramer Junction in southern California. The peak output of the array is about 150 megawatts. The mirrors appear deep blue from this point of view because they are reflecting the sky.

Let the Sun Shine In. The friendly star whose neighborhood we inhabit sends planet Earth a steady energy flux of 175 billion megawatts, which is equivalent to the output of a few hundred million nuclear power plants. Now, admittedly, a third of that energy is reflected back into space before it ever reaches the ground, but there's still plenty left over if we could only figure out how to collect it. But, as of 2000, U.S. utility companies were gathering only about 5,000 megawatts of solar energy.

There are two quite different types of solar power plant. In a solar-thermal system, sunlight is simply a source of heat. Photovoltaic plants generate electricity directly from light, with no moving parts.

The simplest solar-thermal technology uses flat-plate collectors, which work a lot like greenhouses. Pipes carry water or some other fluid through a glass-covered box tilted toward the sun; the pipes and the inside surfaces of the box are painted black to absorb as much heat as possible, and the glass cover retains the heat. It's simple and reliable, but the maximum temperature is well below the boiling point of water, so

you can't produce steam to turn a turbine and generator. Most flat-plate collectors are rooftop installations used for water heating and space heating.

To reach higher temperatures, you have to concentrate the sunlight, collecting over a wide area, and focusing it on a smaller patch. In principle, lenses might be used to do the focusing, but in practice it's always done with mirrors. The ideal shape for a reflective solar collector is a parabola, because this curve has the property that parallel rays of sunlight striking the mirrored surface are all reflected to the same focal point.

One style of collector is a long trough with a parabolic cross section, mirrored on the upward-facing surface so as to focus sunlight on a tube that runs parallel to the trough at just the right position to receive all the concentrated light. The biggest installation of parabolic troughs in the United States is at Kramer Junction, a cross-roads in the southern California desert near the city of Barstow. Each trough is about 15 feet across and 150 feet long, assembled from 300 curved glass panes. Altogether there are 546,000 panes, with a total area of one square kilometer (about 250 acres). At the focus of each trough is a receiver pipe housed in a glass vacuum tube to retain heat. The receiver pipe is painted black for best absorption, but when the plant is operating and the sun is shining, the pipe glows bright white, like a fluorescent lamp.

The receiver pipes are filled with oil, which is heated under pressure to more than 750 degrees Fahrenheit. The hot oil is pumped to a heat exchanger, where it generates steam at 670 degrees; the steam then runs a fairly conventional turbine and generator. The Kramer Junction solar array is divided into five independent plants, each capable of producing 30 megawatts of electricity in full summer sun. It would be more efficient to run one big unit rather than five small ones, but at the time the plant was built, tax incentives for solar power were limited to plants of 30 megawatts or less.

Each trough collector at Kramer Junction has a parabolic cross section, which concentrates the sun's rays on a tube installed at the focus of the parabola. The tube is painted black, but it glows white when the collector is operating. Oil pumped through the collector tubes gathers the solar heat and generates steam to run a turbine.

The troughs at Kramer Junction are aligned on a north-south axis, and during the course of the day they tilt, facing east in the morning, then directly overhead at local solar noon, and finally turning toward the west at sunset. The tracking is done automatically with a sensor that tries to keep the focused image of the sun centered on the receiver pipe. Standing near a trough, you can hear it adjust itself every few seconds.

With 250 acres of glass in the middle of a dusty desert, washing the mirrors is a full-time job. It's done at night, with high-pressure hoses. It takes about two weeks to wash off the entire field of collectors; then the washing starts over again.

Even with a large parabolic trough, the temperatures are not as high as power-plant engineers would like to see for maximum efficiency. To reach still higher temperatures, the trick is to focus the sun's light not on a line of pipes but on a single point. One way to do this is with a mirror in the shape of a paraboloid, like a satellite-dish antenna or a radio telescope, but building a really big paraboloidal mirror that can tilt to track the sun is an engineering challenge. A better idea is to set out lots of small mirrors, called heliostats, which can be adjusted individually so they all reflect sunlight onto the same point. According to legend, the principle was invented by Archimedes in 212 BC, when he had a troop of Greek soldiers at Syracuse use their bronze shields as heliostats to burn the ships of an invading Roman fleet.

Heliostats were the basis of the Solar One project at Daggett, another town near Barstow in southern California. Some 1,800 flat mirrors, with a total area of 17 acres, were continually adjusted so that they all reflected the sun's image onto a black receiver at the top of a 300-foot tower. Water pumped through the receiver boiled to produce steam at about 900 degrees Fahrenheit, which then drove a turbine and generator at the base of the tower. Solar One operated as a pilot project in the 1980s. Later, a new receiver was fitted to the tower and the plant was recommissioned as Solar Two. Instead of boiling water directly, the solar energy was now absorbed into

The Solar One and Solar Two projects at Daggett, California, achieved higher temperatures by focusing the sun's light on a point rather than a line. The last of the California experiments was shut down in 1999, but the program continues in Spain.

molten salt, which could be stored for a few hours so the plant could continue generating electricity even after sunset. Solar Two was shut down in 1999. A similar plant called Solar Tres is under construction in Cordoba, Spain.

Photovoltaic technology has almost nothing in common with solar-thermal power beyond the basic fact that both rely on sunlight as the ultimate source of energy. A photovoltaic device dispenses entirely with boilers, turbines, and generators; it converts light directly into electricity in one step. The transformation is accomplished with no moving machinery. All the complexity is hidden in the microscopic structure of the photocells, which are high-tech products of the semiconductor industry.

The photoelectric effect was first noticed more than 150 years ago, and the first good explanation came from Einstein in 1905 (that's what he won his Nobel Prize for—not for relativity theory). The key idea is that light comes in packets, or particles, called photons, each of which carries some definite energy. If a photon's energy is great enough, it can kick an electron out of its stable orbit inside an atom, making the electron available to carry an electric current. These forced evictions are happening all the time; a coin sitting in the sunshine is seething with liberated electrons. But in most cases all the activity comes to naught because the electrons just wander around for a while and then fall back into the atomic orbits they came from. A photovoltaic cell is designed to capture the ejected electrons and put them to use.

Most photovoltaic cells are made of silicon, and if you can get a closeup look, you may find them to be quite beautiful objects, with crystal facets like frost on a windowpane, in various shades of blue. Stripes or grids of metal electrodes are laced across the surface to collect the electric current.

The output of a photovoltaic collector is direct current (DC) rather than the alternating current (AC) of the national power grid. Also, the voltage produced by an individual cell is closer to that of a flashlight battery than that of a power-plant generator. Thus, connecting a panel of cells to the utility grid calls for special electronics to boost the voltage and to convert from direct to alternating current.

Then there's the matter of cost. Even though the fuel is free, photovoltaic power remains substantially more expensive than electricity from coal-fired power plants. As a result, you are most likely to see arrays of photocells in places where utility lines haven't reached—powering emergency telephones along highways, powering the lights on marine buoys, powering remote homesteads. And most remote of all are the many spacecraft that have relied on photovoltaic power.

The cost of photocells has been coming down steadily for two or three decades, and interest is finally growing in utility-scale projects. The pioneer in this field in the United States is the Sacramento Municipal Utility District in California, which operates more than eight megawatts of photovoltaic collectors. One big array of photocells is next to a decommissioned nuclear plant, but most of the collectors are distributed around the utility's territory on residential rooftops and in parking lots.

It's sometimes said that to run the country on solar power we'd have to pave the whole landscape with collectors. It's not nearly that bad. According to one estimate,

Photovoltaic cells, which generate electricity directly from sunlight, are made of polycrystalline silicon. The array of cells above is installed at Montgomery College in Germantown, Maryland. A detail, below, shows the grain structure of the silicon material and the grid of metallic conductors laid down over it to collect current.

photovoltaic plants that could meet the electricity needs of the United States would occupy a little less than 12,000 square miles. That's a lot of land, but it's only about one-third of 1 percent of the total land area of the nation. So there's no need to pave over the whole country, just the state of Maryland.

Warmth from the Earth. Why bother burning coal to make steam when you can just drill a hole in the ground, and let the steam come whistling out? This is the idea behind a geothermal power plant. It's a great idea. It dispenses with the whole fuel supply and the furnace and the boiler. The trouble is, it only works at a few places in the world—rare hot spots where the heat of the deep earth bubbles up unusually close to the surface.

The simplest geothermal plants take steam straight from the well and, after minimal processing to remove a few impurities, pipe it to the input port of a turbine to generate electricity. In the earliest plants, the spent steam leaving the turbine was just vented to the atmosphere. That's no longer done, for two reasons. First, the steam carries some obnoxious contaminants, chiefly hydrogen sulfide, that could make a geothermal plant a worse polluter than a coal-fired one. Second, it turns out that the supply of underground steam is far from inexhaustible. To keep it flowing, you have to recycle it, pumping water down recharge wells to prevent the reservoir from drying out. Recapturing the spent steam requires a lot of additional equipment: a condenser, pumps and valves, and cooling towers to get rid of excess heat. The cooling

The A. W. Hoch geothermal power plant *(right),* near Calipatria in the Imperial Valley of southern California, generates electricity from steam and hot brine brought up from several thousand feet below the Earth's surface. The fluids come out of the ground at a temperature of 450 to 500 degrees Fahrenheit. One of the insulated pipelines that carry well fluids can be seen in the foreground. The remaining equipment is needed to separate steam from water and to remove salts and other impurities. At one of the nearby wells *(above)* the pressure gauge reads 110 pounds per square inch.

Geothermal steam at higher temperatures allows a simpler energy cycle at the Geysers in northern California. This "dry steam" can be piped to a turbine with little preprocessing; then it is condensed and reinjected into the earth.

towers tend to be the largest component of the entire plant, and their plumes of rising vapor make them even more conspicuous.

Piping steam from the ground directly into a turbine is a plan that works only if the steam is superheated to a few hundred degrees above the boiling point. It's called dry steam, since it's too hot for water to condense in pipelines or holding tanks. Unfortunately, few geothermal sites produce enough dry steam for commercial power production. One of these areas is Larderello, near Pisa in northern Italy, where power production began almost 100 years ago. The one spot in the United States where dry steam comes out of the ground is the Geysers, a landscape of hot springs and hissing fumaroles tucked among the wine-growing valleys north of San Francisco. At the Geysers geothermal power plants have been running, off and on, since the 1960s.

Although dry steam is a scarce resource, many areas of the world have enough subterranean heat to produce large quantities of mixed hot water and steam. But getting useful power out of these lower-temperature fluids calls for more elaborate machinery. Steam and water have to be separated, and then some of the hot water can be persuaded to vaporize by lowering the pressure in a device called a flasher. In another type of plant, heat from the geothermal fluids is used to boil a more volatile liquid such as butane or ammonia that circulates through a turbine in a closed cycle.

In any geothermal area, one thing you're sure to notice is a network of pipelines that feed steam or water from widely scattered wells to the central power plant. The pipes are often of large diameter, and they look even fatter because of a thick blanket of insulation. At intervals there are big inchworm-like loops to allow for expansion and contraction as the temperature of the pipe changes.

THE POWER GRID

ELECTRICITY IS INVISIBLE STUFF, and yet it makes quite a mark on the modern landscape. The power plants that generate it, the transmission lines that carry it across the countryside, and the substations that dispatch it are all giant structures, hard to hide from view. The smaller-scale wires, utility poles, transformers, and other hardware that distribute electricity to homes and businesses are less impressive individually, but they are seen everywhere on city streets and rural roadways. In built-up areas, it's hard to make a landscape photograph that doesn't take in a power line. The whole country hums with the 60-cycle-per-second note of the electric power grid.

Generating plants are described in the preceding chapter. This chapter covers everything else electrical: power lines, substations, and the local-distribution network.

SOME ELECTRIFYING CONCEPTS

The network that brings power to a nation is not fundamentally different from the electrical system in your home. Wires carry electricity from city to city, just as they distribute it to lamps and outlets. The national network has switches, circuit breakers, and fuses that work much the same as those in your basement fuse box. They're just bigger.

The basic unit of all electrical technology is the circuit. Current flows out from a source (such as a generator), along a conductor to a load (such as a motor or a light bulb), and then along another conductor back to the source. Without the complete round-trip path, nothing happens; breaking the circuit open at any point stops the current. (The opposite of an open circuit is a short circuit, when the outbound and inbound conductors touch, bypassing the load: Pffft!)

The simplest electrical circuits operate on direct current, or DC, in which electricity flows steadily in one direction around a loop. Direct current is what comes

Fluted porcelain insulators, shiny aluminum conductors, posts and frames that lift all the machinery high overhead—high-voltage electrical gear has a characteristic look to it that could never be mistaken for anything else. The equipment on the opposite page is in the switchyard of the Salem Creek nuclear power plant in southern New Jersey. The large devices in the middle of the image are air-blast circuit-breaker switches, designed to interrupt a 500,000-volt circuit in emergencies. The large curved structure in the background is a cooling tower.

direct current

single-phase alternating current

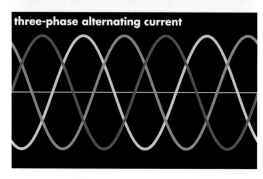

three-phase alternating current

Direct current provides a steady flow of electricity, but in an alternating-current circuit the voltage and current rise and fall rhythmically. Household electricity is single phase, with just one sinuous wave, but power is generated and transmitted with three overlapping waves in a three-phase system.

out of a flashlight battery. The world's power systems are based on alternating current, or AC; the current flows first clockwise around the loop and then counterclockwise, reversing direction many times each second. Nothing ever moves very far in an AC circuit; the current just sloshes back and forth. In the United States and Canada the current completes 60 cycles per second; the power frequency is said to be 60 hertz. In Europe and most of the rest of the world the standard frequency is 50 hertz. (Japan is split down the middle, with both 50-hertz and 60-hertz areas.)

A graph of the voltage in an AC circuit looks something like an ocean wave, with peaks and troughs. From a positive peak, the voltage falls off smoothly, passes through zero, reaches a negative peak, then climbs through zero again to return to the positive peak. The intermittent nature of alternating current has a remarkable consequence: twice in every cycle, there is no current—everything stops. In other words, 120 times a second, the electricity you pay for is slacking off—just sitting there and doing nothing. And the power company is no happier about this situation than you are. Because of all the time spent idling, a simple AC system uses only about 70 percent of the capacity of the wires and other equipment.

The remedy for this waste is three-phase power. Think of three ocean waves following one another so closely they overlap. One wave reaches its crest, and then before its trough arrives, the second wave crests, and then the third. In a three-phase circuit, whenever one phase is at zero voltage, the other two phases are still carrying power, and the system is never idle. Today virtually all commercial electric power is three-phase. Industrial equipment (such as the big motors that drive elevators and large water pumps) is built to run directly on a three-phase circuit. For other uses (including household consumption), a single-phase loop is split off from the three-phase signal simply by forming a circuit between conductors carrying any two of the three phases.

The widespread use of three-phase circuits gives the electric-power infrastructure one of its most distinctive features: just about everything is done in triplicate. A power transmission line is made up of three conductors. At a substation, switches or circuit breakers stand in banks of three. On utility poles, trios of transformers are ganged together. When you start looking around at electrical gear, you begin noticing that almost everything comes in threes.

TRANSMISSION LINES

Steel towers march single file across the countryside, with heavy cables draped from their shoulders. If you follow the line of towers, at one end you are likely to find a generating plant; in the other direction the line will probably lead you to a substation, or switchyard, on the outskirts of a city. Power-company engineers call the transmission lines "feeders," and the energy they supply is indeed a form of nourishment for the way we live today.

The Grid. Hundreds of transmission lines lace together to form a network spanning all but the remotest territories of the North American continent. It's known as the grid. Power flows freely throughout most of this network, finding its own best path from source to consumer. Thus, you can never say for certain where the electricity comes from when you flip a light switch or plug in the hair drier. In Los Angeles a major share of your power comes from the river gorges of Washington State and the coal beds of Wyoming; New York City draws on the hydroelectric resources of Niagara and northern Quebec.

Having many sources of power and many pathways for it to follow improves reliability. If one generating station has to shut down, another can pick up the load. If a transmission line fails, the power it was carrying is instantly diverted onto other feeders. The far-flung grid can also have economic benefits. Utility companies can buy power from the cheapest source, even if it is thousands of miles away, and generating plants in remote areas can sell surplus power to distant cities.

The North American grid is divided into two large zones, the eastern and the western interconnections, which meet at a boundary line that runs along the eastern flank of the Rocky Mountains. Two smaller regions are relatively isolated from the eastern and western networks: Quebec and most of Texas. Inside any one of the four zones, the power system can be looked on as a single gigantic machine. Every generator connected to the grid turns in perfect lockstep with all the other generators, as if they were all attached to the same rotating shaft. Once upon a time, the motors in electric clocks also turned in lockstep with the power system, so that you could

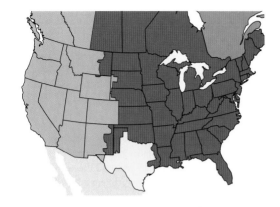

The geography of power: the electrical network in the United States and Canada is broken up into four main regions, called interconnections. The eastern and western interconnections cover most of the territory, but parts of Texas and Quebec are off on their own. Inside each region, electricity flows freely from whoever has it to whoever needs it. Transfers of power across boundaries between regions are actively controlled.

VOLTS, AMPS, AND WATT-NOT

One way to understand electricity is to think about plumbing. Electrons flowing through a wire are like water flowing through a pipe, but the water is something you can see and hear and feel, so its behavior is a little less mysterious.

In the plumbing analogy, voltage is the electrical equivalent of water pressure; it is the force that drives electrons through a wire, measured in volts. Current in an electrical circuit corresponds to the rate of flow in a pipe. In a plumbing system, flow might be measured in gallons per minute; the electrical unit is the ampere, or amp.

Resistance is what opposes the current and dissipates the voltage. When water flows through a pipe, friction slows it and reduces its pressure; the same thing happens to electrons flowing through a wire. In general, the thicker the wire or pipe, the lower the resistance. The unit of electrical resistance is the ohm.

The power passing through an electrical circuit depends on both the voltage and the current. Think about water turning a turbine. A small volume of water under high pressure produces the same amount of power as lots of water at low pressure. Likewise, in the electrical world, high voltage and low current produce the same amount of power as low voltage and high current. The unit of power is the watt.

The volts, amps, ohms, and watts in a circuit are all related; you can't change one without affecting the others. The most important relation is called Ohm's law. It says that current increases with voltage and decreases with resistance. Mathematically, the law is $I = E/R$, where I is the current, E is the voltage, and R is the resistance. In other words, 100 volts across a resistance of 50 ohms produces a current of 2 amperes. The formula for power is $W = E \times I$: the power is equal to the voltage multiplied by the current. Thus, 100 volts and 2 amperes yields 200 watts.

All of these units of measure are named for people—Alessandro Volta, André-Marie Ampère, Georg Ohm, and James Watt. There are a few more electrical eponyms as well. The unit of capacitance is the farad, after Michael Faraday; inductance is measured in henries, for Joseph Henry; and frequency is expressed in hertz, named for Heinrich Hertz.

have on your kitchen wall a tiny rotating device with a direct, second-by-second link to the massive turbines at Niagara or Hoover Dam. (Today, motor-driven electric clocks are antiques. Modern clocks have a digital mechanism, and their timekeeping doesn't depend on the power-line frequency.)

Ties between the four North American zones are looser than the dense network of connections within any one zone. Power is transferred across the boundaries by direct-current links. This allows closer control of the transfers and eliminates the need to keep generators synchronized all the way from coast to coast.

High Voltage. The most important thing about a transmission line is the voltage it carries. It is the voltage that determines the height of the towers, the size of the insu-

V-shaped towers, held upright by guy wires, carry one of the largest power transmission lines in the United States, operating at 765 kilovolts. Branches of the transmission line extend from Michigan and Indiana into southern Virginia; this photograph was made near Ashland, Kentucky.

lators, the width of the right-of-way, and much else. Short-haul transmission lines—up to 40 or 50 miles—often operate at 115,000 or 138,000 volts (more conveniently expressed as 115 or 138 kilovolts). For longer lines the favored voltages vary from region to region. Along the West Coast, in the Mid-Atlantic states, and in the Southeast, the common voltages are 230 and 500 kilovolts. In New England, New York, and the Midwest, most long feeders run at 345 kilovolts, with a few high-capacity lines operated at 735 or 765 kilovolts.

You might think of 115- and 138-kilovolt feeders as the two-lane rural roads of the electric-power system. The 230- and 345-kilovolt lines are like major U.S. highways, and the 500- and 765-kilovolt lines are Interstate routes.

Why are the voltages so high? Electricity comes out of the generator at 10,000 or 20,000 volts, and it reaches your household light socket at 120 volts. So why go to the trouble of boosting the voltage to such a high level, if you then have to reduce it again at the end of the line? The answer lies in the way voltage, current, and power are related. To push more electricity through a power line, you can either increase the current or increase the voltage. But higher currents heat the conductors, just as they heat the filament in a light bulb or a toaster. For any given conductor, there is a maximum safe current. Once you reach that point, the only way to transmit more power is to keep the current constant and go to higher voltages. At the same current, a 345-kilovolt feeder carries more than 6 times as much power as a 138-kilovolt line, and a 765-kilovolt transmission line carries 30 times as much.

Using higher voltages instead of higher currents reduces the waste of energy that goes into heating the transmission-line conductors. A 345-kilovolt line that carries 1,000 megawatts of power might lose 20 megawatts, or 2 percent, in overcoming the electrical resistance of the conductors. In other words, the efficiency is 98 percent, which is pretty good compared with other parts of the power system. Still, the wasted 20 megawatts is enough to run about 40,000 toasters.

If kilovolts is good, maybe megavolts would be better? Why not just keep raising the voltage until you cut the resistive losses to almost nothing? For one thing, another kind of energy loss, called corona, becomes more troublesome as the voltage goes up. But there is an even more important constraint. Higher voltages require taller towers, bigger insulators, and a wider swath of land. All of these factors increase the cost of building the line, and at some point they outweigh any possible energy savings.

How can you judge the voltage of a transmission line from its appearance? When I was a kid, the neighborhood lore was that you just had to count the segments of the insulators, and then multiply by a certain number; the result would be the voltage. But no one could tell me the multiplier. And no wonder. There is no simple and foolproof method to determine the voltage of a transmission line just by looking at it. In particular, lines with the same voltage can have insulators with very different numbers of segments, depending on the design of the insulators.

Still, there are a few things you can look for to make a rough guess about a line's voltage. The best indicator is the distance between the conductors. If the conductors

are less than 15 or 20 feet apart, the feeder probably operates at 200 kilovolts or less. Most 345-kilovolt transmission lines have conductor-to-conductor separation of about 25 feet. The giant 765-kilovolt lines are easy to spot because the distance between conductors is more than 50 feet. (To make a rough measurement of the separation, pace off the distance between the shadows of the wires. You may need to correct for the angle of the sun.)

Another clue to a feeder's voltage is the use of "bundled" conductors. At voltages below 230 kilovolts, each phase of the circuit is almost always carried by a single conductor, but above 345 kilovolts each phase has a bundle of two, three, or even four wires held together by spacers. Circuits at 230 and 345 kilovolts can use either single or bundled conductors.

Transmission Towers. The transmission-line tower everybody knows is an Erector Set latticework of steel girders and diagonal braces. The techniques for designing and building these towers are the same ones used in constructing steel bridge trusses or crane booms. The individual pieces can be made cheaply from rolled steel and then bolted together on the site. This last point is more important than it might seem: transporting a fully assembled tower 100 feet tall is an awkward and expensive business.

Although the steel-lattice tower has long been favored by utility companies in most parts of the country, it is also one of the least-loved objects on the industrial landscape. (Telling people it is built on the same principles as the Eiffel Tower doesn't seem to change their opinion.) And there are lots of other designs. On a Sunday drive you might spot a dozen species.

Many of the alternative towers are single-pylon designs, much like modern street-lighting poles but larger. The pylon is typically four or five feet in diameter at the base and tapers to half that thickness at the pinnacle. The conductors are hung from "boughs" that branch from the mast near the top. Newer single-pylon towers are constructed from hollow steel sections 15 or 20 feet long that fit together like the sections of a fishing pole.

A pylon generally has a smaller footprint than a steel-lattice tower, and so the monopole designs are favored in built-up areas, where the cost of land is high. Out in the countryside, on the other hand, the compactness of the monopole can be a disadvantage. A pylon design requires a deeper and more massive foundation, which means more concrete has to be trucked to the site. This not only raises costs but also does more damage to farmland.

Another drawback of the single-pylon tower is that the conductors generally must be arranged one above the other. Because safety considerations (and government standards) specify the minimum height of the *lowest* conductor, a tower that carries three conductors in a vertical row has to be taller than one with the same three conductors arrayed horizontally. One way of achieving the horizontal layout is with a double-pylon tower, made up of two poles (usually wood) set a few yards apart. The conductors are suspended from a crossbeam near the top.

A gallery of transmission-line towers suggests the variety of forms that can be adapted to the task of holding high-voltage conductors aloft. From left to right and from top to bottom: The red, white, and green steel-lattice tower is in Redipuglia, Italy; the single pylon is in Raleigh, North Carolina; the two-legged colossus carries a 500-kilovolt feeder near Dixon, California; both the X-shaped and the Y-shaped towers are at New Hope, Pennsylvania; and the wood goalpost is in the southern California desert east of Glamis.

Since the basic function of a transmission-line tower is really no more complex than that of a pole used for stringing clothesline—it just has to keep the conductors from touching the ground or one another—many shapes and materials can be adapted to the purpose. There are giant bobby pins and tripods, wishbones and hangman's derricks. Much of the alphabet has been exploited: there are towers in the form of the letters *A, H, I, T, X* and *Y*, as well as in the shape of the Greek letter π.

Often two feeders—a total of six conductors—will be strung on the same line of towers. Doubling up in this way has obvious money-saving potential—towers that carry two feeders cost little more than those that carry one—but it also means that a single lightning strike could knock out both feeders. Sometimes multiple lines of towers run side by side along the same swath of land, presumably because it's easier to assemble one wide power-line corridor than several narrow ones. North of Buffalo, New York, where several power lines from Canada join those coming from the American side of Niagara Falls, nine separate lines of towers run in parallel for a few miles, carrying a total of 16 feeders. It's quite a forest of steel and aluminum.

Not all the towers along a transmission line are identical. Look closely at a tower where the line makes a sharp turn and you will likely find it is wider and beefier than other towers along the route. The added strength and weight are needed to resist the unbalanced pull of the conductors, which might overturn an ordinary tower. These special towers are called deviation or angle towers. Also, at each end of a transmission line there is a heavy-duty termination tower, which serves as an anchor against the tremendous tension in the conductors. And sometimes unusually tall towers are installed where the line crosses a river.

From the air, the pathway of a transmission line stands out as a series of straight lines and abrupt turns, very different from the sweeping bends of roads and railroads. Electricity has no trouble turning sharp corners or climbing steep hills.

If you're driving along a highway parallel to a power line, here's something to watch for: the transposition of conductors, where two of the three conductors exchange places. When three conductors are arranged in a row, the one in the middle has a somewhat different electrical environment; it "feels" the presence of the two flanking conductors, whereas those on the outside each have only one neighbor. On a long feeder, this imbalance can distort the flow of power. The cure is to braid the conductors so that each one is in the middle for about a third of the route. The transpositions usually require a tower with a design different from the rest.

When an engineer lays out a transmission line, one of the basic starting points for the design is the minimum height of the conductors. For a 345-kilovolt feeder, the lowest conductor has to be kept at least 20 feet above the ground; for a 765-kilovolt line the minimum height is 33 feet. The lowest point is usually in the middle of the span between towers. To increase the height of the conductors, the engineer might choose taller towers, or else the towers could be placed closer together, so the conductors don't sag as much at midspan. Another way to reduce sag is to stretch the conductors tighter, but then the conductors, the insulators, and the towers have to be

made stronger to withstand the force. All of these options entail some expense, and in practice the designer seeks the lowest-cost compromise. There are also aesthetic compromises: which looks better—fewer tall towers or more short ones?

And the engineer faces yet another complication: the midspan sag is not a constant. As the conductors heat up (both from hot weather and from the current flowing through them), they expand and droop; as they cool, they contract and stretch tighter. To compensate for these changes, the power rating of a feeder depends on the

A brawny "deviation tower" absorbs the unbalanced forces where a transmission line makes a sharp turn. In this case the deviation tower has a design totally different from that of the rest of the towers along the route. The transmission line, which operates at 500 kilovolts, carries the output of the Mayo generating plant in North Carolina.

Fat cigars in the middle of a conductor are splicing sleeves, holding together two pieces of wire.

Dumbbells hanging from the underside of power conductors *(below)* are installed to absorb wind-induced vibrations that might weaken and fray the wires. This design is known as the Stockbridge damper. Another device with the same function *(bottom)* looks like a snake crawling out onto the wire.

weather. A transmission line with a normal rating of 1,800 megawatts can carry 2,200 megawatts in cold weather but has a limit of 1,580 megawatts in hot weather. The hot-weather "derating" is unfortunate, since the capacity is lowest just when demand is greatest, as everyone turns on the air conditioner.

Conductors. The conductors are the part of a power transmission line that most of us would call wires (although that word seems to be little used by insiders). The job of the conductors is to carry the electric current, and so they should be made of something with very low resistance. Among all substances, the very best conductor is silver, but that is not a very practical choice. The next-best metal, copper, was once the standard material for transmission-line conductors and is still seen in older low-voltage distribution lines and in household wiring. All modern high-voltage transmission lines, however, use aluminum conductors. Aluminum is only about 60 percent as conductive as copper, but it is much lighter and cheaper, and that makes all the difference. To compensate for the lower conductivity, the wires are simply made thicker.

In most household wiring, the conductors are about as thick as a pencil lead. Even the heavy-duty wires that supply a clothes dryer or an electric stove are no bigger around than a crayon. The conductors of a high-voltage feeder can be as thick as a baseball bat. They are made up of many strands of aluminum twisted together, as in a hemp rope. For example, one type of conductor has 61 strands, each about an eighth of an inch in diameter; the complete conductor is an inch and a third thick, has a rated current-carrying capacity of more than 1,100 amperes, and weighs about a pound per foot. Curiously, the various sizes and types of aluminum conductors are named for flowers. The 61-strand example is known as narcissus; there are about 50 other standard types, with names such as sneezewort, valerian, snapdragon, and lupine. The biggest standard size is bluebonnet, which is more than two inches in diameter and can safely carry more than 2,000 amperes of current.

Some transmission-line conductors consist of aluminum strands wrapped around a steel core. Steel is only a mediocre conductor, but it adds mechanical strength. The steel-core conductors are named for birds rather than flowers. The type designated starling, for example, has 26 aluminum strands surrounding seven finer steel strands.

The strength of modern transmission-line conductors was demonstrated dramatically in an ice storm that hit New England and Quebec in the last few days of 1997. In several places the ice-loaded conductors didn't break; instead, they pulled down the steel towers that supported them. This was not the intended mode of failure: rebuilding the towers takes much longer than resplicing a conductor.

On the highest-voltage transmission lines, each phase is carried not by a single conductor but by a bundle of two or three or four subconductors. The reasons for this arrangement are explained later in the discussion of corona discharge. Many 345-kilovolt feeders have two subconductors per phase, separated by spacers about a foot long, so the bundle looks like a spindly ladder with widely spaced rungs. At 500 kilovolts and above, most transmission lines have three or four subconductors per phase.

Conductors for power lines are shipped on giant wooden spools that hold as much as a few miles of wire—but not nearly enough for a long feeder. Where two lengths of wire have to be joined, the splice is not made by twisting the conductors together, the way you might splice household wiring. Instead, the two ends are slipped into an aluminum sleeve, which is then compressed so it tightly grips both conductors. Sometimes the sleeve is compressed with a powerful hydraulic ram; in other cases the compressive force comes from a specially shaped explosive charge, which is set off on the ground before the conductor is raised into position. In either case the resulting splice is easy to spot: it is a cigar-shaped enlargement, making the conductor look like the snake that swallowed the pig.

The conductors of a high-voltage transmission line are bare metal; they have no insulating sheath of rubber or plastic. Adding such a layer would accomplish nothing. If you were to grab hold of a live 345-kilovolt conductor, the thin coating that insulates a lamp cord would not protect you; the voltage would instantly "punch through" the layer. Bare aluminum conductors glint brightly in the sun when they are new, but later they grow a layer of oxide that turns them dull gray.

What are those little dumbbell-like objects attached to the conductors near each supporting tower on some feeders? I wondered about them for years before I learned the answer. They are vibration dampers. In a steady wind, the tightly stretched conductors of a transmission line can begin to vibrate like violin strings (or perhaps more like the strings of a double bass—their natural note is a very deep one). The oscillations cause stress and fatigue in the metal of the conductors and can even shake bolts loose in the supporting towers. Dampers suppress the vibrations much like a thumb held lightly against a violin string. The common dumbbell design is known as a Stockbridge damper, after George Stockbridge, an engineer with Southern California Edison who came up with the device in the 1920s. There are many other styles, including one that wraps a long spiral tube around the conductor—it looks just like a snake winding around the line.

Armor bars are another defensive measure against damage from wind-induced vibrations. Actually they are not bars but reinforcing wires wrapped around the conductor at the point where it is clamped to an insulator. From the ground what you will notice is a bulge in the conductor at each tower.

Still another curiosity you might spot is a series of brightly colored spheres, the size of beach balls, attached to a power line. They are often seen near airports, where they make the conductors more visible to the pilots of low-flying aircraft. At river crossings, similar devices alert the skippers of tall-masted sailboats. In Europe neon tubes and fluorescent bulbs have been hung from power lines to warn off swans and other migratory birds that fly at night. The lights draw their energy from the electric field surrounding the conductor. (I'm surprised no one has thought of exploiting this effect for advertising.)

In the American West, bird problems take another form: eagles like to survey their territory from atop transmission towers, and with their tremendous wingspan can

Armor bars *(above)*, which are extra strands of wire wrapped around a conductor where it is clamped to an insulator, are another mechanism for dealing with fatigue and vibrations. Colorful globes attached to a power line where it crosses a river *(below)* have a quite different function. A bystander guessed that they might be meant to keep the wire afloat if it should fall in the water. In fact, they are a warning to pilots of tall-masted boats and low-flying aircraft.

Double, triple, and quadruple bundles of conductors help to control the sizzling of corona discharge, which becomes more of a problem as the voltage gets higher. From the point of view of an overexcited electron trying to escape, the bundle acts almost like one big conductor.

sometimes bridge the gap between two conductors. Rather than try to drive the raptors away, some utilities have built platforms well above the conductors, where the birds can perch in safety.

Corona. Near the surface of a high-voltage conductor, the electric field can be so intense that molecules of the air are torn apart. Electrons, which have a negative electric charge, are stripped from molecules of nitrogen and oxygen in the air, leaving behind positively charged fragments called ions. When the electrons and the ions recombine, they emit light and radio waves. The light gives the corona its name: *corona* is Greek for "crown" or "wreath," and the discharge takes the form of a faint blue or lavender halo around the conductor. St. Elmo's fire on the mast of a ship is the same phenomenon induced by natural electrical activity in the atmosphere.

To see a corona discharge, find a spot along a high-voltage line well away from streetlights. The ideal conditions are a moonless night with a soft rain or mist. Let your eyes adapt to the dark, then look along the underside of the conductors. The glow is usually fuzzy or diffuse and may be mottled. Sometimes the discharge is noticeably stronger at the clamps where the conductors are held fast to the insulators. Even in daylight, when you can't see the corona, you may be able to hear it. The characteristic note is a hissing or sizzling, like bacon frying, and it may be intermittent. You can also hear the radio-frequency emissions associated with corona: just tune your car radio to the AM band and drive under a high-voltage line. That harsh buzz is the mating cry of atmospheric electrons and ions.

Corona becomes a problem only at very high voltages. At less than 100 kilovolts or so, the discharge is undetectable, but at 345 kilovolts and above, it can waste megawatts of power. Furthermore, the radio emissions—and sometimes even the audible noise—are subject to regulations. And corona generates small quantities of ozone and nitrogen oxides, which are smog-causing pollutants that also attract the attention of regulators. Thus, power companies work hard to minimize corona effects.

The intensity of a corona discharge depends on the shape of the emitting object. Sharp points or edges are strong corona radiators because the electric field has to bend and stretch around them. Accordingly, high-voltage hardware tends to be designed with gentle curves, rounded corners, and smooth surfaces. But nature can spoil the designer's efforts. When a water droplet clings to the underside of a conductor, the electric field distorts the droplet into a sharp spike, shaped like a rose thorn, which becomes an excellent corona source. This is why corona is so much worse in wet weather.

One way to reduce corona emissions is to make the conductor as thick as possible, so that the surface is more gently curved. Expanded conductors have a mat of light, nonmetallic fibers between a steel core and the current-carrying aluminum strands; the mat makes the conductor thicker without greatly increasing its weight or cost. Bundled conductors, as described earlier, have the same purpose. When two, three, or four subconductors run parallel to one another a foot or two apart, the elec-

Three strings of insulators *(left)* are rigged in parallel to sustain the weight and the tension of a transmission line near Three Mile Island in Pennsylvania. Each string consists of 22 disks of translucent green glass the size of salad bowls. The ring at the "hot" or high-voltage end of the insulator helps to suppress corona discharge.

tric fields of the subconductors combine and overlap so that the bundle acts much like a single fat conductor, almost as big as the entire bundle. Corona emissions are greatly reduced.

The troublesome corona hot spots where conductors are clamped to insulators are sometimes protected by corona rings, or shields. These smooth metal rings look just like great handles attached to the end of the insulator, as if they might be used for maneuvering it into place. That is definitely not their function. They work much like a bundled conductor to spread out the electric field and minimize corona losses.

Incidentally, because of corona you will not see birds perching on the energized conductors of a high-voltage transmission line. A bird would not receive a shock from the wire, as long as it didn't touch something grounded at the same time, but evidently the sizzling discharge makes the wire an uncomfortable or unpleasant place to sit. Birds do perch on the aerial ground wires strung above the active conductors.

Insulators. Whereas conductors are meant to carry electricity as efficiently as possible, insulators have the opposite job: blocking the flow of current. They are made of porcelain, glass, or plastic—materials that present an impenetrable barrier to electricity.

The most common insulator on high-voltage lines is an assembly of porcelain segments, called disks, linked together into an insulator string, which looks like some gigantic marine worm, something that might have washed up on a prehistoric beach. One end of the insulator string is hung from the transmission-line tower; the other end supports the conductor. The segments in the string are held together by metal balls and sockets that interlock like the links of a keychain, allowing the assembly to flex as the conductors move in the wind.

A good insulator is one that doesn't leak—one that lets no electricity escape from the high-voltage conductors to the supporting tower and the earth. As a rule, leakage through the body of a glass or porcelain insulator is not a problem; it seldom happens. But sometimes there is troublesome leakage along the surface of an insulator, especially when it is wet or dirty. Designers combat this problem by making the surface smooth and slick so contaminants won't stick to it, and by careful attention to the shape of the insulator. To make the leakage path over the surface—known as the

Conductors can be suspended from a single string of insulators, from a V-string, or anchored to the tower in a "dead-end" arrangement.

creep path—as long as possible, most insulator disks have rings, ridges, or crenelations so that any stray current has to take a circuitous route. The problem of leakage is worst in industrial areas, where soot and grime coat the insulators, and along the seacoast, where deposits of salt build up. Power companies in these places sometimes grease the insulators or else wash them regularly with high-pressure hoses.

AC/DC

One of the great pitched battles in American industrial history was the fight over alternating and direct current. It was bigger than Coke vs. Pepsi, or Ford vs. Chevy. The DC forces were generaled by Thomas Edison, with the AC insurgents commanded by George Westinghouse. Each side argued that its own system was better for everything—with one exception: they both conceded that the other kind of current was more deadly and thus would be better for electrocuting prisoners.

In the end AC won total victory—even for the electric chair. But in recent decades DC

has made a surprise comeback in one area. Some of the longest, highest-capacity power transmission lines carry direct current.

It's easy to recognize a high-voltage DC transmission line. There are just two conductors, whereas all AC lines have three conductors. A DC line operating at 500 kilovolts runs 846 miles from The Dalles in northern Oregon to Sylmar, California, in the suburbs north of Los Angeles. During summer peaks in electricity consumption it carries more than 3,000 megawatts. Another major DC link extends from the hydroelectric plants of northern Quebec to Waltham, Massachusetts. Still more DC ties have been built in Manitoba, North Dakota, and Minnesota, and along a route from Utah to Los Angeles.

The reason DC lost favor a century ago is that you can't build DC transformers to change the voltage level. So how do DC transmission lines work? The power is generated as AC and raised to high voltage by conventional transformers, then it is converted to DC for long-distance transport. At the far end of the line, another station converts the current back to AC again.

The first AC/DC converter stations, built in the 1960s, were based on a device called a mercury-arc valve. It looks and works much like a vacuum tube in an old radio, except that it stands three feet tall, handles 1,000 amperes or more and 100,000 volts, and needs a constant flow of water to keep from burning up. In more recent stations, mercury-arc valves have been replaced by a solid-state

device called a thyristor (just as transistors long ago replaced vacuum tubes in radios). The thyristors are smaller than mercury valves, but they're still enormous as semiconductors go—the same silicon "wafer" that might hold a few hundred microprocessor chips, each with a million transistors, makes a single thyristor for power-grid duty.

Unlike most substation equipment, AC/DC converters have to be kept out of the weather, which means you can't see them from outdoors. The Celilo Converter Station, at the northern end of the Pacific DC Intertie, has a visitor center where you can look into one of the enclosed galleries filled with mercury-arc valves. At Celilo you can also stroll around outside the fenced yards of conventional transformers and switchgear. Some of this equipment even has helpful labels, just like the plaques that identify trees and shrubs in a formal garden or arboretum. (The Celilo station, operated by the Bonneville Power Administration, is on a hill above the dam at The Dalles on the Columbia River.)

Why all this bother to convert from AC to DC and back to AC again? DC transmission has a number of advantages. For a given conductor size, resistive losses are lower. And only two conductors are needed; on a long intertie, this is not a small cost savings. The number of insulators is reduced as well, and since the towers have less weight to support, they can be built lighter. Perhaps even more important, two conductors take less space than three, allowing the use of a narrower right-of-

Often, each conductor is hung from a single insulator string at each tower. In other cases you will notice two insulators per conductor, arranged in a V shape. The extra string is not there to provide more electrical insulation; as a matter of fact, it doubles the leakage current. The purpose of the V-string is mechanical: it prevents the conductor from swinging from side to side. At some towers the line must be

way. And DC transmission lines are sometimes easier to control. With an AC line, the only way to regulate the flow of power is to change the throttle setting at a generating station. With DC, it's just a matter of turning a dial that controls the mercury valves or thyristors.

Still another advantage of DC is that there's no need to synchronize the distant power systems. (With AC, all the interconnected generators must turn in lockstep.) For this reason alone, several utilities have built back-to-back converter stations, where an AC-DC-AC link transfers energy between two nearby power systems that do not run synchronously. All the links along the east-west divide in the United States work this way, and so do many stations along a similar boundary in Europe.

Along a high-voltage DC transmission line you may be able to figure out which conductor carries a positive charge and which is negative. Corona loss is much greater from the positive-polarity conductor. On a dark night, if the corona discharge is visible, the negative conductor will have just a uniform glow, but the positive conductor will give out plumes and streamers. An AM radio may also be able to detect the difference; the positive side produces a stronger rasp of interference. You may even be able to hear the difference in ordinary audible noise. DC transmission lines sound quite different from AC ones. They click and crackle rather than buzz; the DC line sounds just like a Geiger counter. And when you walk under the conductors, the pace of the clicking accelerates, as if you were

radioactive. Your own body is disturbing the electric field around the wires overhead.

The electric fields on the ground near DC conductors are also different from those associated with AC transmission. With DC there is a sustained current of ions flowing from the conductors to the ground—a slow rain of slowly moving charged particles. Standing under the line, you intercept some of this charge. If you are well insulated—wearing

sneakers—and you wait to accumulate a charge, you'll draw a spark when you touch something grounded. Those who regularly work near high-voltage DC equipment report that they can feel the charge: it repels the fine hairs on the nape of the neck and the tips of the ears, producing a "crawling" sensation. It is the same feeling you get when you place the back of your hand near the television screen.

The Iron Cross of insulators: rigid struts *(above)* are increasingly popular with power-line builders. The core of the strut insulator is fiberglass, which gives it the strength to hold up the conductors with arms outstretched. The same fiberglass is also turning up in the newest, slimmest suspension insulators *(below)*.

dead-ended: two separate insulator strings take up the full tension of the conductors on each side of the tower, and often a third string is needed to guide the slack loop of conductor around the tower structure. With very heavy lines or long spans, even more elaborate arrangements may be needed.

For a long time most transmission-line insulators were brown or black. Many of the porcelain ones had a handsome red-tinged glaze that would have looked good on a fine china teapot. But fashions have changed. The color of choice for newer insulators (and for much other electrical equipment) is neutral gray, a shade that one manufacturer calls Skytone. The makeover has no functional significance; it's just somebody's idea of what looks best, or what looks least—as the name suggests, Skytone insulators are supposed to blend into the background when seen from below.

The latest thing in high-voltage insulators is the single-piece rigid post or strut. Instead of a string of jointed disks suspended from overhead, a strong insulating rod juts out directly from the tower. The mechanical strength comes from a fiberglass beam at the core of the insulator, but there is often a glazed ceramic covering to create a smooth surface and cut down on leakage.

Lightning and Other Hazards. A steel tower poking 100 feet out of the landscape is an obvious target for lightning. Transmission towers are struck thousands of times a year, in most cases without damage and without even interrupting the flow of power.

The main defense against lightning is to ground the tower—to connect it electrically with the earth. At the base of a steel-lattice tower you will generally find a heavy copper wire bolted or clamped to each leg. (You can tell the wire is copper because of the green copper-oxide patina.) Each ground wire is attached to a copper rod driven deep into the soil. Metal towers of other designs have similar provisions for grounding, and wood poles have ground wires that run from top to bottom.

The ground wire cannot prevent lightning from striking; all it can do is offer the lightning current a direct and harmless route to the ground—and a route that's supposed to be more inviting than the power-line conductors. The ground wire is analogous to a drain pipe that carries excess water away before it can flood your basement.

Grounding the tower protects the tower itself, but what if lightning strikes the transmission line between towers? The power-line conductors themselves cannot be grounded; after all, the entire edifice of towers and insulators is intended to prevent the conductors from coming in contact with the earth. The answer is to provide a kind of electrical umbrella, a metallic shield at ground potential that extends over the power-carrying conductors. The umbrella consists of aerial ground wires that are strung from the pinnacle of one tower to the next, parallel to the main conductors but above them. The aerial grounds are easy to distinguish from current-carrying conductors because they have no strings of insulators. They are also the highest wires on a tower, and they are thinner than the main conductors.

The role of the aerial ground wires is to intercept a lightning stroke before it reaches the power-line conductors. Aerial grounds work best when they are directly

above the conductors. Where the main conductors are in a vertical arrangement, a single aerial ground can be mounted directly over them. With a horizontal configuration the usual practice is to install two aerial ground wires spaced to provide coverage for all three conductors.

The grounding strategy sometimes fails. A powerful lightning bolt can deliver several million volts, which is far beyond the rating of the power-line insulators. The usual result is a flashover, an arc from the tower to one (or more) of the phase conductors. The arc itself can damage insulators and conductors; a more serious worry is that the voltage surge traveling along the conductors can damage generators, transformers, and other equipment at the end of the power line—including, perhaps, the television set you've plugged into the wall outlet. To deal with these perils there are further lines of defense—circuit breakers, lightning arresters, and surge suppressors at substations and throughout the power-distribution system.

At operating voltages up to about 500 kilovolts, lightning is the only likely cause of flashovers. At still higher voltages the power system can become a hazard to itself. The source of this self-destructive behavior is the energy stored in the magnetic field that surrounds a current-carrying conductor. If the current in the line stops suddenly (for example, when a circuit breaker opens), the magnetic field collapses, and the stored energy is transformed into a high voltage. (The principle is the same one that fires an automobile's spark plugs when the breaker points interrupt the current to the ignition coil.) The transient voltages induced in this way can be three times the normal operating voltage of the line. Protecting against them would be expensive: a 765-kilovolt feeder would have to be insulated as if it were operating at two million volts. A better solution is to design the circuit breakers so that they cannot interrupt the current so suddenly.

In recent years there has been much controversy about possible dangers of exposure to the electric and magnetic fields surrounding high-voltage power lines. Certain rare diseases are alleged to be slightly less rare among people who live near transmission lines and among power-company maintenance workers. Personally, these alarms leave me unconvinced, but I'm not a medical expert. In any case, the casual explorer of the industrial landscape—someone who occasionally spends a few minutes near a power line—should have nothing to worry about: the supposed effects of electric and magnetic fields are attributed only to long-term exposure.

The fields in question certainly do exist. If a conductor 35 feet overhead is at 350,000 volts, the voltage gradient—the rate of change—between the conductor and the ground is 10,000 volts per foot. In other words, if you're six feet tall, then when you stand under a transmission line, your head is 60,000 volts above your feet. How can people walk through such an intense field without being electrocuted? Because air is such an effective insulator, the field remains a static one; no electric current flows through the body in spite of the enormous voltage difference. Anyway, power lines are not the only source of such fields. On a dry winter day you can create a field just as intense by rubbing your feet on a wool carpet.

Aerial ground wire (above), at the very top of the tower, provides a perch for pigeons, which avoid the high-voltage conductors lower down. But the aerial ground is not put there for the convenience of birdlife; it protects the power conductors from lightning strikes. Another form of lightning protection (below) relies on an arrester—the slender rod to the right of the insulator string—to divert voltage surges through the tower and safely into the ground. Arresters are common in substations but rare on power lines; this specimen was photographed in Spartanburg, South Carolina.

An electric personality lights up the night: the electric field emanating from a high-voltage transmission line can be made visible with apparatus no more complicated than a fluorescent tube. Electrons are set in motion by the field beneath the conductors and stimulate the phosphors inside the tube to emit light. The power line used for this experiment operates at 500 kilovolts.

You can detect the electric field under a power line with a simple instrument: a fluorescent light bulb. When I first read about this, I was skeptical, but an experiment on a dark night, at a spot where a 500-kilovolt feeder hangs low over a country road, soon made a believer of me. As I approached the conductors, the 18-inch fluorescent tube suddenly came to life, emitting a mottled blue glow, dimmer than the bulb's full normal output but quite easy to see. Running my hand over the glass, I could chase the glow from one end of the tube to the other. Holding the tube near the ground extinguished the light; it reignited at a height of about four feet. When I stood directly under the middle conductor, the light became feeble and irregular, presumably because the fields from the three phases nearly cancel out there. (*Warning*: If you try this experiment, use a short fluorescent tube—no more than two or three feet long. Holding an eight-footer up over your head could too easily cause *you* to light up. And no baton-twirling!)

SUBSTATIONS

A power-system substation functions like the fuse box in your basement. Power enters the substation on large, high-capacity feeders, and it leaves via smaller distribution lines, much as electricity enters your home through a heavy-duty cable (with a capacity of 100 amperes or more) and is distributed in the fuse box to several 15- and 20-ampere circuits. Large circuit breakers at the substation protect the power system from overloads and surges, just as fuses or smaller circuit breakers at home disconnect circuits if something goes wrong. One point where the analogy breaks down is that

the home fuse box has no equivalent of the substation's transformers, which reduce voltages from transmission levels to more manageable distribution levels.

A good place to look for a large substation is on the outskirts of a city. Often two or three major feeders will converge on the station, with multiple lines at lower voltage carrying the power on to smaller substations closer to town. There the voltage is reduced still further, for the many distribution lines that fan out through the city.

Nine massive transformers are lined up at the Celilo substation in northern Oregon.

A substation "bus" is a set of parallel conductors that carry electric current from one component to the next. Here two buses (with a total of six conductors) are formed out of aluminum tubing held aloft on insulated pedestals. The columnar devices at the left are instrument transformers that monitor voltage on the two buses. The substation is in Redipuglia, Italy.

At first glance, a substation is a bewildering array of hulking steel machines whose function is far from obvious. Ponderous tanklike or boxlike objects are lined up in rows. Some of them have cooling fins or fans; many have fluted porcelain insulators poking out in all directions. Overhead is a clutter of metal scaffolding, studded with still more insulators. There are no moving parts to offer a clue to how it all works; and most of the time, there is no one to be seen tending the equipment; the only sign of life is a steady, droning hum.

If you look closer, you will find there is a logic to this melange of equipment. You can make sense of it. The substation has inputs and outputs, and with a little study you can trace the pathways between them.

Catching the Bus. The organizing principle in almost all substations is the idea of a bus, which is a set of three conductors that run parallel to one another, carrying the three phases of the electric current. Incoming and outgoing transmission lines, transformers, breakers, and other equipment are all connected to the bus conductors. Why is it called a bus? Someone tried to tell me once that it was because city buses, like trains, used to have conductors. It was a bad joke, but there *is* a connection: in both of its meanings, bus is an abbreviation of the Latin word *omnibus*, which means "for

SONG OF THE SUBSTATION

America hums at a slightly higher pitch than Europe and most of the rest of the world. The power grid in the United States runs at 60 hertz—or 60 cycles of alternating current every second—whereas Europe is on a slower 50-hertz beat. If you have a good musical ear, you will be able to tell the difference. (The tones you hear are actually double those of the power frequency—120 hertz and 100 hertz. That's because each half-cycle of the power wave causes the transformer core to vibrate.)

Technically, there is no strong reason to prefer either 50 or 60 hertz; it's just one of those annoying nonstandards that makes life a little more difficult for everybody. In the early years of the twentieth century, 50 hertz was proposed as an international standard, and U.S. utilities adopted it along with everybody else. A few years later, though, American manufacturers of electrical equipment persuad-

ed Congress to mandate a switch to 60 hertz, as a way of protecting the home market against European competitors. But a few islands of 50-hertz service survived until after World War II, the largest of them being the city of Los Angeles. Nearly two million electric clocks, record players, washing machines, refrigerators, and other motor-driven appliances in Los Angeles had to be rewired or replaced before the great switchover on October 26, 1948.

Several other frequencies have had their vogue. The first big hydroelectric plants at Niagara Falls produced AC power at 25 hertz, and a few of them on the Canadian side continue supplying customers who have ancient equipment operating at this frequency. Some electrified routes of the New Haven and Pennsylvania railroads also ran on 25 hertz for nearly a century. The trains themselves have converted to 60 hertz, but at last report

there were still sump pumps deep in the bowels of Grand Central Terminal that required a 25-hertz feed. Many European railroads use power at a frequency of 16⅔ hertz. Lower frequencies have advantages for running large motors, but they won't do for electric lighting because people can detect the flickering. Low-frequency power also requires heavier and bulkier transformers; aircraft use 400-hertz power to save weight.

In one respect, the frequencies chosen for the world's power grids couldn't be worse. If you accidentally touch a live wire, your muscles may seize, so that you can't let go. This involuntary contraction is more severe with alternating current because the pulses of current mimic the firing of nerve cells. It so happens that the most dangerous frequencies are near 50 and 60 hertz. These hertz hurt! (Of course this wasn't known at the time the frequencies were chosen.)

all." Just as the vehicle called a bus carries any and all passengers, the electric-power bus carries currents from all sources to all destinations. (Lately, the word has also become familiar to those who tinker with computers; the parallel conductors that carry signals throughout a computer are also called a bus.)

The conductors of a substation bus are usually rigid aluminum tubes, called bus bars, rather than flexible wires. The tubes are hollow and look much like the ones used in cheap lawn-and-beach furniture, although they are larger in diameter. You might mistake them for the kind of metal conduit that encloses electrical wiring in some buildings, but the bus bars do not have wires (or anything else) inside; it is the tubing itself that carries the current. Some utility companies favor square rather than round tubes, or channels with an L-shaped or U-shaped cross section. Occasionally the bus bars are painted bright colors to make them more conspicuous to workers. In older substations you might still find a few solid copper bus bars, recognizable by their distinctive green patina.

A long run of bus bar is usually interrupted every 50 feet or so by a length of flexible copper braid. This allows for thermal expansion and contraction. In earthquake country it also improves the chance that the substation will survive a shaking without major damage.

Holding the bus bars aloft is the purpose of the metal scaffolding, or superstructure, that towers over the rest of the substation equipment. The classic design relies on beams and girders assembled from the same lacy, Erector-Set framework seen in transmission-line towers, but many modern substations use solid stanchions or poles that create less visual clutter. In either case the bus bars are supported by porcelain insulators similar to the ones that carry transmission-line conductors.

The simplest substation design would have a single set of bus bars running the length of the station. High-voltage feeders might be attached to each end of the bus, and several transformers would draw off current from intermediate taps, reducing the voltage for distribution. Unfortunately, this simple design is not workable in practice.

Multiple buses crisscross to form a confusing gridwork over many substations. At a substation in Durham, North Carolina, the steel members extending from left to right merely support the structure; the conductors—set off on insulated struts—mainly go from front to rear.

The reason is that a single transformer failure, or a lightning strike on either of the incoming feeders, would probably shut down the entire system. So would routine maintenance anywhere in the substation. At home this situation may be acceptable: when the toaster blows a fuse, no great harm is done if the refrigerator and the microwave shut down as well. But in a municipal power system, no one fault should knock out an entire substation.

The solution is redundancy. Most substations have two buses, which can be ganged together or isolated from each other as circumstances require. The incoming power can be directed onto either of the buses, much as trains are switched onto various tracks in a rail yard. Similarly, the transformers supplying the low-voltage distribution system can draw their power from either bus. In the event of a mishap, power can be rerouted around a failed device. And if necessary an entire bus can be shut down for maintenance without cutting off power to customers.

Where space is extremely tight, the buses may run directly over the transformers, breakers, and other equipment, but that plan makes access awkward when something has to be moved or replaced. More commonly, the heavy equipment is off to one side of the buses or in a gallery between them, connected by conductors running at right angles to the main bus bars.

Adding to the visual complexity of a substation, sometimes there is a low-voltage bus, where the outputs of several transformers are gathered for connection to the distribution system. The low-voltage bus bars are generally at a lower height and are supported by smaller insulators. Sometimes the low-voltage circuits are carried by underground cables, to keep them out of the way. Often they emerge from the ground onto pole lines at the perimeter of the substation.

Transformers. The jumbo items at a substation—the biggest boxes on the lot, and also the most expensive—are the transformers. Much of the other equipment is there to protect the transformers in case something goes wrong.

Size is the first clue when you try to pick out the transformers amid all the other gear in a substation. Even a small substation transformer is bigger than the family refrigerator, and some of them would not fit inside a two-car garage. Another identifying sign is the presence of both high-voltage and low-voltage connections, with differently sized insulators. Provisions for getting rid of excess heat are still another distinctive feature. The transformer casing may have fins or tubes, or it could have a complete radiator system, with banks of fans turned on during periods of heavy load. Finally, if you can't identify a transformer by sight, you might recognize it by sound: of all the substation equipment, the transformers hum the loudest.

What's inside the big box? As a boy I cracked open a doorbell transformer and was disappointed to find nothing but a wad of varnished wire and a stack of E-shaped metal plates that I never got back together properly. Nothing inside seemed to do anything. That's because there are no moving parts in a transformer. The mechanism relies entirely on the ethereal throbbing and pulsing of invisible magnetic fields.

The main components of a power transformer are two coils of insulated copper wire, like spools of twine, wound on an iron core called a yoke. The coils are usually concentric, with the lower-voltage one slipped inside the higher-voltage coil. The iron core is made of many thin laminations instead of one thick lump of steel. This method of construction reduces the little whirlpools of electric current in the iron—called eddy currents—that would otherwise waste large amounts of power. The flexing of the laminations as they are magnetized and demagnetized 120 times per second is what produces most of the transformer's hum.

Large transformers at Marysville, Ohio, take in power from long-distance 765-kilovolt feeders and deliver it to a 345-kilovolt network. The bank of transformers seems to violate the fundamental rule that elements of the electrical infrastructure come in groups of three. But of the four transformers here, only three are active at any given time; the fourth is a spare. Also note the walls separating the transformers, meant to contain oil fires.

The whole point of a transformer is to change (or transform!) the voltage and current in a power line. The change is governed by the turns ratio, the relative number of turns in the two coils of wire. Suppose the primary (or input) winding has twice as many turns as the secondary (output) winding. Then the output voltage will be half of the input voltage. When the voltage is halved, however, the current is doubled, so no power is lost (except for small inefficiencies). A transformer that reduces voltage is called a step-down transformer. In a step-up transformer, the secondary winding has more turns than the primary one, so the voltage is increased. The switchyard of a generating plant has step-up transformers to boost the voltage for long-distance transmission; substations near cities have step-down transformers to lower the voltage for distribution.

I know of no way to determine the turns ratio of a transformer simply by looking at it from a distance. But you can make a rough guess by comparing the special insulators (called bushings) that poke out of the top or side of the transformer and guide power conductors through the steel casing. The size of a bushing depends on the voltage it has to handle. If all the bushings on a transformer are about the same size, the turns ratio must be near 1; there is not much difference in voltage between primary and secondary windings. If one set of bushings is much bigger than the other, there is a substantial change in voltage.

What I have described so far is a single-phase transformer. It has one primary winding and one secondary winding. For three-phase power, three single-phase transformers can be ganged together; they are usually lined up in a neat row. An alter-

A MIRACULOUS TRANSFORMATION

The idea is called electromagnetic induction, and it's so important it had to be discovered twice—by Michael Faraday in Britain and by Joseph Henry in America.

Today, electricity and magnetism are known to be close cousins, but their kinship was still a secret 150 years ago. It was Faraday and Henry who established the family connection. They found that whenever an electric current flows through a wire, it creates a magnetic field. Likewise, when a magnetic field moves across an electrical conductor, it "induces" a voltage. The two kinds of energy, electrical and magnetic, are convertible, like two different currencies.

The simplest transformer is just two wires near each other. A fluctuating current flowing through one wire produces a fluctuating magnetic field, which induces a voltage and current in the other wire. As a practical matter, to make the transformer more efficient, the two wires are not simply laid down next to each other. They are wound into coils and wrapped around an iron core, which intensifies the magnetic field.

Faraday discovered something else important: only a changing current or a changing magnetic field gives rise to the induction effect; a steady current or a steady field, no matter how strong, doesn't do it. That's why alternating current is essential for making a transformer work. The continually changing current in the primary winding generates a continually changing magnetic field, which in turn induces a continually changing voltage in the secondary winding.

Personally, I find it a source of amazement that transformers work well enough for all the world's power systems to rely on them. Consider what this means: There is no conducting pathway to carry electricity from a power plant to the lamp by your reading table. Wherever the electricity passes through a transformer—and there might well be half a dozen of them between a generating plant and your home—all the electricity must be converted into the wispy energy of a magnetic field, and then instantly converted back to electric current again. It's like a relay race, where every transformer has to pass the baton from one runner to the next.

In Knoxville, Tennessee, a three-phase transformer comes equipped with a fanny-pack for cooling. The higher-voltage, primary current enters through the taller bushings atop the casing at right; the lower, secondary voltage exits through the smaller bushings at left.

native is to install a three-phase transformer, which is essentially three single-phase transformers inside one big housing. There are six coils inside—three primaries and three secondaries—wound on a three-legged iron core. The three-phase transformer is easy to spot because it has six bushings poking out like porcelain horns—three for the primary windings and three for the secondaries.

One three-phase transformer is cheaper than three single-phase ones, and often it is a little more efficient too (wasting less electricity as heat). So why aren't three-phase units always used? One reason is that the very biggest ones are truly gigantic, and transporting them to the site of a substation can be quite a challenge. Also, keeping a spare unit on hand is less costly with single-phase transformers, where only the faulty phase needs to be replaced.

Modern transformers are amazingly efficient. Of the power coming in the primary side, 99 percent or more comes out the secondary side. But even a loss of less than 1 percent can amount to a few million watts—several thousand toasters' worth—which is a lot of heat to get rid of. For cooling, the transformer casing is filled with oil, which circulates through the windings and carries off heat to fins, tubes, or radiators, much the way circulating water cools an automobile engine.

Isn't there something risky about this—putting hot, live wires into a tank with thousands of gallons of flammable oil? Fire is indeed a worry, and every now and then a transformer does blow up. For a few years it looked like there was an ideal answer to the transformer-cooling problem: fluids called polychlorinated biphenyls, or PCBs, which are good coolants and electrical insulators but don't readily burn. Unfortunately, PCBs turned out to be nasty and persistent toxins, and they have been banned from electrical equipment since the 1970s.

Many transformers are installed above a concrete sump or basin, big enough to contain the entire load of cooling oil in case of a leak. The basin may be filled with gravel or crushed stone to help suppress fires.

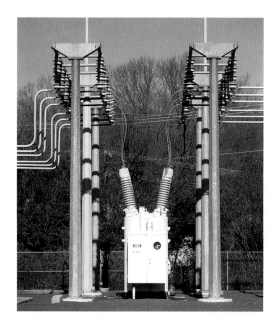

Oil-filled switches, or circuit breakers, are a common sight at American substations. The switches can be used to open a circuit under manual control, but they also operate automatically in the event of an overload or a lightning strike or an equipment failure. The hundreds of gallons of oil in the tanklike enclosures help to snuff out the arc that forms when a large current is interrupted.

The big cylindrical tank on top of many substation transformers is called a conservator, and it holds oil overflows. The oil expands when it heats up, and the conservator gives it a place to go without spilling; when the oil cools again, it is automatically sucked back into the main transformer tank. Sensors continually monitor the state of the oil. Hydrogen bubbles are a sign of arcing within the transformer windings. If a sensor detects a few small bubbles, it will alert maintenance crews to have a look at the transformer. A sudden surge in oil pressure indicates more serious trouble, and immediately triggers a circuit breaker that disconnects the transformer. Transformer hum is another diagnostic sign. Old hands in the switchyard say they can hear the difference when a transformer is about to go bad.

One last thing to look for on a big transformer is a mechanism, called a tap-changer, for adjusting the output voltage. A tap is a connection to one of the transformer windings not at the end of the coil but to one of the interior turns. Changing from one tap to another changes the turns ratio and thus the output voltage. Adjustments of this kind are needed to make sure the voltage reaching customers stays reasonably close to what is promised. Once upon a time, changing a tap meant sending somebody out in the yard to shove a big lever on the side of the transformer casing. Nowadays it's done by remote control, and all you see is a metal box hanging on the side of the transformer; it houses a motor-driven switch. Often the tap can be changed only when the transformer is off-line.

Circuit Breakers and Switches. A switch seems like such a simple piece of hardware. Take two conductors and join them together: The switch is closed, and current flows. Pull the conductors apart and the current stops. This is how a light switch works. But the switches in substations have to handle thousands of amperes and hundreds of thousands of volts, and they are not so simple.

The basic problem in building a switch for high voltages and high currents is that merely pulling two conductors apart does not necessarily stop the current from flowing. Electricity will cheerfully jump across the gap, forming a white-hot arc, which then proceeds to melt the whole mess. If you want your switch to work more than once, you need a way to extinguish the arc. Fortunately for the power engineer, the use of alternating current makes this task easier. Because the current and voltage fall to zero 120 times a second, you don't actually have to stop the current; you just have to wait for the arc to die out at such a "zero crossing" and then cool it quickly enough that it can't reignite.

Most substation switches in the United States use oil to quench the arc—the same kind of oil that cools transformers. The switch contacts are immersed in a deep tank of oil, and when the contacts pull apart, the oil fills the space between them and smothers the arc. Actually, it's not the oil itself that puts out the arc. The extreme heat of the arc breaks down a little of the oil into its chemical components and creates a bubble of hydrogen gas. The hydrogen bubble carries away heat so efficiently that— if all goes according to plan—the arc fails to reignite after the voltage crosses zero.

Usually a few cycles of alternating current—lasting maybe a tenth of a second altogether—are needed before the current stops for good.

I have to admit that when I first heard about oil-filled switches, I thought they made as much sense as dousing a fire with gasoline. Can you really put out an electric arc by covering it with flammable oil and explosive hydrogen gas? It does work: there are thousands of the switches in substations all over North America. On the other hand, they are the most troublesome equipment on the lot. They need frequent maintenance, including an oil change at regular intervals. And when they fail, they fail spectacularly—with a ball of flame and a column of smoke.

You can recognize an oil switch as a tall cylindrical tank with two fluted porcelain bushing insulators poking out of the top at an angle, like horns or an insect's antennae. Since an oil-filled switch and a transformer are both, from the outside,

Air-blast switches rely on a supersonic wind to blow the arc out, rather than a bath of oil. The switches are the nine tall, T-shaped structures. Porcelain insulators form both the vertical stalk and the lateral arms of each T; the switching elements themselves lie in the football-like pod where the vertical and horizontal members intersect. The air blasts come from the tanks mounted directly below each switch. Three switches are rigged in series on each phase of a 345-kilovolt circuit. The installation is outside the Ravenswood power plant in Queens, New York.

"Metalclad" switches use a heavy, inert fluid, sulfur hexafluoride, to smother the arc. Metalclad switchgear is more popular in Europe than in the United States, but these units are in a substation at New Madrid, Missouri.

nothing but a big oil tank, it's possible to confuse them—but look closely and you can tell the difference. Switches are distinguished from transformers by the absence of cooling fins and by the presence of just two high-voltage bushings, which are of equal size. Three identical switches will be lined up in a row, one for each of the three phases. (Three-phase oil switches—with three sets of contacts in one vat of oil—do exist but they're pretty rare. They have six equal-size bushings.)

 The oil-filled switch is the kind you are most likely to spot in American substations, but there are other types. They use other mechanisms to quench the arc, and they are quite different in appearance.

Air-blast breakers blow out the electric arc the way you blow out a candle—except it takes supersonic breath to do it. The switch is a Y- or T-shaped structure, most of it made up of fluted porcelain insulators. The actual switch contacts are inside a small housing at the intersection where the three arms of the Y or T meet. When the contacts open, air under high pressure blows the heated gases away with such force and speed that the arc can't reignite. Air-blast switches eliminate all the problems of oil-filled equipment—not only the explosion hazard but also the danger of leaks and spills and the cost of oil changes. The most serious drawback of air-blast switches is noise. No large switches could be described as quiet, but air-blast switches produce a noise like the report of a cannon. They are not a popular choice for suburban substations with litigious neighbors. You're more likely to find them in the switchyard outside a generating plant.

In another kind of high-voltage switch, the arc is quenched by sulfur hexafluoride, an exotic heavy gas that does the job more effectively than either oil or air, with the result that the switch can be made smaller. Sulfur hexafluoride switches are usually Y-shaped, but they differ from air-blast switches in that most of the housing is made not of insulating porcelain but of conducting metal. (Indeed, sulfur hexafluoride equipment is sometimes called metalclad switchgear.) Sulfur hexafluoride has the big advantage of being nonflammable; on the other hand, it breaks down into sulfur and fluorine, which are nasty and corrosive. In recent years, sulfur hexafluoride has also become fabulously expensive. You're most likely to see metalclad equipment in downtown substations where space is at a premium. It is more common in Europe than in North America.

The circuit breakers in your home can be used in two ways. On an overload or a short circuit, the breaker "trips" automatically, but you can also throw the lever manually if you need to shut down a circuit to make a repair. Substation breakers have the same dual role. Instruments throughout the system continually monitor voltages, currents, and frequencies; when anything goes wrong, the sensors transmit signals that trigger the appropriate breakers. A breaker can also be tripped manually—although "manually" almost never means sending someone out into the yard to pull a lever. The switches are operated from inside a control room, which might be on the grounds of the substation or miles away at a power-company central facility.

The network of instruments and controls for what the power companies call system protection is more complicated than the power network itself. The sensors have to detect faults—such as a lightning strike, or an insulator that has flashed over—in thousandths of a second, so that transformers, generators, and other expensive equipment—not to mention customers—can be disconnected before any damage is done. What makes the problem of protection difficult is that you want to shut off only the faulty segment of the system, not pull the plug on an entire city. Rigging up a set of controls that will trip just the right breakers is a highly respected art in power engineering. For decades it had to be done with nothing but electromechanical relays—switches operated by magnetic coils. Now computers allow faster responses and more

elaborate fault analysis; on the other hand, the power grid itself has gotten more complex and harder to control. A number of major power outages have been traced to malfunctions of the equipment that was supposed to protect against power outages.

Many power-system faults are momentary: a tree limb touches a conductor and then falls to the ground. Accordingly, most breakers are set to automatically reclose after a second or two. If the fault has cleared itself, all you notice at home is a brief dimming of the lights (unless, of course, you are staring at a suddenly blank computer screen). If the fault persists, the breaker immediately reopens and waits a little longer before trying to close yet again. After two or three unsuccessful attempts, the breaker locks out and has to be reset manually. If you happen to be near a substation when all this opening and reclosing is going on, you'll know it: the noise is like a very large machine gun.

The switches in a power network have to be carefully planned and placed so that every piece of equipment can be isolated from the live circuit for maintenance and repair. That includes the switches themselves: you need to be able to switch off power to the switches so that you can work on the switches! But then don't you need switches for the switches for the switches, and so on? It's not quite that hopeless. If you trace along the overhead bus bar feeding a major breaker or transformer, you will probably notice a hinged link, called an isolator or cutout or disconnect, which serves as the simplest kind of switch. You can see at a glance how it works: when the metal gate swings open, current cannot cross the gap; when the gate is closed, the circuit conducts. But the simplicity of the isolator raises another question: if it's so easy to build a switch, why bother with all those huge oil-filled or air-blast breakers? The answer is that an isolator cannot be opened or closed when the circuit is under power; trying to do so would produce a fireworks show and a demonstration of the principle of the arc welder. Before opening an isolator, you first have to shut down the circuit by opening the breaker; once the isolator is open, the breaker can be reclosed to restore service to other areas. A substation has scores of isolators, arranged so that every item of equipment can be taken off line independently.

More Substation Sights. Transformers and circuit breakers are the big-ticket items at a substation, but there are also other kinds of equipment.

Bushings. The bushing-type insulators that poke out of the top of every transformer and breaker are worth a closer look. Superficially they are much like other insulators: a ridged column of glazed ceramic. But the bushing has a harder job to do. Other insulators merely block the flow of all current; a bushing must conduct current through its core while preventing leakage from the inside to the outside.

A high-voltage bushing is not just a hollowed-out ceramic insulator with a copper conductor through the middle. It has a complex internal structure, with alternating layers of conductor and insulator that help to distribute electrical stresses equally. Most bushings are oil-filled. Older ones have a glass globe at the top for monitoring the quantity and quality of the oil. Just like the oil in your car's engine, the

The special insulators called bushings are an essential (and often troublesome) component of many substation devices. Like the rubber grommet that guides a power cord through the casing of a toaster or a vacuum cleaner, a bushing carries electricity along an insulated passage through the metal housing of a switch or transformer. These are spares, photographed in the switchyard at the Fontana Dam hydroelectric plant in western North Carolina. The glass globes at the top allow inspection of the insulating oil that fills the interior.

oil in a bushing insulator turns from tawny to black as it ages. But the cause of these changes is slightly different: it is not dirt and mechanical wear that does the damage, but chemical breakdown caused by microscopic electrical discharges.

Bushings are a common trouble spot in substation operations. Because replacements might be needed on short notice, you are likely to see spares kept in open racks somewhere on the substation grounds. The spares may allow you to get a look at the lower half of the bushing, which is normally concealed inside the transformer or breaker housing.

Lightning arresters. You might think of a lightning arrester as a kind of antifuse. An ordinary fuse is a weak link that stops conducting when the current gets too large. A lightning arrester works the opposite way: it starts conducting when the voltage gets too high, thereby closing a circuit and shunting the excess voltage into the ground. Under normal operating conditions a lightning arrester is an insulator; when exposed to a lightning surge, it briefly becomes a good conductor.

The typical arrester is yet another tall and slender column of ribbed porcelain insulators, stacked up in alternation with smaller, disklike objects. Passing through the middle of each insulating segment is a high-resistance conductor that limits the maximum current passing through the arrester. In older arresters the disks are spark gaps, which arc over when the voltage gets too high. In newer arresters the disks are blocks of the semiconductor material zinc oxide, which switches abruptly from being a good insulator to a fair conductor at a certain threshold voltage. (The same zinc-oxide device, by the way, is the working mechanism of the "surge suppressor" that protects your home computer or television from bad juju coming down the power line.)

Look for lightning arresters where each feeder enters the substation. Additional arresters are often installed close to each piece of expensive equipment. With large transformers, the arrester may even be mounted directly on the case.

Choke coils. What is that washing machine–size, concrete-and-metal basket-like thing at one end of the high-voltage bus? It is another line of defense in the war against faults. When a short circuit develops far out on a transmission line, the fault current is limited by the resistance of the miles of conductors between the substation and the trouble spot. But when the short circuit is nearby, or even inside the substation, the current can climb to dangerous levels so fast that a breaker does not have time to open the circuit before damage is done. A choke coil—known more formally as a current-limiting reactor—chokes off the highest peak currents.

The coil is usually open to the weather, so you can see just how it is built. A few dozen turns of very heavy copper wire or bar stock are wound in a helix two or three feet in diameter. At ordinary current levels, this coil has little effect on the electrical system; from an electron's point of view, it's not much different from a straight wire. But the very sudden surge of current through a nearby short circuit generates a tremendous magnetic field in the coil, which saps energy from the current and thereby limits it. It's the suddenness that makes the difference. The faster the current tries to grow, the more the choke acts to limit it.

Lightning arresters are installed near every transformer and other items of expensive hardware in a substation. The arrester is the tall strut in the foreground, with the skirt at the top. Ordinarily, the arrester acts as a simple insulator, but when voltage on the system rises to extreme levels, the arrester becomes a conductor, shunting the excess current into the earth. The strange lampshade at the top is an arc ring, meant to limit the damage if a lightning bolt "flashes over" from the arrester to the ground.

Choke coils provide another form of protection from lightning and current surges. The choke coils are the six brown cylinders suspended from the steel framework at this Ohio substation. Under normal operating conditions the coils have little effect on the current flowing into the substation. But when lightning or a malfunction causes an abrupt surge, the coil chokes it off, like a constriction in a garden hose.

You'll note that the coil is built very sturdily, with the conductors threaded through notches or holes in a heavy concrete frame. The structure needs to be strong because the magnetic field that chokes off the fault current also exerts tons of force, tending to collapse the coil. Failures are rare but spectacular.

Some reactor coils give off a distinctive, pure-pitch tone, more musical than transformer hum.

Capacitors. One of the trickier aspects of an AC power system is that the voltage and the current can get out of step with each other. Since voltage is the force that pushes current, you might think the two things would always stay synchronized, but that's true only in the simplest circuits. Voltage and current would remain well matched if no one ever plugged in anything but toasters and incandescent light bulbs. These devices offer an almost purely resistive load—all the electrical energy goes into overcoming the friction-like forces that oppose the motion of electrons. But the power supply must also run motors, transformers, and fluorescent lamps, and these devices have not only resistance but also properties called inductance and capacitance. An inductor stores and eventually dissipates energy in a magnetic field; a capacitor does the same in an electric field. When an AC signal hits a large inductance, the current comes out lagging behind the voltage; high capacitance has the opposite effect, making the current peaks lead the voltage peaks.

In practice, inductive loads are more common than capacitive ones, mostly because so much electricity goes to run motors. Every refrigerator, air conditioner, washing machine, and garage-door opener adds a bit of inductance to the power company's load, and the much bigger motors that drive industrial pumps and machinery add a great deal more. As a result, current tends to lag behind voltage at most places in the

national power grid. This is not a good situation. The amount of work you can get out of electricity is measured as voltage multiplied by current, but in AC systems that simple formula holds true only if the voltage and the current are present simultaneously. When they are out of sync, the work is reduced by a percentage called the power factor.

One way to fix a lagging power factor is to attach capacitors to the system, to balance out the inductive load. Conceptually, a capacitor is two metal plates separated by an insulating layer; the bigger the plates and the thinner the insulator, the greater the capacitance. For the capacitors used in the power system, the plates are metal foils and the insulator is a plastic film. These are rolled up together jelly-roll style and stuffed into a rectangular box the size of a briefcase. Substation capacitor banks are usually easy to recognize because there are dozens or hundreds of these boxes lined up side by side on steel racks, like some big luggage checkroom.

When you look closely at how the capacitors are wired, you'll see "jumpers" going from one capacitor to the next all down the row. Only one box at the end of the row is connected to one of the overhead bus bars; the capacitor at the opposite end of the row is connected to a grounding strap. In other words, the capacitors are wired in series. This arrangement is necessary because a single capacitor could not withstand the full voltage of the power line; it has to be divided up among the 10 or 20 capacitors in series.

Instrument transformers. Voltages and currents have to be measured in a substation, but it would be inconvenient (not to mention dangerous) to bring wires carrying

Banks of capacitors at a substation help to keep voltage and current in synchrony. Evidently no one has figured out how to make capacitors on the same gargantuan scale as substation transformers and switches, so hundreds of smaller ones are stacked like books on a shelf.

Mystery item: What are the bright red perforated pie plates? I noticed them and puzzled over them at a number of substations, until a worker at this Louisiana installation explained: squirrel shields.

500,000 volts or 10,000 amperes into the control room. The solution is to use much smaller signals that are proportional to the actual voltages and currents. For example, a power-line current that can range from 0 to 1,000 amperes might be represented by a current between 0 and 1 ampere; an instrument that monitors the small current can be calibrated to read out in units of the larger current. Voltages are handled in the same way: the North American custom is to reduce all voltages to a range of 0 to 110 volts.

The reduced-magnitude signals needed by measuring instruments come from current transformers and voltage transformers. These devices work on the same physical principles as the hefty power transformers at the heart of a substation, but there's no chance of confusing them.

A current transformer is often built directly into the bushing insulator that enters a circuit breaker or a power transformer. The primary "winding" of the current transformer is not a winding at all but simply the high-voltage conductor that bores straight through the core of the bushing; it counts as a winding of one turn. The secondary winding consists of many turns of fine copper wire spun around the core of the bushing. Reducing the current by a factor of 1,000 calls for a secondary winding with 1,000 turns. From the outside you can't see much of a bushing transformer. The telltale sign is small-gauge wiring coming out of a bushing near its base. The wire will be housed in a small metal conduit, which typically either disappears underground or leads to a steel cabinet.

Some recent current transformers are easier to spot. They are doughnut-like collars that surround a bus bar in free air. The collar is in fact a coil of fine wire, which monitors the current passing through its interior.

A voltage or potential transformer is designed the opposite way. Its primary winding, which gets connected to the transmission-line voltage, has thousands of turns of fine wire, and the secondary has just a few turns. The two windings are usually housed inside what looks like an unusually fat porcelain insulator, mounted on a pedestal and connected at the top to a bus bar.

Batteries. An electric-company substation is the last place in the world you might expect to find equipment running on battery power. But it makes sense. Think of all of those relays and computers that protect the system against malfunctions: the moment when they are most needed is the very moment when the regular power supply is most flaky. As a consequence, much of the substation is powered by a roomful of lead-acid batteries. These are the same kind of batteries that start your car, although they don't look it. Each cell of the battery is a heavy glass vessel the size of a large dictionary, and 50 or 60 of these cells are lined up on wood racks like books on a bookshelf. A "trickle charger" keeps them topped up.

Grounding. Near the foundation line of any substation equipment, you'll likely find a heavy strap of braided copper or a thick copper wire entering the earth. With so much high voltage around, good grounding is a major preoccupation of substation designers and operators. If the outer casing of a transformer were not grounded, then

a fault in one of the windings could charge the casing with thousands of volts; this would be a very unpleasant discovery for the next person to touch the case. Grounding provides two kinds of protection. First, as long as the ground strap is intact, the casing simply cannot have a voltage very different from that of the ground. Second, if a high-voltage conductor should come in contact with the casing, the fault current flowing into the ground would trip a breaker. (These are the same arguments that favor grounding kitchen appliances and home power tools—this is why you're not supposed to defeat that annoying three-pronged plug.)

Even the chain-link fence that surrounds the substation may have special grounding provisions in case a conductor should fall on it. The fence may also have insulating links at certain points around the periphery of the station, which have the effect of dividing the fence into electrically isolated segments. In this case the aim isn't protection against the high voltage of a fallen conductor. Instead, the segmenting insulators limit the voltage induced into the fence by the electric and magnetic fields surrounding the conductors.

The grounding straps attached to equipment and fences connect to an array of long rods driven into the ground on a grid pattern, or to a network of rods or pipes that were buried before the station was built over them. In dry and rocky soil, where getting a good ground can be difficult, the ground electrodes may be bonded to the steel casing of wells drilled down into a deep aquifer. The reason for all this trouble becomes apparent when a substation absorbs a really heavy fault, such as a direct lightning strike. Currents through the ground near the point of entry can be intense enough to light grass fires.

Hot sticks. Somewhere on the grounds of a substation there should be some long fiberglass poles with metal fittings at one end. They are not for pole-vaulting over the bus bars! These hot sticks are used for opening or closing isolators, and in emergencies for clearing fallen debris from equipment. A common place to stash them is in a length of plastic pipe attached to a building wall or the perimeter fence.

DISTRIBUTION: POWER TO THE PEOPLE

The final segment of the electric-power system carries current from the substation to your home. The distribution network has two main parts. Primary distribution lines cover distances of several miles, and in rural areas may stretch to 50 miles or more. They commonly operate at voltages of 2,400 to about 25,000 volts, or occasionally as high as 46,000 volts. Secondary distribution circuits bridge the last few hundred yards from the utility pole or the underground cable to the electric meter on the wall of your house. The secondary voltages are the familiar ones you find at household outlets: 120 volts and 240 volts in North America.

Sometimes there is no need for a distribution network. A large customer, such as an aluminum smelter or a municipal pumping plant will buy the entire output of a

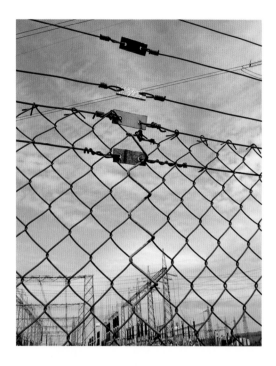

Even the chain-link fence that surrounds a substation has its points of interest. The fence above has a twine thread running through it to block currents induced by power lines crossing overhead. The fencepost below has multiple grounding straps to drain away any stray voltages.

substation. But most substations serve a town or a neighborhood of several hundred or a few thousand homes and businesses. Distribution lines radiate from the substation to all parts of this territory.

Poles. Most distribution lines are carried on wood poles, which have thus become one of the most common sights of the modern roadside and streetscape. We live in a forest of these leafless trees. In North America there are 100 million of them—almost as many as there are houses or cars. Yet we seldom notice them.

When I was growing up, we called them telephone poles. The term has historical merit—telephone service came before electricity in most areas, and so the first pole lines were put up to carry telephone wires. Today, though, the majority of poles are owned by power companies, which lease space to telephone companies and other utilities such as cable-TV operators.

The poles are most commonly cedar, yellow pine, or Douglas fir—trees that grow tall and straight. They are treated with creosote or other preservatives to stave off decay and insects. (Nevertheless, rot is the eventual fate of most poles; they tend to

Power lines fanning out to neighborhoods of Durham, North Carolina, are draped over the crossbars of more than a dozen utility poles.

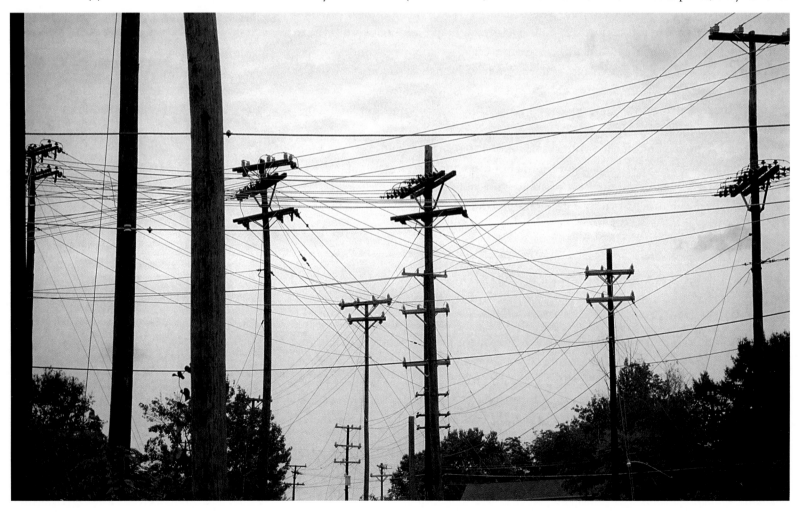

go soft at the ground line.) On high-voltage transmission lines, you might see wood poles as much as 90 feet long, but the poles set along the curb of a suburban street are more likely to be 30 or 40 feet, with 5 or 6 feet of this length below the ground. That doesn't seem like much of an anchorage, but I have never seen a pole uprooted; when they come down in a storm, they are more likely to break than to overturn. Poles carrying an unusually heavy load are braced with guy wires anchored in the soil or to another pole.

Many utilities tag their poles with inventory numbers. The identifying tag is an aluminum or plastic plate at about eye level—if you can find it among all the lost-dog notices and yard-sale posters. When you report a power outage, you might gain a bit of credibility with the dispatcher if you can give the pole number of the trouble spot. Another tag on many poles gives the date of installation or last inspection.

Primary Distribution Circuits. Primary distribution lines are invariably the topmost wires on a utility pole, following the general rule that higher voltages call for higher elevations. In the arrangement seen most often, the three primary conductors are arrayed horizontally on a crossbar near the top of the pole, but there are many variations. Instead of a single horizontal crossbar, there might be an X- or K-shaped arrangement of beams, or the conductors may be hung in a vertical row from insulating struts bolted directly to the pole. Sometimes metal brackets hold the three conductors in an equilateral triangle—one conductor above the top of the pole and the others on either side. If a line of poles carries two or three primary circuits, there will be six or nine conductors near the top of the pole, usually on multiple crossbars.

In recent construction, the conductors themselves are bare stranded aluminum, perhaps half an inch in diameter. But there are still many distribution circuits in service with solid copper conductors.

The conductors are usually supported by porcelain insulators that perch on top of the crossbar (unlike the underslung suspension insulators of high-voltage transmission lines). Pin-type insulators are threaded onto a wood or metal pin set into the top of the crossbar. They have a distinctive shape: a broad skirt flares outward from the waist; underneath it are circular ridges that naturally enough are called petticoats. Post-type insulators bolt directly to the crossbar; the insulator is a tall cylinder with corrugations around its circumference. As with transmission-line insulators, the purpose of all these skirts and petticoats and ridges is to increase the creep length of the insulator—the distance that a leakage current would have to travel over the surface.

Both pin-type and post-type insulators have a groove along the upper surface to receive the conductor. With binoculars you may be able to see how the conductor is fastened into this groove. Various kinds of clips and clamps are made for the purpose, but there are also still plenty of distribution lines where the conductors are lashed on with twisted lengths of wire. Creating neat and secure lashings has traditionally been an important element of the lineman's craft and training. It's particularly impressive to see it done with eight-foot hotsticks when the worker is repairing a live line.

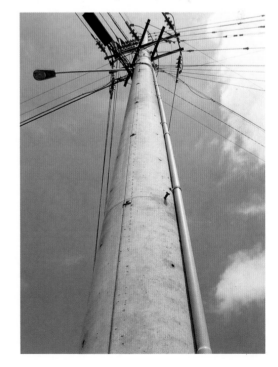

Not all poles are wood: This one, in western Virginia, is made of cast concrete. The thin copper wire running down the surface is a grounding strap.

There is no easy way to judge the voltage of a primary distribution line, although larger and more elaborate insulators usually signify higher voltages. Distribution voltages have been climbing steadily over the years as the system has been stretched to supply greater loads over longer distances. The most widespread voltage was once 2,400 volts; then many systems migrated to 4,160 volts, 7,620 volts, and 13,200 volts. Utilities give careful thought before going beyond this level because it is near the limit for doing live-line maintenance with rubber gloves; procedures for maintaining higher-voltage lines are more awkward and expensive. Nevertheless, more and more distribution systems operate at 23,000 or 34,500 or even 46,000 volts.

THE INDUSTRIAL ECOLOGY OF A UTILITY POLE

In the tropical rain forest, ecologists study the communities of plants and animals that live in vertical zones, from the leafy canopy of the tallest trees down through the understory to the rotting litter on the forest floor. Utility poles also have a series of vertically stratified habitats, each with its own characteristic inhabitants. From top to bottom, here are some of the species you might observe in the utility-pole ecosystem:

• Primary distribution lines for electric power. These are the topmost wires. They are usually hung on a crossarm, and they come in groups of three, mounted on big insulators.

• Switches, fuses, and surge arresters. These connect to the primary distribution lines.

• Transformers. They are mounted below the primary distribution lines but above the secondary ones, with connections to both.

• Secondary distribution lines. Just below the transformer level, they are rubber-sheathed conductors carried on spool-type insulators or twisted around a steel messenger cable.

• Street-lighting fixtures. They draw their power from the secondary circuits.

• Traffic signals. These too are powered by the secondaries. The signal lights are often hung from a steel cable stretched between utility poles.

Everything from the top of the pole down to this level is the domain of the power com-

pany. Below is the realm of communications lines, which operate on lower voltages and therefore don't need to be kept quite as far out of reach.

• The municipal zone. Cities and counties that allow poles to be set on public land sometimes demand a bit of utility-pole real estate in compensation. This zone was once used for the wiring of fire alarms and police call boxes; today those signals are more likely to go over leased telephone lines.

• Cable television feeders. These may be finger-thick coaxial cables, in either a black plastic sheath or a bare metal jacket. In newer systems the trunk lines that carry signals over longer distances are fiber-optic cables.

• Telephone cables. Often the thickest of all the wires strung on a pole, they are actually bundles of dozens or hundreds of pairs of fine copper wires. Fiber-optic cables also show up at this level.

Still lower—indeed, reaching the ground—are some wires that ought to have no voltage at all on them.

• Guy wires. Their function is strictly mechanical; they help to hold the pole up. There may be an insulator inserted into the guy wire for safety, in case a power conductor should touch the upper part.

• Grounding lead. A pole with a transformer generally has a copper grounding wire that runs down the side of the pole and into the ground.

Finally, at eye level, comes the bottommost ecological stratum of the urban or suburban utility pole:

• The yard-sale zone, where the wood bristles with a thousand rusty staples.

On a crossbar carrying a primary distribution line, you might notice small plac-ards labeling the three conductors *A*, *B*, and *C*. Or there may be three colored tags, usually red, yellow, and blue. These markers are meant to identify the three phases of the electric current. Why would anyone care which is which? In a household with single-phase service, it makes little difference which of the three phases you tap into. But a factory that draws on three-phase power cannot be so indifferent. If a lineman were to accidentally transpose two of the phase conductors, every three-phase motor on that circuit would suddenly begin running backward.

So far, a primary distribution circuit looks just like a miniature transmission line, but there is a fundamental difference, apart from the lower voltage levels. In the United States most distribution lines are actually four-wire systems, in contrast to the three-wire standard that prevails throughout the transmission network. The fourth wire is lower down on the pole, less conspicuous, set off on smaller insulators, and usually shared with the secondary distribution system. The presence of this fourth conductor calls for a bit of explanation.

It is in the nature of an electrical circuit that all the power that flows out has to flow back in; indeed, that's why it's called a circuit. In a three-wire, three-phase sys-tem, there is no problem balancing inflow and outflow as long as the loads on the three conductors are equal. But in distribution circuits, individual customers draw power from different phases. If you happen to be burning the midnight juice while your neighbor is sleeping, the loads will be imbalanced. The fourth conductor, called the neutral, carries any leftover currents created by the imbalance. Because the power company tries to equalize loads on the three phases as best it can, the currents in the neutral conductor are generally small.

There are distinctive national and regional customs in the engineering of power-distribution systems. In North America, each branch circuit that supplies electricity to houses and small businesses is wired between one of the phase conductors and the neutral conductor. In Europe single-phase service is usually wired between two of the phase conductors. You can tell the difference by looking at the connections to the pole transformers (see section below). Also, three-phase service is much more common in Europe. In the United States only industrial or large commercial build-ings are likely to have a three-phase connection, but in several European countries many homes have three-phase service for running the motors in refrigerators, air conditioners, and other large appliances. The three-phase motors are more efficient, although also more expensive.

Along rural or suburban roads in some areas you may see a primary distribution line that violates the sacred rule-of-three: instead of three primary conductors at the top of the pole, there is just one, along with a neutral somewhat lower down. Even in these areas the basic power system is still three-phase. It's just that the power com-pany has saved some wiring cost by running only one of the phases along a given road, and feeding all the houses there from it. Other streets draw their power from the other phases, so that the overall load remains balanced.

Post-type insulators *(top)* are heavy porcelain castings bolted to the crossbar. Glass pin-type insulators *(bottom)* are now collector's items, but a few are still in service. Here the conductors are also vintage equipment: The green patina shows they are copper, whereas all new power lines are strung with aluminum.

Pole-Mounted Transformers. The big steel can bolted high overhead on a utility pole—I've heard workers call the thing a "pole pig"—is the last stage in the long series of up-and-down transformations that finally delivers electricity at household voltages. This is where the voltage drops from the primary distribution level (typically a few thousand volts) to the tamer and more familiar 120 or 240 volts. (Some big customers take their jolts at 480 volts.)

Like the larger transformers at substations, the pole-mounted transformer has windings of copper wire wrapped around a steel core. The windings are immersed in oil for insulation and cooling. Larger units are festooned with tubes or fins to radiate away excess heat. Filled with oil, a transformer can weigh 500 pounds or more.

LIVE-WIRE GUYS

Electricity has become such a 24-hour-a-day necessity that the power company has a hard time shutting down transmission lines for repairs. As a result, a great deal of maintenance is done on live wires.

For distribution circuits up to about 12,000 volts, the work can be done with rubber gloves. These are nothing like the supple latex gloves of the surgeon, or even the gloves you might use to wash dishes. They are as heavy and thick as galoshes, and they come with sleeves that go all the way up your arms and link together behind your back. You test them by blowing them up like a balloon; no leaks are allowed, since electrons will wiggle through even the tiniest hole.

You say you wouldn't touch a live wire with a 10-foot pole? That's exactly how it's done at higher voltages. The poles are called hot sticks, and they are made of fiberglass or plastic, materials that provide excellent insulation as long as they stay clean and dry. Hot sticks have various kinds of tools and hooks on the business end so a work crew can do jobs such as replacing insulators while staying a safe distance from the energized conductors. But doing any kind of close manipulation with a long pole is pretty awkward; it takes practice, skill, and strength. And the higher the voltage, the longer the pole needs to be.

The most impressive kind of live-wire work is also the simplest. It is called bare-hand maintenance. It's possible because electricity flows only when there is a difference in voltage. If you touch a transmission-line conductor and a grounded steel tower at the same time, you're toast. But you can safely hold onto either one if you keep your distance from the other. The bare-hand worker leaves the grounded world behind and lives at high voltage.

The challenge is getting there. Sometimes a bucket truck with an insulated boom will reach high enough. Or workers can climb a transmission-line tower and then make their way to the live conductors using insulated fiberglass ladders. Still another choice is to lower a worker from a helicopter.

Strictly speaking, bare-hand work isn't done with bare hands. You wear a conductive suit, with booties, gloves, and a hood covering everything but your face. The suit is made of a fabric woven with silver or carbon threads, and it has a "tail" that you clamp onto the live wire, to ensure that your body is always at the same voltage as the circuit. Those who have done such work tell me there is a strong tingling sensation at the moment you bond onto the conductor, but it fades in a second or two.

Who could fail to admire the mettle of these workers? Even when you understand the physics of the situation—even when you believe beyond question—it is surely a test of faith to grab onto a sizzling 500,000-volt power line.

Hoisting it up the pole with block and tackle was once heavy labor for the lineman; now the lifting is done by the hydraulic boom of the bucket truck.

When you are standing on the sidewalk looking up at a pole transformer, the first things to check out are the "high-side" connections—the wires running from the top of the transformer to the primary distribution conductors overhead. If you're in a residential neighborhood in North America, you'll probably see just one high-side wire, connected to one of the three primary phase conductors. This is the signature of a single-phase transformer connected between a "hot" primary conductor and the neutral, or ground, conductor. Another kind of single-phase transformer has two high-side leads, connected to two of the overhead conductors. These are rare in North America but common in Europe. Finally there are three-phase transformers, which have wires going to all three of the primary conductors. Three-phase transformers tend to be larger, and they mostly serve commercial or industrial customers—a supermarket with a lot of refrigerators to run, or an office building with a heavy-duty air conditioner. Sometimes a three-phase load is served by three single-phase transformers, all hung on the same pole. If you look closely, you'll see that each transformer is wired to a different primary conductor.

Each of the high-voltage connections enters the transformer through a porcelain insulating bushing on the lid of the can. But often there is other miscellaneous hardware up there as well, such as fuses and lightning arresters (see below), so that it can be a little confusing to count the bushings.

The "low-side" wiring generally connects to lugs on the side of the casing, without any need for big porcelain insulators. In the standard American configuration for single-phase power there are two hot wires and a neutral. Usually the hot leads are heavily sheathed in black rubber insulation, whereas the neutral is bare metal. These wires are the ones that loop from the utility pole to your home.

American practice in most residential neighborhoods is to use lots of fairly small pole transformers, each feeding just three or four houses. Thus, when you drive down a suburban street, you may see a transformer hung on nearly every pole. European utilities use fewer but bigger transformers, each of which might supply 50 or 100 homes.

How can you tell the power rating of a transformer? More often than not, the rating is stenciled on the side of the case: it's a number such as 25 or 37.5 or 75. This is the transformer's power-handling limit in kilovolt-amperes, which is the number you get by multiplying the maximum current in amperes by the operating voltage in thousands of volts. Kilovolt-amperes are roughly equivalent to kilowatts, and so a 25-kilovolt-ampere transformer could run 25 toasters that consume a thousand watts apiece. The transformers you see on poles can be as small as 10 or 15 kilovolt-amperes and occasionally as large as 300.

Fuses, Arresters, Switchgear. Like their big siblings back at the substation, pole-mounted transformers need to be protected from overloads, short circuits, voltage surges, and lightning. Moreover, the rest of the power system (including the part that

Pole-mounted transformers—also known as "pole pigs"—are such commonplace elements of urban street furniture that we tend not to notice them at all. In residential neighborhoods, there is a transformer for every few houses. The six above supply power to a small shopping district. They are single-phase transformers ganged up in threes to handle three-phase power.

Distribution transformers are now highly standardized items, but there is still room for occasional variation. The ice-cream cone atop the street lamp above, in Livermore, California, is a disguised pole transformer. European distribution grids often have fewer but larger transformers. The Slovenian one below is a mini-substation, built into a concrete tower.

runs through the walls of your home) has to be protected from malfunctions inside the transformer. Both kinds of protection are provided by fuses and surge arresters.

The kind of fuse you'll most likely see near the top of a power pole is a cardboard tube the size of a large cigar. The tube has a thin metal rod running through it, and the circuit is arranged so that current has to pass through the rod on its way from the primary conductors into the transformer. In the event of an overload or a short circuit, the rod melts, interrupting the current. But that's not the end of the story. The melted fuse element is replaced by an electric arc, which won't necessarily die out anytime soon; it has to be extinguished somehow. That's where the cardboard tube comes in. It's a low-tech solution to the arc problem, but a surprisingly effective one. What happens is that the heat of the arc vaporizes some of the organic material on the inner surface of the tube. The hot gases produced in this way are expelled from the ends of the tube with enough force to blow the arc out. The force is also enough to create quite a loud bang. People often report that a transformer has blown up when the noise they heard was actually an expulsion-tube fuse blowing.

Expulsion-tube fuses occasionally fail, allowing current to continue flowing even after the fuse has blown. To deal with this problem some utilities install backup fuses in series with the regular fuses. The current has to get through both fuses in order to reach the transformer. One style of backup fuse looks something like a large tube of caulk, or perhaps an automobile shock absorber. Inside, a silver filament is encased in pure quartz sand; when the silver melts, the surrounding sand is baked into a form of glass that quenches the arc. Silver-sand fuses are bigger, heavier, and more expensive than expulsion-tube fuses, which is why they are used mainly for backup. They are designed not to blow during an ordinary brief fault; their protection kicks in only when the primary fuse fails.

Switches placed at intervals along a primary distribution line are mainly an aid to maintenance and repair. If a tree gets fouled in the wires near the end of a long line, shutting it down from the substation will leave everyone along the way in the dark. If a repair crew can open up an intermediate switch, power can be restored to at least some customers.

The simplest switches are knife-blade switches, similar to the isolators seen in substations. Unlike so much other electrical equipment, you can see at a glance exactly how they work: when the switch is closed, current flows through the blade to complete the circuit; when the switch is open, the break in the conducting path is obvious. This property is important to the lineman who wants to make sure that the equipment he's about to fix is shut down.

Knife-blade switches generally can't be used to open an energized circuit. The idea is to shut the circuit down with the breaker at the substation, then open the appropriate switches to isolate a section of the line, then close the breaker again so that power is restored to the rest of the system. Opening most knife-blade switches requires climbing the pole or going up in a bucket truck. A few of the switches have a lever or crank at ground level.

The more elaborate and more automated switches in the distribution system are called reclosers. They are similar to the high-voltage circuit breakers at substations, but small enough to hang on a pole. Like a substation breaker, a recloser can be operated by remote control or triggered automatically when it senses an overload or short circuit. And like a breaker it has an arc-quenching medium—oil or vacuum or sulfur hexafluoride—so it can interrupt both normal load currents and fault currents.

In external appearance a recloser is a nondescript box or cylindrical can, painted the usual gray. You might mistake it for a transformer. Even utility workers get confused, which is presumably why at least one power company stencils the word *RECLOSER* on the case of every one. If your local utility is not so helpful, the best way to identify a recloser is to look carefully at the connections and think like an electron. A recloser will be wired so that current can get past it only by going through it. (A transformer, in contrast, offers an optional path to the current, like an exit ramp from a highway, which you can either take or pass by.) A single-phase recloser has two equal-size bushings on top; a three-phase recloser has six equal-size bushings.

It's called a recloser because it not only interrupts a circuit when something goes wrong but also automatically tries to restore service. Because many faults are only momentary, this strategy can save a lot of service calls. Typically the recloser is set to try three times. When it first detects a fault, it opens the circuit and then immediately tries to close it again. If the fault is gone, then fine. If the fault is still there, the recloser opens again, and this time it waits a few seconds before closing the switch again. If the fault persists, the recloser waits a little longer before trying a third time. At this point if the fault has not cleared itself, the recloser "locks out," and a lineman has to come out and reset it manually. When the lights rapidly flicker on and off during a thunderstorm, that's a recloser in action.

The last items of miscellaneous hardware to look for near the top of a utility pole are capacitors. Like those in substations, their role is to correct power factor—to bring the alternating waves of voltage and current into sync when big motors or other such loads get them out of balance. The individual capacitors are oblong boxes, the shape of a telephone book and just a little larger. They are usually mounted in racks that hold three, six, or a dozen lined up side by side. Short circuits inside capacitors are not rare, and so there is likely to be a fuse in the lead that connects the capacitor to the primary distribution conductor. Or there may be a tiny recloser, which not only protects against faults but also allows the capacitor bank to be switched in and out of service as the power factor on the line changes.

Secondary Distribution Circuits. Secondary circuits carry power across the last few feet to your home. It's easy to tell them from the primary circuits. They are strung lower down on the utility pole, below where transformers are mounted (although above telephone and other communications wiring). The secondary circuits may also be sheathed in insulating rubber, unlike the bare conductors seen almost everywhere else in the electric-power system.

Fuses protect power-line equipment just as they do home appliances. On the pole above, three expulsion-tube fuses are bolted directly to the crossbar; they are wired in series with three silver-sand fuses farther to the left, which serve as a backup. On the pole below, a helpful placard reveals that each phase has two 80-ampere fuses wired in parallel, for a capacity of 160 amps.

Cut-out switches *(above)* have such a simple and open mechanism, it's easy to see how they work: put a hook in the loop at one end of the blade, and pull to break the circuit. A recloser *(below)* is a pole-top version of the switches and circuit breakers found in substations. Reclosers can be hard to recognize. The three cylindrical objects below might be mistaken for transformers—but note the two equal-size bushings on the top of each can. (The smaller devices behind the bushings are lightning arresters.)

In the United States, homes are wired for dual voltages: 240 and 120 volts. The two voltage levels are supplied by three wires—two hots and a neutral. The water heater and electric stove and perhaps a few other big appliances are connected to both hot leads, which supply 240 volts. All the other household circuits are fed by one or the other of the hot conductors and the neutral, a combination that yields just half the voltage, or 120 volts.

The secondary conductors are strung from pole to pole (or sometimes along the back walls of buildings) on spool-type insulators. The spool is just like the one that holds sewing thread but larger (the size of a coffee mug), and is made of porcelain. It is mounted on a steel skewer that passes through the hole down the middle of the spool. In one common arrangement, three spools are mounted one above the other to carry the two hot conductors and the neutral. In another style of wiring, the two rubber-coated hot lines are twisted around the bare neutral, and they are all lashed to a single spool. (The bundled conductors are called triplex.)

Why are the secondary conductors, which carry only a couple hundred volts, thicker and heavier than the primary conductors at several thousand volts? Because the size of a wire depends on the amount of current it carries, not the voltage. And for a given amount of power, the lower the voltage, the higher the current.

Secondary wiring practices vary widely. In a downtown commercial district, there may be a secondary bus or grid: Lots of transformers all pump power into the same set of secondary conductors, which in turn serve many buildings. This way no building goes dark if a single transformer fails. On the other hand, a severe fault might knock out the entire secondary grid.

In the suburbs, each transformer generally feeds its own set of secondary conductors. Often, the wires run straight from the transformer to each of the houses being served. If there's a problem with a transformer, only those few homes will lose power.

Electric Meters. A glass bubble filled with brightly polished, spinning gears is fastened like a suction cup to the side of nearly every house in America. Would a visitor from Mars ever guess what that glass pod is there for? Maybe, indeed, the transparent globe is some kind of Martian surveillance device?

The purpose of the pod is indeed surveillance—though the information gathered goes not to Mars but to the local utility company. Technically speaking, the mechanism inside the glass is a watt-hour meter; power-company insiders know it as a revenue meter; to the rest of us, it's just the electric meter. It is the device that reports your energy consumption to the electric company so they can send you a bill at the end of the month.

The meter mechanism is a special kind of electric motor, calibrated so its speed is directly proportional to the amount of power flowing through the meter. The spinning disk whose edge is visible through the glass is the core part of this motor. The clock dials that record energy consumption just count the revolutions of the disk. The reading on the dials continually increases, rather than being reset each month.

Your bill is calculated by subtracting this month's reading from last month's. That way, even if the meter reader makes a mistake and you are overcharged one month, you automatically get it back the next.

The latest electronic watt-hour meters have no moving parts and display their reading in odometer-style numbers instead of on a series of clocklike dials. And some

LIGHT UP THE NIGHT

These days, street lighting is nobody's idea of a high-tech, high-growth industry, but a century ago it was a driver of technological development. The impetus for building the first municipal electric systems was not the urge to make better toast and coffee; the systems were built exclusively for street lighting. Only years later did electricity find its way indoors.

Street lighting was a tough problem for a long time. In eighteenth-century New York, every seventh house was required to hang out a lantern on moonless nights. Later, whale-oil lamps were installed along the streets, and then came the gaslight era. (The gas industry, too, got its start as a street-lighting utility.)

The first electric street lights were carbon arc lamps, working on the same principle as the World War II searchlights that still see duty when a used-car dealer decides to advertise a sale with a roving beacon in the sky. In an arc lamp, two carbon electrodes are briefly brought together and then held at just the right distance to maintain a brilliant, blue-white arc. In Los Angeles in the 1880s arc lamps were mounted atop 150-foot-high towers, remarkably similar to the towers that now illuminate major interstate highway interchanges. Fifty thousand fascinated Angelenos clogged the streets when the lights were first turned on. (It's been a while, I think, since new street lights drew a crowd in southern California.)

Arc lamps were a high-maintenance item. Someone had to come around every night to get the arc started. The tungsten-filament incandescent bulb finally put the lamplighter out of work; an engineer at the central station could throw a switch and light up an entire town.

Incandescent street lights have been supplanted by more efficient designs. Fluorescent tubes had a vogue in the 1950s and 1960s, but now most street lights use either mercury-vapor bulbs or sodium-vapor bulbs. Both types work by passing a current through a heated gas of metal atoms (either mercury or sodium).

Compared with a tungsten-filament lamp, the metal-vapor lamps convert more energy into light and less into heat.

Mercury and sodium lights are easy to tell apart. The mercury ones put out an icy blue-white light that turns your lips purple. Light from sodium-vapor lamps is pink or orange. Actually, there are two kinds of sodium-vapor lamps. The low-pressure ones are more efficient (and also favored by astronomers worried about "light pollution") but the color is a deep orange. The high-pressure lamps emit a broader range of colors but use more electricity.

The colors of mercury and sodium lamps are exaggerated when you see them from a great distance, because the light is filtered by the atmosphere. If you look down on a cityscape from an aircraft at 10,000 feet or so, sodium-vapor lamps look almost buttery yellow. Mercury-vapor lamps remain white or blue.

Recent street-lighting fixtures have a color-coded label that indicates the kind of bulb to be installed—blue for mercury, yellow for sodium, and red for a less common type called the metal-halide lamp. A number on the label encodes the wattage: 10 for 100 watts, 25 for 250 watts, 40 for 400 watts, and 1K for 1,000 watts.

When a metal-vapor bulb is nearing the end of its life, it may begin to cycle off and on every minute or two. What happens is that the vapor grows too cool to maintain the light-producing electric discharge, but once the lamp goes out, a starter filament automatically comes on to heat the vapor again. The whole process then repeats, usually with a fairly stable period. If your timing is good, you may be able to persuade your more gullible companions that by mental effort alone you have the power to turn street lights off and on.

Electric street lighting has eliminated not only the lamplighter's evening rounds but even the central-station engineer throwing the switch at dusk. Almost all street lights are now controlled by a photocell that detects the fall of night. The detector is the little cap or turret mounted atop each light fixture.

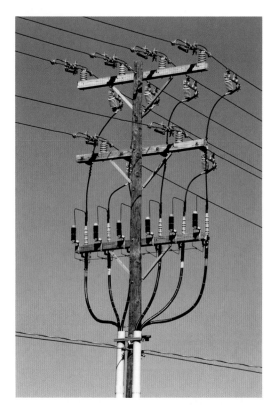

Two three-phase power lines duck underground in bundles of inch-thick insulated cables. The cables can be isolated by cut-out switches where they join the primary lines overhead; they are also protected by lightning arresters—the six dark-brown cylinders mounted on the lower crossarm. Also note the red, blue, and yellow bands wrapped around the six cables; the colors identify the phases.

new meters can be read remotely, without sending out a worker to walk from house to house. Some of them communicate with their home base directly over the power lines; some have a connection to the telephone system; some transmit radio signals.

Underground Distribution. Wander around an upscale new housing development and you will see no power lines strung from pole to pole. Electrical wiring and other utilities run underground, either because the developer believed that would help to sell houses or because a zoning board required it.

You might get the impression that underground electric service is a rather new idea. That's not the case at all. In the downtown core of the largest American and European cities, power and communications lines have been buried under the street almost from the beginning of the wired age. In New York, for example, overhead lines were outlawed following the blizzard of 1888 (when a great many of the lines came down in the streets). Los Angeles enacted a similar law in 1896.

In another sense, the suburban underground systems of recent years are indeed a new technology. They differ in several ways from the older city designs.

Urban underground power systems are all about manholes and buried conduits. Transformers and oil-filled circuit breakers are installed in the manholes, or in larger underground chambers called vaults. Some of the equipment is housed in special sealed casings so it can continue operating even if the manhole floods. Insulated cables for both primary and secondary distribution are pulled through the conduits from one manhole to the next.

The newer suburban systems are built in a different style and by different methods. There are no manholes; transformers and switchgear are housed in metal cabinets (usually painted a grassy green, and perhaps hidden among the azaleas) that sit on concrete pads at ground level. There are also no conduits underground. Instead, the power cable is buried directly in a narrow slit trench.

The big issue in underground distribution is insulation. With overhead wiring, insulation is a lot easier because the conductors can be kept several feet away from anything grounded. That's obviously not possible underground, and so the conductors have to be swaddled in elaborate layers of high-performance insulation.

In older underground circuits, each phase of the three-phase service is carried by a separate cable, which is insulated by wrapping the conductor with oil-soaked paper and sealing it in a lead sheath. Splicing these cables requires the electrician to master some of the plumber's skills, since the lead jacket has to be closed with molten solder.

New installations use a more flexible cable with polyethylene insulation; no lead sheath is needed to keep water out, although sometimes the cable is armored with steel rods for protection against the rogue backhoe (principal foe of all underground utilities). Instead of three separate cables, all the conductors are bundled into a single fat cable. In the latest designs, the three phase conductors are given a pie slice–shaped cross section so that they fit together neatly, and a fourth, neutral, conductor is wrapped around the assembly as a concentric sheath.

When a distribution system is entirely underground, there is not much to see. But in many cases the primary distribution lines run underground only for part of their route. On the boundaries of a downtown district or along a main road near subdivisions with underground service, you might see a number of risers, where a line makes a transition between overhead and underground. The actual transition between insulated cable and bare conductor takes place inside a device called a pothead. (Honest. That's what they call it. I wouldn't kid you.) The pothead seals the end of the cable to keep moisture out. There may be three separate potheads (one for each phase) or a three-phase unit with three bushing-type insulators poking out of the top like the legs of an inverted stool. The potheads are mounted near the top of a pole; the cable descends into the earth along the side of the pole, often through a plastic conduit. There will usually be fuses and surge arresters on the lines feeding the potheads, and tags or tapes to label the phases (*A, B, C* or red, yellow, blue) since it's all too easy to get them mixed up in an underground cable.

Why aren't all distribution lines underground? The power companies say it's a matter of money. Burying a line can cost five times as much as stringing it up on overhead poles, largely because the insulated cable is much more expensive than bare wire for overhead use. The cost penalty has been coming down in recent years, but it may never be reduced to zero. Proponents of underground service argue that the premium is worth paying because it not only eliminates an eyesore but also improves reliability, since lines are not brought down by ice, tree limbs, or auto accidents. Power companies counter that when an underground conductor does fail, finding and fixing the fault takes much longer.

Underground utility service is not totally invisible; equipment that might otherwise hang on poles squats on concrete pads at curbside. Here an electrician has opened up the green box that houses a power transformer.

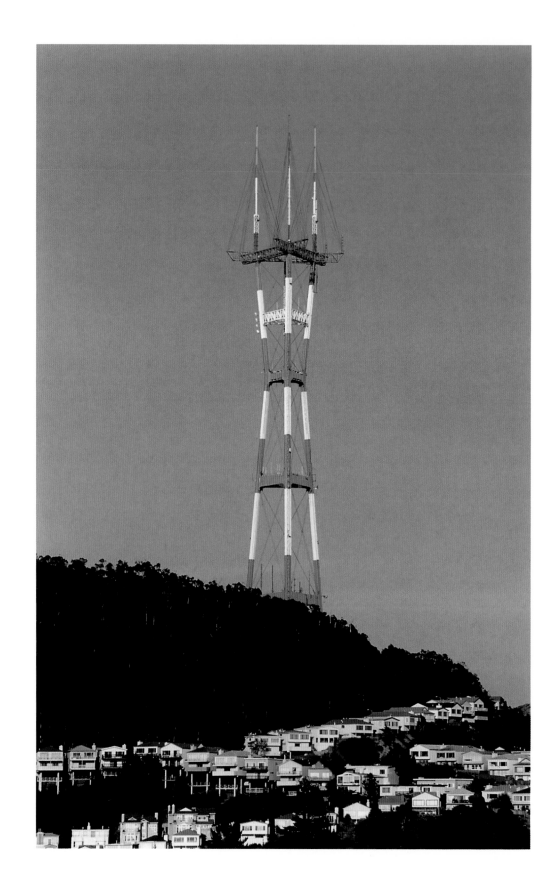

The three-legged, three-headed, red-and-white-striped colossus looming over this hillside neighborhood is the Sutro broadcast tower, where most of the television signals in San Francisco come from. Broadcast towers are among the tallest of all man-made structures, but other elements of the communications infrastructure are less conspicuous and seldom noticed. (This particular tower, of course, could hardly be more conspicuous.)

COMMUNICATIONS

EVERYBODY'S WIRED. If we're not talking on the phone, we're checking our e-mail. Everybody's wireless too. We carry cell phones and pagers, and we browse the Internet while lounging on a park bench, connected to the farthest corners of the world through an untethered laptop computer.

No aspect of the industrial infrastructure has been changing more rapidly than communications technology. In North America it took a hundred years for the web of telegraph and telephone lines to spread over the whole of the continent and extend tentacles into every inhabited place. Then, starting in the 1950s, many of those long copper wires were ripped out and replaced by other kinds of communications channels, chiefly microwave and satellite links. And now the whole continent has been restrung yet again, this time with a filigree of glass fibers carrying digital signals. What took a century the first time was completed in little more than a decade.

THE TELEPHONE

In the Hollywood version of telephone history, Alexander Graham Bell is about to run next door to fetch his assistant when it suddenly dawns on him, Hey, I just invented the telephone! And so he speaks the fateful words into the brand-new instrument: "Mr. Watson, come here. I want you." Maybe it really happened that way. In any case, sometime in March of 1876 Bell did have a working assemblage of batteries, wires, magnets, and other electrical doodads that could transmit the human voice from room to room.

Today the idea of getting sound from electricity no longer excites much wonder, but building a telephone system with the technology of the nineteenth century called for genius. Bell's bright idea—which also occurred to his rival Elisha Gray—

Open-wire telephone circuits are rare today; the photographs on the opposite page were made in 1999 along an open-wire line that runs about 50 miles through southern Nevada, from Hawthorne to Tonopah. The line has eight pairs of wires, stretched very taut on closely spaced poles to avoid clashing. The middle photo shows a transposition, where the two conductors in a pair trade places, which helps to avoid crosstalk between adjacent circuits. There are also transpositions where entire circuits exchange position. The bottom photograph shows the Tonopah end of the line. Each pair of conductors is connected to a small aluminum box with four cables plugged into its underside. Presumably the boxes are multiplexors, and each open-wire pair of conductors can carry four circuits. Thus, the system would accommodate 32 simultaneous conversations.

was a method for using sound as a kind of throttle valve to control the flow of an electric current. Parts of Bell's basic scheme are still in use, and telephones that are 50 or 60 years old work perfectly well when you plug them into the modern network. On the other side of the plug, however, in the central offices and the long-distance networks, everything has changed.

The Local Loop. Your personal link to the global telecommunications infrastructure consists of two slender copper wires, which run from your home to the nearest telephone switching center, known as a central office. The wires have names. They are called tip and ring, terms that come from the two parts of the plug that switchboard operators once used to make a connection in the days before dial-it-yourself telephones.

The local loop really is a loop: it's an electrical circuit. When you lift up the telephone handset from its cradle, you close a switch that completes this circuit, allowing a current to flow through it. In the old days, that current had a very simple effect: it lit a small light bulb on the operator's switchboard at the central office. The operator then plugged into your line and asked who you wanted to talk to. What happens today is a little more complicated and requires no human intervention, but it's still the closing of the circuit and the flow of the current that tells the phone company you're ready to make a call.

Open Wires. In the earliest telephone systems the wires that connected subscribers with the central office were bare copper conductors attached to glass insulators on the crossarms of utility poles, much like the electric power lines of today. Those old-fashioned open wires gave excellent voice transmission, carrying signals farther and with less interference or distortion than modern wiring can. But there was a prob-

GETTING A LOOK

A 1923 pamphlet published by the Bell System offered this hearty invitation: "Subscribers and the public generally are always welcome at the telephone central office where they never fail to be impressed by the intricate apparatus and equipment and the efficient and systematic way in which the telephone traffic is handled." Sad to report, the public generally are no longer quite so welcome at the local switching center. By all means, ask your telephone company for a tour, but don't be surprised if you are turned away.

Major nodes in the communications network tend to be high-security areas these days.

Some of them have adopted the strategy of "security through obscurity"; rather than

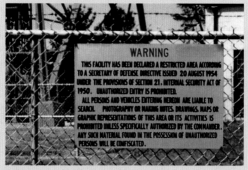

putting the facility in a guarded or fortified installation, it is housed in a nondescript office

building, and the address is not made public. Other parts of the infrastructure, however, are out in the open and in full view. Wires are strung up everywhere, and antenna towers would be very hard to hide.

The intimidating sign in the photograph at left is posted outside a small building with a large antenna tower in Tyson's Corner, Virginia, on a hilltop a few miles from downtown Washington. A worker assured me that although the sign prohibits photography inside the fence, it would be all right to photograph the sign itself. But he told me nothing of what goes on inside the facility.

lem. The wires had to be kept a foot apart so that they wouldn't clash together in the wind. At this spacing a single crossarm could carry only five pairs, and a pole with 10 crossarms was limited to 50 pairs. If a modern city were wired in this way, whole forests would have to be cut down and replanted as utility poles.

Open wiring reached its limits early in the history of the telephone system. By the end of the nineteenth century, many downtown districts were enveloped in a dense web of telephone and telegraph wires, like a growth of jungle vines overhead. There were poles on both sides of every street, and sometimes on rooftops as well. In New York, the blizzard of 1888 brought down hundreds of poles, creating an impenetrable tangle of copper on street corners. The city soon ordered all utility lines underground.

Open wires lasted longer in rural areas, where they have advantages on very long loops. But even in the remote countryside almost all of them have now been replaced by multiconductor cables.

Multipair Cable. The key to squeezing more wires into a smaller space is insulation, so that the wires can be bundled together without shorting out against one another. Early insulation was paper or cloth; now it is a thin coating of plastic. Cables are made with up to 4,200 pairs of conductors—that's 8,400 separate wires. Cables with 2,700 pairs are fairly common; they're as thick as the business end of a baseball bat.

But the cable is not just a bunch of wires lashed together; quite a lot of thought and artistry go into its construction. Two problems need to be solved. First, when long wires run parallel to each other, signals leak from one pair into another, and a private conversation begins to sound like a crowded party. This crosstalk can happen even if the insulation on the wires is perfect; the separate circuits are coupled by their magnetic fields, which go right through insulation. The solution is to take each pair of wires and twist them around each other, so wires from different pairs won't run parallel to each other over long distances. For added protection the various pairs are twisted at different pitches, some as tightly as one full turn every two inches, some as loosely as one turn every six inches.

The second problem is how to find a specific pair of wires in a cable with thousands of pairs. Color coding is the answer, but there's more to it than that. Even if you could give each wire a different color, and even if the eye could distinguish all those thousands of hues and shades, riffling through 8,400 wires in search of the right color would be exquisite tedium. To preserve the sanity of line workers, the cables are organized hierarchically. The twisted pairs of conductors are wrapped in bundles of 25; then as many as 24 bundles are formed into a superbundle of up to 600 pairs; the largest cables are assembled from multiple superbundles. With this arrangement, every wire can be individually identified by looking for combinations of just 10 colors. The plastic insulation on each wire has a solid background color and a contrasting "tracer"—a stripe that spirals around the insulation. The two conductors within a pair use the same colors, but with the background and tracer reversed. For example, in pair no. 1 the tip conductor has a white background and a blue tracer, where-

A telephone lineman in southern Virginia *(right)*, working from a bucket truck 15 feet over the roadside, separates pairs of wires while installing a new splice case. Many older multipair cables *(below)* have an outer sheath of lead rather than plastic, and splice cases have to be soldered in place. The two thick cables are lashed to steel messenger wires, which support their weight.

as the ring conductor has a blue background and a white tracer. Dividing the set of 10 colors into two contrasting groups of 5 allows for exactly 25 combinations with one color from each group; thus, each pair in a bundle can be uniquely colored. A similar color code is applied to the ribbons that bind together all the pairs in a bundle, and to those of the superbundles. The result is highly festive! A specific wire might be identified as the blue-red conductor within the orange-black bundle within the brown-yellow superbundle.

According to old repair manuals, line workers remember the 10 colors by memorizing two five-word sentences. The color series blue, orange, green, brown, slate is encoded in the slogan "Bell operators give better service." For the series white, red, black, yellow, violet, the mnemonic device is "Why run backwards you'll vomit." But I fear these traditions have faded. When I tried the latter sentence on the repairman who came to fix my phone line, he gave me a very quizzical stare.

All those party colors are hidden away inside a rather dull outer wrapper. Older cables have a lead sheath, gray or dun in color. Newer ones have a shield of corrugated aluminum to protect against electrical interference, with a black plastic cover to keep out the weather. When big cables are strung on poles, they are lashed to a steel messenger wire, so that the conductors don't stretch under their own weight.

Multipair cables can also be installed underground, either buried directly in the soil (the usual suburban practice) or pulled through conduits (the preferred method in cities, since the cable can be repaired or replaced without digging). Underground

cables are usually pressurized, so that if any small holes develop in the outer casing, air will leak out, which is better than having water leak in.

The largest multipair cables are slowly disappearing from the landscape. Whenever one of them fails, the replacement is unlikely to be another cable with thousands of copper wires. A route with that much communications traffic is a candidate for some other high-capacity medium, such as an optical fiber.

The Splice Case. The big lump in the overhead telephone cable, which looks like the snake that swallowed the pig, is called a splice case. It is where the telephone wires from your home join others on their way to the central office. In a large cable, a splice case can be the size of an automobile muffler. Under the hinged cover are hundreds of connectors for joining wires. In current practice the individual splices are made with plastic crimp-on connectors that linemen call chiclets. Of course the trick is connecting the *right* pairs.

For underground service, splices are made in a pedestal, which generally pokes up out of the lawn near the curb. It's a metal box a foot or two high, usually painted an insipid green—as if one might mistake it for shrubbery.

Don't Everybody Talk at Once. In the local loop described earlier, the pair of wires running from your telephone to the central office is reserved for your exclusive use 24 hours a day. No one else can talk over those wires. Since you're probably not on the phone all day and all night, this arrangement doesn't make very efficient use of

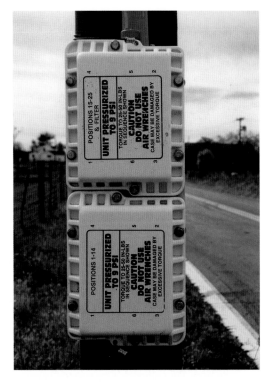

Many buried telephone cables are pressurized to keep out moisture, which means that splice cases and other connection points must be pressure-tight. These cases are on a rural road in eastern Kentucky.

Burning up the wires? Of course it is merely a trick of the light at sunset, which makes the multipair cables draped from the poles along this roadway more conspicuous than they would otherwise be. Heavy cables like these, filled with hundreds of copper wires, are gradually fading from the scene, replaced by fiber-optic technologies. The photograph was made near Greensboro, North Carolina.

equipment. If someone else could use those wires when you don't need them, everyone would save a little money. Schemes for sharing equipment or other resources are known in the communications business as multiplexing.

The grandmother of all multiplexing methods was the party line. (My own grandmother, as a matter of fact, was a party-line participant.) Think of the way extension phones work: if you pick up the phone in the kitchen while someone else makes a call from the bedroom, you both become parties to the conversation. That's the way a party line worked too, except that the extension phones were in someone else's house. All the parties shared a single pair of wires going back to the central office. Actually, a party line was not quite the same as an extension phone: there were special ringing arrangements so you would know which incoming calls you were sup-

—•• ——— — —•• •— ••• ••••

Western Union, the telegraph empire, once had an opportunity to buy the rights to Alexander Graham Bell's telephone. They passed up the chance, saying the telephone would never amount to much.

Perhaps the company's arrogance is understandable. At the time, the telegraph was everywhere and seemed indispensable. The wires crisscrossed the continent and even reached overseas. Large cities had hundreds of operators tapping out messages. Urgent or important news always came by telegram. It was part of the culture.

The technology of telegraphy was remarkably simple, which may explain why it worked so well and lasted so long. The sending key was merely a switch that could open or close a circuit. At the other end of the line, the telegraph sounder had an electromagnet that attracted an iron bar when current flowed; when the circuit opened again, a spring pulled the iron away from the magnet. The result was a pair of clicks for each press of the sending key. The telegrapher heard these rapid-fire clicks as the dots and dashes of Morse code. Reminiscences of former operators tell of attaching a Prince Albert tobacco can to the sounder to make the clicks louder.

The system was not supposed to be so simple. The first telegraphs were equipped not

with a sounder but with a contraption called a register, which automatically recorded the dots and dashes on paper tape. The idea was that a clerk would read the dot-dash markings from the tape and transcribe them for delivery to the customer. But the register apparatus was temperamental, and the clerks listening to its ticking were soon writing down the message before the paper tape emerged. (Years later, though, the telegraph companies were using teletypewriters, which tapped out readable messages directly on paper, leaving not even the job of transcription to the operator.)

Samuel F. B. Morse, the American portrait artist whose name is so closely linked with the telegraph and its code, sent his first message in 1837 in his studio on Washington Square in New York. The famous "What hath God wrought!" message from Washington to Baltimore was sent in 1844. (The second message on that line was less bombastic and more gossipy: "Have you any news?") A transcontinental link was open by 1861; it took just four months to string wire from Chicago through Omaha and Salt Lake City to San Francisco. The charge for sending 10 words was $6, which was then equal to a laborer's weekly wage. Nevertheless, it put the Pony Express out of business.

The telegraph has a prehistory that goes

back long before Morse. Signaling from hilltops with flags or bonfires or mirrors was an art known centuries ago; a relay of fire beacons, supposedly, brought the news of the fall of Troy back to Greece.

The first communications infrastructure to be called a telegraph was begun in France in the 1790s by Claude Chappe. The system flourished throughout the Napoleonic era and remained in use until 1853. Chappe's telegraph had no wires or electrical gadgets. The signaling was done with mechanical semaphores, flaglike arms that could be set at various angles to indicate numerical codes or letters of the alphabet. Semaphore stations were set up on high ground at intervals of several miles. The attendant at each station would observe the next station down the line with a telescope, write down the symbols transmitted, and then relay them to the next station. At one time there were more than 500 stations, stretching from Paris to Venice.

In the United States a few semaphore stations operated briefly. Telegraph Hill in San Francisco gets its name from such an installation, and Massachusetts has seven places named Telegraph Hill.

As for Western Union, the name survives, but the last telegraph operations in the United States were shut down in 1989.

posed to answer, and which ones you were supposed to secretly eavesdrop on. The party line may not have had much to recommend it technically, but it was an interesting social institution.

Since my grandmother's day, telephone companies have found other ways to share local-loop circuits. In a suburban housing development you may spot a metal cabinet labeled *MUX* or *MXU*. I've heard line workers refer to these boxes as muxes or muxers (short for multiplexors); they're also known as concentrators. Their function is to funnel all the local loops from a neighborhood into a smaller allotment of wire pairs.

Suppose the houses in the neighborhood have 100 telephone lines overall. In a conventional system, 100 twisted-pair circuits would run all the way to the central office, but with multiplexing these wires extend only as far as the mux cabinet. From there, cables carry a smaller number of circuits—perhaps 30. When a neighborhood resident wants to place a call, equipment in the mux cabinet connects the resident's local loop to one of the 30 central-office circuits. When the call ends, the circuit is made available for someone else. What happens if more than 30 people try to use the phone at once? Somebody is out of luck; the 31st call won't go through.

Multiplexing is not a new idea. There were multiplexed telegraph lines in the 1880s. Another early form of multiplexing carried Muzak into offices and factories. (It's a remarkable fact that there was elevator music almost as soon as there were elevators.) In the telephone system, multiplexing has long been used in the "trunks" that run between central offices and in the long-distance network, squeezing hundreds or thousands of conversations into a single circuit.

The Central Office. Even in smaller towns, the telephone exchange building is likely to be a substantial edifice, not ostentatious like the county courthouse or the First National Bank, but built of solid brick and two stories tall. Why the imposing building? In part, no doubt, the phone company just wanted to make an impression. But it also needed a lot of space. In the early years the space was for people: dozens of switchboard operators, clerical employees, and maintenance staff. Later it was for machinery: the long racks of click-clacking mechanical switchgear that displaced the operators. Now, much of the space is simply empty. The electromechanical switches have been hauled away to the scrapyard, replaced by electronic ones that are getting steadily smaller, like everything else in the world of computers. A machine the size of a pizza box can handle a town's telephone system.

The offices are lonely as well as empty. Quite a few are "lights-out buildings"— entered so infrequently that they are left dark. The machines run themselves and call for help when they need it.

Telephone offices have so much extra room now that they're taking in boarders. It's called colo space, short for colocation. If you sell Internet access, for example, or provide paging services, the basement of a telephone central office is a convenient and secure spot to put your equipment, where it can be wired directly into the telephone network.

The orange labels identify the contents of this cabinet as "MXU 10" and "MXU 11." The multiplexing units provide telephone service to an outlying neighborhood without stringing a separate pair of wires all the way from the central office to each house.

A telephone-exchange building is distinguished by what it lacks: windows. This switching office, in Durham, North Carolina, evidently once had windows, but they were bricked in.

Something else to notice about central offices and other telephone switching facilities: they are windowless. Recent offices are built without windows; older ones have had the window openings bricked up. These blank-faced fortresses are said to be a legacy of the Cold War; windows were eliminated to improve the odds that communications links would survive a nuclear attack. (I have been unable to find any documentary evidence of regulations or legislation mandating this policy.)

For a few years, the highest-priority telephone facilities were given even more serious blast hardening. A nationwide defense network built in the 1960s and 1970s had its switching offices built in underground concrete bunkers, with the machinery hung from the ceiling on springs. These buildings were supposed to survive a 20-megaton explosion two miles away. Most of them are now abandoned or converted to less dramatic uses. A bunker in western Massachusetts serves as overflow book storage for the Amherst College library.

Batteries Included. All the on-line equipment in the central office runs on batteries. As a matter of fact, the entire global telephone network is battery-powered.

As you might guess, the batteries look nothing like the ones you put in your flashlight. They are lead-acid batteries, the same kind that start your car, but are housed in glass or plastic cases about the size and shape of a "jerry can" for gasoline. Each case is a single lead-acid cell that produces a little more than 2 volts of electricity. The cells are wired together in groups of 24 to generate 48 volts overall. (A car battery has six cells in series, producing 12 volts.) Depending on the size of the office, there

may be several banks of 24 cells each. They are usually arranged on tiered shelves, lower in the front and higher in the back, like gymnasium bleachers.

Why batteries? Why not just plug into the power grid? Part of the reason is historical. The telephone entered American life before electric power was widely available. And once the telephone system was designed to operate on the 48 volts of direct current supplied by batteries, it was too late to change over. But in any case batteries also have an advantage in reliability. After all, they are what you use when the lights go out!

The batteries have to be continually recharged, and the energy for that does come from the national power grid. If the utility power fails, the batteries have enough capacity to keep the phones working for several hours. Long before the batteries run down, an emergency backup generator kicks in.

Rings, Beeps, and Buzzes. When you pick up the telephone, it hums at you. At other times it goes "brng-brng-brng" or "bzzz-bzzz-bzzz." Remarkably, we all seem to understand these weird noises and respond to them appropriately. At the other end of the line, the switching equipment in the telephone office apparently understands the signals *we* send, such as the melodic "touch tones" of the push-button dial.

The dial tone is actually two tones—two audible frequencies sounded at the same time, like a chord on the piano. In North America the standard frequencies are 350 and 440 hertz, which correspond closely to the notes F and A above middle C. The busy signal consists of 480 hertz and 620 hertz, interrupted 60 times per minute. There's also a "fast busy," which uses the same tones but switches on and off 120 times per minute; you get this signal—formally known as "all-trunks busy"—when

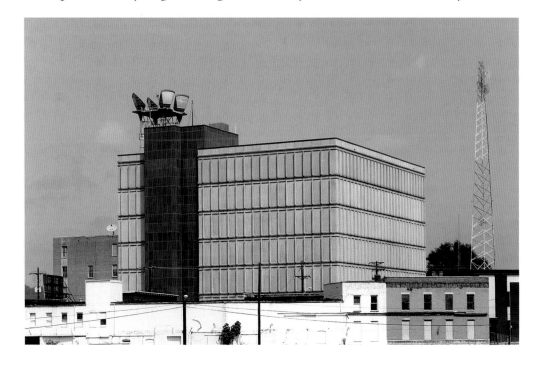

In Knoxville, Tennessee, a much larger and somewhat newer central switching office was windowless from the outset.

THE NORTH AMERICAN NUMBERING CRISIS

Of all the shortages that might afflict us, who could have guessed that we would be running out of numbers? Surely that's an inexhaustible resource if ever there was one. But numbers of one particular kind—telephone numbers—have been desperately scarce for the past decade.

A North American telephone number has 10 digits: three for the area code, three for the central-office code, and four for the local-line number. The 10-digit format seems to allow for 10 billion telephone numbers, starting at 000-000-0000 and running through 999-999-9999. That ought to be plenty—enough to give every person on the continent 20 or 30 telephones. But it turns out that not just any 10-digit number can be a phone number.

For example, no area code or central-office code begins with a 0, because 0 is reserved for dialing the operator. Forbidding an initial 0 wipes out 100 possible area codes and a million potential phone numbers within each area. For slightly different reasons, a first digit of 1 is also not allowed, eliminating another 100 area codes and another million phone numbers per area.

In part, the prohibition of initial 0 and 1 digits is a legacy of Almon Strowger's switching mechanism from the 1880s. That device and its descendants were "step-by-step" switches, meaning that the machine had to decide what to do with the dialed digits one at a time, without waiting to see what might come next and with no opportunity to go back and change its mind. When you dialed a 0, you were immediately connected to the operator, so any subsequent digits would be ignored.

For many years there were lots of other restrictions on the format of a valid phone number. Central-office codes avoided 0 and 1 in the second digit as well as the first. This was because the codes used to have names, such as BUtterfield 8, and on the telephone dial 0 and 1 have no alphabetic equivalents. Area codes, in contrast, always had a 0 or 1 as the middle digit, to help distinguish them from central-office codes.

In 1947, when the system allowing people to dial their own long-distance calls was first set up, North America was divided into 86 area codes, with another 50 codes held in reserve. The plan allowed for a total of 696,320,000 phone numbers. This is only 7 percent of the 10 billion that would fit in a 10-digit format, but it seemed like more than enough. There were fewer than 50 million numbers in use then, and

planners thought the scheme would last for 300 years. They didn't count on cell phones, fax machines, beepers, and modems. And they assumed that the system would continue to be controlled mainly by one company, not shared with hundreds of competitors, each demanding its own allocation of numbers.

By the late 1980s the miscalculation had become obvious, and the planning authorities began scrambling to find more numbers. They relaxed the restrictions on the format of both central-office codes and area codes. Only the prohibition of 0 and 1 as the leading digit remains inviolate; apart from that, anything goes. The effects of this change have been most dramatic in the case of area codes. Instead of 136 available codes, there are now 800 possibilities (all three-digit numbers from 200 to 999). Some of these codes are reserved for special purposes (for example, 411 and 911), leaving 675 "assignable" codes. More than 350 of these are in use, and most of the rest are spoken for in one way or another. Only 29 uncommitted area codes remain.

The proliferation of area codes has changed the telecommunications landscape. In 1947, area code 213 covered all of southern California from Santa Barbara down to the Mexican border and from the ocean to Arizona. Today, 213 designates a few blocks of central Los Angeles, and the rest of its former territory has been carved up into 14 new area codes, with more to come.

For a while, it looked like we would run out of numbers completely within just a few years, but conservation measures have eased the crisis somewhat. The most important steps have prevented telephone companies from stockpiling numbers they don't actually need. Under the old system, numbers were allocated in blocks of 10,000, even if a company needed only a few; the new rules avoid this waste.

The latest projections suggest that we'll use up the last available 10-digit telephone number sometime in the 2030s. At that point American telephone numbers will have to grow beyond the traditional 10 digits—unless by then we have figured out how to dispense with telephone numbers altogether.

the system can't find an available circuit to reach your destination. Then there's the receiver-off-the-hook tone, the obnoxious noise that grates on you when you forget to hang up. It is made up of four discordant frequencies (1,400, 2,060, 2,450, and 2,600 hertz) at a higher volume than other signal tones. An earlier off-the-hook warning was a siren-like "howler" that was so loud there was a risk of ear damage.

The most deceptive signal is the sound of the telephone ringing at the other end of the line, while you're waiting for a call to be answered. It's a total fake. Yes, it sounds just like a muffled version of a standard telephone ringer, but think a minute. Telephones these days make all sorts of strange bleeps and warbles, but the "ring-back" tone you hear while waiting for a call to be answered is always the standard old Bell System ring. In fact, the sound does not come from the other telephone at all; it is generated in your own central office. It consists of tones at 440 hertz and 480 hertz, cycled on for two seconds and then off for four seconds.

The classic old pre-electronic ringer had some lore associated with it. There were two brass gongs, tuned to slightly different pitches, and a steel clapper that bounced back and forth between them. As the story goes, the Bell System wanted to encourage people to answer the phone more quickly, since a ringing phone ties up equipment but generates no revenue. An advertising campaign helped, but what was most effective was a ringer so loud, unpleasant, and imperious that no one could stand to ignore it. As the story came down to me from a long-time Bell employee, "Americans are the only people in the world who will interrupt sex to answer the telephone."

Dialing and Switching. Some of the latest, fanciest telephones boast of a voice interface. There's no need to dial a number; you just pronounce out loud, "Call the office." My grandmother's phone had the same feature more than 50 years ago. It didn't even have a dial. She just picked up the receiver and said, "Hello Jenny. Would you ring Noretta for me, please? Thank you, dear."

Jenny and 250,000 other operators (nearly all women) comprised the principal switching mechanism of the telephone system in that era. What Jenny did in response to my grandmother's request was take a plugwire connected to my grandmother's local loop and insert it into a jack connected to my Aunt Noretta's line. The possibility of automating this process had been explored as early as the 1880s. The story has a quality of legend about it: Almon B. Strowger was an undertaker in Kansas City who worried that calls meant for him were being diverted to one of his rivals. It seems the operator for the local telephone exchange was the rival's wife. Strowger found a technological fix: a machine for directing calls without human intervention.

The heart of Strowger's mechanism was a 10-position selector switch, with a pivoting central arm that could rotate to connect with any of 10 electrical contacts arranged in a semicircle. The pivoting arm was moved by an electromagnet, with the help of various springs and ratchets. Each time the electromagnet received a pulse of current, it advanced the arm by one position. In the first network to try Strowger's idea, the customer operated the switch by means of push buttons. To dial a 7, you

pressed a button seven times, thereby sending seven pulses of current to the electro-magnet. The push buttons were soon replaced by a rotary dial, which automated the counting of pulses.

Strowger's electromechanical marvel made "chun-ka-chunk" noises as the rotary switches stepped through their 10 positions. In some later technologies the circuit switching was done by tiny magnetic reeds sealed inside glass vials; they made a quite different sound, like the rustling of insects. Now switching is done by computers that have no moving parts, and all you hear is the whoosh of ventilating fans.

Whatever the switching technology, the control panel that puts you in charge of all this machinery is the dial or the number pad on your telephone. The original dial—the round one with 10 finger holes—works by repeatedly opening a switch and thereby briefly interrupting the local-loop current. The scheme behind touch-tone dialing is known officially as DTMF signaling, for Dual Tone Multiple Frequency (a phrase that seems vaguely redundant). Each button produces two musical notes at the same time; the central office listens for these sounds to identify the number you are dialing. The pair of tones assigned to a button depends on the button's position in the keypad: each row and each column has a separate frequency. For example, all the buttons in the top row share a frequency of 697 hertz, and all those in the left-most column have a frequency of 1,209 hertz; when the central office hears these two frequencies in combination, it knows that a 1 has been dialed, since that is the button at the intersection of the top row and left column.

Four rows and three columns provide 12 combinations of tones—two more than the 10 digits needed to dial a telephone number. The two extra combinations are the * and # keys. What do you call the # key? You might hear it referred to as the pound sign, the sharp sign or the hash mark, but according to official standards of the International Telecommunications Union it is the square. Insiders, however, like to call it the octothorpe. And you thought the phone company had no sense of humor!

LONG DISTANCE

The technology of the local loop really is *local*. There is no boosting or amplifying of the voice, which limits the range to a few miles. Trying to communicate coast-to-coast with local-loop technology, you might as well just shout. Because of this problem, telephone companies were essentially disconnected islands for 40 years. The first coast-to-coast telephone call wasn't made until 1915.

What made long-distance service possible was the vacuum-tube amplifier, which works much like the audio amplifier in a home stereo system. A faint signal comes in one end, and a much stronger copy of the signal goes out the other. Of course, amplifiers don't use vacuum tubes anymore; the glass bulbs with their glowing filaments have long since been replaced by transistors and integrated circuits. (It's not a coincidence that the transistor was invented by three employees of the telephone company.)

To keep the volume up, long-distance calls have to be boosted again and again as they make their way cross-country. Some lines have an amplifier every eight miles; in others the amplifiers are as closely spaced as one per mile. When you place a call over one of these wires, your voice decays and is restored to full volume hundreds of times along the signal path.

In some areas of the country, telephone amplifiers are a commonplace—though generally little noticed—element of the industrial landscape. They are typically housed in bright stainless-steel canisters, about the size and shape of a five-gallon paint bucket, mounted on a post or attached to a utility pole a few feet above ground level. The cables—one coming and one going—enter through the bottom of the housing. Both the cables and the canister are pressurized to keep out moisture, so there's an airtight seal and a label warning the repair crew to bleed off the air pressure before unlocking the latches, or else the lid might fly into the air.

An amplifier is a one-way device. Like a telescope, it has a small end and a big end, and you don't want to put a signal through it backward. For this reason amplified lines can't work the same way as local loops, where signals flow in both directions over a single pair of wires. The long-distance network is made up of "four-wire" circuits, with a separate pathway for each half of the conversation. It's quite possible for the two parts of a connection to follow different routes. If you're in Philadelphia and talking to someone in Seattle, what you say might go through Omaha, whereas what you hear might come back via St. Louis or Dallas. This is why you sometimes get a connection that's noisy in one direction but clear in the other.

Aluminum "lobster pots" by the roadside house amplifiers for long-distance telephone lines. The amplifiers and the cables they connect to are pressurized to keep out moisture.

Multiplexing. Although it's called a four-wire connection, that doesn't mean there are four physical wires between your telephone and the one you're calling. As a matter of fact, you don't even get one wire all to yourself. The phone company squeezes thousands of calls onto each wire. This is another form of the process called multiplexing, discussed earlier in the context of the local loop. Multiplexing for long-distance lines works a little differently. There are several variations.

Frequency multiplexing is much like radio broadcasting. A radio receiver picks up different stations when you turn the dial; all of those signals are in the air at the same time, but they don't get mixed up because they occupy different frequency bands—different places along the electromagnetic spectrum. Telephone frequency multiplexing is just the same. Each conversation rides piggyback on a different frequency band, then all the frequencies are mixed together and transmitted over a single wire. At the far end, the jumble of signals is delivered to a bank of receivers, each of which is tuned to a separate carrier frequency. In effect, each receiver listens for just one voice, extracting it from the hubbub of other conversations.

A second form of multiplexing is even more bizarre. When you're talking on the telephone, you have the impression that there's a continuous, uninterrupted connection between your phone and the one at the other end of the line. But on most out-of-town calls, this is an illusion. The truth is that your voice is being chopped up into thin salami slices so your conversation can be shuffled together with hundreds of others. You actually hold the line for only one eight-thousandth of a second before you have to surrender to the next party waiting for a chance to talk. In one common scheme for time multiplexing, 24 calls are sliced and diced in this way.

There's more. The long-distance network has gone digital. It's not your voice that goes out over the telephone line. It's not even an electrical analog of your voice. It's just numbers, bits, on-off pulses like those that race around inside computers. To digitize your voice, the telephone system measures the height of the sound wave 8,000 times per second, then transmits the measurements in the form of binary (base-2) numbers. The stream of data for one digitized conversation amounts to 64,000 bits per second. At the far end of the line, the numbers are read off the line and used to reconstruct your voice. Digital phone calls can be multiplexed even more easily than analog ones. You just need to send more bits per second.

Coaxial Cable. When telephone engineers first tried squeezing multiple conversations onto a single circuit, they used ordinary twisted pairs of copper conductors. They soon ran into problems.

Multiplexing conversations produces high-frequency signals—not just the few thousand hertz that the local loop was designed for, but a million hertz or more. High frequencies do weird things to electrical circuits. We tend to think of a wire as a kind of pipe for electricity: The electrons flow through the solid metal the way water flows through the interior of a pipe. The electrons are kept from leaking out by the insulating material that surrounds the wire, just as the water is confined by the walls of

the pipe. But the metaphor breaks down with high-frequency signals, which are more like waves than flowing water. The waves want to spread out in space like ripples that spread across the surface of a pond. As a result, less of the wave reaches its destination; the energy is dissipated.

Running two closely spaced conductors side by side, as in the twisted-pair arrangement, reduces the energy loss to some extent. The waves are partially confined to the space between the two conductors. In effect, each conductor shields one side of the other conductor. The scheme would work even better if you could shield all sides of a conductor. That's what coaxial cable does, by a clever trick: it puts one conductor inside the other. The conductor in the middle is a thin solid wire, which runs through a hollow metal tube. The two conductors are separated by an insulat-

BREAK GLASS, PULL LEVER

The bright red metal box on the street corner that summons the aid of the fire department is a living fossil among communication systems. The technology goes back to the middle of the nineteenth century, when the telegraph was still a novelty and the telephone undreamt of. The first alarm turned in by a call box was transmitted on April 26, 1852, in Boston.

The simplest fire-reporting system would run a separate pair of wires from a central station to each alarm box. Pulling the lever on a box would close a switch and thereby light a signal identifying the specific box. But wire was expensive in 1852, and large cities needed thousands of boxes. To economize, the designers of the early systems devised a means of sharing wires, a technique that would now be described as code-space multiplexing—the latest trick in cell-telephone systems.

In a typical alarm system, one pair of wires runs through many call boxes in series. Under normal circumstances, a current is always flowing through this circuit. Any box can signal an alarm by interrupting the current, but the system needs some way of identifying which box was activated. The answer is a marvel of nineteenth-century clockwork ingenuity. Inside the box is a spring-driven wheel, with a distinctive pattern of notches on the perimeter. When you pull the lever, you set this wheel

in motion, and each notch momentarily opens a switch that interrupts the circuit and thereby rings a gong at the station. Firefighters are expected to learn the pattern of bells that identifies each call box, although from early years there have also been devices that record the box number, usually on paper tape.

The wheel for call box 312 would have three closely spaced notches, then a blank space, a single notch, a second space, and two more notches. At the station, the box would sound the pattern: "ding ding ding—ding—ding ding."

Having the alarm mechanism interrupt a normally closed circuit (rather than close a normally open one) is a fail-safe feature. A break in the wiring is immediately apparent. On the other hand, the alarm system is susceptible to some other kinds of failure. If two alarm boxes are pulled at the same time, the signals they transmit will be garbled. (In the vocabulary of modern networking, the system has no protection against collisions.)

Now that telephone service is nearly universal, many cities have dismantled their fire-alarm networks, but others are maintaining them. New England is particularly rich in bright red boxes, but they can also be found in San Francisco, Atlanta, and Buffalo.

In many places, the boxes now communicate with headquarters over conventional telephone wires or even over radio instead of private circuits; and instead of ringing a gong, the alarm signal may go directly into the computer that runs the municipal 911 system.

Interestingly, the company that built the first alarm-box network in Boston in 1852 is still in the business of making and selling emergency-reporting equipment. It is the Gamewell Company of Ashland, Massachusetts. Its emblem can be found on most of the surviving alarm boxes, including the one in the photograph, spotted in Concord, New Hampshire.

On a hilltop in western New Jersey a tower holds aloft
a dozen of the sugar-scoop antennas that are the trade-
mark of microwave relay links used for telephone traf-
fic. The antennas come in pairs, one for transmitting
and one for receiving. Each pair is carefully pointed at
a matching pair of antennas on another tower 20 or 30
miles away. The configuration of the antennas on this
tower suggests that it lies at the intersection of three
microwave relay routes. At each of three levels, a set of
four antennas is arranged to pass signals along a route
aligned on a different compass direction. Hundreds of
towers very much like this one were built all across the
United States starting in the 1950s; most of them are
still standing. A few microwave towers are different in
appearance although identical in function. The upper
photograph on the opposite page shows a tower in
Greensboro, North Carolina, which was given a fancier
architectural treatment. Many European microwave
towers are even more elaborate. The one in the lower
photograph on the opposite page stands in the Italian
city of Verona. The large windowed structures below
the antenna level hold transmitting and receiving equip-
ment that in American practice would be installed in a
separate building at ground level.

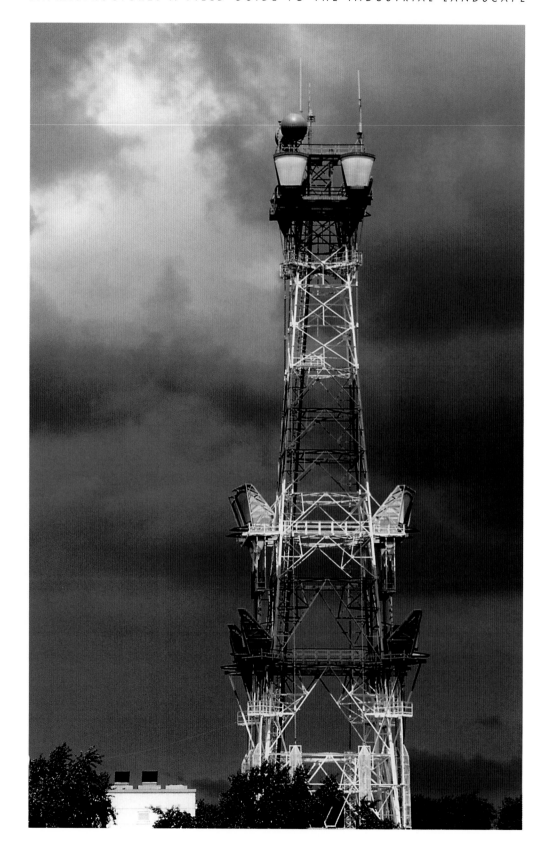

ing layer, which keeps the inner conductor precisely centered, or coaxial—hence, the name. To those who work with the stuff, it's just coax (pronounced with two syllables—it's what Aristophanes thought that frogs say).

Coaxial cable was an invention of the 1940s. The first uses were not for ordinary voice telephone circuits but for high-grade broadcasting links, such as a connection in New York City between Radio City Music Hall and transmitters at the Empire State Building, some 15 blocks away. The first long-distance telephone link on coax ran from New York to Philadelphia and Washington.

MICROWAVES

Pushing on to still higher frequencies, even coaxial cable reaches its limits. In the gigahertz range—a billion cycles per second or more—the metal conductors of coax begin to disrupt the waves more than they conduct them. Rather than try to confine the signal to a wire, it would be better to send it through the open air as a radio wave. This is the idea of microwave transmission. Microwaves get their name because their wavelength is short compared with other radio signals. Wavelengths go from a few inches down to less than an inch.

Microwaves have a built-in economic advantage. When you lay down thousands of miles of coaxial cable, you have to pay not only for the cable itself but also for a right-of-way across the land. Microwave technology is wireless. You need to buy sites for the towers that transmit the signals, but there's no charge for the use of the air between towers.

Out in the countryside, the microwave relay towers of the long-distance network are hard to miss. They are those giant "ears" or "sugar scoops" pointing in various directions atop a steel tower on high ground. They have to be high because microwave links require a clear line of sight between transmitter and receiver. The waves can't go around corners or follow the curvature of the earth or tunnel through a hill that happens to be in their way.

The sugar-scoop antennas are properly called horn-reflector antennas, and they work somewhat like a megaphone to efficiently transmit and receive waves. The distinctive shape is designed to form the microwaves into a well-defined beam at the transmitter end, and to gather in such a beam at the receiver. (The transmitting and receiving antennas look the same; there's no way of telling them apart.) The business end of the antenna is at the bottom, where the horn tapers down to a narrow throat. There the signal from the transmitter at ground level is fed into the antenna and travels upward, spreading out through the conical or pyramidal part of the horn. At the top the waves bounce off a surface tilted at a 45-degree angle. This surface is not a flat plane but a section of a paraboloid—a surface that, like a flashlight reflector, creates a beam of parallel rays. The beam emerges through the opening at the front of the horn. But the opening usually isn't really open: a cover of fiberglass or fabric

An especially handsome conical horn antenna, made of satin-finished metal, is installed on a low tower in Yuma, Arizona. The fabric cover over the face of the horn is transparent to microwaves. The pipe joined to the bottom of the horn is a wave guide for conducting microwaves; a second wave guide next to it continues up the tower to another antenna. On the opposite page, a small tower at a television station in Norfolk, Virginia, *(upper photograph)* sprouts multiple microwave antennas in a different style: they are dishes, drums, domes, and cones. Antennas of this kind are used mainly for private communications links. (The triangular structure in the middle of the tower is not a microwave antenna but a cellular-telephone antenna.) Another drum antenna *(lower photograph)* has lost its cover and hence provides a rare glimpse of what's inside the drum. The device on the central stalk emits and receives microwaves, which are then focused by the paraboloidal surface at the back of the drum. The side walls are lined with a fabric that absorbs microwaves, in order to suppress stray reflections. The antenna is on a hilltop communications tower in Trieste, Italy.

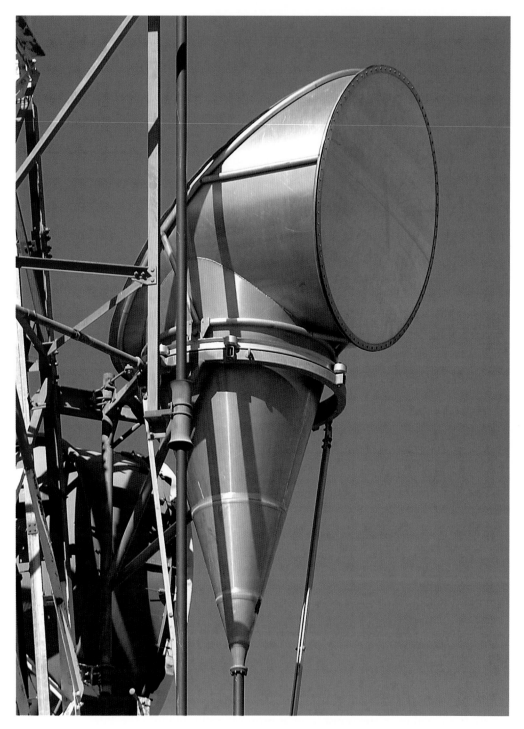

keeps out nesting birds and insects. The cover material is transparent to the microwaves, and so they escape the horn as if through a window.

At the receiving end, the waves follow the opposite path. They enter through the window, bounce off the paraboloidal backplane, and are funneled into the throat of the horn; from there the signal travels down to the receiver below.

In most horn reflectors the horn is a four-sided upside-down pyramid, with a square cross section. But some later antennas have a conical horn, with a circular cross section. They are known as circular cornucopia antennas—or simply as ice cream cones.

Telephone microwave links are one-way circuits, so the horn antennas usually come in pairs to create a two-way channel. (Sometimes a third antenna is added as a spare.) The beams going in opposite directions along the same route are operated at slightly different frequencies so as not to interfere.

What limits the distance between towers is not just the need for an unobstructed line of sight. The main limit is "rain fade." It turns out that water is a good absorber of microwaves, and so a heavy shower could interrupt transmission. The problem gets worse as the operating frequency goes up. The telephone system uses frequencies near the bottom of the microwave spectrum, at four gigahertz and six gigahertz, where the maximum practical distance is about 30 miles.

Some microwave towers are built next to (or on top of) telephone switching offices, but out in the countryside they tend to stand in lonely isolation on hilltops. At the base of the tower you'll find a small building, or hut, that houses the transmitters, receivers, and other electronics. Between the antennas at the top of the tower and the hut at the bottom, microwave signals travel through wave guides. As noted earlier, wires are not much good for carrying signals in the microwave range; they act more like antennas than conductors. A wave guide is an inside-out wire—a tube with metal walls and a hollow interior. In other words, it's a lot like a pipe, and people who work with microwaves refer to wave-guide systems as plumbing. But the way a wave guide works is more complicated than the way a pipe carries water; it's more like an organ pipe than a water pipe. The wave guide has to be just the right shape for the microwave signal, typically with an inside diameter of half the wavelength. Since the four- and six-gigahertz signals of the telephone system have wavelengths of two or two-and-a-half inches, the wave guides are about an inch in diameter.

Wave guides generally go straight up the tower, without making any bends or turns. Every change in direction can cause disruptive reflections. Indeed, wave guides have a reputation for being finicky and temperamental—a dent can be disastrous, and even a tiny spot of corrosion or a loose joint can block transmission.

The lowest few yards of each wave guide are sometimes surrounded by metal-mesh shielding, which is meant to absorb any leaks. The shielding protects employees working close to the wave guide. High-intensity microwaves are nasty; they're what cooks a meal in the microwave oven.

Microwave communication has a surprisingly long history. A microwave link between Dover and Calais carried phone calls (one at a time) as early as 1931, and in the 1940s a similar connection provided phone service to Catalina Island, off the coast of California. But microwave entered the mainstream only in the 1950s, with the completion of the first coast-to-coast chain of relay towers. There were 106 of them.

Telephone companies have no monopoly on microwave communication. Private microwave links usually have antennas that are nothing like the giant sugar scoops of

the phone company's towers. Instead, they are mostly circular structures—dishes, domes, drums and cones. The parabolic dish is the basic form: a concave metal reflector that gathers parallel rays and brings them to convergence at a focal point. Dish antennas are better known for their use in satellite communication. Whereas satellite dishes are usually built at ground level and point upward, dishes for terrestrial microwave circuits are mounted on towers or rooftops and are aimed horizontally at a similar unit a few miles away.

The drums, domes, and cones are really dishes in disguise. In each case the rear surface of the antenna is a parabolic reflector, with the same shape as a dish. The rest of the structure just offers protection. Private microwave links generally operate at a higher frequency than those of the telephone system—in bands at about 18 and 23 gigahertz. Rain fade is a more serious problem in these bands. Users put up with that nuisance because the antennas are smaller and cheaper.

In some dish antennas the reflector is not a solid surface but is made up of tubes or wires arranged in a grid pattern. Looking at this sievelike structure, you might think the microwaves would just slip right through, but in fact the porous surface makes a good reflector as long as the grid lines are closer together than half a wavelength. Grid reflectors are popular because, while stopping microwaves, they allow the air to sift through. As a result they create less wind load than a solid reflector, and can be mounted on a flimsier tower.

One of the strangest of microwave antennas sits at ground level below a more conventional microwave tower in Monrovia, Maryland. The antenna is made of concrete and formed part of a communications network design to survive a nuclear attack.

FIBER OPTICS

All the historical shifts in telecommunications technology through the past 100 years—from copper wires to coaxial cable to microwaves—are wholly overshadowed by the revolution now under way. Optical fiber reduces every other transmission medium to insignificance. The first fiber-optic telephone link was installed in Chicago in 1976. Within 20 years, more than 90 percent of all long-distance voice circuits were fiber.

Light Pipes. The optical fiber itself is simple to describe, although making the stuff is a high art. The fiber is a fine strand of glass with two concentric layers—the core in the middle and the cladding surrounding it. Both layers are glass, but they differ slightly in chemical composition and physical properties. In particular, the core has a higher index of refraction, which measures the ability of the glass to bend light rays as they enter or leave. The difference in refractive index has an extraordinary consequence: light can become trapped inside the core of the fiber, with no leakage through the cladding. In effect, the core is a pipe for light, a pipe whose inner surfaces are perfect mirrors.

An ordinary mirror is far from perfect—and the problem is not just that you always look 10 pounds heavier and 10 years older than you might prefer. At the reflecting surface of the mirror, some light is absorbed, some is transmitted, and some

gets scattered in the wrong direction. But inside the fiber an effect called total internal reflection eliminates all these losses. Light rays in the core bounce off the cladding without losing any of their intensity. Nothing leaks out. Total internal reflection sounds like magic, but it's really not so exotic. If you swim underwater and look up at the surface, the silvery patches are areas of total internal reflection, where light cannot escape from the water into the air.

The glass in optical fibers has to be extraordinarily transparent. When you look through a window pane, it seems clear enough, but you are looking through a thickness of only an eighth of an inch. If you try looking into the edge of a sheet of glass, all you will see is a deep, dark emerald green. Not much light gets through from edge to edge, even over a distance of only a foot or two. Trying to push light through thousands of miles of such glass would be totally impractical, like rowing a boat across dry land.

It is the development of exceptionally clear glass that has made long-distance fiber optics possible. The fiber is not only more transparent than window glass; it is also more transparent than the clearest water or the clearest air; the only thing clearer is nothing at all—the vacuum of outer space. The key to achieving this clarity is to eliminate all impurities from the glass, since even rare contaminant atoms absorb light. The impurity that's hardest to get rid of is water, which slowly sneaks into the fiber throughout its life, no matter how carefully the glass is sealed. Apart from absorption, another problem is scattering, where light strikes tiny imperfections and bounces out of the fiber. Both causes of attenuation have been reduced to such a low level that light can go through 50 miles of fiber and lose only 1 percent of its strength.

Almost all optical fiber today carries digital rather than analog signals. At the input end of the fiber a laser flashes on and off to represent the ones and zeros of the digital signal. At the other end is a detector somewhat like the photocell in a burglar alarm. The basic idea is not fundamentally different from the flashing lights used to communicate between ships at sea. But the laser flashes off and on far faster than any sailor could blink a signal light. There are fibers capable of carrying 129,024 separate voice channels all at the same time—the telephone chatter of a large city funneled into a single fine thread of glass. In terms of digital data, that's about 10 billion bits per second.

Fiber on the Landscape. Delicate glass fibers are not just strewn across the countryside bare. They come packaged in many layers of protection. One effect of all this casing is that fiber cables don't look at all glasslike. They look so much like ordinary copper wires that many phone companies put special colored tags or markers on them to warn repair crews, lest a hasty snip of the lineman's pliers cut off a hundred thousand phone calls to Cleveland.

Bare optical fiber is as fine as a hair, but stiffer and more fragile. The outer diameter of the glass cladding is 125 micrometers, which is about one two-hundredth of an inch. The core is only 8 or 9 micrometers in diameter. The first protective layer surrounding the fiber is a clear coating of hard acrylic plastic applied as the fiber is

Fiber-optic cable strung on utility poles often bears yellow or orange warning tags *(above)*. The warning does not signify that there is anything dangerous about the cable; rather, it is meant to warn off repair crews who might mistake optical fiber for an ordinary copper cable. The sheath of a fiber-optic cable may also offer identifying information *(below)*, including distance in feet or meters.

Spare hanks of fiber-optic cable, stored on various overhead fixtures, have become one of the most visible signs of the rewiring of the communications infrastructure. The extra cable is left in place to facilitate repairs or minor rerouting of the line. It has to be stored in a manner that avoids sharp kinks, which could damage the fiber. The most common fixture is the eyelet or thimble seen in the top photograph; usually, a pair of thimbles is installed, and the spare cable is looped around them in a figure-eight arrangement. There are also various other loops and coils.

being manufactured. The plastic helps to keep out moisture and to prevent nicks or breaks. It brings the thickness of the fiber up to either 250 or 500 micrometers—a quarter or half a millimeter.

The coated fiber is encased in a still thicker layer of plastic called a buffer. Then multiple buffered fibers are gathered together in groups to make a fiber-optic cable. Cables include as many as 96 fibers, along with a steel or Kevlar strength member, all surrounded by plastic and fabric sheaths. In some buried cables there's also a layer of steel armor. The armor affords some protection against the hazards of pick and shovel, but its main purpose is to discourage the scourge of all buried wiring: gophers.

A fiber-optic cable with 96 fibers represents a communications flow of Amazonian dimensions. Each fiber carries signals in only one direction, so there are 48 pairs for two-way communication. If all 48 pairs were running at the maximum rate in common use today, they could carry more than six million conversations; in other words, nearly everyone in the city of New York could talk to someone in Los Angeles, all over a single cable roughly an inch thick.

Fiber gets installed both underground and up on utility poles. The buried cable, of course, is out of sight, but there are often clear signs on the surface to indicate what's below—more so than with many other kinds of underground infrastructure. Most telecommunications companies put up plastic posts or pylons along the route.

Sometimes the cable is buried directly in the soil by digging a slit trench with a power tool that looks like a jumbo chainsaw. In built-up areas the fiber cable may run through a conduit, which is often a colorful, flexible plastic pipe, about two inches in diameter. The conduit offers a bit of extra protection, but its main purpose is to allow the cable to be replaced or supplemented with additional cables without having to dig up the entire right-of-way. Small concrete vaults every few hundred feet provide points of access where the fiber can be fished out if needed.

Companies take pains to protect fiber cables against damage. Fiber is immune to the rain fade that afflicts microwave communication, but it is vulnerable to "backhoe fade." When a single strand of glass can carry such an enormous volume of traffic, a cable cut can be very disruptive.

Fiber-optic cables strung up on utility poles look remarkably like ordinary copper wires, but there are several ways to distinguish them if you know what to look for:

• At each pole there is likely to be a little slack to allow for thermal expansion and contraction. Sometimes the steel messenger wire that supports the fiber-optic cable is clamped directly to the pole, but the fiber itself is allowed to droop a few inches below it.

• Lengths of extra cable are coiled like a garden hose near a pole or looped around teardrop-shaped eyelets, called thimbles. The thimbles are grooved metal guides that ensure the cable bends smoothly, without sharp kinks. Why are lengths of extra fiber hung all along the route? After all, you never see such spare footage in a metallic cable. The motive for providing extra fiber is to minimize the number of splices. Without a spare loop, a break in the cable would have to be repaired by inserting a patch, which

Another way that fiber has brightened the industrial landscape is with multicolored coils of plastic conduit. When several fiber cables are all buried in the same trench, each runs through a differently colored conduit as an aid to identification. These reels of conduit were photographed in a supply yard in Toledo, Ohio.

means adding two splices. If some slack cable is available nearby, it can be pulled through to close up the break with just a single splice.

• A fiber-optic cable has no sharp turns or kinks. Because glass is more brittle than copper, a sharp turn can break the fibers. As a rule of thumb, the minimum bending radius is 10 times the diameter of the cable.

• Some communications companies mark fiber lines with a bright orange or yellow tag attached to the cable or the pole, mainly for the benefit of repair crews.

• The cable itself has identifying markings printed on the outer jacket by the manufacturer. The letters *OF* are part of the product code for all the common fiber types. Some manufacturers also mark the cable with sequential numbers every foot or every meter, to help in tracking down breaks or defects.

Fiber-optic routes tend to follow existing rights-of-way, such as highways and railroads, pipelines and power transmission lines. One long-distance communications company began as an operator of natural-gas pipelines. In the 1980s they found themselves with several thousand miles of pipe that was no longer adequate for carrying high-pressure gas. Rather than refurbish the pipe, they decided to run fiber through it and become a wholesaler of communications bandwidth. (Later, they returned to pumping gas.)

Several electric-power utilities have also gotten into the communications business. Typically they employ a special cable with fiber inside a conductive aluminum sheath. The cable is hung at the pinnacle of electric transmission towers, above the main power conductors, where it does double duty: carrying communications traffic and serving as an aerial ground wire to protect the power system from lightning strikes. Amazingly, this hazardous electrical environment has no effect on the stream of data bits and chitchat flowing through the glass fiber.

Getting Spliced. With copper wiring it's very easy to make connections. You can just twist two conductors together, and the electrons will find their way through. With fiber optics, in contrast, splicing is a delicate and difficult operation, calling for the kind of microsurgery that rejoins severed blood vessels or nerves.

Marker pylons warning of buried optical fibers have sprouted like weeds along many roadsides.

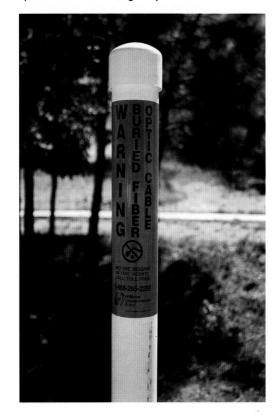

First, you have to separate the individual fibers, removing the outer sheath and the buffer tube. Then, the acrylic plastic coating on the glass fiber has to be stripped away. This is a tricky step because any nicks or scratches in the fiber could leave a weak

IN SEARCH OF CYBERSPACE

Surely the most remarkable recent development in communications—both technologically and sociologically—is the sudden blooming of the Internet. A few years ago it was the exclusive playground of an academic and government elite, but now almost everyone makes frequent visits to the place called cyberspace. But where is that, exactly? The Internet has become a part of our daily lives, and yet there's scarcely any physical evidence of its existence. There are plenty of telephone wires and cable-TV wires out there, but where are the Internet wires?

The short answer to this question is that most Internet traffic goes over the same hardware facilities as telephone conversations. The high-capacity fiber-optic cables that crisscross the continents don't know or care what kind of information they transmit. It's all just bits—telephone calls, Web surfing, e-mail, downloading an MP3 file. The same fiber carries it all.

At a higher level, however, the Internet does have an infrastructure distinct from that of the telephone network. The telephone system is described as a circuit-switched network. When you dial a phone number, a circuit is dedicated to your conversation for the duration of the call. The Internet is different. It's a *packet-switched* network. Messages of all kinds are broken down into small bundles of bits called packets, which navigate the network independently from source to destination. Each packet has to be individually addressed so that the nodes of the network, called routers, will know where to send it. From this point of view, the best answer to "Where is the Internet?" might be "Wherever the routers are."

The biggest concentrations of large routers are at sites called *peering points*, where inde-

pendent networks come together to exchange packets. When an Earthlink customer sends e-mail to an America Online subscriber, the message crosses from one network to another at a peering point. But where are the peering points, and what do they look like?

A decade ago, when the Internet was first cut loose from government control, a few peering points were set up as cooperative undertakings by major network operators. In 1997 I had an opportunity to visit one of those places, called MAE-East, which was then the largest peering point in terms of traffic volume. On maps of the Internet, MAE-East appeared as a grand hub, where dozens of lines converged from all directions. When I got there, I found that this center of the virtual universe was a cinder block enclosure in one of the underground levels of a parking deck in suburban Virginia. Inside, floor-to-ceiling steel racks were lined up like library bookshelves, and great multicolored rivers of cable—orange and yellow fiber optics, creamy thin coax, gray-jacketed bundles of copper—flowed through overhead trays and cascaded down the sides of the

racks. Harsh light poured out of fluorescent tubes; the air was filled with the white noise of a hundred computer cooling fans and a hint of battery fumes.

Soon after my visit, MAE-East moved out of the parking deck, and today most peering points occupy digs that are decidedly upscale. The machinery is installed in the kind of look-but-don't-touch glasshouse where corporate mainframe computers were kept on exhibit for so many years. A raised floor keeps all that beautiful wiring out of sight. But in one respect nothing has changed: Peering points and other major Internet facilities still try to be inconspicuous. If they don't actually conceal their whereabouts, they don't advertise it either.

Nevertheless, a few clues sometimes give away the secret. If you discover a nondescript office building with two huge diesel generators for backup power and dual air-conditioning systems, that building may well house more than clerks in cubicles. The windowless concrete building in the photograph below may or may not be a peering point in Pennsauken, New Jersey.

spot. Next, you need to create smooth, flat, clean surfaces on the ends of the fibers to be mated. The fiber is snapped to get a fresh surface, then polished using techniques that go back to the centuries-old art of grinding lenses and mirrors. Now comes the most crucial stage: aligning the fibers. It's not enough just to line up the outer edges; what counts is that the fiber cores meet head-on. The actual fusion of the two mated fibers is fairly simple. It's done with an electric arc that heats the glass to near the melting point so the fibers grow together in a permanent bond.

Fiber-optic splicing is not the kind of job you would want to do while hanging from a climbing belt 20 feet up a utility pole. It's an operation that needs a clean room and a microscope. It's usually done in a specially equipped truck or trailer.

OVER AND UNDER THE SEA

Surprisingly, the first transatlantic telegraph cable was laid in 1857, four years before the United States was spanned by a coast-to-coast overland line. On the other hand, the first transatlantic telephone cable was not completed until 1956, a century later.

For the first undersea telegraph cables the key problem was insulation. The plastics used today were unknown, and rubber wouldn't last in seawater. The material of choice turned out to be gutta-percha, a reddish resin extracted from an Asian tree. Gutta-percha still has uses in dentistry but is not a major commodity in world trade. However, the plan to wrap 2,300 miles of copper wire in gutta-percha created quite a bonanza for the small London company that imported the stuff. They did not let the opportunity slip through their fingers: the Gutta-Percha Company is now known as Cable & Wireless, one of the largest international telecommunications companies.

The first transatlantic cable was the project of Cyrus Field, an American entrepreneur of legendary patience and persistence. When the 1857 cable was completed after many trials and mishaps, there was a jubilant celebration on both sides of the ocean, with an exchange of messages between Queen Victoria and President Buchanan, and newspaper editorials predicting a new era of perpetual world peace. Then, a month later the cable stopped working. The failure was not from natural causes; the cable had been operated at too high a voltage and essentially had burned out. Rumors circulated that it had never worked in the first place—that the whole project had been a hoax and a stock-market fraud. Nothing daunted, Field raised more capital to lay another cable and, after many more adventures, finally established reliable transatlantic telegraph service in 1866.

Because of the great length of an intercontinental cable, the crisp on-off pulses transmitted from one end were smeared out into mushy, gradual changes in voltage by the time they were received at the far end. This degradation of the signal limited the sending speed to about 10 words per minute. Prices were high, roughly $10 a word. And yet there was no shortage of customers; Field's cable paid for itself within a year. By 1928 there were 21 telegraph cables across the Atlantic.

Telegraph cables worked without repeaters or amplifiers, but telephone signals could not travel nearly far enough without being boosted, and so undersea telephone service could not begin until vacuum-tube amplifiers were available. Furthermore, amplifiers working on the bottom of the ocean would have to be extraordinarily reliable; you couldn't just send out a repair crew every time a tube burned out.

Because of the amplifier problem, the first overseas telephone service was based on radio rather than submarine cables. For transatlantic traffic a transmitter station was built in 1927 at Rocky Point on the north shore of Long Island; the corresponding receiving station was at Houlton, Maine (nowadays the end of the road for Interstate 95). These were "long-wave" stations, operating at frequencies below the modern AM broadcast band. They could accommodate just one call at a time, and only if the weather and sunspots cooperated. A three-minute call to England cost $75. The transmitting and receiving installations had to be separated so that the signal from a nearby transmitter wouldn't swamp the receiver and drown out the fainter signal from across the ocean.

A few years later high-frequency (or short-wave) radiotelephone service began. For transatlantic links, all the stations were in New Jersey, with transmitters at Lawrenceville and Ocean Gate and receivers at Manahawkin and Netcong. On the

A Stonehenge of long-range radio antennas at Dixon, California, seems as mysterious—and perhaps as antiquated—as the stone circle on Salisbury Plain. The site once handled all telephone calls to Hawaii and Japan. Outgoing signals were transmitted from Dixon; the incoming half of the conversation was received at Point Reyes, California, 50 miles away. It is now a Navy installation.

West Coast, for calls to Hawaii and Japan, a great complex of powerful transmitters was built at Dixon, California, near Sacramento, with receivers at Point Reyes, north of San Francisco.

An early test of amplifiers for submarine telephone cables was made on the route from Key West, Florida, to Havana, a distance of 125 miles. The cable, laid in 1950, remained in service until 1989, overcoming political as well as technological odds.

The first truly oceanic telephone cable was TAT-1, which crossed from Newfoundland to Scotland and was completed in 1956. It could handle 35 simultaneous two-way conversations. The technology of TAT-1 was analog transmission over coaxial cable, with amplifiers spaced every 40 miles. The 102 submerged repeaters had a total of 1,608 vacuum tubes—not one of which failed in the first 10 years of operation.

Fiber optics has transformed undersea telephony just as it has the dry-land kind. No one will ever again build a coaxial submarine cable. The first transatlantic fiber cable, TAT-8, was commissioned in 1987 and is still operating. It runs from Tuckerton in southern New Jersey out across the Atlantic and then branches to Widemouth, England, and Penmarch, France. The cable includes three pairs of fibers—two for active service and one as a spare. The cable also has metallic conductors to supply power to 125 amplifiers, spaced every 31 miles.

The latest TATs—numbers 12 and 13—have just a single pair of fibers, but their capacity is five billion bits per second, equivalent to almost 80,000 voice channels. Furthermore, they are all-optical systems, with no metallic conductors. Instead of conventional electronic repeaters, they have optical amplifiers, which use a special laser to boost the light signal without converting it to electronic form.

By their very nature, submarine cables don't offer the industrial sightseer much to look at. After all, they're at the bottom of the ocean. And where the cable comes ashore, it does not just emerge from the waves on a bathing beach. Cable companies look for a sheltered landing site. On the East Coast of the United States, most cables come ashore in bays behind sandy barrier islands.

CELLULAR TELEPHONES

Along with the Internet, the cell phone has totally changed the way we communicate. But whereas the Internet is almost invisible on the landscape, the cellular telephone system has changed the skyline as well as the airwaves. The antenna towers stand out—most of them are 150 or 200 feet high—and they are everywhere. Cellular service in the United States began in 1983. Twenty years later there were more than 100,000 towers. A whole new infrastructure has mushroomed up.

The key to the cellular-phone system is not just building a wireless telephone; the portable two-way radio—the walkie-talkie—has been around for decades. The trick is to build a wireless phone system that can handle millions of calls simultaneously, and do it with no more than a few hundred available frequency channels.

The cellular telephone tower, an artifact utterly unknown to the public a little more than 20 years ago, has become so commonplace that visual clutter from the towers is a contentious issue. The classic specimen shown here was photographed in 1996, when it was sparkling new, against the pure blue sky of Albuquerque, New Mexico. The white pods are directional antennas, pointed toward three cells that meet at the site of the antenna tower. The tall masts extending above the rest of the structure are omnidirectional antennas used for tracking the location of cell phones.

Mobile-telephone service was offered in large cities long before the cellular idea came along. Each city had one central station with a tall antenna tower, and every mobile phone in the area communicated through that tower. To keep calls separate—so you could hear only your own conversation—each call had to be assigned a separate frequency. Actually, a call needed two frequencies, one for the uplink (mobile to base station) and the other for the downlink (base to mobile). In each city there were only 24 frequencies available, so there could never be more than 12 calls active at any one time. This situation put a severe constraint on the number of mobile-phone subscribers. In all of New York City there were never more than a thousand—and most of them spent more time waiting for a free channel than they did talking.

Adding more frequencies is an obvious way to ease this problem, and the Federal Communications Commission (FCC) has allocated several new blocks of frequencies for mobile service. But the electromagnetic spectrum is a finite resource; there are simply not enough channels to go around. What's needed is some kind of multiplexing so that many phones can use the same frequency. The solution adopted in the cellular-telephone system amounts to spatial multiplexing: a region is divided into many small territories called cells, each with its own base station and antenna. Then the same frequencies can be used in many different cells. Two mobile callers can occupy the same channel provided they are not too close together. As long as you can keep subdividing the cells to cover smaller and smaller areas, there's really no limit to the number of simultaneous calls. And the process has an incidental benefit: since the transmission range is shorter, the mobile phone doesn't need as much power, allowing it to be made smaller and lighter.

The price for this expanded capacity is an increase in complexity. When someone calls you on your cell phone, the system has to figure out where you are at the moment so it can route the call to the appropriate cell tower. If you are cruising down the freeway as you chat, the phone may leave the initial cell and cross over into another. At that point the system has to complete an intricate handoff operation, transferring control to a new base station, instructing your phone to switch to another pair of frequencies, and rerouting the land-line part of the call. The whole hand-off process takes place in milliseconds; it can happen in the middle of a word and you'll never know it. (It can also fail totally, dropping the call without warning.)

The Cellular Honeycomb. If you were going to map out an array of cells for telephone service, your first impulse might be to carve up the territory into a neat grid of squares, like city blocks. But nature offers a better model: the hexagonal honeycomb pattern. Why is it better? With a tower in the middle of each cell, you want to reach the farthest corners but intrude as little as possible into neighboring cells. Hexagons produce much less overlap than squares.

Real cells are seldom perfectly regular hexagons, because hills and buildings distort the radiation pattern. Besides, you can't always put a tower exactly where you want one. And because the population of mobile-phone users is not spread uniformly

A gallery of cellular telephone towers *(opposite page)* suggests both the variety of designs and the features they have in common. The main common element is the prevalence of triangular motifs: most of the towers have antenna elements pointing toward three adjacent cells. From left to right and top to bottom, the towers were photographed in Richmond, California; Philadelphia, Pennsylvania; Berlin, New Jersey; Lincoln, Nebraska; Vaughn, Ontario; and Durham, North Carolina. The tower in Lincoln is notable in that it has seven decks of antennas, the most I have seen on a single tower.

Cellular towers incognito *(opposite page)* have become more common as opposition to the proliferation of antennas has grown more adamant. The preternaturally tall pine tree is in Cary, North Carolina, although there are a number of others like it. The stand of saguaro cactus (one is the real thing, two are fakes housing antennas) is in Fountain Hills, Arizona. The mast with the swept-wing weathervane is part of an artwork by Tom Grubb commissioned by the city of Fayetteville, North Carolina. Many other cellular antennas take advantage of tall structures that would poke up into the skyline in any case. Shown here are antennas fitted to a shopping center sign near Media, Pennsylvania, a smokestack in Chelmsford, Massachusetts, and a water tower in Metuchen, New Jersey. Of the disguised towers, only the cactus would be likely to fool anyone. I found the illusion convincing, but the detail photo below reveals the truth.

over the country, some places need many small, closely packed cells whereas other areas can get along with just a few big ones. In spite of all these distortions, though, a map of cell boundaries is still recognizable as a honeycomb, even if it's one made by slightly drunken bees.

There's one more trick to the geometry of the cellular system. Putting a tower in the middle of each cell obviously calls for one tower per cell. It may seem like there's no way to reduce this investment, but in fact it's possible to do much better, provided you can use directional antennas, which radiate preferentially on one side. A tower erected at the point where three hexagonal cells come together can carry directional antennas pointed toward each of the adjacent cells. This strategy requires only one-third as many towers on average. Most cell-phone systems are laid out in this way, and that's why you usually see a triangular array of antennas atop a cell-phone tower, a sort of three-cornered hat.

Because of overlap and interference, nearest-neighbor cells cannot use the same sets of frequencies. Some careful planning is needed to make sure that all the cells get their fair share of the available frequency space. You can think of the problem as one of coloring the cells on a map so that no two adjacent cells have the same color. Mathematicians know this can always be done with no more than four colors, meaning that the complete list of frequencies could be split up into just four sets. A given hexagonal cell might use only the red frequencies, and it would be surrounded by cells assigned green, blue, and yellow frequencies. The nearest cells also using the red frequencies would be one cell beyond the innermost ring of neighbors. In practice, designers try to ensure even greater distance between cells that share a frequency assignment. Common plans use either seven or nine colors—that is, they divide the available frequencies into seven or nine groups.

The Base Station. Cell-phone towers come in all descriptions. Some are built out of a steel lattice like the stanchions of electric power transmission lines. Some are tubular pylons, like oversized streetlight poles. Most are about 150 feet high.

The antennas at the top of the tower come in two basic types (with dozens of minor variations). Directional antennas are large, rectangular, podlike structures typically hung on the outside of a triangular framework as if they were bumpers fending off stray cars and trucks. They radiate or receive in the direction they are facing, usually covering a fat wedge equivalent to one-third of a full circle. The other antenna type is a pole, or whip, roughly 10 feet high and most often sticking up above everything else. This is an omnidirectional antenna; it radiates or receives equally well in all directions. In form it's not much different from a car radio antenna.

A given cellular tower could have either directional or omnidirectional antennas or, very frequently, some of both. One arrangement has three directional antennas at the vertices of an equilateral triangle, with three omni antennas poking up above them. The directional antennas are for receiving, and the omni antennas for transmitting. Another plan uses a total of nine directional antennas, with three bolted to

each side of the triangular frame. Two of the antennas facing in each direction are for receiving, and the third (usually set slightly apart from the others) is for transmitting. Towers that use directional antennas for both transmitting and receiving usually have at least one omnidirectional antenna as well. It's used for measuring the signal strength of mobile units in the area; information from this antenna tells the system when to hand a call off from one cell to another.

If you look carefully at the directional antennas, you may notice that they point downward slightly, as if they've drooped a little. This is not sloppy installation. The angle is carefully calculated, along with the antenna height and the signal strength, to cover the entire cell as uniformly as possible while minimizing overlap with other cells.

Sometimes you can see the cables that feed the antennas. They are thick, black-sheathed coaxial cables, one for each antenna, forming a bundle that can be more conspicuous against the sky than the structure of the tower itself. At ground level the cables lead into a hut or equipment cabinet housing the transmitters, receivers, and other gear. The equipment racks in the hut hold one transmitter and one receiver for each channel the cell can handle. Also in the hut is the equipment needed to connect the cell-phone system to the land-line network. The cell needs the equivalent of one voice circuit for each two-way radio channel. Most often the connection is made over a buried coaxial cable or optical fiber, but an alternative is a microwave link. In the latter case you'll see a drum-type or dome-type microwave antenna mounted partway up the tower.

The base station is a cell phone's point of contact with the global network, but it's not where the intelligence of the system resides. All the cells in an area funnel their calls through a central facility, called a Mobile Telecommunications Switching Office, where computers keep track of phone locations and choreograph the handoffs. They also handle the billing.

Cellular Proliferation. Most of the telecommunications industry is glacially conservative and shy of novelty. Conventional telephone hardware has evolved so slowly that you can still plug in a 1940s desk telephone and expect it to work with the modern network. But the cellular-telephone business has not inherited this philosophy of compatibility in perpetuity. In the cell world, every shiny new idea gets its chance. No one seems to be afraid of making last year's model obsolete.

The oldest cellular-phone system, called AMPS, or Advanced Mobile Phone Service, uses analog communications technology. In the United States it is the lowest-common-denominator system that most phones fall back on if they are out of range of other services.

AMPS has a total of 666 frequency channels, but 42 are used for control rather than communication—for tracking phones as they move among the cells—which leaves 624 frequencies for voice circuits. Since two frequencies are needed for each conversation, a system consisting of one big cell could serve 312 callers at once. Dividing the complete set of frequencies into seven groups leaves each cell with the

Even the base station that provides service to mobile phones can itself be made mobile. These truck-mounted cellular antennas were on duty in Lower Manhattan in November 2001.

capacity to handle 40 to 45 two-way conversations. This is the number that governs the size and the layout of the cells. Each cell should be small enough that it will seldom contain more than 40 or 45 subscribers who want to use their phones at the same time. In a busy downtown neighborhood, a cell just a few blocks across might strain this limit. Out in the countryside, you might go for miles before meeting a cowboy holding a phone to his ear.

Most of the newer cell systems transmit digital signals between the mobile phone and the base station, unlike the analog AMPS system. They also add new kinds of multiplexing to increase the number of simultaneous calls per cell. One multiplexing scheme is called Time Division Multiple Access, which allows up to 24 telephones to share the same band of frequencies. It's done by compressing your voice into a burst of data that lasts about half a millisecond; then your phone waits politely while the other 23 take their turns before sending the next burst. Another multiplexing process called Code Division Multiple Access spreads the voice signal over a wide band of frequencies, then reassembles it at the other end of the connection.

For a decade, several incompatible digital cellular systems in the United States have been competing for subscribers as well as for places to erect their antenna towers. Although the contest is not quite over, it looks like the winner will probably be GSM, a standard imported from Europe—although the American implementation is incompatible with equipment in the rest of the world.

Base stations for many of the newer digital cellular systems are more compact than older ones. The antenna mast may be just as high, but the antenna elements are smaller and more closely spaced, and the equipment hut at the base of the tower looks more like a file cabinet and less like a house trailer.

Another new development is a kind of miniaturized cellular system meant specifically for use in center-city neighborhoods, with "microcells" that cover only a few hundred yards. The microcell antennas are square, book-size panels being hung on streetlight poles and traffic signals and tucked away under the eaves and cornices of buildings. Keeping the antenna below rooftop height helps limit the signal to a small radius. These ventures are in flux, and it's not clear whether they will succeed in competition with the larger cell systems.

The sudden blooming of cellular antenna towers all over the landscape has provoked a backlash. Neighbors organize to oppose new towers, and local zoning boards resist granting permits. In response, the telephone companies have become a little more creative about where to put their antennas.

A first approach to reducing the number of new towers has been to double up and triple up antennas on the existing towers. Many towers have been fitted with extra decks of antennas so several competing companies can use the same tower for different kinds of cellular service. Towers built for other reasons have also been pressed into service. Television and radio broadcast towers, for example, which are much taller than the typical cellular mast, have acquired a girdle of cell-phone antennas at the appropriate height.

Microcell networks take the cellular concept a step further, with base stations whose range is just a few hundred feet. The white box suspended from a streetlight bracket in Berkeley, California *(above)*, is an antenna unit for a microcell network called Metricom, focused on data communication. On the brick facade of a building in Midtown Manhattan *(below)* the four large green tiles are merely decorative elements, but the two smaller appendages are microcell antennas.

At the base of an 1,800-foot-high broadcast tower, looking up at the top puts you at risk of falling over backward. The tower is a new one, built to transmit Public Broadcasting Service signals in Columbia, North Carolina, near the Outer Banks. The tower could not stand without the three sets of guy wires that lend it stability. The red balls attached to some of the guy wires are intended to make them more visible for low-flying aircraft. Soon after the tower was completed, several of these balls were damaged in a storm. An engineer on the site explained how they were replaced: a worker ascended the tower (there's an elevator part way) and then descended by sliding down the guy wire, removing broken balls as he went and leaving new ones behind.

Indeed, almost any tall structure has become a likely spot for a cell site. Hundreds of municipal water tanks are festooned with antennas, and so are high-voltage power-line pylons. (Apparently, interference from the power-line currents is not a problem.) Antennas are bolted to the walls or roofs of hotels, offices, and apartment buildings. I have seen them attached to smokestacks, church steeples, flagpoles, billboards, stadium lighting stanchions, and those long-legged gas-station signs at rural interchanges on Interstate highways.

And there are cell towers in disguise. The most popular costume seems to be the phony pine tree. It's a hundred feet tall, the apotheosis of the artificial Christmas tree, with plastic branches that fit into sockets in a brown steel trunk. When you look at the base of the trunk, there's an oval access panel where heavy black cables spill out. Near Phoenix there's a fake saguaro cactus that hides an antenna.

RADIO AND TELEVISION

The telephone is a point-to-point, one-on-one technology. Radio and television broadcasting are different: The signal goes everywhere, and anyone is welcome to tune in. Also, for the most part, radio and TV are one-way channels; talking back is difficult.

Broadcasting has a presence on the visual landscape because of the antennas that transmit and receive the signals. Transmitter towers are the tallest of all man-made structures. In the United States the legal limit is 2,049 feet, or just under two-fifths of a mile; hundreds of towers approach that ceiling. The limit is set by the Federal Aviation Administration, which takes the position that there has to be *some* height where the land ends and the sky begins.

Towers for broadcast antennas come in two basic types: free-standing and guyed. The free-standing ones have a tapered form, like the Eiffel Tower, with a broad base so the tower won't topple over in the wind. Guyed towers are so slender they could not possibly stand up without wires to support them. The tower is like a pencil standing on its point; indeed, the base of a guyed tower often narrows to a single point and is attached to the ground through a hinged bearing deliberately designed not to resist any bending motions. The reason for this design is that the tower inevitably sways in the wind, and rigid joints would be vulnerable to metal fatigue.

The choice between a free-standing and a guyed tower is a matter of engineering trade-offs. A free-standing tower needs more steel, but a guyed tower may need more land to make room for the cable anchorages. As height increases, guyed structures become more common. In any case, for new construction a large tract of land may be needed regardless of tower type. Many insurance companies now insist that a tower be sited where it can't hit anyone else's property if it ever falls over. Thus, a 2,000-foot tower needs a circular plot of land almost a mile in diameter.

Several European cities have made broadcasting towers into a showplace or landmark, building not a minimalist steel structure out in the countryside but a sculptural

At its base the Columbia tower narrows to a point, so that it stands like a dancer rising on one toe. The rationale for this design is that the tower will sway in the wind no matter how strongly it is braced; if the legs were rigidly anchored to the ground, the swaying might cause metal fatigue. The single-point connection to the earth functions like a hinge, allowing the tower to sway without damage.

spire right in the middle of town, often with an observation deck and other tourist facilities. This custom hasn't caught on in the United States, but Canada has taken it to new heights—specifically to 1,815 feet 5 inches. The CN Tower in Toronto broadcasts the signals of seven television stations and nine FM radio stations, but that function seems secondary to the tower's role as tourist attraction.

Every tall tower in the United States has a small plaque that says something like "FCC ID# 1006708." The plaques are there mainly so that people can report burned-out light bulbs in aviation-warning lights. If you call the FCC, they will tell you who owns a tower with a given registration number; the information is also available on an FCC Web site.

Antennas. An antenna's role is analogous to that of a loudspeaker or a microphone. Just as a loudspeaker takes alternating currents flowing through an electric circuit and converts them into sound waves, a transmitting antenna converts currents into electromagnetic waves. And just as the microphone converts sound waves into electric currents, a receiving antenna produces measurable currents from electromagnetic waves. The analogy can be taken a step further. Loudspeakers come in various sizes to reproduce sounds in different frequency ranges—big woofers for the long basso waves and tiny tweeters for the small treble waves. Antennas also need to be matched to the wavelength of the signals they send or receive. In fact, the size of the antenna is your best clue to what kind of signal it conducts.

The archetypal antenna is the dipole: two conductors extending in opposite directions, with a connection to the transmitter or receiver where they almost meet in the middle. (They don't quite meet; there has to be a small insulating gap.) Each half of the dipole is a quarter wavelength long, so the complete antenna is half a wavelength. Thus, if you can recognize a dipole antenna and estimate its length, you'll know approximately what wavelength it is transmitting or receiving.

A dipole directs most of its energy to the sides—perpendicular to the axis of the two conductors. If the antenna runs north and south, it will radiate strongly only to the east and west. The pattern is the same when the antenna is used as a receiver: it is sensitive to signals striking the conductors broadside but not to those coming from the ends. The rabbit ears antenna that was once a common sight on top of the living-room television cabinet is an example of a simple dipole. If you're old enough to remember that quaint device, you will also remember the struggle to get better reception by adjusting the length of the two ears and by turning the antenna to face in various directions. Those attempts relied on the antenna's wavelength resonance and its directional radiation pattern.

AM Radio. AM stands for Amplitude Modulation, which describes the way an audio-frequency signal is impressed on the radio-frequency carrier wave. The amplitude is the height of the radio wave, and so an AM signal varies in height with the music. But all this has little to do with what gives an AM radio broadcasting anten-

na its distinctive features. What matters more is the band of frequencies and wavelengths assigned to AM broadcasting in North America. The AM band runs from 535 to 1,610 kilohertz, which means the wavelengths range from about 1,800 feet down to about 600 feet. The antennas are correspondingly large.

Often the broadcast antenna for an AM radio station is not mounted on a tower; the antenna *is* the tower. In other words, the entire structure is electrically "live" and radiating energy. The best way to recognize such an antenna is to get a look at the base of the tower: it will be insulated from the ground. This takes quite a hefty insulator. In the case of a free-standing tower, an insulator is installed at the base of each leg; a guyed tower has a single big insulator at the foundation. The insulators are impressive ceramic structures, anywhere from a few inches to 10 feet high. The guyed tower also needs insulators in the guy wires.

The height of an AM broadcast tower is usually one-fourth of the wavelength of the station's signal. That works out to 450 feet for stations at the bottom of the dial and diminishes to 150 feet at the other end of the band. (Increasing frequency corresponds to decreasing wavelength.)

An antenna tower with an insulated base amounts to half of a dipole. Where is the other half? It is created with mirrors. Out of sight in the earth surrounding the tower are buried conductors that form what is known as a ground plane or counterpoise. For radio waves, the ground plane acts as a reflector, just as silvered glass does for light. The presence of this radio mirror creates a "virtual image" of the tower that effectively turns the antenna into a dipole. The ground plane is constructed by laying down a sunburst of copper conductors one-fourth of a wavelength long, radiating away from the foot of the tower. AM broadcast antennas generally have 120 buried radials, placed every three degrees around the compass.

A vertically mounted antenna radiates equally in all horizontal directions. For many purposes, that's exactly what's wanted—equal signal strength all around—but some AM stations are required to beam their broadcasts preferentially in one direction or another, to avoid interfering with other stations. One way to steer the beam is to transmit with a horizontal dipole, which emits more strongly broadside than at the ends. The horizontal dipole is a wire suspended between two towers spaced half a wavelength apart. The feed line from the transmitter is hung from the midpoint of the antenna.

Another scheme for directional broadcasts requires multiple vertical antennas. Several towers in a line are spaced half a wavelength apart. The directional effect comes from interference between the waves emitted by the separate towers. Where the waves are in phase (peaks lining up with peaks and troughs with troughs), they reinforce each other; where they are out of phase, they cancel. Unfortunately, you can't tell just by looking at the towers which way the beam is being steered. That depends on the phases of the signals being fed to the various towers. If all the phases are identical, the antenna array has a broadside pattern; if the signals are of opposite phase, the array produces "end fire."

The red and white stripes required on many tall towers have to be repainted from time to time. In the photograph on the opposite page a painter toting red and white buckets is working down from the summit of a tower in Spartanburg, South Carolina.

Most broadcast towers are utilitarian structures, but the CN Tower in Toronto is obviously more than the bare minimum needed to elevate an antenna. At 1,815 feet it is "the world's tallest building and free-standing structure," according to the owners.

Porcelain insulators installed in the feet of a tower *(right)* allow the entire structure to function as an antenna; this is the usual practice with antennas for AM radio stations, where the waves are hundreds of feet long. On the insulated base of another antenna *(below)* the metal bracket shaped like a backward J is a spark gap for lightning protection. It allows the antenna to remain electrically isolated during normal operation, but shunts a lightning surge into the ground.

Television. Television channels are at frequencies a hundred times higher than those of AM radio, and so the wavelengths are 100 times smaller. It follows that television transmitter antennas are comparatively small—only a few feet long instead of hundreds of feet. It's ironic, then, that the television broadcasters put up the tallest of all towers for their antennas. But they don't do it just for the bragging rights.

Higher frequencies allow smaller antennas, but they have another effect as well. The lower frequencies of AM radio can bend around the curvature of the earth to some extent, and they bounce off a layer of electrified particles high in the atmosphere (the ionosphere) to reach distant receivers. But at television frequencies, communication is line-of-sight. A transmitter can reach only those receivers that are directly in view. To cover an entire metropolitan area, the antenna has to be way up in the air.

Television transmitter masts are merely mechanical supports meant to elevate the antenna, which is a much smaller structure at or near the top of the tower. Hence, there is no need to insulate the base of the tower or the guy lines. On the contrary, the builders go to some trouble to make sure the framework is securely grounded.

The antenna elements at the top of the tower come in a variety of shapes and sizes. There are dipoles and crossed dipoles (two dipoles arranged like a plus sign), often mounted in front of a reflecting panel. Batwing antennas take their name from the scalloped trailing edge of a bat's wing. They are also known as turnstile antennas, which is a better description of what they look like when mounted on the mast. Helical antennas spiral around the mast like a morning-glory vine climbing a fence post. The pitch of the helix—the vertical distance traversed by a conductor in making a single 360-degree turn around the pole—is equal to the signal wavelength. Slot antennas are a kind of inside-out dipole: instead of forming a metal rod or wire into

the shape of a dipole, you cut a hole of similar dimensions into a continuous metal panel. Often there are lots of these dipole slots, giving the active part of the antenna the look of a cheese grater or colander.

The size of the antenna elements depends on the signal wavelength and, hence, on the station's channel assignment. Channel 2, at the lowest frequencies, has a wavelength of about 18 feet; channel 13, at the top of the VHF (very high frequency) allotments, corresponds to a wavelength of less than 5 feet. The UHF (ultra-high-frequency) wavelengths are still shorter, down to about a foot for channel 83. Thus, the basic building block of most antenna elements—a quarter-wavelength conductor—can range in length from four-and-a-half feet down to about three inches.

Sites for new broadcast towers are in short supply. As a result, existing towers have become electronic high-rise tenements for the communications industry. A tower may have been erected by a broadcaster primarily for its own use, but over the years it will have accumulated dozens of other tenants, who lease space for a variety of purposes. A TV tower can also be home to an FM radio station. (The FM broadcast band lies just above television channels 5 and 6, so the antenna structures are similar.) Lower down on the tower there may be a girdle of cellular-telephone antennas. Paging services, police and emergency dispatchers, and truck and taxi fleets all seek a few feet of tower space. At a tower in the western suburbs of Washington, D.C., I was able to count 31 separate antennas, and I probably missed a few.

One problematic challenge is putting two TV transmitting antennas on the same tower. Sometimes a single antenna can be designed to broadcast both signals, and sometimes two antennas can be stacked up totem-pole style. But when those approaches don't work, another option is available; it creates some of the world's most unusual tall towers. Two or more antennas can be mounted side-by-side on a beam or platform that gives the tower the form of a candelabrum. Such forked towers are distinctive landmarks in the cities that have them, including San Francisco, Boston, and Washington.

The market for tower space has gotten even tighter with the introduction of digital high-definition television (HDTV). For years to come, most stations will be broadcasting both HDTV and the traditional analog signals.

Antennas for receiving radio and television signals are not nearly as big or elaborate as the transmitter antennas, but they can still have a visual impact because there are so many of them. Television receiver antennas once grew on every urban roof. Cable and satellite TV have now made them rare in most places, but they are not quite extinct. The type seen most often is known as the Yagi or Yagi-Uda. It has several parallel dipole-like rods, of graduated length, all attached to a metal boom. If you look carefully, you'll find that only one of these parallel elements is actually connected to the cable that leads to the receiver; all the others are passive elements that improve the antenna's sensitivity and make it highly directional. For optimum reception the dipoles must be broadside to the approaching waves; in other words, the antenna boom has to point toward the transmitter.

Television antennas are mounted on a tower but remain electrically isolated from it. The tower above, in Pleasantville, New Jersey, has a single antenna with five elements. (They all broadcast the same signal.) The tower below, in Oklahoma City, is a candelabrum structure with separate antennas for at least three stations.

A directional antenna is important for television reception because of the problem of "multipath" and "ghosts." When a broadcast signal bounces off a building, your set may receive both the direct and the reflected waves, an instant apart; both signals display on the screen, displaced slightly along the horizontal axis of the picture because of the delay. A directional antenna exorcises these ghosts by excluding the reflected waves. Incidentally, ghosts allow you to turn your TV into a distance-measuring instrument. On a 19-inch screen, measure the distance in inches between the primary image and the ghost; two-thirds of that number is the approximate extra distance in miles traveled by the reflected signal.

CABLE TELEVISION

Among communications technologies, cable TV is a strange amalgam. It shares some characteristics of the telephone infrastructure, in that millions of homes are wired to central offices. But cable also has much in common with broadcasting, since the communication is almost entirely one way, and the same signal—more or less—goes to everyone.

Cable TV began in the 1950s as "community antenna television," and occasionally you still see the abbreviation *CATV*. The idea was hatched in "fringe" areas, 50 or 100 miles from the nearest big city, where reception was difficult but not quite impossible. Those who happened to live high on a hillside, or who were willing to invest in a tall antenna tower, could get at least a marginal picture, but others were blacked out. And so the idea arose to put up one super-duper antenna at the best site

Rooftop television antennas have disappeared from most American neighborhoods, but there are still forests of them in some European cities. The grove of antennas at right is in the Italian city of Monfalcone. Most of the antennas point toward Venice and the rest toward Trieste, the two nearest places with television broadcasts.

in town, and distribute the signal over cables to all who wanted it and were willing to pay a share of the cost.

There's some dispute over where CATV was invented, but there's no doubt early systems were built both in northwestern Oregon (just out of reach of both Portland and Seattle) and in the coal-mining region of northeastern Pennsylvania (beyond the range of New York and Philadelphia stations). The first Oregon system, in Astoria at the mouth of the Columbia River, was set up by an appliance dealer to help sell television sets; in 1949 it was bringing Seattle's sole TV station to 25 subscribers. Could anyone have looked at that modest beginning and foreseen CNN and HBO?

The world of broadcasting didn't have to evolve in just this way. To some extent, the birth of cable TV was an accident of history, or of federal regulation. When small towns in fringe areas began to demand access to television, there were other possible responses. The entrepreneurs who built CATV networks might have started up local, independent television broadcasting stations. Or the big-city stations might have set up additional transmitters or repeaters in outlying areas as a way of extending their territory. As it happens, neither of these options was available. From 1948 through 1952 the FCC had a moratorium on new television broadcasting licenses while it reconsidered how best to allocate the available spectrum.

Modern cable TV systems have two main components: the head end, where programs are received, and a distribution network. Typically the distribution network has a treelike structure: trunks lead from the head end to major hubs, then feeder lines spread out through neighborhoods, and drop lines carry the signal to individual homes.

The Head End. What began as a community antenna has evolved into a major communications ganglion, with tentacles reaching out to regional, national, and international networks. In many communities you can spot the head end from a distance because of the tall tower bristling with antennas pointed in various directions, and on the ground a cluster of bright white satellite dishes, much larger than the kind you see in backyards. For a small system all this equipment may be clustered around an unattended cinder block hut. A big metropolitan head end will be a more substantial establishment, possibly including a studio for local television productions.

Signals from local TV stations are plucked out of the air by antennas that aren't much different from ordinary rooftop models, but they're built more robustly and can be tuned to pick up a particular channel. More important, they are mounted high enough to have a direct line of sight to the transmitting antenna, and they are carefully pointed toward it. For each channel there's a separate receiver. The head-end tower may also support microwave antennas. Some television stations supply their signal directly to the cable system over a microwave link.

Satellite communication was not part of the original plan for cable TV, but today it's the main source of programming. Multiple dishes are needed because each dish can lock on to only one satellite at a time, and typically can receive no more than two channels simultaneously. There's probably a spare dish too, just in case.

The head end of a cable system in Fairfax, Virginia, includes a multitude of satellite dishes as well as a tower with antennas for broadcast signals and microwave links to local stations.

Hardline is a metal-sheathed form of coaxial cable popular with operators of cable TV systems. In the photograph at right, rivers of hardline make smooth bends around a utility pole. The U-shaped dips in the hardline allow for thermal expansion and contraction. The photograph was made in Fairfax County, Virginia, which seems to have a notable abundance of cable infrastructure.

A "strand-hung amplifier" for the cable system (upper photograph) is powered by voltages transmitted over the cable itself. The amplifier is the large metal box with ribs or fins for cooling. Customer taps (lower photograph) connect individual houses with the distribution cable. Bar-coded tags identify subscribers.

Distribution. The early cable distribution networks were quite simple. All the channels were fed into a coaxial cable, which was strung up on utility poles. Each subscriber had a tap, which was pretty much what the word suggests: a device that cut into the cable and diverted a share of the signal, just as a tap on a pipe drains off a fraction of the water flowing through it. To maintain the power level, booster amplifiers were needed every quarter or half mile. Even with the boosters, cable runs could not be longer than a few miles.

All of this infrastructure remains in place, but new levels of interconnection have been built on top of the local distribution network. In recent cable systems, signals are dispatched from the head end in "supertrunks" of optical fiber. The supertrunks fan out to regional hubs, where they connect with trunk lines, which can be either coax or fiber. The trunk lines continue to neighborhood distribution nodes, where the local coax with taps for subscribers takes over.

The hardware of the cable system hangs on the same crowded utility poles that carry electric-power and telephone lines, but cable TV equipment is pretty easy to recognize. It usually runs just above the telephone wires and well below any power lines. In many communities the coaxial cable for cable TV is a distinctive type called hardline, which has no insulating outer sheath (unlike the coax favored in the telephone system). The silvery, satiny aluminum cable stands out amid all the black plastic on a typical utility pole. Most hardline is half an inch in diameter; inside the metallic outer conductor is a layer of foam plastic insulation, surrounding the copper wire that forms the center conductor. Because of the foam, the cable is surprisingly light. Hardline is fairly stiff and can't survive sharp bends or kinks, so you'll see it formed into careful, large-radius turns wherever a change of direction is necessary.

Taps into the local coax line are small boxes, the size of a pack of cards, with individual household droplines attached. The drops are made of a narrower, more flexible coax, which carries the signal the last 50 feet or so to the television receiver.

Amplifiers are somewhat larger boxes, sometimes bolted to a utility pole but more often suspended from the messenger wire that supports the cable. (Cable guys talk about "strand-hung amps.") A modern cable amplifier is the size and shape of a half-gallon milk carton. The cast metal housing is sealed against moisture with a rubber gasket and often has fins molded into the lid to carry off heat. Amplifiers are powered by 60-hertz current transmitted through the coaxial cable along with the video and audio signals.

SATELLITE COMMUNICATIONS

The first communication satellite, Echo I, was an oversize version of those shiny Mylar Happy Birthday balloons. For launch it was folded up in the nose cone of a rocket and then inflated in orbit to a diameter of about 100 feet. The name Echo was well chosen: high-frequency radio signals simply bounced off the metallized surface.

A communications satellite of the current generation is a different animal altogether. If you go look at a hilltop microwave relay station and then imagine all that hardware lofted into the sky, your image won't be far off. Like a relay tower, a satellite receives microwave signals, amplifies them, and then rebroadcasts the same information. Many of the satellites work in the same four- to six-gigahertz band employed by terrestrial microwave telecommunication links.

Satellite communication has had a roller-coaster history. The idea was dreamed up in 1945 by Arthur C. Clarke, who was then an engineer working for the British Post Office; he went on to become better known as an author of science fiction (*2001: A Space Odyssey*). Some experimental satellites were launched in time to beam broadcasts of the 1964 Tokyo Olympics back to the United States. The first commercial satellites came just in time to put the Vietnam War on the nightly news. By 1970 satellites had more transoceanic capacity than cables. Many observers took it for granted that orbiting relay stations would be the mainstay of all future long-distance communications. Then along came optical fibers, which soon surpassed satellites in both bandwidth and glamour coefficient. The contrary view took hold: the communications satellite was seen as a has-been technology. But it hasn't turned out that way either. Today there are more satellites than ever before.

Geostationary Satellites. The communication satellites with the longest history are geostationary satellites, known in the trade as GEOs. They seem to hover stationary over a fixed point on the earth's surface. Actually, they've got to keep running at almost 6,000 miles an hour just to stay in the same place.

A satellite's orbital speed depends on its altitude. In the lowest possible earth orbit, just above the edge of the atmosphere, a spacecraft completes one revolution in about an hour and a half. A satellite 240,000 miles up (namely, the moon) takes 28 days, more or less, to circle the earth. Somewhere between these extremes there must be

At a sports stadium, 10-foot dish antennas are equipped both to transmit and receive broadcasts via satellites in geosynchronous orbit, 22,250 miles above the Equator. The curious metal brush atop the dish is apparently for lightning protection.

an altitude where a satellite completes exactly one revolution per day. If the satellite is orbiting eastbound above the Equator at just this altitude, it will appear to stand still in the sky, the way the sun did for Joshua. The geostationary altitude turns out to be 22,250 miles above the earth's surface, a region of space now called the Clarke Belt. (In Arthur C. Clarke's vision, however, the satellites would have been manned space stations. Evidently it was easier to imagine putting a crew in orbit than it was to imagine an electronic communications system that would run unattended.)

What makes a GEO ideal for communication is that it's always in the same spot. Other satellites rise and set; a GEO stays put. But GEOs also have drawbacks. Because the orbit is so high, boosting tons of electronic gear to the geostationary altitude costs much more than launching a satellite into a lower orbit. Communication with the satellite also takes a lot of power because of the distances to be crossed. And the distance has still another disturbing effect. Even traveling at the speed of light (186,000 miles per second), a signal takes about a fourth of a second to make the round trip from an earth station to a GEO and back down to another earth station. If the signal happens to be your voice saying "Hello hello hello" on an international phone call, it will be fully half a second before you hear the other party's answer, which makes conversation awkward. For this reason, GEOs have lost favor for voice telephony. The delays are not a problem in television transmission or in some kinds of data communication.

The business end of the satellite is a cluster of dish antennas and a package of electronic transponders. Each transponder is a receiver-amplifier-transmitter built as a kind of bucket brigade; anything that comes in the receiver's antenna is immediately amplified, shifted to a new frequency, and rebroadcast back to the home planet. The frequency shift is needed so that the high-power signal from the satellite's transmitter will not leak into its sensitive receiver. Most of the satellites now flying have either 24 or 36 transponders, each working on a slightly different pair of frequencies.

The electricity to run this equipment comes from solar power. In spring and fall, each satellite passes through the earth's shadow once a day. During this eclipse, which lasts up to 70 minutes, the transponders are kept alive by batteries. Spring and fall also bring a different kind of astronomical interference. As viewed from the earth, each GEO passes in front of the sun, cutting off the downlink for about 20 minutes.

There are over 150 active satellites in geostationary orbit, forming a jeweled belt around the planet's midriff. The major limit facing the technology is that there's no room for more. Not that the satellites are jostling against each other in orbit; distributing 150 satellites at equal intervals around the Clarke Belt would leave almost a thousand miles between them. The problem is that with closer spacing, ground stations can't focus their beams narrowly enough to select just one satellite. "Slots" in geostationary orbit are some of the most prized real estate not on earth.

Earth Stations. The control centers for a major satellite system have multiple antennas, including some very large ones, roughly 100 feet in diameter, much like the radio telescopes that NASA uses to track interplanetary probes. At a somewhat smaller

scale, a television station or the head end of a cable system may have dishes 30 feet across, equipped both to transmit and to receive. The 10-foot dish antennas that dot rural and suburban neighborhoods are receive-only devices. So are the even smaller plastic dishes clamped to windowsills and high-rise apartment patios.

Other things being equal, size determines performance. In receiving mode, a dish with a larger area can collect more of the energy emitted by the satellite, just as a larger net catches more fish. When transmitting, a bigger dish has higher "gain": it concentrates more of the transmitted power in the desired direction.

Even with a large signal-gathering reflector, however, detecting the satellite's distant whisper is still an impressive feat. One strategy is to mount the first stage of the signal-processing electronics—a device called a low-noise amplifier—right on the antenna, so it can immediately boost the signal to a higher level. A converter also shifts the received signal down to a lower band of frequencies so it can be carried back to the receiver by a coaxial cable instead of a wave guide.

If you're ever lost in the wilds of suburbia, backyard dishes can be a useful aid to getting your bearings. In the Northern Hemisphere they always point generally southward (although the exact compass bearing could be anywhere from southeast

BELLS AND WHISTLES (AND SIRENS)

There was a time in the history of Western civilization when the most important instrument of mass communication was the church bell. In part that's why the village church had a tall steeple—not just to satisfy architectural or spiritual sensibilities but to get the bell up in the air where people could hear it. It was rung routinely to mark the passing of the hours, as well as in celebration and alarm.

For a time the factory whistle took over the role formerly played by the church bell—a transition that did not go unremarked by those who believe we have all sold our souls to the company store. But church bells still ring, whereas I don't know of a single town where people continue to live by the factory whistle.

If anything has taken the place of the church bell in modern life, it is the siren. The great age of siren building was the 1950s and 1960s. The sirens were meant to warn of incoming Russian bombers and missiles, telling us it was time to duck and cover. Some of those Civil Defense sirens can still be seen on

rooftops, although they're looking a bit rusty, and I wouldn't count on them for early warning of Armageddon.

But lots of other sirens are well maintained and tested. On coastal plains sirens warn of hurricane floods. Near nuclear power plants they sound the alarm when radiation escapes. On the prairies they warn of tornadoes. The

siren in the photograph at left summons volunteer firefighters in Bryson City, North Carolina.

The traditional siren is a disk or cylinder with a pattern of holes around the periphery. When the siren turns, air blowing through the holes is periodically interrupted, producing the characteristic wail. The noise can be extraordinarily loud, rivaling a jet engine. Another design works more like an automobile horn than a siren: a stack of vibrating diaphragms puts out basso profundo blasts of sound that you feel in your gut.

For some years I lived in a Minnesota town equipped with tornado sirens, including a unit just a few hundred feet from my house. The sirens were tested every Wednesday at 1:00 p.m. I quickly got used to this routine. When the siren started up, I would glance at my watch, note that it was one o'clock on Wednesday, and go back to my business. But I would also wonder, every week, how the town was going to warn us if a tornado was spotted at 1:00 p.m. on a Wednesday.

The little gray dishes, the size of a pizza pan, sprout everywhere from yards, balconies, rooftops, windowsills. The cluster above is in Zagreb, Croatia.

to southwest). At the present time, all of the 18-inch dish antennas in the United States point to a pair of satellites at 101 degrees west longitude, which is the meridian of North Platte, Nebraska. If you're anywhere in the Midwest, any of those mini-dishes is a pretty good pointer toward due south. Sometimes a dish can also give a rough estimate of latitude. When a dish is pointing south, its angle with respect to the vertical is approximately equal to your latitude.

The very idea of home satellite receivers was not something that the marketing wizards of the communications industry dreamed up. The first home dishes were built by tinkerers, and the early commercial models were bought by people too isolated to get either broadcast or cable TV. The subsequent evolution of the technology has followed a familiar but interesting path. Just as cable started in small towns and then caught on in the city because it offered more than the local broadcast stations, so satellite receivers are now competing with cable for urban and suburban customers. Indeed, there are now several satellites aloft that do nothing but beam programs directly to home receivers.

Low-Earth-Orbit Satellites. Since the 1960s, giant, high-power GEOs have dominated the satellite communications industry, but another scheme altogether has lately gone into orbit. The new type of communications satellite is called a LEO, which stands for *low earth orbit.* LEOs are only a few hundred miles up—just high enough to avoid being brought down by the drag of the earth's atmosphere.

LEOs have some big advantages over GEOs. They don't suffer the long delays of waiting for a signal to bounce back from 22,000 miles away; when you communicate via a LEO, the signal path is not much longer than it would be on the ground.

The prototype of the satellite dish antennas that are now being molded in plastic by the millions was the radio telescope, with a history going back almost to the beginning of radio itself. The dish antennas in the photograph at right are 82 feet across; 27 of them are arranged on a plain near Socorro, New Mexico, to form a listening post of exquisite sensitivity.

Launching the satellites is cheaper because they don't have to go as high. And communicating with a satellite that's buzzing by at treetop level (relatively speaking) doesn't require a high-power transmitter and a high-gain antenna.

The LEOs also have an obvious drawback: they do not hover over a fixed point on the ground. On the contrary, they zoom by quite rapidly, staying within sight of any one spot on the earth's surface hardly long enough for a phone call. But even this trait can be turned to advantage. The key idea is to adopt the philosophy of the cellular-telephone system. You put up lots and lots of LEO satellites, just as a cell-phone operator has lots of antenna towers. Each satellite communicates with those callers who happen to be directly below at a given moment, then hands off the call to another satellite. The orbits are arranged so that there's almost always at least one satellite over any point on the earth.

The pioneer among LEO projects was the Iridium system. The original proposal called for a "constellation" of 77 satellites, which gave the system its name—the element iridium has atomic number 77, meaning that an iridium atom has 77 orbiting electrons. Before the satellites were launched, the plan was scaled back to 66 active satellites, but no one wanted to change the name to Dysprosium. Maybe they should have. Iridium was a technical success but a commercial disaster, at least initially. It was rescued from bankruptcy with support from the Department of Defense.

TALK OF THE FUTURE

Over the course of the past century, the communications infrastructure has been evolving faster than the infrastructures of other industries. A Victorian engineer would have no trouble understanding a modern municipal water supply, or the Interstate highway network, or most of the hardware for electric-power distribution. Railroads have changed all too little, and even airplanes have looked much the same for 40 years or more. But fiber optics, cellular telephones, the Internet, and satellite communication—these technologies did not exist when I was born. They have developed so rapidly that they are now utterly commonplace and unremarkable. Children use them every day. And the evolution is by no means over.

One of the most puzzling aspects of the communications world is the continual interplay of wired and wireless technologies. For a while it seemed clear how they would sort themselves out. Broadcasting, or one-to-many communication, would occupy the airwaves, while telephony, or one-to-one communication, would be confined to copper or glass. This division seemed to make the most efficient use of resources, balancing the scarcity of frequencies in the electromagnetic spectrum and the cost of laying down wires.

But the continuing success of cable TV and the growing importance of cellular telephones and wireless computer networks argues that some other principle is at work here. I don't know how it will come out in the end.

ON THE ROAD

K EEP YOUR EYES ON THE ROAD" is familiar advice, but almost no one does it. We may be alert to traffic, or we may take in the scenery along the roadside, but we seldom pay much attention to the road itself. And what's to see anyway? It's just a paved-over path, isn't it, a featureless strip of concrete, a way of getting somewhere else—and the quicker the better, please.

Learning a little something about the design of highways has given me a new perspective when I sit behind the wheel. Road building is an engineering profession, hemmed in both by codes of practice and by economic constraints, but there is also a strong tradition of aesthetic sensibility among highway designers. They pay a lot of attention to the visual experience of the driver. The placement of turns and grades is partly prescribed by the terrain, but highway engineers still have a lot of latitude in working out the details of how the road will unfold before you as you follow the dotted line at 60 miles an hour.

Of course, this aesthetic judgment is based mainly on what the road looks like from the driver's seat. To critics of highway policy, that's just the problem: road builders consider only the needs of motorists, and the remedy for any problem is always to build another road or else to widen one—until the countryside is sliced to ribbons by multilane highways, the cities are choked with cars, and we all choke on the fumes.

There's no question that the automobile has altered our landscape and our lifestyle more than any other technology of modern times. It's not just a matter of the acreage paved for roads and parking lots, or the number of hours per day we spend in the car. Even more important, the automobile has transformed where we live and work. For more than a century, cities have been sucking people out of the rural countryside, but the cities themselves have also been sprawling beyond their boundaries. Especially in North America, many cities have become hollow at the core while they

Over, under, and through: The ramps of the highway interchange have become such a familiar everyday sight that they go largely unnoticed. We forget that before the age of the automobile, the idea of making a left turn by going right, right, and right (as in a cloverleaf intersection) would have seemed quite bizarre. The steel-girdered ramps and overpasses on the opposite page were photographed at an interchange in Durham, North Carolina.

A network of streets and highways spans the United States, although it gets sparse in the Mountain states. The image below was created by dividing the area of the nation into squares 1 kilometer on a side, then assigning a color to each square according to the total length of roads found in the square, as recorded in a geographic database compiled by the Bureau of the Census. The colors range from black (for areas with no roads at all) through blue and red to yellow (for areas with the highest density of roads and streets). Although no individual roads are plotted in the image, only the density per square kilometer, it is easy to pick out some major thoroughfares. In the detail at right, the route of Interstate 70 passes through Denver, Colorado, which is the brightest of the major blobs. To the north, Interstate 80 forms another prominent east-west corridor through Cheyenne and Laramie, Wyoming. The original image was created by Patrick J. Hayes of the National Geophysical Data Center; the color coding is by the author.

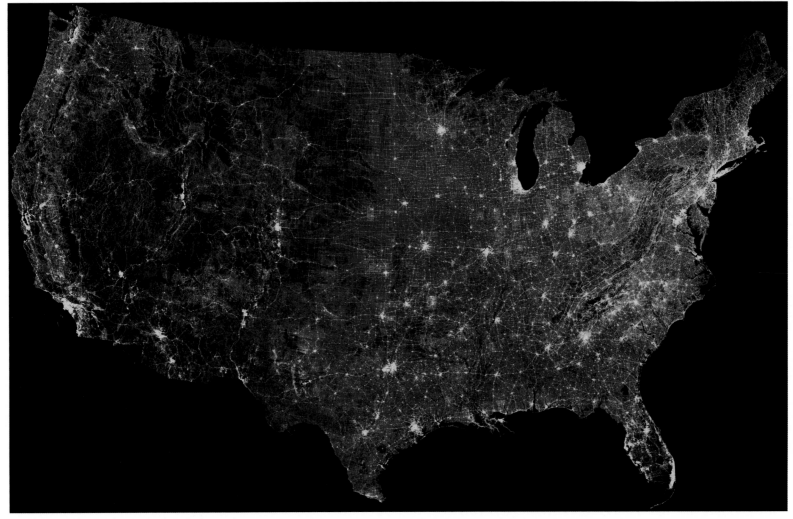

grow concentric rings of suburban housing developments and shopping malls. It's as if everyone wants to be *near* the central city, but nobody wants to live *in* it. We value the amenities of city life—the restaurants, the shopping, the big-league sports and museums—but we prefer a house set on a woodsy acre or two. Strategies for reconciling these conflicting urges tend to rely heavily on the automobile. The house on the woodsy acre has a two-car garage.

Still, the image of the automotive juggernaut is sometimes exaggerated. It's true that we're continually surrounded by cars and car culture, but that's partly a result of our own choices. If you go everywhere by road, then you'll find roads everywhere you go. A few unpaved patches of the planet still exist, but you have to get out of the car to see them. It's also worth noting that the exodus from the central cities was not initiated by the automobile. It began well before World War I, when few families had cars. The first suburbs were streetcar developments—in many cases designed, financed, and promoted by the owners of trolley lines.

THE PUBLIC THOROUGHFARE

Some of the roads we travel today have been well-worn paths since antiquity. A few of them predate humanity altogether; they were animal tracks before people discovered them.

A number of the oldest roads are ridgeways: routes that stay well above valleys, following high contour lines, typically on the warm, sunny side of ridges. It's easy to see why a road would tend to follow a contour line (that is, a line of constant elevation). There's no sense in climbing up and down hills if you can avoid it. But why did those earliest trailblazers choose higher rather than lower contours? The answer, apparently, is mud: The valley bottom tends to get soggy and overgrown with vegetation. "Ye'll take the high road and I'll take the low road, and I'll be in Scotland before ye," says the old ballad of Loch Lomond—but the historical evidence suggests the high road has usually been the quicker way.

In Europe and Britain there are ridgeways known to be 7,000 years old. One of many surviving examples in Britain is a 90-mile trail called the Rudge, which meanders from Avebury to Ivinghoe in Buckinghamshire. In America, the Natchez Trace is a ridgeway route that supposedly began as a track created by buffalo migrating from the lower Mississippi to salt licks near the present site of Nashville, Tennessee. New York Route 5 between Buffalo and Albany follows the route of the Iroquois-Mohawk Trail, a ridgeway footpath 400 miles long.

Later roads were not all fitted so carefully to the contours of the land, either high or low. The clearest counterexample is the gridded road system of the American Midwest. The Land Ordinance Act of 1785, passed at the initiative of Thomas Jefferson, divided new territories beyond the 13 original states into townships six miles square. Each township was divided in turn into 36 one-mile-square sections,

and each section was divided into four quarter-section farms. The owners of property on either side of a section line were required to donate a 33-foot-wide strip to public use, creating a 66-foot-wide right-of-way at one-mile intervals. (Why 66 feet? That's the length of the Gunter's chain, a standard measuring device in surveying.) Roads in much of the Midwest still conform to this rectilinear pattern: they run straight and parallel and are aligned to the cardinal points of the compass, heedless of hills and valleys and all but the most extreme obstacles. And yet the lattice of squares within squares isn't quite perfect. It can't be because even an act of Congress can't repeal the law of mathematics that says a lattice of squares won't map smoothly onto the surface of a spherical planet. Because lines of longitude converge toward the poles, the parallel north-south roads have to deviate to the left or right every few townships.

City planners also have a fondness for the rectilinear grid—perhaps understandably since the nucleus from which many cities have grown is a crossroads. In some cases the determination to have streets intersect at right angles is powerful enough to triumph over the most difficult terrain—and even over common sense. The obvious example is San Francisco, where straight-arrow streets climb hills so steep that the sidewalks have to be carved into stairways.

City blocks, unlike rural quarter sections, are seldom actually square; houses fit together more efficiently if the grid consists of skinny rectangles. But whatever the shape of the blocks, organizing space into columns and rows makes it easier to find your way, especially when the naming or numbering of the streets follows some decipherable system. (There are a few American cities where you'll have an easier time navigating if you know the chronological sequence of presidents.)

Among real-estate developers, straight lines and right angles went out of fashion sometime in the middle of the twentieth century. If you look at a town or a residential neighborhood laid out since then, you are more likely to find sinuous, serpentine roads—whether or not the topography offers any excuse for the curves. Many of these roads go nowhere: they are loops that bring you back to where you started, or they are cul-de-sacs. Making it easy to find your way through the network of streets is obviously not a high priority. This is an interesting development in urban geography: having redesigned the city to accommodate the automobile, we now search for ways to discourage people from driving on the streets.

OCEAN TO OCEAN

In North America, thousands of miles of roads date to the colonial era. On the West Coast, the Spanish built El Camino Real (The Royal Road) from Mexico City to Sonoma, north of San Francisco. Parts of that route now form the main drag in dozens of towns and cities from Santa Barbara through Silicon Valley. (Much of El Camino Real coincides with U.S. Highway 101.) On the East Coast, the British built

The geometry of city life underwent a major transformation sometime in the twentieth century. Where earlier generations of planners drew ruler-straight streets, all crisply intersecting at right angles, later architects have favored soft curves, loops, and cul-de-sacs. Both of the aerial photographs on the opposite page were made by the U.S. Geological Survey in their Urban Areas series. The upper image shows Oak Park, Illinois, and adjacent neighborhoods of Chicago, an area built up in the years just before and after 1900. Although Oak Park, as the birthplace of Frank Lloyd Wright, is famous for its "prairie style" houses, the layout of the streets is relentlessly urban. The lower image shows part of North Highlands, California, a community near Sacramento where building began in the 1950s. As in Chicago, houses are lined up side by side along all the roads, but the texture of the landscape is quite different because the designers have twisted and warped the grid so that almost all the roads have smooth curves.

their own version of the Royal Road, and a few sections in New England are still called the King's Highway; elsewhere it is the Boston Post Road and U.S. Route 1.

After American independence, there was a spurt of road building directed toward the interior of the country. The Lancaster Turnpike, running 60 miles west from Philadelphia to Lancaster, Pennsylvania, was an entrepreneurial venture in the 1790s. Private investors laid claim to a network of preexisting roads and began charging tolls to travelers; at each gate a carriage drawn by two horses paid 25 cents—not a small sum in those days. The many nearby roads that still bear the name "Shunpike" attest to the ingenuity of toll avoiders.

Another project of this era, the National Road, was publicly funded under an act of Congress passed in 1806. The original plan was for a wagon road running west from Cumberland, Maryland, across the Appalachians to Wheeling, West Virginia. Once this section was completed, however, the builders plunged on enthusiastically across Ohio and Indiana, aiming for the Mississippi River at St. Louis. But the road never quite got there. Federal funding ran out early in the 1840s, when the National Road had reached Vandalia, Illinois, about 50 miles short of the goal. Here's another fascinating fact about the National Road: until 1850 it mandated driving on the left, long after the rest of the country had settled on the keep-right convention.

The failure of the National Road was not an isolated incident. Road building had lost much of its luster by the middle of the nineteenth century; public interest had shifted to canals and then to railroads, which seemed more glamorous. It was almost a century later before the automobile finally brought roads back to the top of the public-works agenda. A revival of interest in the 1920s and 1930s led not only to the construction of new roads but also to the paving and widening of older ones. And what was most distinctive about this era was the fad for mapping networks of designated long-distance routes. The National Road became part of this movement, incorporated into a route with an even more grandiose name: the National Old Trails Ocean-to-Ocean Highway. Its main rival was the Lincoln Highway, championed by Carl Fisher, whose other notable promotions were the Indianapolis 500 and Miami Beach. Fisher's Lincoln Highway followed a more northerly route than the Ocean-to-Ocean Highway, incorporating the old Lancaster Turnpike and passing through Chicago instead of St. Louis. Eventually both of these routes were co-opted or superseded by the system of numbered U.S. highways. Roughly speaking, the Ocean-to-Ocean Highway became U.S. Route 40, and much of the Lincoln Highway was designated U.S. Route 30.

But that's not quite the end of the story. U.S. Routes 30 and 40 and other roads of the same era—such as Route 66 in the Southwest—remained the primary choices for long-distance travel through the 1950s, but with the coming of the Interstate system, they reverted to back-road status. For example, Interstate 70 runs parallel to U.S. 40 for well over a thousand miles, and in many places the roads are only a mile or two apart. The older road is often designated the business route or the scenic route or the historic route—all ways of saying, "Don't go that way if you're in a hurry." In the West,

A milestone on the old National Road in Ohio is one of hundreds still surviving. The legend at the top gives the distance west of Cumberland, Maryland. On the right flank, westbound travelers see they have 29 miles to go to Columbus, Ohio; eastbound travelers, approaching from the left, are 98 miles from Wheeling, West Virginia, and 24 miles short of Zanesville, Ohio. The places listed as "H." and "J." were evidently too small to bother spelling out; they probably refer to the villages of Hebron and Jacksontown, Ohio.

several sections of U.S. 40 were completely obliterated, because the Interstate was built over the same right-of-way. Indeed, U.S. 40 no longer reaches the West Coast; it simply disappears, like a river drying up in the desert, somewhere in Utah.

A final note on the rivalry between highways and railroads. In the 1880s, when rails of steel were still the epitome of high technology and the darlings of the stock market, the railroad barons began planning a new rail route from Philadelphia and New York to Pittsburgh. Over the next 50 years they invested millions in acquiring land, surveying the route, and digging several tunnels through the Allegheny Mountains. But by the 1930s the plan had collapsed. What was eventually built along the right-of-way (and through the tunnels) was not a railroad but a highway: the Pennsylvania Turnpike.

GEOMETRIC DESIGN

To build a road, you obviously have to know where you're going and how to get there, but beyond these basic decisions about the route, there are countless smaller-scale choices to be made: How sharp will the curves be? When hills get in the way, will the road go over, around, or through them? What is the width of the lanes? Issues like these are known in the road-building trade as elements of geometric design. They have a major effect on what it will feel like to drive the road.

Typically, a highway layout begins with the choice of a design speed. The curves, hills, lanes, and so on are then designed so that you'll feel safe and comfortable driving the road at this speed. Incidentally, the design speed has nothing to do with the

A highway cut in western Maryland creates a groove 800 feet deep where Interstate 68 passes through the ridge known as Sideling Hill. The cut allows the highway to take a direct route over (or through) the ridge; U.S. 40, the earlier road that follows the same corridor, takes a three-mile detour to climb over the crest.

Two styles of roadbuilding can both be found in the Interstate highway system. In one approach the designer begins by laying out long straight segments and then connects them with relatively short curves. The alternative is to start with sweeping curves and join them with short straight sections. The contrast is apparent here between a view of Interstate 10 near Casa Grande, Arizona *(this page)*, and Interstate 77 near Odd, West Virginia *(opposite page)*. A feature to note in the latter photograph is the coordination of vertical and horizontal profiles. A "sag," or valley bottom, in the foreground coincides with the apex of a curve. Designers today increasingly favor the curves-first method, but of course the terrain has much to say about the options available. A series of wiggling S-bends on the flat expanse of southern Arizona would look silly.

legal speed limit; engineers base their calculations on surveys of how fast people actually drive on roads of various kinds, not on what the sign says they should do.

Curves and Grades. Highway engineers measure curvature in a curious way. They pace off 100 feet along the centerline of the road and measure the road's change in direction over this distance. If the direction of travel changes by 5 degrees in the course of 100 feet, that is called a 5-degree curve. The larger the degree, the sharper the curve. As the design speed of a road increases, the maximum allowable curvature has to decrease. For ordinary roads the usual limit is 5 to 7 degrees, but highways designed for high-speed travel will have curves no sharper than 3 degrees. Following a 3-degree trajectory, you have to travel more than half a mile along the road to make a full left or right turn of 90 degrees.

Limits on slope, or grade, also depend on the design speed, because heavy trucks cannot maintain speed while climbing steeply. (They cannot "make the grade.") For a design speed of 30 miles per hour, the maximum grade is 12 percent (that is, the road gains 12 feet of elevation for every 100 feet of horizontal travel). At 70 miles per hour the steepest allowable grade is only 5 percent.

What does the engineer do when a road with a 5 percent grade limit needs to cross a series of hills with 20 percent slopes? One choice is to trade horizontal for vertical distance: take a roundabout route, spreading out the climbing over a longer span. The ultimate expression of this strategy is the Alpine switchback roadway. A second approach is called cut-and-fill. You excavate a trench through the hilltop, and use the earth removed from these cuts to fill in an embankment across the adjacent valleys. If you plan it just right, cut and fill are balanced: there's just enough material

removed from the cut to bring the filled embankment up to the level of the road-way. A final option replaces cuts and fills with tunnels and bridges, which call for less earthmoving but more engineering.

Limits on curvature and grade are easy to express in numerical form, but many other criteria for highway design reflect aesthetic principles as much as engineering rules. For example, many designers believe that if a road has long straight sections, they should be connected by long sweeping curves, not short sharp ones. Designers also try to avoid "broken-back" curves, where two curves in the same direction are connected by a short straightaway. On the other hand, "reverse curves"—the two parts of an S curve—should be separated by a straight section of a few hundred feet to ease the transition. The idea is to induce a smooth rhythm for drivers on the road, with changes in direction coming at a predictable pace.

Another guiding principle suggests that the point where a road's slope changes direction (the peak of a hill or the bottom of a valley, called a sag) is also a good spot for a change in horizontal direction (the apex of a curve). On the other hand, a turn at the crest of a hill can't be too sharp, because the driver cannot see the turn coming, especially at night. The maximum vertical curvature at a crest is often determined by minimum sight distance: if the road bends too sharply over the brow of the hill, a driver cannot see far enough ahead to maintain the design speed. Near intersections, too, the road should be as straight as possible in both the horizontal and the vertical dimensions to improve sight distances.

Until recent years most roadways were designed by first laying out a series of straight lines across the landscape, then adding curves as transitions. On two-lane

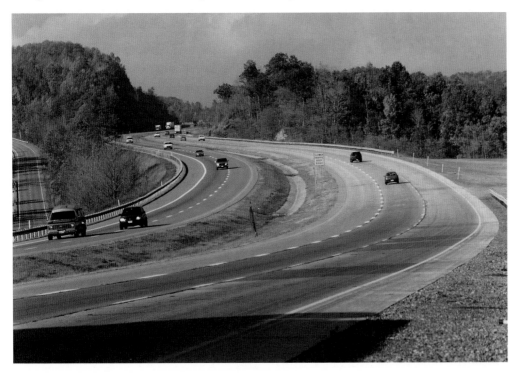

The fashion for curvaceous roadways affects even access ramps. This elevated serpentine connector carries traffic from a bridge across the James River just south of Richmond, Virginia.

roadways this style of layout is still favored because the long straightaways maximize unobstructed vision for passing; unfortunately, they also create an unpleasant sensation of veering first in one direction and then another. On roads with four or more lanes, sight distance for passing is of less concern, but nevertheless the same design rules prevailed for many years on these highways too. Modern practice, however, increasingly favors laying out a series of long sweeping curves, which are then connected by short straight segments or spiraling transitions.

I had been driving both kinds of roads for years without ever being aware of such stylistic distinctions. Once a highway engineer pointed them out, however, the differences were unmistakable. On a road built in the older style of long straightaways, where you can often see a mile or two ahead, a trip is experienced something like a slide show: a sequence of still photographs, shown one after the other. On a newer road with sweeping curves and short straights, your destiny unfolds gradually and continuously, so travel is more like a movie.

By the way, a persistent legend says that one mile out of every five in the Interstate highway system has to be straight so the roads can be used as runways for military aircraft in time of emergency. If you think this might be true, keep a record during your next highway trip of how many mile-long sections are free not only of curves but also of signs, light poles, overpasses, and other obstructions to aviation.

In the vertical dimension, a road is laid out as a series of straight lines—in other words, sections of unchanging slope—connected by either circular or parabolic curves at the peaks and sags. The parabola has been the traditional choice, because it has an especially simple equation, making it easy to calculate the elevation of the road at each point. A parabola is also the trajectory followed by a free-flying projectile—

such as Evel Knievel jumping his motorcycle over a line of flaming cars and buses. But take it for granted that if you get airborne at the crest of a hill, you're going well over the design speed. With the computers available to designers today, the calculational convenience of parabolas is no longer an issue, so maybe the shapes of hills and valleys will change.

PARKWAYS AND FREEWAYS

In the design of most roads built before the twentieth century, the reference vehicle was an ox cart, and the design speed was a few miles per day. There was a brief but shining interlude, around 1900, when the needs of bicyclists came to the forefront in roadway engineering, but since then cars and trucks have been totally in command. Many of the biggest roads exclude all other traffic.

In the United States, the first roads specifically meant to expedite automotive traffic were parkways built in the years just before and after World War I. These roads had most of the features that we now associate with freeways and expressways: a dual-carriage highway with a barrier or a grassy median strip to separate counterflowing streams of traffic; entrance and exit ramps; overpasses and underpasses for intersecting roads; some of them even had rest stops with gas stations and restaurants. Most of the parkways also had at least two lanes in each direction, which is probably the most important innovation for the driver. Because you can pass without waiting for a break in opposing traffic, a four-lane road has much more than twice the capacity of a two-lane road.

You can still see and drive on parkways built in the style of the 1920s and 1930s. Perhaps the best place in the world to do this is New York City and its suburbs, where many of the early parkways seem to be frozen in time. The first link in the system was the Bronx River Parkway, built between 1916 and 1923; it runs from the Bronx Zoo up to White Plains in Westchester County. The success of this road inspired several more of similar design in Westchester (the Saw Mill River, the Sprain Brook, the Hutchinson River) and the Merritt Parkway in nearby Connecticut. Later, Robert Moses—the master planner of much of New York's road network—built half a dozen more parkways on Long Island. A biography of Moses remarks, "He deliberately made the overpasses low to keep out the buses used by the lower class non–car owners." It's also worth noting that Moses himself, while paving over so much of New York, never learned to drive a car; he had a chauffeur.

Parkways of similar design in other parts of the country include the DuPont Highway in Delaware (1924), a "superhighway" in Michigan from Pontiac to Detroit (1925), and Lake Shore Drive in Chicago (1933).

Driving these roads is an interesting experience for anyone who has grown accustomed to the scale and rhythm of the later Interstate highway system. Although the parkways have the appurtenances of modern freeways—ramps, overpasses, and so

Older parkways and newer Interstate highways are built out of the same basic elements, but they create a different impression from behind the wheel. On the Garden State Parkway near Toms River, New Jersey *(upper photograph)*, the narrow right-of-way and the massive concrete overpass encroaching on the shoulders creates a slightly claustrophobic feeling. The more recent Interstate 75 near Wapakoneta, Ohio *(lower photograph)* offers wider vistas.

on—all these elements seem to be in miniature because the design speed is lower. The lanes are narrower, the curves sharper, the grades steeper, the sight distances shorter.

The first roads that approached modern standards of freeway practice were the early autobahns in Germany and the autostradas in Italy. The first such American road was probably a stretch of the Meadowbrook State Parkway leading to Jones Beach State Park on the south shore of Long Island, opened by Robert Moses in 1934.

Today most U.S. freeways are part of the Interstate highway system. This network of roads was first conceived in broad outline during the Roosevelt administration in 1937; the authorizing legislation was enacted in 1944; but construction was not started until 1956, after much prodding by General Motors, the road-building lobby, and other interested parties. The system now has more than 45,000 miles of highway open to traffic. If you had a big enough gas tank (and a big enough bladder!), you could drive coast to coast without ever getting off the highway. But the dream of a transcontinental freeway was a long time in coming—much longer than transcontinental rail and air service.

The Interstate system is all about standards and uniformity. The red-white-and-blue shield is supposed to certify that the road was designed and built according to federal rules that spell out the minutiae of geometric design—from the width of traffic lanes to the height of underpasses—as well as construction details such as the strength of the concrete in the roadbed. In other words, the road you drive on is just as uniform and predictable as the motels and the burger joints clustered at every exit. Or at least that was the plan. In practice, road builders have been allowed a lot more leeway than fast-food franchisees. There are two-lane Interstates; there are Interstates with traffic lights; there are even Interstates with drawbridges and railroad crossings.

The impulse to renovate and rebuild the road network seems to be a recurrent one. There was the early spurt of work at the time of the National Road, then we created the U.S. highway system in the 1920s, and just a few decades later started all over again with the Interstates. Is it safe now to sit back and say, "We'll never have to do *that* again"? Maybe. In 2001 the governor of California dedicated a new section of Interstate 210 that he said would be the last freeway ever built in the state. Whether or not that prediction comes true, it seems unlikely that today's superhighways will be replaced by some whole new network of ultrahighways, built to even higher standards and design speeds. But a major upgrading of the roads is not unthinkable.

Route Numbers. There's a method to the numbering of the Interstate routes. If you wake up with amnesia at a truckstop where two Interstates cross, you should be able to figure out what part of the country you're in just from the road signs. The basic idea is that east-west routes get even numbers, and north-south roads odd numbers. The major, long-distance routes are assigned numbers less than 100, ending either in 0 (for the east-west highways) or 5 (for north-south). The numbers increase from south to north and from west to east (which is the opposite of the scheme that was adopted for the U.S. highways in the 1920s). Thus Interstate 10 runs along the coun-

Which way is up? It will come as no surprise to learn that this section of road, which seems to go north and south at the same time, is actually aligned east-west.

The simple idea of the crossroads has evolved into elaborate geometric fantasies like this intersection in Maryland. The lowest roadway is the Baltimore Beltway (Interstate 695) and the uppermost level, crossing at right angles, carries Interstate 70, just a mile or two from its eastern terminus. The diagonal lanes threaded through the intersection are ramps carrying left-turn traffic. The structure is known as a directional interchange.

try's southern flank, whereas I-90 crosses the northern tier of states; I-5 is the principal thoroughfare along the Pacific coast and I-95 along the Atlantic. So far so good. But in between, the grid gets a little twisted. For example, I-85, which ought to be a north-south route west of I-95, veers off on a diagonal and has intersections with I-65, I-75, and I-95.

Route numbers higher than 100 are reserved for shorter spurs, loops, and connectors. Where the leading digit is even, the road is supposed to be a loop or bypass; an odd first digit indicates a spur that does not return to the main road. For example, I-495 in Maryland and Virginia is the Washington Beltway—the famous loop road outside of which the rest of the country lives. I-395 in the same area is an expressway to downtown Washington. But these rules too have not been universally observed, and a community of "road geeks" delights in collecting exceptions. Their favorite example seems to be I-238, a short connector south of Oakland, California. According to the approved numbering scheme, this ought to be a loop off of I-38, but in fact it runs between I-580 and I-880. There is no I-38.

INTERSECTIONS

The crossroads is one of those ancient artifacts nobody had to invent; it just happened. But later generations of highway engineers have worked some elaborate variations on the basic idea.

Years ago, two unpaved country roads might have had a totally "unprotected" meeting. Drivers were left to their own resources to negotiate who had the right-of-way. As traffic got heavier, stop signs might have been added to the intersection, and

A diamond-type interchange is the usual choice where a major highway crosses a lesser road. Cars enter and leave the larger route via high-speed ramps, but at the smaller road the ramps terminate in conventional surface intersections, where drivers must contend with cross-traffic. The interchange shown is on Interstate 40 south of Raleigh, North Carolina. All of the photographs on these two pages are from the U.S. Geological Survey Urban Areas series.

then signal lights. If the streets continue to get busier, it might be worthwhile to add dedicated lanes for right and left turns, so that through traffic doesn't pile up behind vehicles waiting to turn. Still another step in this evolution could be to "channelize" the intersection, installing raised islands or concrete barriers to separate streams of traffic going in different directions.

On a two-dimensional surface, it's an obvious fact of life that whenever two roads cross, there must be some patch of pavement that belongs to both of them. One way

AROUND AND AROUND WE GO

A quite different approach to managing traffic at intersections is the device known variously as the rotary, the traffic circle, or the roundabout. Over the years it has come in and out of fashion several times; at the moment it seems to be on the upswing again.

Some of the grandest traffic circles serve as hubs for the spokelike boulevards of cities such as Paris and Washington, D.C. (shown at right is Dupont Circle in Washington). A circle is perhaps the only plausible choice for such an intersection, where cars are entering from a dozen directions and might choose as many destinations. A conventional system of traffic lights and turn lanes would be a nightmare.

Early in the twentieth century, circles were also popular for rural intersections in some states. They seemed more efficient than signal lights or stop signs because cars could zip

through without pausing. But many of the circles were found to have a high accident rate, and traffic in them would sometimes freeze

solid; no one could get in because no one could get out. By the 1960s, state highway departments were ripping out circles.

The European experience was better, at least in part because the rules of the road were slightly different there. Under the American system, drivers inside the circle were asked to yield to those trying to enter, but European practice gave the right-of-way to those already in the circle. Apparently, this change greatly lowers the chance of a packed-solid traffic jam. European designers also avoided multilane circles, so drivers would not get trapped on the inner ring.

Starting in the 1990s, there has been a new wave of circle building in the United States, adopting European designs and traffic patterns. To emphasize that the new circles are something different, they are usually called "roundabouts," which seems to have quaintly British connotations. Whether they'll work for American drivers remains to be seen.

or another, the roads have to allocate access to this shared territory, usually by having drivers take turns. The only way to eliminate this contention for the right-of-way is to escape into the third dimension. This was already apparent to Frederick Law Olmsted in the 1850s, when he built overpasses and underpasses for the carriageways in New York's Central Park. Today such "grade-separated intersections" are a defining feature of freeways everywhere. Actually, the overpasses and underpasses are the easy part; they merely carry the through traffic. Where things get tricky is in the arrangement of ramps that allow drivers to make turns—to switch from one road (and one level) to the other. There are dozens of designs for such highway interchanges, but they fall into three main categories, known as the diamond, the cloverleaf, and the directional intersection.

The diamond is the usual choice where a small, low-traffic road connects to a freeway. Cars leave the freeway via a high-speed exit ramp, and they enter it by merging from another ramp. At the smaller road, however, the ramps terminate in conventional right-angle intersections, usually equipped with stop signs or signal lights. A troublesome feature of this arrangement is that drivers on the minor road who are turning left to enter the freeway have to cross oncoming traffic. As a result, the strip

The cloverleaf, with its four symmetrical leaves, is the classic among highway interchanges. This example, squeezed between shopping malls and a residential neighborhood, is on Interstate 405 in Kirkland, Washington. A drawback of the cloverleaf configuration is that cars taking an exit have to weave their way through traffic entering the highway just upstream. Here this transition has been eased on the major north-south route by giving the weaving zones a separate roadway. Also note that traffic entering the southbound lanes is being "metered": Cars are lined up behind signal lights on the ramps and are released one at a time.

The directional interchange *(this page)* is by far the most elaborate scheme for getting one road over another. At the center, the two levels of through lanes are surmounted by two more levels of "flyover" ramps, carrying left-turn traffic. This is the intersection of Interstate 20 and Texas Route 360 in Arlington, Texas, between Dallas and Fort Worth. The area visible in this image is more than half a mile on a side. On the opposite page is another series of intersections a few miles north along Texas Route 360, with variations and elaborations on both the diamond and the cloverleaf, and a complex braiding of local and express lanes. Both photographs are from the U.S. Geological Survey Urban Areas series.

of road between the ramps can become badly congested. But the diamond is more compact than the other designs and thus costs less.

Cloverleaf is a word sometimes used for any highway interchange with ramps, but it really refers to a specific four-leaf geometry. The first true cloverleaf intersection was built by Eugène Henard in Paris in 1906, and there are even earlier precedents in railroad practice. In 1928 a cloverleaf was built in Woodbridge, New Jersey, not far from the present site of the massive interchange between the Garden State Parkway and the New Jersey Turnpike. The 75-year-old Woodbridge cloverleaf is still carrying traffic today where U.S. Route 1 crosses New Jersey Route 35. Nearby, someone has built a cemetery called Cloverleaf Memorial Park.

A cloverleaf eliminates all cross-traffic movements; no one ever has to make a left turn across oncoming lanes. As a matter of fact, no one ever makes a left turn at all. To turn 90 degrees left at a cloverleaf, you turn 270 degrees to the right. (Two wrongs don't make a right, but three rights make a left!) If you need to make a U-turn, you'll do a full 540 degrees, and you may be a touch dizzy at the end.

One drawback of the cloverleaf design is that the loop roads have to turn rather sharply, and so cars—and especially trucks—have to slow down substantially. Even more troubling, each entrance loop disgorges cars onto the roadway just upstream of the point where other drivers are trying to reach an exit. Thus, the stretch of road between the loops becomes a "weaving zone" where the entering and exiting threads of traffic cross over. Amazingly, drivers seem to handle this tricky maneuver quite well, but it can make for a nervous moment.

The directional interchange is the brute-force solution to all these problems. As you drive toward the intersection from any of the four directions, you get a choice of lanes going left or right or straight through. All of the exit ramps depart the main roadway before you reach the crossroads, and they merge back into the main road downstream of the crossing, so there are no awkward weaving sections. Or one might say that the ramps do the weaving, rather than the drivers. Also, no one has to turn 270 degrees in order to make a 90-degree turn; as a result the curvature of the ramps is much less severe, and speeds are therefore higher. But the price for all this convenience is a further leap into the third dimension: In a directional interchange, even the overpasses and underpasses have overpasses and underpasses. Threading the left-turn ramps through the crossover requires two more levels of roadway, making a four-level structure altogether. The resulting concrete confection is visually very impressive when seen from the air; many directional interchanges have the elaborate symmetries of a fancy bow tied on a birthday package. Driving through the structure on the flyover ramps can be like an amusement-park ride. But erecting a four-tiered tower of highways is expensive, and the interchange sprawls over an expanse of real estate bigger than many entire towns. The design is seldom considered except where major freeways come together. One other minor annoyance: There's no way to make a U-turn.

THE ROAD SURFACE

For at least a couple of thousand years, paved roads were built as if they were walls laid flat on the ground, with large stone blocks or bricks fitted together like tiles to make a level surface. This style of paving was supposed to be a cure for ruts and mud, which made dirt roads impassable to wagons during spring rains. But often the paving stones didn't do their job very well. Wheels cracked them or jarred them out of alignment; frost heaved them up; floods undermined them; people stole them.

The road builders tried setting the stones flat (flagstones) or setting them on end (cobblestones); they tried bedding them in sand and bedding them in mortar. Nothing made much difference. It was a bumpy ride every time. This is one of the more dramatic instances of a technology going a very long way down the wrong alley. Running heavy loads over big paving blocks was just the wrong answer.

The first hints of a better way came from Pierre-Marie Jérôme Trésaguet in France and from Thomas Telford (a famous bridge builder) in Britain. Trésaguet and Telford

both suggested covering a foundation of carefully set stone blocks with finer pebbles to provide a smooth surface. But the real innovation was the idea of John Loudon McAdam, who dispensed with the foundation of stone blocks. McAdam showed that a well-compacted layer of fine, broken stone can be strong and durable when laid down directly on the natural roadbed. (The paving material named for McAdam is generally spelled *macadam*, but the change is of no great consequence. McAdam wasn't the man's real name anyway. He was born John Loudon. As the story goes, he added McAdam to poke fun at Scots who boasted of their clans' great antiquity; McAdam was laying claim to an ancient ancestor of his own.)

McAdam's idea of paving roads with crushed stone seems too simple. Surely some enterprising Roman road builder would have figured it out, if only by chance. In fact, there is a trick to McAdam's method. The key is the use of stone crushed to the size of pebbles, *not* naturally occurring pebbles. Crushed stone has sharp, angular edges and corners that dig into each other and form a strong, interlocked network. Natural gravels (and especially river stones) tend to be more rounded, and they do not bind as tightly. On McAdam's road crews, men, women, and children sat on stools by the wayside breaking stone with hammers. This was not an economically sound

TRAFFIC CALMING AND THE ASPHALT REBELLION

Highway engineers work to make traffic flow faster and smoother. They straighten and widen roads and remove bottlenecks. But in many places they are now being asked to do just the opposite: To make streets narrow and crooked, and to build obstructions that compel drivers to slow down. It's called traffic calming.

Many of us are a little ambivalent about ideals of road design. When we're sitting behind the wheel, impatient to get home from work, we see things differently than when we're sitting on the front porch watching cars speeding down our own street. Traditionally, highway departments have tended to take the driver's side in this conflict. But the balance of power is tipping, and transportation officials have begun giving greater weight to neighborhood concerns.

Advocates of traffic calming have come up with quite an array of sedative devices. There are speed bumps, speed humps, and speed tables; there are bulbs, knuckles, islands, chokers, and miniroundabouts; there are chicanes

where the driver must negotiate a slalom course (see photo). In many cases, all it takes to slow down drivers is narrowing the traffic lane by having cars park diagonally rather than parallel to the curb. Sterner measures include closing off

one end of a street—or even closing both ends to make a pedestrian mall.

Fire-and-rescue services often oppose traffic calming, but residents have voted for the measures even when warned of slower responses to emergencies.

Traffic calming is mainly an issue in residential neighborhoods, but similar questions

are also being asked about larger roads. Is widening and straightening a highway always an improvement? Does building more and bigger roads accomplish anything but attracting more and bigger cars?

A signal event in the debate over these issues came in 1973 with the collapse of the West Side Highway, an elevated expressway along the Hudson River in Manhattan. (Ironically, what brought the structure down was a cement truck, on its way to make overdue repairs.) Plans for replacing the highway with an even bigger road, called the Westway, had the backing of city, state, and federal officials. Nevertheless, the Westway was never built. Persistent community opposition derailed the project, and today a more modest surface road occupies the space, along with parks and a bicycle path.

Resistance to repaving has not always been as successful elsewhere, but no road-building project today gets approved without the scrutiny of environmental and community groups.

method, to say nothing of its inhumanity. Widespread macadamizing of roads came only with the advent of the steam-powered stone crusher.

The original macadam had no adhesive binder to hold the stones together. Today, however, the term *macadam* usually refers to asphalt, in which sand and crushed stone are mixed with bitumen, the tarry residue of petroleum refining. Freshly laid hot asphalt is a gooey semiliquid that remains somewhat flexible even after it cools. This resilience helps the pavement resist damage, but it also means that any defects or irregularities in the underlying foundation will "telegraph through" to the surface. Also, asphalt softens in hot weather. At a busy intersection it can be pushed up into waves like ocean swells by the horizontal forces of cars stopping and starting. Where the winters are severe, it erodes into giant potholes.

The alternative to asphalt is concrete, which is also made of crushed stone and sand mixed with a binder, but the binder is portland cement instead of bitumen, and the final product is very different. Where asphalt is like taffy, concrete is hard candy. Under stress, asphalt yields slowly and smoothly, but concrete forms a rigid slab that resists bending. Stiffness and strength sound like they would be virtues in a paving material, and generally they are, but they are not without drawbacks. If a weak spot develops under a roadway, asphalt paving will gradually slump into it, producing a smooth dip. A concrete slab forms an impromptu bridge over a void in the foundation, concealing the flaw entirely, but when the concrete finally reaches its breaking point, it cracks open, leaving a hole with jagged edges.

A distinctive feature of a concrete roadway is the series of tar strips that set up that annoying "thump-thump-thump" as you drive the highway. People generally call these strips expansion joints, but most of them are contraction joints. Here's the difference: An expansion joint has a built-in gap so two blocks of concrete can move either toward or away from each other; a contraction joint is closed solid when the roadway is first poured. The contraction joint starts out as a groove sawn partway through the slab and filled with something pliable (hot-poured tar is the traditional choice) to keep water out. During the first winter, as the roadway cools and contracts, a controlled crack forms through the rest of the thickness of the slab. There may be dowels in the joint to distribute traffic loads across it. Expansion joints, which go all the way through the slab, are generally found only at intersections or other breaks in the pavement. (Bridges also need expansion joints, sometimes elaborate ones.)

Whatever you choose to call the thumping joints, they are typically spaced every 15 feet or 22½ feet or 30 feet. If you are really bored on a long trip, try counting the thumps between mileposts; there will be about 350 if the spacing is every 15 feet, 235 if the spacing is 22½ feet, and 175 for every 30 feet. I have read about roads built with joints at staggered intervals (for example, 13, 19, 18, and 12 feet) to avoid creating rhythmic stresses in the concrete, but I've never found one.

Roadways built with embedded steel reinforcing bars need fewer contraction strips. They still crack upon cooling but on a finer scale, and the steel holds the faces of the cracks close together. Some concrete pavement has no expansion or contrac-

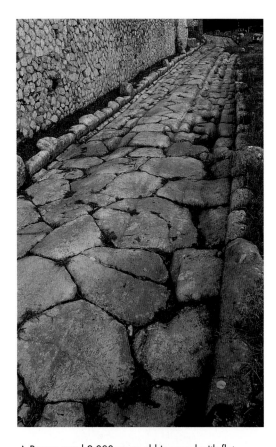

A Roman road 2,000 years old is paved with flat stones carefully fitted together like jigsaw puzzle pieces. The deep grooves, seen in many such ancient roads, look as if they might have been dug by generations of passing oxcarts. But in fact they were probably carved deliberately with axes or chisels, creating tracks to guide carts that had no steerable wheels. The preserved segment of road is in the city of Cassino, halfway between Rome and Naples.

Construction details: Above, steel dowels protruding from the end of a newly poured concrete slab will form a rigid joint with the next slab in the roadbed. Below, a contraction joint is formed by sawing a groove partway through the slab and filling it with a plastic cord and liquid sealant. A crack propagates the rest of the way through the slab.

tion strips at all over long intervals—a mile or more. This continuously reinforced concrete was first tried shortly after World War II, but it didn't begin to catch on for freeways until the 1970s. By now some 13,000 miles of U.S. roadway has been built this way. The key to making it work is installing steel reinforcing bar throughout the slab, and welding it together into a mat or mesh. Even with this internal skeleton, the concrete still cracks under thermal stress, but the cracks form every few feet, and they remain small. You can see them if you get down on your hands and knees at the roadside, but at highway speed they are undetectable.

One final note about paving: what really makes it work is the pneumatic tire. If the vehicles running over modern roads still had iron-rimmed wheels, both asphalt and concrete would quickly be reduced to rubble. But the pneumatic tire has a remarkable ability to spread load evenly over the road surface. Within limits, the pressure that a pneumatic tire applies to the road depends only on the inflation of the tire, not on the weight of the vehicle. It's hard to believe, but a skinny-tired bicycle can apply more pressure to the road surface than an 18-wheel tractor trailer. How can this be? The key factor to keep in mind is that we're talking about pressure, not force. The total force exerted on the road is simply the weight of the vehicle, which is perhaps 150 pounds for a bicycle and rider and 50,000 pounds for a big rig with its cargo. Pressure, on the other hand, is force per unit area—pounds per square inch. The bicycle produces more pressure because all of its weight is supported on two tiny spots, whereas the truck rests on broad patches under each of its 18 wheels. If the bicycle tires are pumped up to 150 pounds per square inch, then they will squish where they meet the road just enough to distribute the 150-pound load over an area of 1 square inch. If the truck tires are inflated to 50 pounds per square inch, then they will spread the 50,000-pound load over a total of 1,000 square inches.

A paving crew lays down a new topcoat of hotmix macadam on a residential street *(right)*. The paving machine and the dump truck up ahead, carrying a load of hotmix, move along in tandem. The two workers on the platform at the rear of the machine regulate the width and the thickness of the layer. The fresh pavement is soft enough to show footprints.

Roadside Hazards. In an alpine pass where a car that leaves the road will plunge hundreds of feet into a ravine, guardrails make obvious sense. Elsewhere, the situation is not so clear-cut. A rail so stiff it stops you dead in your tracks may do more harm than good. And a rail that sends you bouncing back into traffic or flips your car on its roof also does you no favors.

The guardrails you see out on the highway are designed with these considerations in mind. The most common kind of rail is the W beam, mounted on wood, steel, or concrete posts. The W-shaped cross section gives the steel beam just the right resiliency: it yields somewhat, absorbing energy, without letting a car crash through. The metal beam is generally mounted on wood or rubber supports that hold it several inches away from the posts so that even after the beam has crumpled somewhat, a car will not snag the posts themselves.

Guardrails made of wire cables work the same way, restraining the car without stopping it too abruptly. Sometimes there are springs built into the anchorage of the cables to add resiliency. Barriers of solid concrete or rigid steel pipes are used mainly where the consequences of going through the guard rail would be truly dire, such as on high bridges.

In recent years there has been a major effort in some areas to modify the upstream ends of guardrails so they do not "spear" a car that strikes them. Sometimes the end of a W beam is bent into a big curl like a violin scroll. Sometimes it is tapered away from the road or buried in the ground. Still another solution is a blunt cap mounted on an energy-absorbing rail.

Barriers in the median strip of a dual-carriage highway are much like other guardrails, but it is especially important that vehicles not climb over them since that could turn a one-car accident into a multicar crash. The concrete New Jersey barrier (which probably originated in California) has a tapered cross section designed specifically to turn vehicles back without flipping them over.

Crash cushions are installed where cars might smash into unyielding obstacles such as bridge abutments or retaining walls, and in the aptly named gore area where an off ramp splits away from a main road. One type of cushion consists of barrels filled with sand or water and restrained by cables. Old tires also work well. These days you even see crash cushions on the trucks used by construction or maintenance crews. The cushion is typically a low metal or plastic box, about six feet square and two feet thick, slung from the back of the last vehicle in a convoy.

Signposts that need to be installed close to high-speed traffic usually have a breakaway link in their structure. If you look near the base of the post, you will see a joint held together by bolts or rivets that have a thin neck, deliberately designed to yield under impact. Many of the metal masts that support streetlights also have such a weak link; when you hit one, it is supposed to shear off and tumble safely behind your car. This is emphatically not true of standard wood utility poles, whose use is therefore discouraged along major arteries unless they can be set well back from the traffic lanes.

Guardrails are meant to protect motorists who stray off the road, but in recent years much attention has focused on the hazards presented by the protective structures themselves. Above, a guardrail is fitted with an energy-absorbing blunt face plate, so that cars will not be impaled on it. Below, Jersey barriers are stockpiled near a construction zone.

A trucker on Interstate 40 considers taking the runaway ramp *(below)*, but then veers back on the roadway despite clouds of blue smoke from the brakes and tires. The escape ramp is one of three on the eastern flank of Mount Pisgah on the border between Tennessee and North Carolina.

Breakaway signposts on the margin of a highway are mounted on bolts with a built-in weak point, designed to shear off if hit by a car. The sign is in Sparrows Point, Maryland, on Route 158.

The pedestrian-crossing sign in the foreground sports the new fluorescent-yellow-green color, in contrast with the older orange sign in the background.

For really big signs, such as the overhead minibillboards that announce major Interstate exits and intersections, the breakaway strategy won't work. The signs and supporting structures are so heavy that a weak link would break under the weight of the sign itself, or from wind loads. The alternative is to move the supporting pylons back from the roadway out of harm's way. The sign is hung on a long cantilevered boom or on a bridge stretched across the traffic lanes.

Runaway truck lanes are built on long mountain downgrades where trucks may overheat their brakes and be unable to stop. The lane is filled with loose sand or gravel and sometimes slopes upward. Once a truck is in there, it may have to be towed out, but that's a minor nuisance compared with the alternative.

SIGNS AND SIGNALS

The stop sign, the red light, the one-way arrow—these are among the most recognizable icons in the civilized world. Yet not one of them existed a hundred years ago.

Even directional signs—those indicating where a road goes—were rare until the 1890s. At that time it was the bicyclists who took the lead in posting signs, followed shortly by the Automobile Club of America. State governments didn't take over until the 1920s and 1930s. Some international standards for highway signs were adopted in 1931 by the League of Nations (and you've been wondering all this time just what the League of Nations did!). The abstract white bar on a red disk, meaning "do not enter," was one of the signs created at this time.

In the United States, both the shapes and the colors of road signs are supposed to carry meaning. The stop sign is the only octagon; the yield sign is the only equilateral triangle; and a diamond is always a warning sign. Among background colors, red is reserved for imperative messages, such as stop, yield, do not enter, and wrong way; yellow is for warnings; and orange, for temporary construction or maintenance signs. On the Interstates and many other roads a green background is used for directional signs (officially known as guide signs) and for mileposts; brown signs mark points of interest such as parks; and blue signs advertise food, fuel, and lodging. Recently, an intense, fluorescent yellow-green has been introduced for signs warning motorists of pedestrian crossings and bicycle lanes. A few other colors have been reserved for future use: purple, light blue, and coral.

Most signs are printed on an aluminum panel by the silk-screen process—the same way that tee-shirts are printed. In cities and at major highway interchanges, the signs may be lighted at night, but elsewhere they rely on retroreflective coatings. These materials don't simply reflect light the way a mirror would; they have the special property of bouncing the light back in whatever direction it came from. Thus, when your headlights are shining on a sign, the letters look bright to you but not necessarily to someone in another car. Up close, you may be able to see glass beads or prisms embedded in a translucent plastic layer. (If you're ever feeling down on your-

self, park your car with the lights shining on a retroreflective sign and then stand in the beam. You'll see the shadow of your head surrounded by a dazzling halo.)

Special reflective coatings are also an essential part of the stripes that mark lanes, centerlines, and road edges. The stripes were once merely painted on the road sur-

THE SCIENCE OF TRAFFIC JAMS

You're cruising along at the limit when suddenly the road ahead becomes a red glare of brake lights. For the next mile or two, you inch forward impatiently, wondering what's responsible for the tie-up—an accident? a stalled car? construction? But when you finally come out the other side of the jam, there's nothing there to explain it. The traffic reporter on the radio merely says the highway is "congested," as if it had a stuffed-up nose.

No one enjoys getting stuck in traffic, but if you're forced to endure a major jam, you might take some satisfaction in knowing that these spontaneous slowdowns are interesting enough to attract the notice of physicists and mathematicians. It seems that traffic patterns can be explained by the same kinds of laws that govern waves in fluids and transitions between the phases of matter, such as the freezing and boiling of water.

The idea that traffic flows like a fluid stream goes back at least 50 years. And yet there has to be more to it: A freeway at rush hour is not just a river of cars. For one thing, when a river swells with the spring flood, it doesn't slow down; the velocity increases along with the volume of water. We can only wish that highways worked the same way.

A clue to what makes automobile traffic such a peculiar fluid comes from looking at events on a microscopic scale. In a river, individual molecules of water jostle against one another. When a fast-moving molecule happens to come up behind a slower one, they collide and bounce off in new directions like billiard balls. On the highway, the molecules are cars, and they behave a little differently. Even the most aggressive drivers don't routine-

ly bump slower cars out of the way. They may flash their high beams and beep the horn, but they also take their foot off the gas.

Physicists who study traffic flow distinguish three main modes of motion, roughly analogous to the gaseous, liquid, and solid phases of matter. With very sparse traffic, drivers are free to change lanes and pass at will, so cars move almost like independent molecules in a gas. As traffic gets heavier, there is a transition to a more liquid-like state called synchronized

flow. Under these conditions drivers are forced into follow-the-leader mode, keeping a constant distance from the car in front. They can't readily change lanes. The final transition to the solid phase comes when cars get so closely packed that traffic comes to a complete stop, at least momentarily. In this bumper-to-bumper state you are like a molecule locked in the rigid lattice of an ice crystal.

Once traffic slows down, the jam has a life of its own, persisting long after the original cause disappears. At the trailing edge of the jam, cars slow down and then inch their way through the stop-and-go zone. At the leading edge, they are suddenly liberated and speed away onto the open road, but in the meantime others have piled up behind them. Depending

on the rate of new arrivals, the wave of congestion can remain stationary or propagate back up the highway.

We tend to think of a traffic jam as the inevitable result of too many people trying to travel the same route at the same time, but that's only partly true. In many cases, the road could accommodate all the cars trying to squeeze in, if only they could be organized a little more efficiently.

Traffic engineers monitor the throughput of a road, which is the number of vehicles multiplied by their average speed. When traffic is light, throughput is low: cars move freely, but there are few of them on the road. When traffic is jammed up, throughput is low again: although there are lots of cars, their average speed is low. Evidently there's a condition of optimum throughput somewhere between these extremes.

Field studies suggest that if 20 percent of the roadway is covered by cars, congestion threatens. At 30 percent, traffic slows significantly, and throughput begins to fall. Thus, if you see the average distance between cars dropping below three or four car lengths, you may be about to experience the physics of congested flow.

One device meant to prevent unnecessary jams is the metered entrance ramp on freeways, which dribbles cars onto the highway at a controlled rate. Drivers held up by metered ramps tend to resent the delay. Why should they have to wait when those already on the highway are allowed to pass without interference? But if the scheme is working, and the road stays closer to the condition of maximum throughput, then everyone gets home sooner, even those who have to pause on the metered ramp.

At an intersection near Albany, New York (right), near-ly 40 signal-light units and about a dozen signs direct traffic. Below is a new signal head with light-emitting diode (LED) elements, in Chapel Hill, North Carolina.

face. Now they are coatings of plastic, melted and extruded onto the roadway in a thick ribbon. To form a reflective layer, glass beads are spread over the surface while the plastic is still warm and soft. For a few days after the striping is done, parts of the road may have a glorious iridescent coating of tiny glass beads.

The plastic reflectors embedded in the centerline and edges of many roadways are known as Botts Dots, after Elbert Dysart Botts, who invented them. Because they protrude above the road surface, the Dots are much more visible in heavy rain than any pavement stripes could be. For the same reason, they provide not only visual guidance but also a tactile and auditory warning if you stray out of your lane. The Dots come in amber and white, and many of them also have a red rear face that shows up if you should go the wrong way up an exit ramp. The occasional blue Dot identifies fire hydrants.

The key problem in developing Botts Dots was not creating a plastic reflector that could survive being run over by cars and trucks; it was finding a way to attach it to the road. Spikes and bolts driven into the road surface didn't work; the reflectors would break off, leaving a tire-threatening shaft protruding from the pavement. The answer turned out to be an adhesive, applied after sandblasting a patch of road to clean it. But there's one thing even modern adhesives can't stand up against: a snow-plow. Hence, Botts Dots are less common in the north than in areas with mild winters. Even in warm climates they are a high-maintenance item. In California I once watched a road worker inspecting the Dots with a go-cart. Riding an inch or two off the road surface, he tested each Dot by banging it with a rubber mallet. The ones that rattled or moved were marked for replacement.

Traffic Lights. Whereas roadside signs and symbols are passive markers, signal lights take an active role in regulating traffic. The familiar system of red, yellow, and green traffic lights evolved from earlier railroad practice, although the meanings of highway and railway signals have diverged somewhat.

A single assembly of traffic lights is called a signal head; a set of lights shining in a specific direction is called a face; each light within a face is an optical unit. The col-

ored glass lens is known as a roundel—a word I find curiously sonorous. Older signal heads were made from iron or steel, but the modern ones are either aluminum or polycarbonate plastic; these lighter materials make it easier to hang the signals over the lane they regulate, rather than mounting them on a post beside the road. The heads used to be painted black or dark green, but now the standard color is something called federal yellow.

Optical units come in two common sizes—both of which are bigger than they look from the driver's seat. The smaller ones are 8 inches in diameter and the larger ones 12 inches—the size of a dinner plate.

Traditionally, the lamp inside the unit was an ordinary incandescent light bulb (60 to 150 watts) mounted in a reflector, like a flashlight bulb. But since the late 1990s, light-emitting diodes (LEDs) have been replacing light bulbs everywhere. Instead of one big bulb filtered by colored glass, the LED signal has a geometric array of small, semiconductor devices that emit various distinct colors of light. Red LEDs entered the market first; amber and green ones weren't quite bright enough. The LEDs still cost more than light bulbs, but they last longer. Moreover, if one small emitter burns out, the rest continue to shine.

The controller for a set of signal lights is usually in a nearby metal cabinet, which can range in size from a medicine chest to a small refrigerator. Running the signals is more complicated than you might think. A major crossroads with turn lanes and pedestrian signals might have 15 to 20 signal faces and perhaps 50 optical units. The simplest control algorithm goes through a cycle with just four phases, but when left-turn signals and pedestrian crossings are included, the control cycle can easily grow to six or eight phases, and occasionally more. Sensors that detect the presence of a vehicle and change the signal only when needed add another complication.

For decades, traffic-signal controllers were clockwork mechanisms that operated much like a music box or player piano. A motor slowly turned a set of disks that had

A traffic-signal control cabinet holds the electronic modules that determine the timing and sequence of signals. Each signal face and other device (such as the pedestrian push-button in the foreground) has a plug-in module in one of the racks in the lower part of the cabinet. The programmable master controller, at the top, governs overall timing. The cabinet was opened for the installation of a module controlling audible signals for pedestrians.

Newly installed vehicle-detection loops are buried in the pavement at an intersection in San Diego. The conductors sense the presence of the large mass of metal in a car and trigger a traffic signal.

That's not a bird perched on the boom next to the signal light above. It's a sensor pointed toward oncoming traffic; when it detects flashes from a strobe light mounted on fire engines and other emergency vehicles, it overrides the normal signal sequence and gives the emergency vehicle a green light. The tower below carries video cameras for traffic surveillance. Both of these devices are in Nashua, New Hampshire.

bumps or depressions in their perimeter. Switch levers riding against the edge of each disk would turn the signal lights on and off in a fixed sequence. Some of these electromechanical controllers are still in operation; if you stand close to the cabinet, you can hear the clock motor humming or grinding and the switches clacking. The mechanisms are very reliable, but they are also inflexible.

Controllers based on digital electronics began to appear in the 1970s, and a standard for computerized systems was published in 1992. By now the majority of controllers have been converted to digital technology. The task of issuing commands to traffic lights is one for which computers are obviously well suited, but it is also a task where a computer bug could be fatal. The big no-no in traffic signaling is allowing two conflicting streams of traffic to both get a green light at the same time. The main line of defense against such a disaster is a conflict-monitoring unit built into every computerized controller cabinet. The module directly monitors the electrical currents flowing in the wires that feed the signal lamps so it can tell with certainty if two greens are incorrectly lit at the same time, no matter what caused the error—a subtle programming bug, a stuck switch, or crossed wires.

Traffic lights are at their most annoying when they stop cars on the main thoroughfare at a moment when there's no one to take advantage of the green on the cross street. The remedy for this situation is a sensor that triggers the cross-traffic green only when a car shows up to use it. Quite a variety of sensor technologies have been tried, including microwave, ultrasonic, infrared, video, and magnetic. By far the most common type, however, is the induction loop, which is a coil of wire buried in the roadway. The loop carries an electrical signal at audio frequencies (a few thousand hertz). The frequency is altered by the presence of the metal in an automobile. The same principle is at work in the metal detectors that beachcombers use to find lost coins and bottle caps, as well as in the metal detectors in the security checkpoint at the airport.

Most of the loops are installed by sawing a groove in the pavement, laying in the coil of wire, and then sealing it with a watertight epoxy. The standard inductive detector has a single loop for each lane of traffic; this arrangement is called a dipole detector. Sometimes you might notice a lane that has two loops side by side in a squared-off figure-eight geometry. This double loop is called a quadrupole detector. It provides greater sensitivity—detecting motorcycles and some bicycles as well as cars—and has a sharper cutoff at the edge so that a left-turn-lane signal, for example, will not be triggered by traffic in an adjacent through lane.

Another kind of detector allows fire engines and ambulances to preempt normal traffic signals; the light automatically turns green for the emergency vehicle as it approaches the intersection. Public-transit buses may also be equipped to trigger the green light. These systems rely on an optical sensor—a photocell mounted on a stalk—that detects distinctively timed flashes from a strobe light on the vehicle. I've heard people claim to have triggered the preemption by flashing their high beams at just the right frequency. The manufacturers tell me it's just not so.

In most cities today, traffic signals are linked together in a citywide network controlled from a central command post. This is not a new idea. In Manhattan a central traffic-control station was operated as early as 1929. Of course the modern systems are more highly automated, but this is not to say that a central computer takes direct charge of individual faces or optical units at each intersection. As a safety measure, the local controllers continue to operate the signals and monitor for conflicts. The central station merely issues orders to alter the timing of the various signal phases.

During rush hours, many cities time their signal lights to form "platoons" of cars on one-way arterial streets. Waves of alternating green and red signals propagate along the street at a set speed. Vehicles traveling at the same speed are not stopped by red lights; the vehicles spontaneously form platoons that stay together from one signal to the next. It turns out that smoothing the flow of traffic in this way maximizes the capacity of the roadway.

SMARTER ROADS? SMARTER CARS? DUMBER DRIVERS?

At the 1939 World's Fair in New York, a General Motors exhibit called Futurama offered a famous vision of travel by automated automobile. In the morning you would read the newspaper as your car navigated the freeways and took you to work. After a long day at the office, you could nap as the car drove you home.

Several technologies that might be seen as steps toward such a future are already on the scene. Cars and trucks can be equipped to receive navigational signals from Global Positioning System satellites and show the vehicle's location and velocity on a moving-map display. In-car radar units that monitor distance to the vehicles ahead and behind are becoming available. Cruise control has long since ceased to be a novelty. Most major highways have embedded sensors and video cameras to monitor the density and speed of traffic. (Chicago's freeways, for example, have some 1,600 built-in sensors.) Computer software can calculate a route between any two addresses.

In 1997 the idea of the smart car was given a test drive on a section of Interstate 15 in San Diego. Two lanes of the highway—separated from normal traffic by concrete barriers—were studded with "magnetic nails" driven into the pavement every 100 feet. Eight cars were equipped with sensors for these magnetic lane markers, as well as radar, video cameras, and a computer to interpret all the sensor readings and drive the car accordingly. On the first day the system was tested, a platoon of cars sped down the highway in close formation (see photo), with the drivers—or rather the nondrivers—

holding both hands out the window, like kids showing off on a roller coaster.

It's a long way from this test of the concept to routinely entrusting human lives to a supposedly intelligent vehicle. Skeptics ask, What happens when the car has a flat tire? When the roadway floods? When a child runs into the street? Most of all they point out that one of the things computers are known for doing is crashing. The proponents of the system say they are well aware of these hazards and believe they can be overcome. Given that auto accidents in the United States kill 40,000 people a year, it might be argued that human drivers aren't doing such a hot job, and the machines deserve a chance.

Another argument for smart cars is that they could use highways more efficiently, at least in the case of roads open exclusively to vehicles under automatic control. Because the cars could coordinate all their movements, they could drive with much closer spacing at any given speed, effectively increasing the capacity of highways. This leads to an economic argument for the technology: even if retrofitting roads for smart vehicles costs $100,000 per mile, that is far less than the cost of building new freeways.

The social and psychological effects of adopting such a new mode of travel are harder to fathom. In a society where what you drive and how you drive have so much to do with self-image, surrendering the car keys to a computer will call for some major readjustments. (Photograph courtesy National Automated Highway System Consortium.)

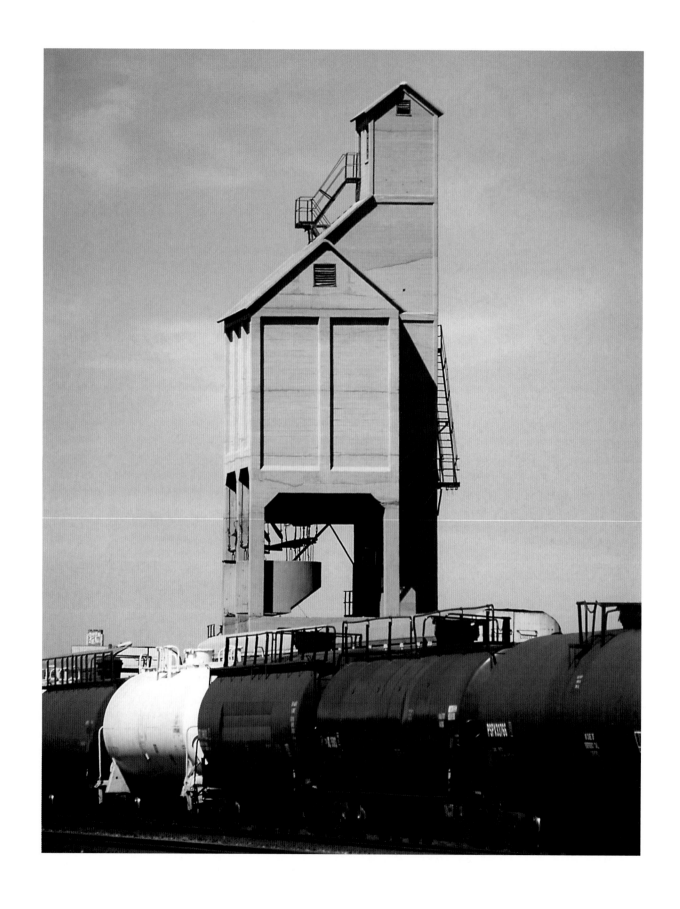

THE RAILROAD

IF YOU WANT TO SEE NINETEENTH-CENTURY technology still hard at work today, just go stand by the nearest railroad crossing and watch the past roll by. Not that you'll see steam locomotives and antique rolling stock; most of the equipment on today's railroads is quite up-to-date. But the industrial style and culture of the railroads have changed remarkably little in the past hundred years. Railroad technology was the centerpiece of the nineteenth-century Industrial Revolution, and it still is.

The basic idea of railroading is steel wheels on steel rails. Compared with rubber tires on a concrete roadway, this metal-to-metal contact allows heavier loads to be carried with less friction. Reduced friction makes for greater energy efficiency, but it has other consequences as well, not all of which are entirely welcome. For one thing, less friction means the driving wheels can't get much of a grip on the rails, so trains have only limited acceleration. You'll never see a locomotive make a drag-race start. And for the same reason, trains don't stop very well. The engineer's nightmare is seeing a catastrophe waiting to happen a mile up the track, and knowing the train can't be stopped before getting there.

Running on rails has another obvious effect: You don't have to steer. In fact, you can't steer, even if you want to. The train's motion is one-dimensional; it goes wherever the track goes. The switching point where one track branches away from another may look something like a highway exit ramp, but it works differently. On the highway, it's the driver who decides whether to go straight or turn off, but on rails it's the switch in the track that makes the decision.

More than other modes of transport, railroads attract amateur enthusiasts and collectors of lore and memorabilia. Some of these "railfans" have nostalgic or antiquarian interests, commonly focused on the Age of Steam or the era of luxury passenger trains in the 1920s and 1930s. Others are fascinated by modern train operations; they monitor railroad radio frequencies and keep life lists of locomotives spotted, in the

A lonely looking grain elevator looms over train tracks in the Texas panhandle *(opposite page)*. Although much of the drama of nineteenth-century American railroad building focused on the effort to span the continent and link the two coasts, the major role of the railroads for many years was carrying farm produce from the nation's interior to urban markets. But the grain elevator seen here is shuttered, and the train passing by below consists entirely of oil tank cars.

manner of bird-watchers. The information collected in this chapter does not reach the level of detail that would satisfy a railfan. My aim is merely to present some of the commonplace elements of railroad infrastructure.

TRACKS

The steel rail is an artifact whose form has been carefully optimized. This gradual refinement of the design was done not by a single brilliant engineer but by more than a century of industrial evolution. The rail was never meant to be an object of beauty, but its cross section has all the elegance of fine typography.

It is called T rail, but it's a pretty oddly formed T. The upper part, which actually supports the wheels of the train, is called the head; the horizontal piece at the bottom is the base; the vertical member connecting these parts is the web. Not all rails have exactly the same shape, and the dimensions and weight vary considerably. High iron rails—those that carry the heaviest loads at the highest speeds—weigh 130 to 140 pounds per yard and stand more than six inches tall. The rail on a small branch line might weigh less than 100 pounds per yard and be less than five inches tall.

Every rail is embossed with its date of manufacture and some other information that's not hard to decipher. Look on the web of the rail on the side facing away from the center of the track. You'll see markings something like these:

<div align="center">1360 RE CC CF&I 1975 IIII</div>

Here, *1360* indicates a rail weight of 136 pounds per yard, *RE* designates the particular shape of the rail cross section, *CC* stands for "control cooled" (a manufacturing step that enhances strength), *CF&I* is the manufacturer (CF&I Steel Corporation of Pueblo, Colorado), *1975* is the year the rail was rolled, and *IIII* indicates April, the fourth month of the year.

A standard rail in the United States is 39 feet long and weighs somewhat less than a ton. Why 39 feet? Because the cars on which the rails are transported are 40 feet long.

Except on high-speed and high-load main lines, the 39-foot rails are bolted together end to end through connecting steel links called fishplates. The fishplates nestle into the web area between the base and the head of the rail so that holes in the fishplates line up with holes predrilled in the rails. Either four or six bolts pass through the sandwich of fishplate, rail, and fishplate to hold the assembly tight. A few railroads have been experimenting with boltless joints; the fishplates are glued to the rails with an industrial adhesive.

In a fishplate joint, the two rails are not butted tightly against each other; a gap allows for expansion and contraction of the rails as the temperature changes. In cold weather the gap can open up to a quarter inch or more. Each of these gaps is felt as a distinct bump in every rail car passing over it. This is the source of the unmistakable "clickety-clack" of a train in motion.

Rail markings *(below, enhanced with chalk)* show that this section of 132-pound rail was rolled in November of 1985. A cross section sawn from a rail *(opposite page)* is shown at actual size. It's a hefty slice of steel, and yet when you think that the half-inch thickness of the web has to support locomotives and freight cars weighing hundreds of tons, it seems quite delicate.

Rail joints: In a standard fishplate joint *(top)*, 39-foot sections of rail are bolted together. An insulated joint *(middle)* has the same structure, but a plastic insert keeps the two rails electrically isolated. Welded rail *(bottom)* eliminates the rhythmic ka-chunck ka-chunck. Note the empty bolt holes, suggesting this is probably recycled rail.

At most fishplate joints a copper wire is attached to the rails on both sides of the joint. This bonding wire ensures electrical continuity of the track, which is vital to the operation of signaling systems. At boundaries between signaling blocks, two sections of track have to be electrically isolated, and so special joints are installed, without a bonding wire and with plastic insulators between the fishplates and the rails.

On most of the main-line track in the United States, joints have been eliminated by the installation of continuously welded rail. The idea of welding rails together into one long ribbon of steel was once considered totally unworkable. When the rail heated up on a sunny day, the forces of expansion would be irresistible, people thought, and the rails would buckle. Or in cold weather the rails would contract and straighten out curves, or else the rail would simply fracture. These fears turned out to be unfounded. If a welded rail is anchored to the ties in the normal way, the ties and their foundation can resist the thermal forces. When you see welded rail on a cold day, it is under great tension, stretched taut like a giant violin string. On a hot day the rail is under tremendous compression, like a spring ready to uncoil. But the stresses are held in check by the mass of wood and stone in the roadbed.

Welded rail is constructed from the usual 39-foot sections. Most of the welding is done in a shop or yard with a flash-welding machine that passes electric current through the joint. Welded sections of rail 1,440 feet long are then loaded on special rail cars and transported to the prepared roadbed. In transit, the rails bend as the train carrying them goes around curves. But the train cannot accommodate more than about 40 rails at a time, or the spring tension might snap the train off the tracks.

Additional welds are done in the field as the track is laid. One method uses thermite: a mixture of powdered aluminum and iron that burns hot enough to weld the heavy steel rails. The track laying has to be done when the weather is moderate, to avoid extremes of expansion and contraction.

A glance at the surface of the rails can tell you whether a track is in frequent use. On main lines the load-bearing surface has a bright polish or luster. Depending on the weather, it takes anywhere from 24 hours to two weeks for a fine layer of rust to coat the polished steel. Something else to look for on the rail surface are corrugations caused by braking forces or friction on turns. You might also spot the occasional locomotive burn, a concavity where an engine has spun its wheels.

Gauge. Take a tape measure to your nearest railroad tracks. If you find the distance between the inner edges of the two rail heads is anything other than 4 feet 8½ inches (plus or minus a quarter-inch or so), then either you are not in North America or you have discovered a great rarity. Why that specific distance? It goes back to the very beginning of railroading: 4 feet 8½ inches was selected by George Stephenson, the English engineer who built several of the earliest steam railways, starting in the 1820s. Stephenson probably based his choice on the wheel spacing of English trams and wagons. (A popular legend traces the ancestry of Stephenson's gauge back to Roman war chariots, but experts are skeptical.)

Not all railroad builders followed Stephenson's lead, and the standard gauge of 4 feet 8½ inches was not always universal. When railroads were a growth industry in America, in the 1830s and 1840s, gauges ranged from 4 feet 3 inches up to 6 feet. Much of the American South adopted a gauge of 5 feet—a choice that lives on today, strangely enough, in Russia, because the first Russian railroads were built by G. W. Whistler, an American southerner (and the father of the artist James McNeill Whistler). For rail lines that were not interconnected, the variations in gauge didn't matter much, and sometimes there was a competitive advantage in being nonstandard. The Atlantic and St. Lawrence Railroad, between Montreal and Portland, Maine, was built with a gauge of 5½ feet to discourage shippers from sending their goods on to Boston over standard-gauge lines.

There were also disadvantages to incompatibility. A nonstandard railroad might hold its own customers captive, but it also gave up all hope of gaining a share of the competitors' business. In the years after the Civil War, more and more railroads were forced to conform or perish. By 1886 virtually all tracks in the United States had been converted to standard gauge.

But that's not the end of the story. The question remains why other aspects of railroad operation have resisted the forces of standardization. For example, electrified railroads do not all use the same voltage and frequency. In Europe, a train going from Sweden to Portugal would pass through six or seven different voltage regimes. Signaling systems and even such ordinary items as speed-limit signs also vary from one railroad to another (unlike the highly standardized signs and signals of the highway).

A hypothesis that might explain these oddities is that a uniform gauge works to the advantage of individual railroads by allowing rolling stock to move freely throughout the North American system. At the same time operational idiosyncrasies mean that crews and locomotives are not interchangeable. Each railroad thereby preserves its exclusive right to haul trains over its own tracks. It will be interesting to see if more uniformity develops following the series of mergers that have been consolidating the American railroad industry in recent years. Even 125 years after the first transcontinental rail journey, it is still not possible to cross the country on tracks owned by one company.

For rail cars to be interchangeable, it's not just the wheel gauge that has to match up. The load gauge defines the maximum dimensions of rolling stock, so trains can pass through tunnels and under bridges, and so trains on adjacent tracks do not scrape against each other. In North America the maximum height is 15 feet, the maximum width 10 feet 8 inches. (European trains are generally smaller.)

Ties and Ballast. What I find most remarkable about a railroad track is not the steel rail, whose strength is readily apparent, but the timber crossties and crushed-stone ballast that support the rails and maintain their alignment. For carrying loads of 10,000 tons at a time, I would have expected something more elaborate, something more carefully constructed—perhaps a bed of reinforced concrete or a framework of

A bonding wire ensures that electrical signals for train control can pass through rail joints. At this Connecticut train station someone has helpfully sprayed a bonding wire bright green, for ease of identification.

steel beams. But the timber-and-stone foundation, despite its impromptu appearance, works remarkably well.

If you stand near a track and watch a slow freight lumber by, you can see the rails sag under each set of wheels. The sag is typically about a quarter of an inch. This flexing of the rail absorbs energy; just like a wagon wheel sinking into soft sand or mud, it makes the train harder to pull. American railroad engineers say the energy cost is worthwhile because it saves wear and tear on equipment. A more rigid substrate would be battered to pieces by the pounding of the trains.

European and Japanese railroaders evidently see things differently. They build many of their rail lines with steel-reinforced concrete crossties. At least part of the reason for this difference is surely that wood is scarce and expensive in Europe and Japan, whereas North America is infested with trees.

Incidentally, what Americans call ties or crossties, the British call sleepers—a term evocative of bodies in their graves.

Timber crossties are most often oak, sawn to a width of about 8 inches and a length of eight or nine feet. They are treated with creosote or coal tar to prevent rot, a process that has made a tremendous difference in track maintenance. Before the preservative treatments were introduced, ties had a lifetime of 5 years; now they last 30 years or more. Typical spacing from one tie to the next is 20 inches, which works out to about 3,000 ties per mile.

Rails are fastened to the crossties through steel tie plates, sometimes with an added cushioning pad of plastic or rubber. The rail and tie plate are spiked into the tie by a machine. John Henry, the steel-driving man who raced a steam hammer, lost that race a long time ago.

Track-laying machinery refurbishes a section of Southern Pacific right-of-way near the Arizona-California border. The plow on the opposite page spreads and smooths the ballast stone; the tamper below seats the ties in the ballast and aligns the rails.

The ties rest in a thick, heaped-up bed of crushed stone: the ballast. (How this rather nautical word entered railroad usage is not entirely clear, but it goes way back. Ships used to be ballasted with stone, so perhaps that is the source.) On most American tracks the ballast is either limestone, which is light gray when it's clean, or the darker, heavier and harder traprock, crushed into angular fragments an inch or two across. As in macadam paving (see Chapter 8), the angularity of the crushed stone is important; it helps the pieces interlock to form a stable and cohesive mass.

The rails, ties, and ballast work together as a structure to support and guide a moving train. The weight of the train—the simple, downward gravitational force—is distributed by the rails over a few nearby ties. When a wheel is centered over a tie, 40 to 50 percent of the wheel's weight goes directly to the tie below, with almost all the remainder taken up by the two nearest adjacent ties. The tie in turn distributes the weight to the ballast stones below. Lateral forces (when the train is turning) and longitudinal forces (from acceleration or braking) also get transferred from the rails to the ties to the ballast. And it is the ties, providing a rigid link between the rails every 20 inches or so, that maintain the proper gauge between the rails, resisting the train's tendency to split them apart.

The rail-tie-and-ballast structure is not maintenance-free. Rails get out of alignment—wiggling in both the horizontal and the vertical dimension. Wood ties rot, and concrete ones crumble. The ballast gets worn, filled with mud, and loses its resiliency. It used to be that track was restored by crews of gandy dancers, who would muscle the rail back into position with long iron levers. Now track-maintenance machines lift the very rails they are running on, replace worn-out ties, snuffle up the ballast like a harvesting combine, and clean it—all in one operation.

Track Layout. Rules for the geometric design of a railroad are far more stringent than those for highways. Trains simply cannot make sharp turns or climb steep hills, and the track has to be laid out accordingly.

Limits on curvature are about the same as those for a high-speed freeway: designers try to avoid curves any tighter than 3 or 4 degrees (meaning that if you pace off 100 feet along the curve, you will turn by an angle of 3 or 4 degrees). Sharper curves require trains to slow down or risk derailment. But there are further constraints; even at a crawling pace, there are turns too sharp for a train to negotiate. In practice, the limit is a 13-degree turn, which is equivalent to a radius of about 450 feet. If a turn were any more severe, the middle of a long car would hang over the inside of the curve so far it could interfere with trains on an adjacent track or with other trackside obstacles. (On the highway, some truck trailers carry warning stickers that read: "This vehicle makes wide right turns." That goes double for a rail car long enough to hold two trailers in tandem.)

Like high-speed roads, some railways have banked, or superelevated, turns. The banking allows higher speeds and reduces the danger of tipping over. Equally important, it reduces wear. Without banking, wheel flanges tend to rub hard against the outside rail of a curve, to the detriment of both the rail and the wheels. Banking equalizes the wear.

Worse than curves are hills. A 2 percent grade that a car would hardly notice, and that even a bicyclist could easily climb, can make a 7,000-horsepower locomotive look like a real wimp. On main-line tracks designers try to keep the grade to 1 percent or less. And the standards for high-speed railroads, like the French Train à Grande Vitesse, call for grades no steeper than 0.25 or 0.35 percent. (At a slope of 0.25 percent, gaining 100 feet in altitude takes seven and a half miles of horizontal travel.)

Leveling the landscape for a railroad requires more cuts and fills, more bridges and tunnels than grading for a highway does. Driving a road parallel to a railway, you are likely to see the tracks at one moment below you and at the next above you, as the road follows the undulations of the terrain, while the rails maintain their steady course. The absence of hills also explains why abandoned railroad lines are so popular as bicycling paths.

Carefully designed railways have compensated grades: Where an ascending track enters a curve, the grade is made a little less steep, since friction on the curved rails tends to hold the train back. When turning and climbing are properly coordinated, the train maintains a steady speed.

The steepest grade on U.S. main-line track is at the small town of Saluda, on the Norfolk Southern line between Spartanburg, South Carolina, and Asheville, North Carolina. The grade goes on for three miles at a slope of 4 or 5 percent. Some trains get over the top with the assistance of extra engines, linked either as helpers at the front or as pushers at the rear. Other trains must "double" over the hill: the crew breaks the train in two at the bottom of the grade, hauls half the cars to the summit, then goes back for the rest, finally reassembling the train at the top.

Looking up the steepest hill in American railroading: the Saluda Grade, on a route from Spartanburg, South Carolina, to Asheville, North Carolina, climbs at a slope of 4 or 5 percent. For trains descending, the track is equipped with a switch that diverts any train going faster than eight miles per hour onto an emergency escape route, which works much like a runaway truck ramp on the highway.

Whereas major highways always have multiple lanes, it is not unusual for long segments of a main-line railroad to consist of just a single track, shared by trains moving in both directions. I once spent a day watching trains go by on the Southern Pacific in New Mexico, which is a single-track line for many miles. All through the morning the eastbound trains barreled through, with no westbound traffic at all. I began to wonder if this wasn't a one-way thoroughfare, with another track elsewhere to carry trains heading west. Then, in the afternoon, after an idle period of an hour or so, a westbound freight appeared, then another right behind it, and another. Finally I caught on to the plan. (It's embarrassing how long it took me.) On a long single track, you can't very well alternate trains going in opposing directions. The trains run in platoons, all eastbound for a while, then all westbound.

Even with the platoon system, any long stretch of single track requires sidings, or sidetracks, so that opposing trains can pass when they meet. For that matter, double-track lines also need sidings to allow faster trains to overtake slower ones.

A siding branches off the main line and then rejoins it farther on, so a train being passed doesn't have to back up. The siding has to be long enough to hold the entire train, which can mean a mile or more for long-haul freights. The dead-end branch tracks that serve factories and other freight customers are often called sidings, but the proper term for one of these tracks is a *lead*. Another important track feature is a wye—a three-way intersection with turnouts leading from each track to both of the other tracks. By turning at a wye, then flipping a switch and backing up, then flipping another switch and going forward, a train can turn around and come back the way it was going. (It's like pulling into a driveway to make a U turn.)

Switches, Frogs, and Diamonds. A railroad switch is one of those mechanisms that has no secrets. All of its parts are out in the open for anyone to see. There is no mystery to how it works. And yet sometimes I find it hard to believe that it does work—that merely by shifting a lever, thousands of tons of train can be deftly diverted from one track to another.

First, a minor matter of terminology. Americans call them switches, but the British call them points. Or at least that's what I always thought. But it turns out that real railroad workers both in the United States and abroad seldom call them either switches or points; they call them turnouts.

Whatever the name, the heart of the mechanism is a pair of flexible, tapered rails (sometimes called switch points, as if matters weren't confused enough already!). These special rails have the normal T-rail cross section at one end (the heel of the switch) and thin down almost to a sharp point at the other end (the toe). Consider a turnout where the main line goes straight ahead and a siding veers off to the left. The righthand rail continues undisturbed straight through the switch structure, while the lefthand rail curves to the left. The two tapered switch points—also known as tongue rails—lie between these stock rails. When the switch is closed, the lefthand switch point nestles up against the lefthand stock rail and carries a train approaching

The tapered tongue rails *(above)* are shifted left and right to determine a train's direction. But the most critical part of the switch is the frog *(below)*, where two rails cross over.

In a freight yard in Greensboro, North Carolina, a worker throws a switch by hand to turn a locomotive into a side track.

the switch straight through on the main line. When the switch is open, the lefthand switch point is moved out of the way, and the righthand switch point lies up against the righthand stock rail. Now a train coming into the switch is diverted to the left, onto the siding.

To get a train through the turnout, another problem has to be solved: whichever way the switch is set, one rail has to cross over another. The crossover is a problem because the flanges of the train wheels can't climb over a rail. (That's why the flanges are there—to keep the train on the tracks.) The solution is to install a massive steel forging where the two rails cross, with grooves cut to allow the flanges to pass. This X-shaped steel structure at the point of intersection is called a frog—a term whose source is an utter mystery to me.

Because of the grooves cut through the frog, wheel flanges have nothing to bear against there, and a train could slip off the rails. To prevent this, guardrails are installed

just inside the two stock rails opposite the frog. The guardrails prevent the opposite wheel from drifting inward.

The switch-point rails are shifted from one position to the other by steel rods that pass under the rails between the ties. Originally, switching was done by a member of the train crew, who climbed down from the engine cab or the caboose and moved a clublike lever that had a heavy iron weight on one end. The weight was there to ensure that the lever would turn a full 180 degrees, flopping from one horizontal position to the other. (A switch left in an intermediate position has a good chance of derailing any train that tries to go through it.) Today, manual switches are seen only on low-traffic branch lines and leads. Other switches are moved by electric motors under remote control from a signaling tower or a dispatcher's office, which might be hundreds of miles away.

Turnouts come in various sizes, which differ in how sharply the tracks veer away from each other. A No. 6 turnout measures a little less than 50 feet from the toe of the switch to the middle of the frog. On a No. 20 turnout this distance is almost 100 feet. The long turnouts take up more space, cost more to build, and require more force to change position, but trains can negotiate them at higher speed.

A turnout is not the only way for two rail lines to intersect. There is also a diamond, where the tracks merely cross, without any provision for trains to switch from one line to the other. A diamond is basically made up of four frogs, arranged at the four points where pairs of rails intersect. If the tracks cross at right angles, the diamond is actually a square; as the angle between the tracks gets shallower, the diamond grows more elongated. Standard frogs are manufactured for angles down to 3 degrees, which yields a diamond 180 feet long.

Changing the switch position requires flexing the steel tongue rails, and thus it calls for a good deal of force. The heavy weight at the end of the lever arm helps ensure the switch is always locked positively in one position or the other, not floating in the middle.

A diamond is an intersection of railroads where no turns are allowed: The tracks cross, but all paths go straight ahead.

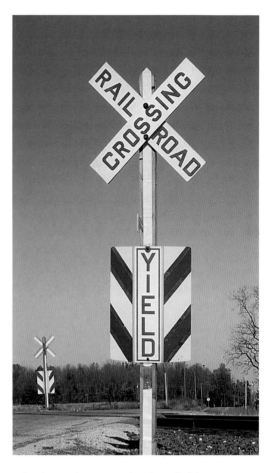

What the car driver sees *(above)*: a colorful variation on the traditional crossbuck sign. What the train driver sees *(below)*: a "paddle" whose markings make sense when you listen to the pattern of horn toots at a crossing.

Highway-Railway Crossings. Two worlds meet at the highway-railroad crossing, and not always on friendly terms. In the first place, there is the physical problem of getting flanged steel wheels across an asphalt roadway, and rubber tires over steel rails. More challenging is the task of getting trains and cars to cross paths safely.

The main trick in constructing the crossing itself is to build up the roadway so that it's flush with the top of the rails, but to leave a groove for the train's wheel flanges. The filler between the rails can be ordinary asphalt or concrete, but these materials have to be jackhammered out when track maintenance is needed. An alternative is to lay timbers parallel to the rails. (Bicyclists hate these.) And there are prefabricated solutions with rubber or polyethylene mats of various kinds.

Warnings to motorists range from a simple crossbuck sign to blinking lights to gates. In general, the signs and signals are erected by the railroad rather than by the highway department—and it shows. Stylistically, railroad gear is just different from anything else you see along the highway. For example, the visor that shades the blinking red lights from the sun is much longer than the one on a standard traffic signal.

The train detectors that activate automatic warning lights and gates have to be fairly elaborate and sophisticated. The crossing needs to be closed well before the train reaches the roadway—perhaps a mile away for high-speed trains—but it's not acceptable just to close the gate whenever there is a train within that range. Drivers expect to cross as soon as the end of the train has cleared the road. But what if the train goes through the crossing, then stops, and backs up? Dealing with all such contingencies requires complicated logical calculations, and for many years it was all done with nothing but relays, motors, and timers—no computers.

Gate-protected crossings usually block off only two quadrants of the roadway—in the United States, the righthand lanes of each approach to the crossing. The rationale for not blocking the entire roadway—apart from saving the cost of extra gates—is to allow an escape route for any vehicles that might be trapped between the gates when they come down. But drivers also use the openings to sneak around the gates and try to beat the train. For those who don't make it, the consequences are grim. Experiments are now under way with four-quadrant gates and with median barriers to discourage the overeager driver.

(Hint: if you should ever get trapped behind a gate in a railroad grade crossing, just drive through it! The gate is just a thin slab of wood, and it will snap right off. You may well scratch the paint on your hood, but that's nothing compared with the damage an oncoming train will cause.)

Under federal regulations, trains are supposed to sound their horn or whistle when approaching a grade crossing. The standard signal is two long blasts, a short toot, and a final long wail that lasts until the engine has cleared the intersection. (This pattern of long, long, short, long is the Morse code symbol for the letter Q, although as far as I know that's mere coincidence.) There are many other whistle signals that train crews use to communicate among themselves. One short toot means stop. Two long blasts means release the brakes, or go. The generic acknowledgment signal, equiva-

A double-gated, four-quadrant barricade with flashing lights and clanging bells guards a grade crossing in San Diego.

lent to a wave of the hand, is two short toots. If you make the whistle-pulling gesture to a friendly engineer, this is usually what you'll get.

Every grade crossing in the United States has an inventory number assigned by the Department of Transportation and the American Association of Railroads. The number is often posted on a small plaque somewhere near the crossing. It can be useful if you need to report an emergency, such as a car stuck on the tracks.

More Track Hardware. Here are a few more items you might notice along the rails.

Derailers. Given all the efforts railroads expend to prevent derailments, you might be surprised to find tracks fitted with devices whose very purpose is to dump rolling stock on the ground. The derailers are installed on industrial leads to prevent parked freight cars from wandering onto the main line and causing an accident. They are also set up on a temporary basis to protect work crews. The derailer is a steel wedge or ramp clamped to one rail; it lifts up the wheel flange and drops the wheel outside the rails; the wheel on the opposite side simply falls into the space between the rails. A lever much like the one that operates a turnout raises the derail wedge into position or retracts it out of the way. A marker or signal light shows train crews whether the device is set.

Bumpers. This is a case where one glance at the object tells you what the function is. A bumper or bumping post is erected at the end of the line to keep cars or trains from rolling off the end of the rails. They tend to be very sturdily built—as you might expect for something meant to stop a train—but even so, they cannot stop a high-speed runaway. The padded head is about three feet above the rails and a little to the right of the track centerline, so it meets the coupler.

Rail lubricators. Railroad cars have solid axles, which means the two wheels on opposite ends of an axle always turn at the same speed. That's fine on a straightaway, but on a curve the wheels travel different distances, and something has to give. One result is a terrible screeching as the wheels scrape over the rails. Another result is excess wear, especially on the outer rail. To reduce the wear, a railroad may install a lubricator near the entrance to a problem curve. The lubricator pumps grease onto the wheels of passing trains, and the wheels spread it out over the rail.

SIGNS AND SIGNALS

End of the line. Both the energy-absorbing bumper *(below)* and the derailer *(above)* are never supposed to see service. They're a safety net for trains that wander where they shouldn't.

On the highway, stop signs and traffic lights are mainly intended to prevent cross-traffic collisions at intersections. Some railroad signals also address this problem, but the majority are meant instead to forestall head-on and rear-end collisions—accidents that truck and automobile drivers are mostly expected to avoid on their own. Trains need more help in this respect because of their tremendous momentum. It takes so long to stop a train that by the time you see another train up ahead on the same track, it's probably too late.

When railroading began in the 1830s, signaling was difficult for a fundamental reason: the train itself was the fastest thing around, and so no message could outrun it. Under these conditions, the safest way to operate was to run by a strict timetable. (Modern commuters may long for a return to that regime.) If trains are started at

hourly intervals and they all move in the same direction at the same speed, they should remain safely separated at all times. In practice, of course, speeds vary, some trains make local stops while others are express, and a single track may have to carry traffic in both directions—all factors that greatly complicate the timetable. Worse, a breakdown or accident can leave a train helplessly blocking the track, regardless of where the timetable says it's supposed to be.

A better way to run a railroad emerged as soon as the electric telegraph allowed communications to speed ahead of the train. The protocol that evolved is called block signaling, and variations on it are still in use today. The idea is to divide the track into segments, or blocks, with a signal at the entrance to each block. In the simplest version of the system, only one train at a time is allowed in a block. The entrance signal gives a "do not enter" warning to any approaching train if the block is occupied.

Initially, block signals were set manually by workers in trackside towers who telegraphed word of train movements up and down the line. Many of these towers still exist, although few are in routine use. They are typically two-story wood or brick buildings erected every few miles along the track.

As early as 1872 automation came to block signaling, using the track itself as an electrical circuit. Within each block a voltage is applied across the two rails. In the absence of a train, no current flows between the rails because they are insulated by the wood crossties. When a train enters the block, the wheels and axles close the circuit, allowing a current to flow. The current betrays the presence of the train and sets the block signals accordingly. To avoid interference between adjacent blocks, the track joints at block boundaries have to be insulated.

Although the concept of automatic block signaling is simple, making it work is quite a trick. A steel rail is not an ideal conductor of electricity, and wood ties and stone ballast are not the best insulators. Thus, the task of the signaling system is not just to detect whether or not a current is flowing but to discriminate between a background leakage current that is always flowing and the slightly larger current that indicates the presence of a train. Detecting the train gets harder still when the rails are rusty or the track is wet. And on an electrified railway, with the enormous currents that power trains flowing through the rails, signaling is still more problematic.

The early automatic block systems relied on simple direct-current (DC) circuits—basically, a battery was connected across the rails, and a relay measured the current flowing between them. Later came block systems based on alternating current (AC), then on high-frequency alternating current (the frequency is a few thousand hertz, or cycles per second), and finally on pulsed currents, where the current is turned on and off a few times per second. These more complicated circuits detect trains more reliably and reduce interference between adjacent blocks. Assigning different frequencies or different pulse rates to nearby blocks can even eliminate the need for insulated rail joints at the block boundaries. In this case the current leaks through from one block to the next, but the sensors in each block are tuned to respond only to one frequency or one pulse rate.

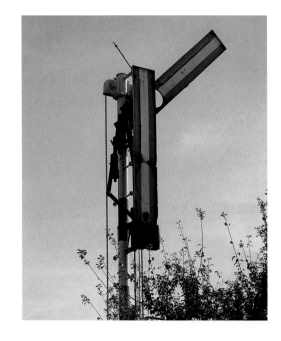

Trackside signals can be perplexing to those who know only the red-yellow-green routine of the highway. The semaphore above says "go slow"; the lights below mean "stop."

A double loop of wire detects the passage of a train across this block boundary and sends a signal to the dispatcher.

If you happen to see the boundary between two blocks (it is often marked with a sign, naming the blocks), you may be able to tell whether the block signals are based on DC or one of the AC technologies. With DC blocks, there is an insulated rail joint with no electrical connection between the rails on opposite sides. With the other methods there is usually some sort of iron box buried in the ballast or attached to the ties between the rails, with heavy wires feeding into it from the two blocks. In either case, there is likely to be a closet-sized metal cabinet nearby, where the train-detecting hardware is housed.

With some of the more elaborate block-control methods, the two rails cannot carry all the information that has to be passed up and down the line from block to block. Railroads therefore erected lines of poles along the side of the track, typically carrying a dozen or so bare copper wires on crossarms. (The copper weathers to a bright green.) The poles tend to be lower than those set up by telephone and electric utilities. Lots of these pole lines can still be seen, and some of them are still in use. On the other hand, many railroads have also laid fiber-optic cables along their right-of-way, which have a vastly larger capacity. The fiber-optic cables are buried, but there are usually markers at grade crossings to show where they are.

In an automatic block system, adjacent blocks communicate with each other, but there is no one point from which the positions of all trains can be monitored. A technology called centralized track control, or cTc, provides just such an overview. Under cTc, every block throughout a wide area reports its status—whether or not it is occupied by a train—to a control room, where dispatchers set the signals and switches that govern all train movements. Railroading with cTc is something like flying with air-traffic control. A single control room can manage a railroad spread out over several states. For example, all dispatching for the CSX Railroad, which stretches over the entire eastern United States, is done from Jacksonville, Florida.

The big advantage of cTc is that it allows traffic to be planned and coordinated better. Whereas other block systems allow movement when they know where a train isn't (namely, in the next block), cTc bases its actions on knowledge of where a train is. As a result, trains can be run closer together without risk of collision. The increase in traffic density is big enough that installing cTc has allowed some railroads to scale back sections of double track to single track.

The most visible parts of a railroad signaling system are the signals themselves. I find it highly curious that after 150 years of railroading, there is still no national standard (much less an international one) for how signals ought to be displayed and interpreted. On the highway, red, yellow, and green look the same and mean the same thing everywhere, but each railroad has its own set of signals, and even within a railroad there's room for variation.

The earliest railway signal in the United States was a white ball hoisted up a mast to indicate the way is clear—hence the phrase "giving a train the highball." In New England a few of these ball signals remained in use as late as the 1950s, but they are all gone today.

Another early signaling device was the semaphore, which was adapted from the optical telegraph systems that preceded the electric telegraph. The semaphore is an arm pivoted on a mast so it can be turned to various positions. Generally, a horizontal semaphore arm means stop, a vertical arm means go, and an arm at the 45-degree intermediate position means caution—proceed at reduced speed. Under federal regulations semaphores are still legal railroad signals, but I haven't seen one in use for years, except in railroad museums.

The signals in widest use today encode their meanings in the colors of lights, much like highway traffic signals. In the first version of this system, a red light meant stop, a green light meant caution, and a light with no colored lens meant go. The problem with this scheme was that a broken or missing lens would be interpreted as a go signal, perhaps with fatal consequences. The alternative of red, yellow, and green lights, now so familiar to motorists as well as locomotive engineers, was introduced in 1899 on the New Haven Railroad.

But the similarity of highway and railroad color-light signals is slightly deceptive. In the first place, the railroad signals apply to an entire block, not just to an intersection; a yellow light, for example, restricts a train's speed throughout the block. Also, a red light often has a meaning more like a stop sign than a highway red light; the train stops but can then continue slowly. What's most confusing (to automobile drivers) is that many railroads install multiple lights at each signaling position, to provide finer control over train movements. You might well see red and green lights shining at the same time; this would be nonsense on the highway, but to the train driver it means proceed but expect to stop at the next signal.

An entirely different system of signal lights was favored on the former Pennsylvania Railroad and still operates in parts of its old territory. In these signals, meaning comes from the position or orientation of the lights, not their color. In essence, the signals mimic a semaphore arm. A horizontal array of lights signals stop; a diagonal line of lights means proceed with caution; a vertical line of lights is the go signal. On the Pennsylvania tracks, all the lights were the same amber color. The Baltimore and Ohio combined the position and color systems, using horizontal, diagonal, and vertical arrays of red, yellow, and green lights.

Although signal lights are still installed at trackside everywhere on American railroads, many trains no longer need them. They have signal indicators in the locomotive cab that pick up messages transmitted through the track from a central dispatching station. A big advantage of cab signals is that when a block signal goes from red to yellow or from yellow to green, the train does not have to wait until it reaches the next trackside stanchion before speeding up.

The rules for interpreting and obeying signals are stricter for passenger trains than they are for freights. A freight train can cross into a block occupied by another freight at low speed (20 miles an hour is the limit), but passenger trains are absolutely prohibited from occupying the same block as another train. Some signals on commuter railroads and other passenger lines are equipped with a trip-stop mechanism: When

Railroads have strung up thousands of miles of signal wire along their rights-of-way. In recent years much of the signal wire has been abandoned, but this stretch in New Mexico appears to be well-maintained.

the signal goes red, a lever arm is raised next to the track. If a train tries to pass, the trackside lever snags another lever on the train, which automatically applies the brakes.

A technology that has had a big impact on railroad operations is two-way radio. Once upon a time, dispatchers could communicate with train crews only by telegraphing ahead to signal stations, where written orders were passed up to the train (or grabbed on the fly from a hoop). Later, telephones were installed at intervals along the road so a train crew could stop and report trouble. Now radio keeps the train crew in constant touch with the dispatcher. Furthermore, handheld radios allow communications among members of a crew on the same train. No longer does the engineer signal by blowing the whistle, and brakemen need not wave flags or lanterns.

LOCOMOTIVES AND ROLLING STOCK

The American locomotive has undergone a strange evolution. The starting point is the nineteenth-century steam locomotive, which had a long barrel-like firebox and boiler in front, and a boxy cab for the engineer and other crew members at the rear. Putting the driver at the rear of the engine has an obvious drawback: you can't see where you're going. In most steam locomotives the only forward visibility is through narrow slits on each side of the cab. There was a reason for this arrangement in the age of steam: the cab had to be adjacent to the coal tender, which was hitched directly behind the engine so the crew could stoke the fire.

When the diesel locomotive came along in the 1930s, the coal tender disappeared. Diesel fuel is carried in tanks slung under the frame of the locomotive itself. Nevertheless, the first diesel engines reproduced the layout of the steam locomotive. They are known as rear-cab switching engines, and they again put the driver in the backseat, looking around or over a long hood that covers the power plant. (They are switching engines because they were mainly used in switching yards, for pushing cars around while making up trains.)

In 1945 the Electromotive Division (EMD) of General Motors introduced a dramatically different diesel locomotive called the F series. It put the crew in front of the motor, with forward visibility through a windshield much like that of a car or truck. The F series was a tremendous success, especially for pulling passenger trains. More than 3,000 of them were built, and perhaps a hundred are still in service today. They define a classic era in American railroading.

And yet the F series, in spite of its success, was not the model for the design of most later locomotives. Instead, the engines that pull most American trains today evolved from the rear-cab switchers.

There have been several steps in this evolution. An early development was the addition of a small pod behind the cab, which typically housed a steam generator for

heating passenger cars. With this addition, the cab was no longer at the very rear of the locomotive, but it still had 40 or 50 feet of engine in front of it. Moreover, the visibility problem grew worse as the engines became more powerful—the long hood in front of the engineer was continually getting taller and broader and longer, once again leaving only slits for forward visibility.

Then someone had a bright idea. (Why did it take so long?) If you simply turned the locomotive around, it could run with the cab near the front. The small pod that used to trail behind the cab now became a low "snoot" out in front. This is now the most common design for locomotives pulling freight trains in North America, as well as many passenger trains outside the denser metropolitan corridors.

The switcher-derived locomotives tend to be very boxy or blocky, with no attention to aerodynamic efficiency. This is another historical anomaly. The F series of the 1940s was a much cleaner design aerodynamically, and for that matter there were celebrated steam locomotives of the 1920s and 1930s that had elaborately streamlined cowling. (They sure looked fast, whether or not the lines of the bodywork had any scientific basis.) So it's not the case that railroaders were unaware that the earth has

Three diesel locomotives—they are General Electric C44-9Ws—haul a fast freight along one of the main rail corridors from Texas to Southern California.

The evolution of the American locomotive has followed a curious path over the past 50 years. The pre-eminent stylistic statement of the 1950s was the General Motors F series; the two examples at right, after a long career with various Canadian railroads, were observed sitting idle and forlorn on a siding in Keokuk, Iowa. Although the F series was highly successful, it was not the proto-type for most of what followed. Instead, the model was the humble switching engine, whose usual job is to move freight cars short distances at low speeds. The switching engine in the photograph below is a 50-year-old General Motors product, now marshalling freight cars in southern Nebraska for Cargill, the grain merchant. A modern long-haul freight locomotive, the General Motors GP50 (bottom photograph) owes more to the boxy switching engine than to the streamlined F-type. In essence, the GP50 is a switching engine turned end-for-end, so that the cab is at the front.

an atmosphere that trains must push their way through; it seems they chose to ignore that fact. Some of the latest locomotive designs, such as the General Electric Genesis (often seen hauling Amtrak passenger trains), are noticeably slipperier.

Under the sheet metal, all diesel locomotives work on much the same principles. The long hood covers a massive diesel engine with 12 or 16 or even 20 cylinders, which turns at low speed but produces as much as 4,000 horsepower. The engine turns a DC generator, which is wired directly to electric motors in the axles of the drive wheels. Why the roundabout conversion from mechanical energy to electrical and back to mechanical? Why not just use a drive shaft and gearbox, the way power flows from motor to wheels in a car or truck? Because no one has built a gearbox that can withstand the loads of railroad service.

One of the best places to get a look at a locomotive is from an overpass, where you can see the top deck. In addition to the diesel exhaust stack (you may not want to be directly overhead), there are fans for cooling the engine. In many cases a second set of fans cools the regenerative-braking equipment. Regenerative braking works something like slowing a truck on a mountain descent by shifting into low gear; the locomotive's electric motors operate as generators, producing power instead of consuming it, and in the process they slow the train. The electricity generated is dissipated in a network of big wire-wound resistors, which are kept from melting by the roof-mounted fans.

Rolling Stock. There are more than a million freight cars rolling around North America. The kinds you see depend a great deal on where you are. Coming out of Wyoming or West Virginia, you might see trains with a hundred open hopper cars

hauling 10,000 tons of coal. In the Midwest the hoppers are covered and filled with grain. In California, trains of refrigerated boxcars ("reefers") carry produce to points east. Along the Gulf Coast the black tank cars are filled with petroleum products or molten sulfur. In other areas you might notice trains carrying ore, lumber, automobiles, cattle, steel, crushed stone, or chemicals.

Despite all these differences, most of the cars are the same under the skin—they run on essentially the same undercarriage. There's a swiveling two-axle truck, or bogie, at each end so that the car is supported by eight wheels overall. As noted earlier, wheels on opposite sides of the car are connected by solid axles, which is what causes that awful squealing on a tight turn: one wheel or the other has to slip since they are being asked to rotate at different speeds. A loaded car, groaning and swaying down the tracks, can press down with a weight of 30 tons on each axle, which is close to the maximum weight that a tractor and semitrailer distributes over its 18 wheels.

For many years all freight cars had journal bearings: the axles turned in a lubricated bronze sleeve. If a journal bearing is overloaded, runs out of lubricant, or becomes contaminated with grit, it can overheat and fail spectacularly. It glows cherry red and then shoots flames that could ignite a wood boxcar floor. Train crews had to be constantly on the lookout for such "hotbox" problems. Newer cars have roller bearings or ball bearings, like those in bicycle wheels and roller skates; the cars are said to be Timkenized, after the Timken Company, a manufacturer of the bearings. Roller bearings are much less troublesome. When you're sitting at a grade crossing watching a slow freight trundle by, it's easy to tell the two kinds of bearings apart.

A few of the million-plus freight cars that roll across North America await sorting at a marshalling yard in Roanoke, Virginia.

Standard hardware mounted at the end of this hopper car *(above)* includes an air reservoir for the pneumatic brakes and a crank wheel for the hand brakes. All newer cars have roller-bearing wheels *(below left)*, but a few journal bearings are still on the road *(below right)*. Note that only gravity holds the wheels onto a rail car; if you were to lift up the body, the wheels would stay behind.

Look at the center of a wheel, where it's fitted to the axle. A journal bearing lies within a steel box that has a flap door on the front, hinged at the top so lubricant can be added. The roller bearing has a round or triangular cap on the end of the axle, which turns with the wheel.

While you're checking bearings, you might also listen for flat wheels. Even a very small flat spot—created when the brakes lock up—produces an unmistakable thumping as the train lumbers by—"kabunk kabunk kabunk."

The suspension system that supports the car usually includes two sets of coil springs on each side of each bogie. Small springs are placed between the axles and the frame of the bogie; large ones, between the bogie and the body of the car. Some boxcars also have pneumatic dampers, which work like the shock absorbers of a car.

The couplers that join American freight cars work like two hands clasping with fingers curled; railroad workers call them knuckles. The official name is the Janney coupler, after Eli H. Janney, a nineteenth-century inventor whose design won out over hundreds of rivals. The fingers link automatically when the cars are pushed together; then a pin drops down by gravity to lock them into place. The air hoses that operate the brakes still have to be connected manually; the fittings on the mating ends of these hoses are called gladhands. To uncouple a car, a crew member pulls a lever at the side of the car, which lifts the pin and allows the coupling to open. The brake cables automatically pop apart as the cars separate.

Each coupler has half an inch to an inch of slack, and so a train with 100 cars might be six feet longer when stretched out than when bunched together. The slack serves a purpose: if a train is bunched up when it comes to a stop (as it usually is), then when it starts up again, the locomotive doesn't have to start every car moving at the same instant. It's marvelous to stand near a stopped freight and listen to the wave of noise propagating from end to end as the train gets under way. It can take several seconds after the engine starts moving before the last car feels the pull.

The couplers on European freight cars are quite different from those used in North America. The coupler itself is a simple arrangement of two loops or eyes that are joined when a worker inserts a pin manually. But what's most distinctive are the buffers—spring-loaded bumpers that keep the coupler under tension, so the train almost never bunches up.

The railroad braking system is a remarkable technological fossil. The scheme that George Westinghouse sketched out in 1869 is still in almost universal use, hardly changed after a century. The brakes are pneumatic, powered by a compressor in the locomotive that pumps air into a reservoir. The standard working pressure is 70 pounds per square inch, or about twice the inflation pressure of an automobile tire. If you stand near an idling locomotive, you'll likely hear the compressor cycling on and off, and the pressure-relief valve occasionally releasing a snort of excess air. These sounds, punctuating the steady drone of the diesel itself, can make a locomotive seem a living thing, with not just horsepower inside but something like horses.

The air pressure is carried from the locomotive throughout the train by an air pipe that runs under all the cars, and is joined by rubber hoses at each coupling. In the earliest air-brake systems, the brakes were applied by letting pressure into the pipe; the pressure moved pistons, which in turn forced brake shoes against each wheel. There is an obvious problem with this scheme: if the compressor quits or the air pipe springs a leak, the train is left without brakes.

Westinghouse turned the whole idea upside down. The brakes are activated not by applying pressure but by releasing it; that way, most malfunctions will lock up the brakes and stop the train—an inconvenience, but preferable to not stopping. When the train is first assembled, the locomotive pumps air through the brake pipe to reservoirs mounted on each car; it is the air stored in these canisters that will be used to power the brakes. As long as the brake pipe remains pressurized, however, a valve on the car holds back the air in the canister, keeping the brakes off. To slow the train, the engineer moves a brake lever that releases a little air from the main brake pipe; the lowered pressure slightly opens the valve on each car, pressing the brake shoes lightly against the wheels. In an emergency, all the air is dumped from the brake pipe, and the brakes grip hard. A major leak or the uncoupling of a car has the same effect.

The main parts of the braking system are readily visible on most freight cars. The brake shoes are wedged between the wheels on each truck, although they may be partly hidden by the frame and springs of the truck. The air canister—which looks like a heavy-duty beer keg—is usually tucked up under the body of the car, but on some hopper cars and tank cars there's more room for it at one end or the other. The brake pipe runs along the underside; the hoses that connect the pipes of adjacent cars hang down a little below the coupler. At one end of the car there is a handwheel that operates a manual brake, which can be used when the car is not attached to a train and the air system is empty.

Perhaps the reason the Westinghouse air brake has endured so long is simply that it works. On the other hand, the brakes are not perfect. A particularly nasty problem

A Janney coupler, or knuckle, links rail cars with a grip like two hands with hooked fingers. The black lever extending toward the left foreground pulls the pin that uncouples the cars. In addition to the coupler itself, the brake hoses must be connected.

Four types of rolling stock: a boxcar, a doublestack well car carrying two freight containers that fit directly on truck chassis, a gondola for bulk goods such as scrap iron, and a tank car for liquids.

comes up when trains get very long—100 cars or more. It can take 10 seconds or longer for a change in air pressure to propagate all the way through the brake pipe from the engine to the last car. As a result, when the brakes come on up front, the rear of the train is still moving with full momentum. This is not a comfortable situation; in extreme cases the train could jackknife and derail, creating accordion folds of freight cars snaking back and forth across the track.

A remedy for this problem has been known for decades: continue to rely on air pressure to power the brakes, but use an electrical signal rather than a pneumatic one to trigger brake action. Since electricity can cover the length of a train in microseconds, all the brakes respond to the engineer's command almost instantaneously. Electropneumatic brakes have been tested on a number of integrated trains, in which a set of identical cars and locomotives are permanently coupled and always haul the same commodity (usually coal) over the same route. Equipping other trains with electropneumatic brakes runs into a logistical problem because older cars cannot be converted all at once, and the two kinds of brakes would have to coexist for a long time, even within a single train. Nevertheless, the first steps toward creating a new electropneumatic standard for train brakes have already been taken.

Types of Freight Cars. There are freight cars specially equipped for carrying everything from railroad rails to pickles, and I can't possibly catalogue them all here. What follows are notes on a few of the common varieties.

Boxcars. For a long time the 40-foot boxcar was the mainstay of rail freight, able to carry anything that would fit through its sliding doors—sacks of flour, crated machinery, livestock, mail, even automobiles. There are still hundreds of thousands of boxcars in circulation, but demand has slacked off. Some commodities have gone to more specialized rail cars—flour to hopper cars, automobiles to automobile carriers—and much of the rest now goes by truck or intermodal flatcar. Meanwhile, the standard size is no longer so standard. There are 50-foot boxcars and even 90-footers. All of the newer ones have steel bodies; anything made of wood is a genuine antique.

Incidentally, the markings on boxcars can also be antique. As railroads have busily merged and reorganized, they have tried to keep their locomotives up-to-date in paint and signage, but many of their boxcars still carry the identifying logos of long-defunct companies. If you feel nostalgia for the Southern Line, the Norfolk and Western, or the Milwaukee Road, just stand by the tracks and watch the faded glories roll by.

Flatcars. A platform on wheels, these are the simplest of rail cars—or at least that's how they began. In recent years, flatcars have begun to mutate and interbreed with other kinds of rolling stock, creating a strange profusion of flatcar subspecies.

The bulkhead flatcar has tall barriers at each end, to keep cargo from shifting longitudinally. The centerbeam flatcar has a spine down the middle as well as bulkheads at the ends. It often carries lumber, sheetrock, or similar materials; the central beam

adds stiffness for these heavy loads. Then there are the various kinds of piggyback and intermodal container flatcars. The piggyback cars carry complete truck trailers, combining the advantages of rail transport (lower fuel costs and labor costs) with delivery from door to door. Containerized freight travels not only by rail and by truck but also by sea. When these intermodal transport systems began in the 1950s and 1960s, trailers or containers were simply lashed down on standard flatcars. Now there are flatcars specially adapted for the purpose—and they are not particularly flat. A spine car is a skeletonized flatcar, with just a beam down the middle and platforms at either end to support the wheels and the hitch of a semitrailer. A well car has a deep recess between the wheels so containers can be stacked two-high. (But these doublestack cars require a 22-foot overhead clearance, so they have to be restricted to main-line tracks that can pass them. Most tunnels are out of bounds.)

Gondolas. Among the earliest railroad freight cars, gondolas have a flat bottom, an open top, and low sides. Like boxcars, they are not as commonplace as they once were, but some can still be seen hauling scrap iron destined to be remelted in steel mills, or wood chips on their way to paper mills. (The name of this very utilitarian rail car apparently does come from that of the romantic, $100-an-hour Venetian boat; the two vessels don't have much else in common.)

Tank cars. The first tank cars were built to carry petroleum, but today pipelines have taken over that trade, and little crude oil is hauled by rail. Nonetheless, tank cars are in great demand. They transport an impressive variety of liquids and gases: refined petroleum products, propane, alcohol, ammonia, carbon dioxide, hydrogen, chlorine, liquid nitrogen, liquid oxygen, molten sulfur, acids, dyes, detergents, ink, corn syrup, honey, chocolate, fruit juice, wine and gin and whiskey. A list of some 1,600 tank-car commodities even includes an entry for "cucumbers in brine," although I have never seen a rail car claiming to carry that cargo. On the other hand, I did once have a look at the receiving dock of a bubblegum factory, where spills during the unloading of tank cars had left oozing mounds of pink goo.

The typical tank car is a cylinder with rounded end caps—a form known as a bullet tank. Atop the middle of this cylinder is a "dome," which covers up valve gear, gauges, and usually a hatchway through which workers can enter the tank for cleaning and inspection. There may also be valves and nozzles under the tank so it can be unloaded by gravity flow. Fluids such as asphalt have to be heated before they will flow, and so tank cars for these commodities are equipped with steam pipes that coil through the interior volume; the steam connections are generally on the underside. Some tanks are also built in a swayback shape so they are easier to drain through the center outlet.

Because tankers often carry hazardous cargoes, they are subject to more stringent regulations than other freight cars. Capacity is limited to 34,500 gallons; most tanks carry from 10,000 to 20,000 gallons. The actual capacity—along with much other information—is stenciled on the sides and the ends of the tank. Some tanks tell you exactly what they contain, with a marking such as "phosphoric acid only" or "molten

Hopper cars for coal *(above)* and grain *(below)* load from the top and unload from sawtooth hatches in the bottom. In some cases coal hoppers are simply turned upside down to dump them.

sulfur." Even if the contents aren't stated, you may be able to glean a few clues from a hazardous-material placard or other markings. For example, the U.S. Department of Transportation classifies tank cars, and so you are likely to see a number such as DOT 111A60W1. The most interesting part of this legend is the number 60, which is the car's pressure rating in pounds per square inch. If the car is carrying a volatile liquid such as anhydrous ammonia or liquefied petroleum gas, the pressure rating will be higher—perhaps 340 or 500 pounds per square inch. Even the color of the tank can be significant. An uninsulated tank holding a heat-sensitive liquid such as anhydrous ammonia has to be painted a light color over the upper two-thirds of the body.

Tank cars rated for hazardous cargoes have special couplers with a "shelf" above and below the coupler itself. The couplers prevent the car from coming adrift in many kinds of accidents. Some tankers also have bulkheads at the car ends to reduce the chance of puncture in a derailment.

Hopper cars. Since the 1980s hoppers have taken over from boxcars as the most common freight cars. The characteristic feature of a hopper is the series of chutes or discharge gates underneath. The car is loaded from above and then unloaded from below. Open-topped hopper cars haul coal, crushed stone, and ore. Covered hoppers—with a row of loading hatches in the top—carry grain, portland cement, and other bulk commodities that need to be kept out of the weather.

Coal hoppers come in three varieties. The sawtooth hopper has a series of small, angled unloading doors in the bottom; it gets its name from the shape of these doors seen in profile. In the bottom-dump hopper, the entire bottom of the car falls away

A unit train is made up of cars going to a single destination; in this case the train carries coal for a Missouri power plant. The cars are unloaded in the shed at the extreme right by unceremoniously flipping them upside down.

WHERE'S THE CABOOSE?

Every child knows that a train ends with a caboose, just as a sentence ends with a period and a dog ends with a tail. But no more. Trains march shamelessly cabooseless over the American landscape. Already, a working

caboose at the end of a train is almost as rare and quaint as a steam locomotive at the front.

The caboose was a standard fixture of freight trains for more than a hundred years. It provided office space for the conductor and storage for tools, as well as a place for the crew to rest, eat, and get warm. Perched in the raised cupola—or leaning into bay windows protruding from both sides—the crew could keep an eye on the train ahead. There was also a gauge for monitoring brake pressure.

What has replaced the caboose is the End of Train Device, or ETD, also known (I'm not sure how seriously) as the Flashing Rear End Device, or FRED. Whatever you call the thing, it's a small box with a blinking light hung on the rear coupler of the last car in a train. At first glance it looks like nothing more than one of those battery-operated highway flashers that you see in a construction zone, but in fact there's a little more to it than that. If you look closer, you'll see that a brake hose is attached to the box, and there is also a radio antenna. A major responsibility of the ETD is to monitor brake air pressure. The device continually broadcasts pressure reports to a receiver in the locomotive cab, which will alert the engineer of a malfunction. Some ETDs also have a radar unit to monitor train speed and direc-

tion: if the rear end of the train isn't going the same way as the front end, something is probably amiss!

The ETD is one of the few wholly new technologies introduced into railroading in recent years. It seems to do the job, and already some of the more ardent railfans have learned to decode ETD radio transmissions. Still, the little box with the blinking light is going to have a hard time displacing the caboose in railroad lore and romance.

to dump the load. Finally, top-dump cars are overturned bodily for unloading. On these, look closely at the coupler, which has a big disklike rotary joint that allows individual cars to be flipped over without breaking the train.

At a large coal-fired power plant you might be able to see the unloading operations. Typically the cars are moved through the unloader not by a locomotive but by a winch that automatically "spots" each car at the right position. Sawtooth hoppers require a shaker to get all the coal out. This is an electrically driven vibrator; even from half a mile away, you can hear its distinctive buzzing. Unloading a sawtooth hopper can take 10 minutes. Bottom-dump cars unload faster and require no vibrator. The top-dump cars are faster still. Each car rolls onto a turntable, where it is clamped in place; then the turntable pivots around the axis of the car couplings. The cycle takes about 30 seconds; a 100-car train can be unloaded in less than an hour.

In northern climates you may find a thawing shed ahead of the dumping station. It looks something like a drive-through car wash, but instead of spraying the cars with water, it warms them with electric or gas-fired radiant heaters.

Ore hoppers are similar to those for coal, although they have not evolved quite as rapidly. Whereas coal cars have grown steadily larger, iron ore is still hauled in 24-

A series of tank cars wend their way through the fan of tracks in the Norfolk-Southern hump yard at Linwood, North Carolina. As each car navigates through the "fan" of classification tracks, switches are set automatically to direct it to the appropriate track. The car's speed is also regulated, so that it collides with cars already on the track just firmly enough to engage the couplers.

foot hopper cars, because that length matches the spacing of loading pockets and hatches on the Great Lakes ore boats.

Covered hoppers have revolutionized the shipment of grain and especially of flour and other powdered products, which previously had to be moved in bags or barrels rather than in bulk. The flour hoppers are unloaded pneumatically. A conveyor tube four to six inches in diameter is connected to an outlet near the bottom of the hopper, and a pressurizing air hose is connected to an inlet, also near the bottom. The incoming air flows through a perforated deck, fluidizing the powder and driving it out the discharge tube. Pressure-relief valves keep the pressure differential at about five pounds per square inch. At one end of the car you'll see a rupture disk, which prevents the car from splitting its sides in case the relief valve malfunctions. Flour cars hold about 200,000 pounds.

FREIGHT YARDS

Freight railways work on a hub-and-spoke system (which they adopted long before the airlines discovered the idea). Trains from various cities converge on a hub, where their cars are separated and reshuffled into new trains, which then depart for new destinations. The hub is a freight yard—also known as a classification yard.

There are two types of classification yards, called flat yards and hump yards. In a flat yard, a switching engine grabs onto individual cars, drags them to the appropriate track, and shoves them together. It's the obvious solution. The idea behind a hump yard is more daring: it's railroading on a roller coaster. Engines push a line of cars slowly up a hill, ascending at a walking pace. At the crest, the cars are uncoupled one by one and roll down into a "bowl" where many tracks fan out to the left and right. Once uncoupled, the cars roll under gravity power, with no brakes and no one aboard. A series of switch points directs each car to the track for the correct outgoing train.

The arrangement of tracks in the bowl of the hump yard is called a fan. The main track coming over the hump splits into left, middle, and right branches, and then each of these branches splits in turn, and the ramifications continue until there are dozens of tracks spread out in a broad plain at the bottom of the bowl.

Making this humping system work requires delicate control. In the first place, each switch point in the fan has to be set correctly before the free-rolling car reaches it. What's more, the speed of the descending car has to be precisely right. If it rolls down the slope too slowly, it won't have enough momentum to couple with cars already on the classification track; it may even get stuck on one of the switches, where the next car in line could snag it. On the other hand, if the car moves too fast, it will slam into the cars already waiting on the track, possibly wrecking the rolling stock or the cargo. (You'll occasionally see a rail car stenciled "Do Not Hump.")

In the earliest hump yards, a century ago, the control was entirely in human hands. A worker rode along on each car to regulate the speed with the hand brake. This was a famously dangerous job; if you happened to get a car with defective brakes, your

An aerial photograph in the Urban Areas series of the U.S. Geological Survey maps out the hump yard in Blue Island, Illinois, near Chicago. The hump and the fan of classification tracks are at left; completed trains are pulled out to the right. Various bypass tracks allow for traffic that need not or should not go over the hump.

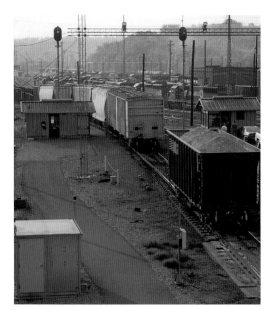

As a brakeman pulls the pin, cars separate at the crest of the hump at Linwood *(above)*. The master retarder *(below)* squeezes the wheel flanges to control the cars' speed. Masts next to the retarder are for speed-measuring radar guns.

choices were few and unattractive. Another group of workers stood by the switch points, flicking great levers back and forth to direct the passing cars.

Later, a mechanism called the retarder eliminated the brakeman's hazardous job. The retarder is a section of track with moveable clamping jaws mounted just inside the normal running rails. As a car passes through the retarder, the jaws are closed to squeeze against the car's wheel flanges. Varying the pressure controls just how much the car is slowed. Originally, workers stood next to the track and gauged each car's speed by eye and ear, then pushed on a lever to cause what they judged would be the right amount of deceleration.

Now the whole process is totally automated. As a car comes over the hump, radar units, much like those used by the highway patrol, measure its speed. The measurement may be repeated at two or three points in order to estimate acceleration and rolling resistance as well. Then a scale built into a section of track determines the car's weight (which has an effect on its momentum). Based on all this information, a computer determines how much to slow down the car, and sends the appropriate signal to a hydraulic actuator in the master retarder, just downhill from the radar sensors and the scales. The computer also sets the switches that steer the car to the appropriate track. And the computer can fine-tune the car's velocity by operating "group retarders" built into tracks farther out in the fan.

Just about the only part of the process that's still done by hand is "pulling the pin" to uncouple cars at the crest of the hill; everything else is controlled from a few computer keyboards in a trackside tower.

Seen from a distance, a humping operation looks like a magical game of pinball. The cars are launched onto their downhill journey in quick succession, each one beginning well before the last has finished, so at any moment there are five or six cars in motion. The switch settings are not visible from far off, and so the cars in the fan veer left and right unpredictably, as if, eerily, they had a will of their own and knew just where they wanted to go. Everything happens in stately slow motion; it can take minutes for a car to roll down the hump and reach the end of a classification track. For the most part, the yard is quiet, but occasionally a resonant boom echoes up out of the bowl when two cars couple hard.

How do the computers in the control tower know where to send each car? The answer is easy to find if you know where to look. Facing either side of any freight car, focus on a point about waist high and a few feet from the right-hand end. You'll notice a small gray lump, which on closer examination turns out to be a bullet-shaped piece of plastic riveted to the car body. It is called an AEI tag, for Automatic Equipment Identification, and it is a miniature radio transponder. Just about every freight car in North America has one (actually two—one on each side). When a scanner by the side of the track emits an interrogating signal, the AEI tag replies, identifying the car. It even manages to do this without batteries, drawing on energy in the received signal to power the response. (Some of the tags you mount in your car for paying bridge or turnpike tolls work the same way.)

What the AEI tag broadcasts is simply the car's identifying number, which is also stenciled on the body. Every freight car in North America has a unique number; it usually consists of two to four letters (identifying the car's owner) followed by four to six digits. The car's number is all that the hump yard's computer needs to know in order to make decisions about the car's fate; the computer consults a database to find out where that car ought to be going.

AEI tags were introduced in the early 1990s. A few freight cars still bear the marks of an older tagging technology. In the 1960s the railroads experimented with an early form of bar codes, similar in purpose to the ones read by grocery-store scanners but very different in appearance. The bar code was a foot-high block of multicolor stripes printed on the side of each car at about eye level. Specially outfitted cameras sensed the stripes as the cars rolled by. But they didn't sense them very well. With grime and graffiti, the system proved unreliable and was abandoned after only a few years. But there are still cars out there with a patch of faded stripes.

After a batch of cars is assembled in a freight yard and hauled to a distant city, they have to be broken apart again for delivery to various destinations. Now another kind of sorting is needed. In this case you can watch the process in any industrial neighborhood where several plants have rail service. As the local freight makes the rounds, it has to drop off and pick up cars from various sidings and leads. Finding the most efficient sequence of deliveries and pickups is a difficult mathematical problem, which the train's conductor is charged with solving. Consider a simple example: A train of five cars is to deposit its first two cars on an empty siding. The train stops short of the siding and uncouples between the second and third cars. Then the engine pulls the front two cars onto the siding and uncouples from them. Now the engine can continue out the far end of the siding onto the main line, back up to recouple with the remaining three cars, and continue on to the next stop. Altogether, the maneuver requires two uncouplings and a recoupling, four changes of switch position, and two changes of locomotive direction. It gets much more complicated if the cars being delivered are not all at one end of the train, if the siding is not empty, if cars need to be picked up as well as left off, or if cars have to be pushed onto a dead-end lead. Sometimes the train must go miles out of its way to find another siding to use as a temporary holding area. The puzzles can get really tricky—noted mathematicians have tested their mettle on them—and I am full of admiration for the railroad workers who solve them routinely day after day.

ELECTRIFIED RAILWAYS

From the point of view of fuel consumption and energy costs, the best way to run a railway is to plug it in. Compared with diesel engines, electric trains also offer lower maintenance costs and less air pollution. And they attain higher speeds. So why are few American railroads electrified?

An AEI tag is mounted on every rail car in North America. When interrogated by a trackside scanner *(bottom)*, it responds with the car's identifying number.

An older boxcar carries both an AEI tag and the tattered remains of an older identifying technology—a multicolor barcode.

The usual answer is that the cost of construction is too high, and the density of traffic too low. In Europe almost all major rail lines are equipped for electric propulsion, but routes are shorter there and more heavily traveled. Higher fuel costs in Europe also put a greater premium on energy efficiency. Finally, the higher speeds attained by electric railways are most important for passenger service, which is a major concern in Europe but of very little interest to most American railroads.

Most of the electrified tracks in the United States radiate from New York City. There's a simple reason: All rail traffic into Manhattan passes through tunnels. Running a coal-burning locomotive through a long tunnel would choke the crew and passengers, or at least blind the engineer to trackside signals. Hence, the New York lines began electrifying their tracks very early on, at a time when large electric motors and the equipment needed to power and control them were brand-new contrivances. The Pennsylvania Railroad began electric operation when it opened the trans-Hudson tunnel to Penn Station in 1910. At first, only the few miles of track in and near the tunnel were electrified. Trains bound for the city traded a steam locomotive for an electric one—or else passengers were shuffled from one train to another—at an isolated station in the Hackensack Meadowlands called Manhattan Transfer. This station in the middle of nowhere is gone now but not altogether forgotten; Manhattan Transfer became the title of a John Dos Passos novel and later the name of a pop singing group.

In the 1930s the Pennsylvania Railroad extended electrification south to Washington, D.C., and west from Philadelphia to Harrisburg. The Long Island Railroad, which enters Penn Station from the opposite side of Manhattan, had begun electrification even earlier, with 38 miles of track powered in 1905. The New York Central and the New Haven Railroads, whose trains tunneled under Park Avenue to Grand Central Terminal, followed soon after.

The network of electric railways in New York and the other cities of the Northeast Corridor is still running, and indeed it has changed little after almost a century. You can still see them working today. As a matter of fact, they are just about the only electric railroads you will see in the United States. After the great spurt of electrification in the early years of the century, the railroads seemed to lose interest. There are a few electrified commuter lines elsewhere, but not long-haul routes.

Because the evolution of electric-railway technology stalled at an early stage, many weird variations have been frozen in place. Even with only a handful of electrified lines, some trains run on alternating current and some on direct current, and at voltage levels ranging from 600 to 25,000 volts. In addition, there are many ways of connecting a moving train to a stationary electric power source.

Direct current was favored in the earliest systems because DC motors are easier to control than AC ones. But direct current also has a serious drawback: there is no convenient way to shift DC voltages up and down (see Chapter 6). Power has to be delivered to the train at the same voltage used by the motor in the locomotive. The common voltages are 600, 750, 1,500, and 3,000 volts. The usual arrangement is to

distribute power up and down the rail line by a high-voltage AC transmission circuit, carried on poles or towers near the tracks. Every few miles along the route, a booster station draws power from the high-voltage line and feeds it to the traction system. The booster station has a transformer that reduces the voltage to the train's working level and a panel of rectifiers that convert AC to DC. This equipment is usually housed in a fenced enclosure next to the track; the rectifiers may be inside a small building or cabinet. The booster stations have to be spaced closely because the low-voltage traction power cannot be transmitted far without unacceptable losses. The lower the operating voltage, the shorter the distance between booster stations.

Much of western Europe relies on DC traction: trains in Italy and Spain run on 3,000 volts DC, Ireland and the western parts of France are wired for 1,500 volts DC, and an area of southeastern England has 750-volt DC service. In the United States, DC power is used mainly by commuter railroads.

AC traction systems run at much higher voltages: 11,000 volts, 15,000, volts or 25,000 volts. At these voltages, power can travel for much greater distances, and booster stations are needed only every 10 or 20 miles. (And, of course, the stations have no rectifiers.) AC can be distributed at these high voltage levels because the locomotive itself includes a transformer that reduces the voltage to a more reasonable level for running a large motor, such as 600 or 2,400 volts.

Much of the equipment for an AC traction system—transformers, circuit breakers, insulators, and so on—is just like the equipment of a utility power grid and can be bought off the shelf from the same sources. But there's a complication. The world's electric utilities have standardized on two power frequencies: 60 hertz in North America and 50 hertz in most of the rest of the world. Electrified railroads are more various. For example, the railroads of Switzerland and Germany run on 15,000 volts at a frequency of 16⅔ hertz. And until recently some American trains required 25-hertz power. Railroads operating at a nonstandard frequency cannot simply draw power from the local utility. They must either run their own coal-fired or oil-fired generating station or somehow convert one frequency to another. On 57th Street in New York, among art galleries and boutiques, there is a basement room where for many years large electric motors took in 60-hertz current and turned dynamos to generate 25-hertz power for trains of the New Haven line running out of Grand Central Terminal. The machinery is idle now; all the trains have finally been converted to 60 hertz.

The trickiest part of designing an electric railway is getting the power from the stationary source onto the moving train. Dragging a cord behind the train is not an option. One answer is the catenary-and-pantograph system. The idea is to string up an overhead wire (the catenary), centered over the tracks at a height of 18 or 20 feet, and suck in the power through a sliding electrical contact (the pantograph). But for this arrangement to work reliably, the overhead wire has to be at a constant height all along the rail line, and that's hard to achieve. No matter how tightly a wire is stretched, it always sags somewhat between its supporting posts. (Indeed, the word

The overhead power supply for a European railroad is held taut by a system of counterweights, chains, and pulleys.

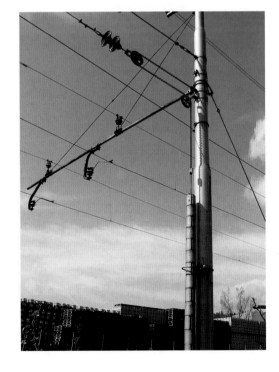

catenary refers to the curve assumed by a sagging wire or chain.) To combat this tendency, the catenary is built as a kind of long suspension bridge with many spans. An upper wire, called the messenger, is hung from posts or frames every hundred feet or so along the tracks and assumes the catenary profile. A lower wire, the trolley wire, is suspended below the messenger and connected to it by struts that vary in length so that the trolley remains nearly level throughout its length. The entire catenary struc-

THE LAST TROLLEY STOP

In 1917 the United States had almost 45,000 miles of trolley tracks and almost 80,000 trolley cars. Trolleys were the main way of getting around in most American cities, and some of them also provided service from one town to the next. By hopping from trolley to trolley to trolley you could get from Maine to Boston to New York, then on to Philadelphia, and as far west as Harrisburg, Pennsylvania. In 1904 newly-weds Clinton and Louisa Lucas took a 500-mile honeymoon entirely by trolley. Los Angeles, now famous as a city where you can't get anywhere without a car, had one of the best streetcar systems in the world, with a thousand miles of trolley tracks and 2,700 daily runs.

So where have all the trolleys gone?

Their demise is generally blamed on the automobile. Indeed, some critics have argued that the trolleys were done in by a perfidious conspiracy of automobile, oil, and tire companies, led by General Motors. In the years just before and after World War II—so the story goes—GM and its allies bought up urban trolley companies deliberately to bankrupt them. They raised the fares and curtailed the service; then, having chased away the customers, they complained that they were losing money. They substituted buses for trolleys. To make sure there was no going back, they ripped up the rails.

As conspiracy theories go, this one is more plausible than most. In 1949 GM and several other companies, including Standard Oil of California, Phillips Petroleum, and Firestone Tire and Rubber, were convicted of illegal anti-competitive practices under the Sherman Antitrust Act. (For a balanced account of the controversy, see David J. St. Clair's *Motorization of American Cities*.)

But a plot by the automobile interests cannot be the whole story. The decline of the trolleys began before any conspiracy could have gotten organized. Track miles were already falling before 1920, and by the end of that decade, the number of passengers was also dropping sharply. Trolleys were reviled and ridiculed even in places where the campaign could not have been coordinated from the GM boardroom. In New York City, for example, two mayors who were champions of public transit (they built most of the subway system)

nevertheless called trolleys "useless junk" and boasted of finally getting them off Madison Avenue. In Britain—far from GM's power base—a royal commission decided trolleys were "in a stage of obsolescence" and recommended that they be allowed to gradually disappear. The commission got its wish. France, too, dismantled its tram systems.

Even when trolleys were new, they were never seen as an elegant way to travel; 30 years later they were considered distinctly old-fashioned, and many of them were simply old. And in trying to upgrade and modernize their operations, the trolley lines faced a tremendous economic disadvantage. Trolleys were run by private companies, which had to lay their own track. Cars and buses, in contrast, could run over roads maintained at public expense. By the 1950s and 1960s, few trolley lines survived, and public transit of all kinds had become a charity case.

Trolleys were no sooner swept from the streets, however, than they began to make a comeback under a new name, with a new image and a new constituency. Where earlier generations knew them as trolleys, trams, or streetcars, the new term is *light rail*. (At least it's not "lite rail.") It is a favorite instrument of those who want to save cities from being strangled by automobiles. In North America more than 300 light-rail systems have been built since 1980.

The technology of modern light rail differs in many details from that of the older systems, but there is also much that a veteran trolley

ture is insulated from the supporting poles with porcelain insulators, much like those used in the electric power grid.

The catenary system has to be kept under constant tension, which can be tricky when the wires stretch and contract with changes in temperature. On some of the supporting stanchions you may see counterweights hung from insulated cables; these weights rise and fall to take up any slack in the overhead wires.

motorman would recognize. The light-rail cars are generally powered by direct current at 600 volts. Higher voltages are considered too dangerous for city streets. The electric current is collected from overhead catenary wires, with return flow through the rails. For trolleys of the earlier generation, the current collector was a small grooved wheel held against the underside of the catenary wire by a spring-loaded pole. This mechanism was not always reliable; most of the recent systems use a pantograph with a broad, sliding current collector.

The driving motors of a light-rail car are mounted in the bogies, close to the wheels. Some designs provide separate motors for wheels on the left and right sides, with no solid axle connecting them. This allows the wheels to turn at different rates and eliminates squealing when the car turns a corner.

The brakes on a light-rail car are different from those on conventional trains. Electrical braking, in which the drive motors are run as generators, slows the cars; the energy recovered in this way can be fed back into the catenary wire if another car elsewhere can absorb it at that moment; otherwise, it has to be dissipated in big resistors mounted under the floor or on the roof. To bring the car to a stop, the driver relies on disk brakes much like those of an automobile, operated either hydraulically or pneumatically. A third type of brake, used only in emergencies, is a heavy steel electromagnet slung between the wheels of each two-axle bogie. When this brake is energized, magnetic attraction pulls it down to the steel rail and

drags the vehicle to a stop *(see photo below)*.

Weaving through the grid of city streets, light-rail cars have to round much sharper turns than standard trains. Many of the newer cars are articulated; they have a big hingelike joint in the middle so they can snake around corners. On the outside of the car, the joint is concealed by a pleated, accordion-like cover; inside, the joint is marked by a large disk in the floor, which the two ends of the car pivot around. (The disk is an irresistible attraction to children, who will invariably be found straddling the floor joints.) Another strategy for dealing with tight corners is to taper the ends

of the cars, which reduces the swept area that has to be kept clear of traffic.

The modern revival of light rail began in Edmonton, Alberta, in 1978. San Diego built a line south to the Mexican border at Tijuana in 1981, and has since extended it in several other directions. Portland, Oregon, has another notably successful system, which carries more than six million passengers a year. Baltimore has a line with some technical innovations, such as AC drive motors *(see photo on opposite page)*. In Los Angeles—allegedly

the main victim of the GM conspiracy—a new light-rail line runs to Long Beach, and there are plans for another 150 miles of track. A few cities never lost their trolleys altogether and have now revamped them; Boston's Green Line, which has handsome new articulated cars, is a notable example.

The resurgence of light rail is rich in ironies. Early trolleys were disliked most of all because they snarled traffic and disrupted street life; now light rail is seen as a solution to traffic congestion and a way to get people out of their cars. In the 1920s there were complaints about the tangle of wires over the street; now cities that have banished all other utilities underground allow new wires to be strung for light rail. Early trolleys charged lower fares than buses and were seen as transport for the working class. Now, light rail is the darling of yuppies, and it's the bus that has taken on the stigma of the poor-person's vehicle.

A final irony is that light rail should come back into American life as a counterbalance to automotive culture and suburban sprawl. It was not the automobile that created the first American suburbs and drew people away from the central city; it was trolley lines. I grew up in such a community—a suburban development begun in the early 1900s and centered on a transit terminal where the subway line from downtown ended and half a dozen trolley lines radiated into the surrounding countryside. When I lived there in the 1950s and 1960s the trolleys were still running. We were a household without a car.

The folding, spring-loaded apparatus that collects power from the overhead wire is called the pantograph.

By the way, the trellis-like steel frames that support the catenary wires on the old New Haven Railroad look remarkably like the toy girders of an Erector Set. The resemblance is not coincidence. A. C. Gilbert's invention of the Erector Set was inspired by looking at those frames while riding the New Haven line.

The pantograph is the spring-loaded apparatus on the roof of the train that slides along the trolley wire of the catenary and collects the electric current. The springs allow the mechanism to follow any fluctuations in the height of the catenary wires. On high-speed trains the pantograph may have specially streamlined sections to keep it stable in the airstream. The strip that actually slides on the trolley wire could be made of brass, but it might also be carbon, which makes good electrical contact without wearing the wire away too quickly. Near the pantograph are insulated porcelain bushings to carry the current through the shell of the car or locomotive. There might also be a lightning arrester.

The alternative to the catenary and pantograph is a third rail and shoe. A heavy conductor mounted about a foot off the ground on insulated posts or brackets runs parallel to the main rails. The shoe is a metal paddle that reaches out from the under-carriage of the train to contact either the upper or the lower surface of the current-carrying rail. Third-rail systems are limited to relatively low voltages, and even so the right-of-way has to be fenced off to keep stray animals and people away from harm. Subways and some other forms of urban mass transit are commonly built with a third rail, but overhead power is the usual choice for other kinds of railroads.

Whatever the voltage, frequency, and method of power delivery, there is just one conductor supplying electricity to the train. How is the circuit closed? How does the current return to the source? It flows down through the wheels into the rails and then continues both through the rails and through the earth. Having large currents

coursing through rails at ground level might seem to pose an electrocution hazard, but the danger is remote precisely because the rails are laid on the ground. Their voltage can never differ very much from ground voltage.

PASSENGER TRAINS

Does passenger rail service deserve to be relegated to a brief appendix at the end of a long description of rail freight—like some quaint and obsolete caboose? American railroads seem to think so. Passenger trains form a minuscule fraction of the total rail

THE FAST TRACK

Making a train go fast takes a lot of horsepower, but getting up to speed is not the biggest challenge. The hardest part is keeping the passengers alive and well as the train goes hurtling down the track. Oh, and making it pay is a challenge too.

The leaders in high-speed rail technology are Japan and France. Japanese railroads have been running trains at 150 miles an hour or more since the mid-1960s. The high-speed routes are known as Shinkansen, which means "new trunk lines," but the rest of the world knows them as bullet trains. The French project is the Train à Grand Vitesse, or TGV. The first line, running southeast from Paris to Lyon, opened in 1981. There are now three more lines, including a connection to Britain through the Channel tunnel. Both the Shinkansen and TGV can take you for a ride at 300 kilometers per hour, or 186 miles per hour.

In their most basic technologies, it's remarkable how conventional the Japanese and the French high-speed railroads are. They both rely on ordinary welded rail. The French rails are laid on concrete crossties and ballasted with crushed stone; most of the Japanese tracks are built on concrete slabs. Power is delivered through an overhead catenary system of conventional design.

Where the high-speed lines depart from usual practices is in the layout of the tracks.

The route is straighter than normal, and curves that can't be avoided are gradual as well as steeply banked. This layout is not meant just to keep the trains from rolling over or derailing; the most stringent limits are set by passenger comfort. At an amusement park, passengers might pay for a ride that throws them violently from side to side, but on the railroad they expect to drink their morning coffee.

The Shinkansen and TGV routes are sealed corridors dedicated exclusively to high-speed passenger service. There are no highway crossings at grade. In most cases the entire route is fenced or walled off to keep out animals and pedestrians.

If you get a look at a TGV train from the side—and if you're able to notice anything before it whooshes out of sight—you'll see the most unusual feature of the design: the arrangement of the wheels under the passenger coaches. Instead of having each car ride on a bogie at each end, the TGV places a single bogie between each pair of cars. This strategy saves weight (the number of bogies is cut almost in half) and reduces aerodynamic drag. Even more important, it improves the stability of the cars and reduces their swaying.

The passenger compartments are slightly pressurized, like the cabin of an airplane, but for a different reason. There's no shortage of oxygen along the rail route, but entering a tun-

nel or passing a train on an adjacent track could cause an abrupt pressure change that might be annoying.

Both the French and the Japanese high-speed trains seem to be commercially successful. Passengers by the hundreds of millions are willing to pay premium fares to ride them. Accordingly, the idea has begun to spread. Germany has built 300 miles of Intercity Express (ICE) high-speed lines, and Spain has plans for a network based on French TGV technology. The French TGV itself may be extended into Switzerland, Belgium, and the Netherlands.

In the United States there have been dozens of high-speed rail proposals, including routes in Florida, Texas, the Chicago area, and on the West Coast. The one place where work has begun in earnest is in the Northeast Corridor between Washington and Boston, but the project there is an upgrading of an old line rather than the construction of a new one. New rolling stock and locomotives run on the existing right-of-way, which has been straightened and regraded here and there, with catenary power extended north from New Haven to Boston. But the track does not approach TGV or Shinkansen standards, and neither do the speeds. The new trains knock about 15 minutes off the travel time from Washington to New York.

industry in North America. There are fewer than 2,000 passenger rail cars operating in the United States, compared with 1.2 million freight cars. About 350 locomotives pull passenger trains, compared with almost 19,000 in freight service.

In any case, the technologies of freight and passenger service are similar. People aren't much different from other cargoes, just fussier. Freight trains and passenger trains run over the same kinds of track, and often over the very same tracks. They respond to the same signals (although the rules are stricter for passenger trains).

One distinguishing feature of passenger trains is that they carry very light loads, as judged by railroad standards. A whole trainload of people, with their suitcases, golf clubs, and presents for the grandkids, weighs less than one loaded coal hopper.

Because of the lighter loads, passenger cars can be built with much softer springs in their suspension than freight cars. As a result, the cars tend to sway from side to side (or sometimes even wallow like a ship in heavy seas) rather than jolt or bounce. If you ride passenger trains frequently, you may well notice differences in their spring rates. You can quantify the differences by counting the number of full side-to-side oscillations in 10 seconds, then dividing by 10 to get the frequency in cycles per second, or hertz. On my last Amtrak trip, the natural frequency of my car seemed to be about 1.2 hertz (that is, I counted 12 cycles in 10 seconds). A New York City subway car, in contrast, swayed at almost twice that rate, about 2.0 hertz.

Another difference between passenger and freight cars is in the couplers. Like tank cars, passenger coaches have couplers with a "shelf" above and below the knuckle so they can't jiggle apart on rough track. They are also machined to fit together more closely, reducing the slack between cars, which makes starting up less jarring.

The Coaster, one of Amtrak's passenger trains along the Pacific coast, has double-decker coaches and a locomotive reminiscent of streamliners from the 1930s. (Under the skin, however, the locomotive is the same F59 that pulls many freight trains.)

Many electrified commuter lines run self-propelled rail cars. There is no separate locomotive; instead, every car has motors in the undercarriage. Every car may also have a cab at each end for controlling the train.

WAITING FOR A SLOW TRAIN

Some closing thoughts: To put it kindly, railroading is a technologically conservative industry. To put it bluntly, railroading is fossilized.

It's remarkable how much of the technology described in this chapter has gone unchanged for 50 years or more. If I had been writing about railroads in the 1940s, I would have discussed welded rail, electric and diesel-electric propulsion, in-cab signal systems, and hump yards with automatic retarders. Automatic block control goes back a full century, and centralized track control was first demonstrated on a working rail line in 1927. It's true, there have been some recent innovations—the end-of-train device, a computer database for tracking freight cars—but mostly one marvels at the longevity of railroad hardware.

And it's not just the equipment that remains unchanged; so does the performance of the railroads. Consider the "Saladbowl Express," a high-priority freight train that carries fresh produce from California to the East Coast in refrigerated boxcars. The train takes a full week to cross the continent from Salinas to New York City, for an average speed of less than 15 miles an hour. Trucks cross the country in half the time.

Through the nineteenth century and on into the 1920s, railroad companies were powerful, innovative, high-growth-rate, glamorous enterprises, a little like computer and biotechnology companies today. The Pennsylvania Railroad ran one of the country's largest industrial research laboratories at Altoona, Pennsylvania. Railroads made the first use of vacuum tubes outside of radio, well ahead of uses in aviation or telephony. The pulse-coded train signaling system was an early precursor of modern digital communication techniques. But the Altoona lab closed long ago, even before the rest of the Pennsylvania Railroad followed it into an oblivion of mergers and bankruptcies.

What follows fossilization, often, is extinction. There was a period when it appeared American railroads might die out altogether. They were losing freight traffic to trucks, and passenger traffic to automobiles and airplanes. The crisis is not over yet, but in recent years the railroads have begun a resurgence. Coal trains shuttling between western strip mines and eastern power plants saved several railroads. Intermodal freight—trains carrying trailers or containers—has also made a difference. And many railroads have found new ways to exploit what has always been their most important asset: their vast land holdings. Running alongside the tracks are power lines, pipelines, and fiber-optic cables; for all of these land uses, the railroad collects revenue.

Still, the recent resurgence in railroading is also a retrenchment. In the 1920s there were 400,000 miles of track in the United States. The total now is less than 150,000 miles; more than half the track in the country has been abandoned.

The state of the American railroad: more than half the track mileage has been abandoned.

BRIDGES AND TUNNELS

ORDINARY ROADS AND RAILROADS smooth our passage over the surface of the earth, but bridges and tunnels create a path where none existed before, either spanning free space or burrowing through the solid earth.

Bridges make connections; they bring people together—a role that has made them a traditional emblem of friendship. Consider the town of Mostar in Bosnia-Herzegovina. When fighting between ethnic factions broke out there in the 1990s, nothing symbolized the social disintegration more clearly than the destruction of a sixteenth-century stone-arch bridge that had linked the two parts of the town on opposite banks of the Neretva River. And the emblem of efforts to heal the divisions is a rebuilt bridge, opened with fireworks and fanfare in July of 2004.

Bridges themselves, in many cases, are looked upon with affection. Whereas many other large engineered structures—refineries, power plants, highways, airports—tend to be seen as a blot on the landscape, bridges are granted an exemption. San Franciscans don't complain that the Golden Gate Bridge is a desecration of the Bay. And Sydney, Australia, considers its Harbour Bridge one of the city's main tourist attractions. Even some tunnels have a bit of romance attached to them, although it tends to focus more on the perilous process of digging the passage rather than on the finished artifact.

BRIDGES

There are 589,685 highway bridges in the United States. I haven't counted them, but the statisticians at the Federal Highway Administration have. They have counted, measured, rated, classified, and tabulated, and 589,685 was the total they came up with in the most recent census, at the end of 2001. Across town at the Federal

From deep beneath the tidal flow
Two granite towers proudly rise
To hold the pendent span aloft—
A harp against the sunset skies.

Each pylon frames, between its shafts,
Twin Gothic portals pierced with blue
And crowned with magic laced design
Of lines and curves that Euclid knew.

It's not the best poetry ever written about the Brooklyn Bridge—the subject has attracted literary notables such as Hart Crane, Marianne Moore, Vladimir Mayakovsky, and Jack Kerouac—but these lines were written by an engineer, David B. Steinman, who designed some of New York's other landmark bridges. Painters, too, have been drawn to the gothic towers and the steel-rope web of the Brooklyn Bridge: It appears in works by Frank Stella, Georgia O'Keefe, and thousands of Sunday afternoon amateurs. And the bridge itself, far more than most industrial artifacts, is taken seriously as a work of art. The photo on the opposite page looks toward the Brooklyn side.

Even small girder bridges built as highway overpasses can make an aesthetic statement. A spare, rectilinear design *(top)* was observed on the Italian autostrada north of Rome. In Maryland between Baltimore and Washington *(middle)*, girders blend into triangular struts. In Arizona north of Phoenix *(bottom)*, a concrete box girder picks up motifs of the desert environment.

At a construction site in western Virginia *(right)*, workers prepare to hoist a girder into place during the widening of a bridge that carries Interstate 77 across the New River. The girder's span between support points is 300 feet.

Railroad Administration no one claims to have an exact count—is it a less obsessive bureaucracy, or just less well funded?—but in 1993 they estimated there were about 100,000 bridges carrying railroad tracks.

When we think of bridges, it is the dramatic and monumental long spans that come to mind first, especially the graceful suspension bridges such as the Golden Gate and the Brooklyn Bridge. But the great majority of those hundreds of thousands of bridges in the United States are not such spectacular structures. They are ordinary overpasses, with spans of 30 or 40 feet, carrying roadways or rails across other thoroughfares or over small streams. You see such bridges by the dozen on any drive down the Interstate. They may be lacking in glamour, but they are most representative of the bridge builder's art.

Girder Bridges. Consider how a three-year-old builds a bridge out of toy blocks: two blocks are set upright, and then a third block is laid horizontally across the top. This is the principle of the girder bridge—although the engineering details of real bridges do extend a bit beyond the preschool level. Often, the role of the upright blocks is taken by reinforced concrete columns, or piers. The toddler's horizontal block is transformed into several long girders called stringers, laid parallel to the axis of the bridge. For a highway overpass, the stringers are likely to be steel I-beams three to four feet deep. Concrete I-beams and box beams are also common. Railroad bridges carry heavier loads than highway bridges, and so the girders are deeper in cross section and spaced more closely.

The deck of a girder bridge is constructed much like the floor of a house. The supporting piers correspond to the building foundation, and the stringers are like floor joists. Laid crosswise to the stringers are smaller beams or planks that correspond to the subfloor of the house, and finally the concrete or macadam road surface is like the finished floor.

When you think about how one of these bridges works, the crucial point is that all the beams and girders are supported at both ends. This means that all the forces from the weight of the bridge (and from the weight of traffic on the bridge) can be transmitted directly downward. If you put a heavy load in the middle of the span, the beams would sag, and eventually they might even break, but the weight would not cause any seesaw-like pivoting or twisting that could tip the bridge over. A structure with this property is said to be statically stable. To a first approximation, the bridge would hold together even if the pieces were just laid on top of each other, with no fasteners. (In practice, of course, fasteners are essential because the bridge has to handle other forces in addition to gravity. The forces come from "live loads"—the traffic moving over the bridge—as well as from the wind and even earthquakes.

One potential failure mode of a girder bridge is familiar to toddler engineers everywhere. If you build a bridge out of blocks and then push horizontally on the top of the deck, parallel to the bridge axis, the structure will collapse by "racking" as the uprights tip over. A defense against racking seen in some tall girder bridges is diagonal bracing between the uprights, but most highway overpasses adopt a simpler solution. Both ends of the bridge are fitted into massive abutments so the deck can't move horizontally.

The engineering and construction of girder bridges are pretty routine these days, but the bridges are not quite standard items you order from the Sears catalogue. The girders, whether of steel or concrete, are custom built for each bridge, then trucked to the site and hoisted into place with a crane. The designer still has scope for variation and creativity, and it shows out on the highway: some overpasses are prettier than others. One advantage of girder bridges over other types is that it's easy to

Not all girder bridges are small and shyly utilitarian. At left the Governor Thomas Johnson Bridge rises 140 feet over the Patuxent River in southern Maryland.

Girder-bridge construction makes use of the same principle over and over: spanning space with a beam supported at both ends. A bridge still under construction *(top)* shows the curvature of the girders, custom-fitted to the site. The "spikes" atop the girders will hold crossbeams supporting the concrete road deck. A similar bridge is seen from the underside *(middle)*; a railroad bridge *(bottom)* needs deeper girders more closely spaced.

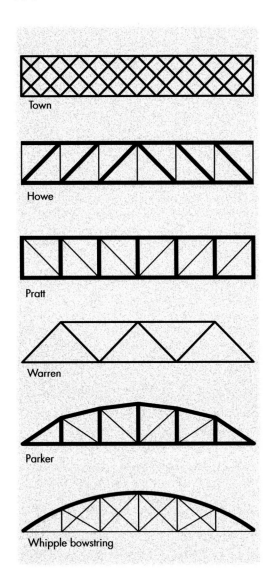

A rough taxonomy of bridge trusses is based mainly on figuring out which elements of the truss act in compression *(heavy lines)* and which are in tension *(fine lines)*. The Town and Warren trusses have elements that can work either way.

design them with a curved or sloping roadbed so they do not interrupt the smooth flow of a highway's alignment.

Truss Bridges. As the span to be crossed gets longer, the stringers in a girder bridge need to be made deeper and stouter, lest they sag too much in the middle. Beyond some critical length, the girders become impractically massive, and it makes sense to start thinking about another kind of bridge. The next candidate is most likely a truss, which you might view as a kind of lacework girder—a structure that's designed to maximize strength and stiffness while minimizing the quantity of material.

If the basic idea of the girder bridge is piling blocks one on top of another, the key idea in a truss is the triangle. Among all the simple geometric figures with straight sides, the triangle has a unique property: natural rigidity. A square can squirm into a diamond shape and then collapse into a flat line, but a triangle can't be bent without breaking something. Bridge trusses are therefore built out of beams, bars, or rods connected in triangular arrangements. A typical truss has two main horizontal elements—called the upper and lower chords—connected by numerous vertical and diagonal elements arranged in rigid triangles.

Bridge-truss designs have been evolving and proliferating for at least two or three centuries. As a result they have a complicated family tree, with several major branches and dozens of minor variations. Learning to recognize and name them is one of those activities, like wine-tasting and bird-watching, where expertise is always threatening to degenerate into mere showing off. But you needn't be a connoisseur of bridges to understand how a truss works. The key is to figure out which elements of the structure are in compression (the ends are being pushed toward each other) and which are in tension (the ends are being pulled apart).

Take a long box or tube of cardboard, lay it across the space between two chairs, and press down in the middle. When the force exceeds the strength of the material, the cardboard will tend to buckle along the upper surface and tear along the bottom. This is a clue that the upper part of the cardboard bridge is under compression and the lower part under tension. Analysis of just this kind is an essential part of truss design: You figure out which parts of the structure will be compressed and which parts stretched. The compression members need to be stiff enough to resist bending or buckling, and so they must be beams of substantial cross section. Tension members can be thin rods or even flexible chains or cables—strength matters in these parts, but not stiffness. The differences between compression and tension components are often conspicuous enough that you can look at a bridge and immediately perceive which are which.

The principles of truss design were first learned in the construction of wood-framed houses and other buildings. The first truss bridges were also made of wood. The form really caught on, however, when iron and steel became economically attractive materials for bridge building. The railroads erected thousands of steel truss bridges, and the basic idea of the steel truss has spread throughout industrial tech-

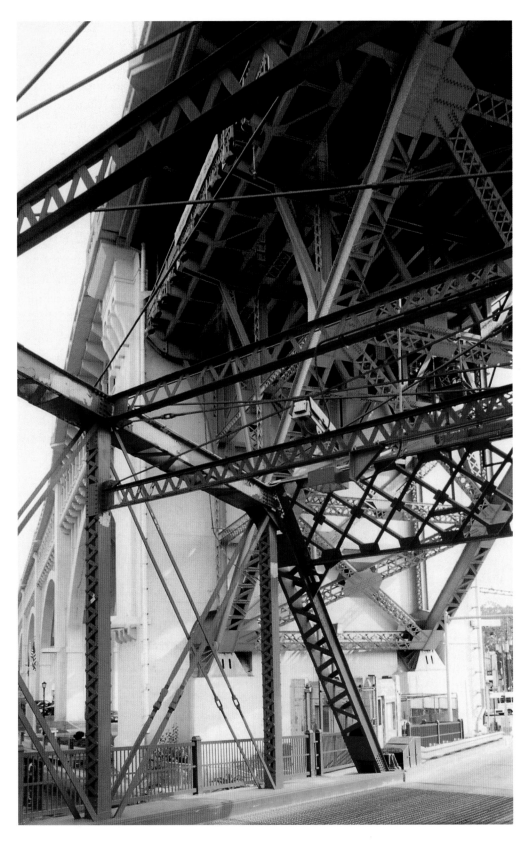

Truss design is a recursive art: Very often the structural members assembled to form the truss are themselves truss-type girders assembled out of still smaller pieces. This principle is visible here both in the red latticework of the Center Street Swing Bridge in the foreground and the gray girders of the larger arch bridge that looms over it—both in downtown Cleveland. The Center Street bridge is a variation on the Pratt truss. The rigid vertical braces operate in compression, whereas the diagonal elements—flexible rods tightened by turn-buckles—carry only tension.

nology. Trusslike structures are conspicuous in many other kinds of bridges and in devices such as cranes.

Here are some notes on a few of the truss types seen in bridges.

The Town truss, invented by Ithiel Town, of New Haven, Connecticut, is a diamond lattice made up entirely of diagonals, all the same size. (Thus, there's no distinction between compression and tension members.) Town received a patent on the design in 1820 and charged royalties for its use based on the length of the bridge— a dollar a foot.

The Howe truss is named for William Howe, a Massachusetts farmer (and uncle of Elias Howe, the sewing-machine guy). It has diagonal compression members and vertical elements in tension. The original design from 1840 called for timber diagonals, with iron rods for the verticals.

The Pratt truss (named for Caleb and Thomas Pratt, father and son, who patented the idea in 1844) is essentially the opposite of the Howe truss: the verticals are in compression and the diagonals in tension. (The doublecross and triplecross variations have diagonals that extend across two or three panels of the truss.)

The Warren truss (the inventor, Captain James Warren, was British) has only diagonal braces, without verticals. The diagonal beams are designed to take both compressive and tensile loads, so they should all be the same size. Although the absence of vertical elements is the definitive trait of a Warren truss, there is also something called a Warren truss with verticals (just to keep you on your toes).

THE OLD COVERED BRIDGE

Why do the back roads of America have so many nineteenth-century covered bridges? Part of the answer is that the bridges were built to last: The whole point of the cover is to keep the wooden framework of the bridge out of the weather, thereby protecting it from rot. But another factor may be even more important: covered bridges have survived because people consider them worth preserving. They are seen as charming and picturesque; they remind us of stories about buggy rides and innocent secret kisses. And so a rescue committee quickly forms whenever a bridge is threatened— whether by physical deterioration or by the bulldozers of the state highway department.

Under the covers, most covered bridges are truss structures. (Among the important early builders were Ithiel Town and William Howe, whose names survive today in the truss designs they patented.) From the inside, you can usually see the truss members and how they are fitted together. Many of the original structures had elaborate joints assembled with wooden

pegs rather than metal nails or screws, but some of that workmanship has been lost in restorations.

The covered bridge was not an American invention, but nowhere was it more popular. Thousands of the bridges were built in New England, down the eastern seaboard, and later in areas of the Midwest. There was plenty of timber in this region, and a wood bridge could be thrown across a river faster and at lower cost than one of masonry. Even after iron became widely available, wood was still the material of choice in many places. The bridge in the photograph, in Philippi, West Virginia, was built in 1852, then modernized in 1938, and restored after a 1989 fire. A rare two-lane covered bridge, it carries heavy traffic on U.S. Highway 250.

Today, old covered bridges are viewed with such fond affection—especially by local tourism boards—that there are not enough of them to go around. And so we're building them again. For example, Ashtabula County, Ohio, boasts of 16 covered bridges, but 4 of them have been built in the past 25 years.

The bridge that carries the California and Arizona Railroad across the Colorado River is a Parker truss, and it is located in Parker, Arizona. Could the town be named after the inventor of the bridge truss? No, it is mere coincidence. The bridge designer was C. H. Parker. The town was laid out by another railroad engineer, Earl H. Parker. But the municipality is named for a third Parker, General Eli Parker.

In all of the truss styles discussed so far, the main horizontal members—called the upper and lower chord—are straight beams. But this geometry is wasteful. The bending stress on the bridge is greatest in the middle but diminishes toward the ends; thus, it makes sense to give the truss its maximum depth and strength only in the center, and to let it taper away near the abutments. This is the idea behind the Parker truss, which has a humped profile, thickest in the middle. As in the Pratt truss, the verticals are compressive and the diagonals are tensile. Many variations on the Parker truss were popular with the railroads.

Another truss that looks a little like the Parker but works in a totally different way is known as the Whipple bowstring truss, after Squire Whipple, yet another Yankee inventor and the first theoretician of bridge-truss design. The bowstring is half arch and half truss. The upper chord is a curved arch held in compression and prevented from spreading out by tension in the lower chord, which also forms the deck of the bridge. Vertical and diagonal braces connect the deck with the arch, all of them working in tension. Some bowstring-truss bridges are beautifully light and lithe, almost like the suspension bridges of a later era.

Apart from this classification based on how a truss is put together, there are also differences in how the truss is arranged with respect to the rest of the bridge. In a deck truss the entire structure is underneath the roadway; in essence, the truss simply replaces the stringers in a girder bridge. A through truss is like a box around the roadway; you drive right through the middle of it. This arrangement allows the truss to be much deeper (and therefore stronger) without reducing clearance under the bridge. It also, incidentally, gives you a better view of the bridge structure as you drive

The cantilever principle is unmistakably at work in the Crni Kal viaduct, seen here under construction in Slovenia. (The viaduct, now completed, carries a highway from the port of Koper to the capital, Ljubljana.) The hollow concrete box girders grow outward in both directions from their supporting columns until they meet. The tallest of the piers is more than 300 feet high.

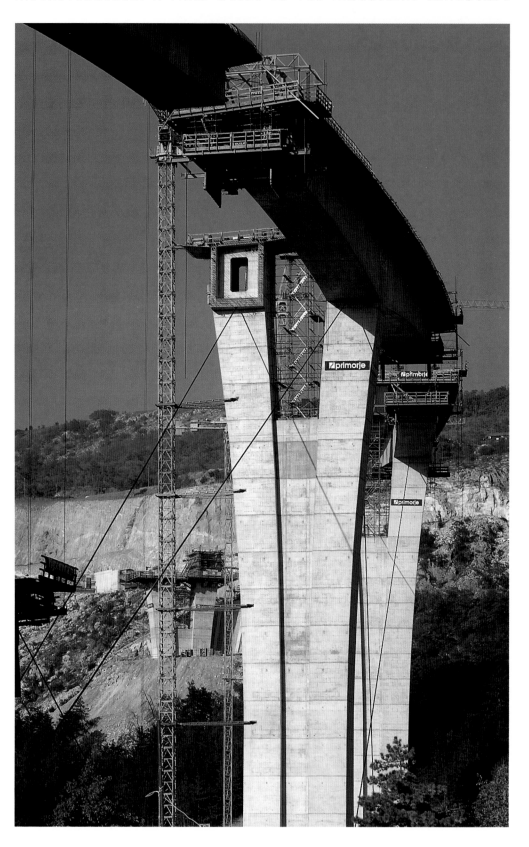

over the bridge—and a correspondingly worse view of the scenery beyond. A pony truss is a through truss without a roof; the truss extends above the roadway on both sides but has no cross bracing to connect the two top chords.

Cantilever Bridges. As the span of a truss bridge increases, the truss has to be built stronger to keep the bridge from sagging in the middle. But a stronger truss is also a heavier one, which causes the bridge to sag more, so the truss has to be made stronger still, which adds even more weight to the load. One way out of this vicious cycle is the cantilever truss, which through some clever geometry makes the center of the span the lightest part of the structure rather than the heaviest.

A girder or a simple truss has to be propped up at both ends. A cantilever is a structural member supported at only one end, like an arm held outstretched; it has to be anchored or balanced somehow so it won't topple over. To return to the world of kindergarten blocks, instead of spanning the gap between two uprights with one long plank, you balance a block on each upright. If you arrange these two balanced blocks to meet in the middle, you have a bridge made up of two cantilevered beams. If you need to extend the cantilevers to span a larger distance, you may need to add counterweights on the opposite ends of the beams, or else pile up mass over the point where each beam rests on its upright anchorage. Both of these techniques are employed in cantilever bridges.

A cantilever truss is easy to recognize. Where an ordinary truss is either uniform in depth throughout its length or has a hump that makes it thickest in the middle, a cantilever truss is thinnest in the middle and usually has its deepest sections directly over the supporting piers. The enlargements at the piers are needed to resist the overturning forces acting on the cantilevered section of the truss. A special property of

The telltale characteristic of a cantilever truss is that it's deepest over the supporting structures and thinnest in the middle (in contrast to a truss supported at both ends, which is often thickest in the middle). Here the twin spans of the Greater New Orleans Bridge—the longest cantilever-truss bridges in the United States—cross the lower Mississippi. The two bridges were built 30 years apart, in 1958 and 1988.

The Queensborough Bridge (a.k.a. the 59th Street Bridge) is the best-known work of Gustav Lindenthal, patriarch of New York bridge builders. The span shown here (there are two more) has cantilever arms reaching out from Queens (on the right) and Roosevelt Island (on the left), shaking hands over the East River.

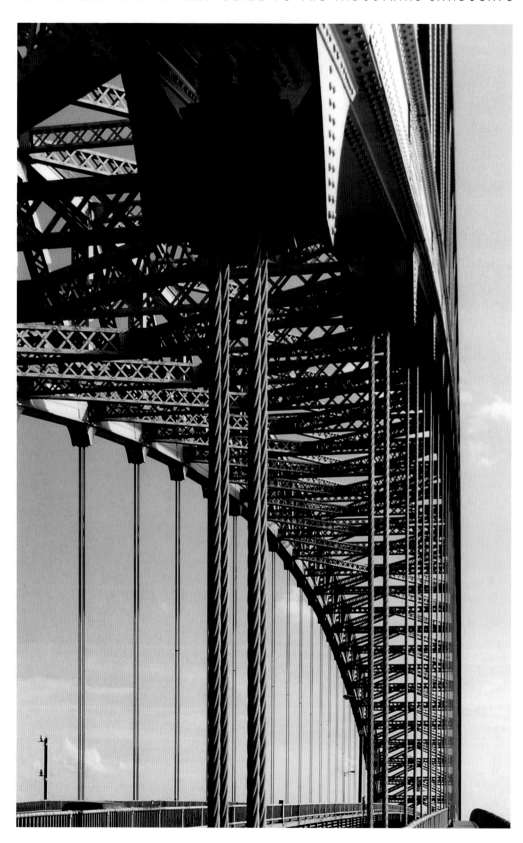

The superstructure of the Bayonne Bridge arcs high above the roadway, which is suspended from the arch by steel cables. The arch may look something like a truss, but its operating principle is different. Instead of relying on stiffness to hold up the bridge, it exploits its geometry to convert vertical forces (the weight of the bridge) into horizontal thrust. The Bayonne Bridge connects Staten Island, New York, with Bayonne, New Jersey, to the north. The span was designed by Othmar Ammann and completed in 1931. At 1,675 feet it is a foot longer than the very similar Sydney Harbour Bridge in Australia, which was built at about the same time.

The tandem steel arches of the Hernando de Soto Bridge carry Interstate 40 across the Mississippi at Memphis, Tennessee. On first glance, this bridge presents a mystery. An arch converts vertical to horizontal thrust, and yet there is nothing but thin air at the ends of these arches to absorb the thrust. Where does it go? The answer is that these are "tied" arches: The road deck, acting in tension, holds the ends of the arches together.

cantilever trusses is that they are stable and self-supporting structures even while they are being built. Typically trusses from opposite sides of a river are extended toward each other piece by piece; when they have almost met in the middle, a short section of ordinary truss is hoisted into place to connect them.

Cantilever trusses set records for bridge spans in the era before the great suspension bridges. Among the cantilevers still standing (and much revered) are the Firth of Forth Bridge near Edinburgh (1890) and the Queensborough Bridge in New York City (1910). In recent decades cantilever bridges built of concrete rather than steel trusses have also become popular, especially in Europe.

Arch Bridges. An arch works on a different principle from a girder or a truss. Instead of transferring loads straight down into the foundations, an arch deflects vertical forces to the sides. The arch is among the oldest of bridge technologies; the Romans were practiced in its construction. It has the distinction that the structure operates entirely under compression; this is an advantage if you are building with materials such as stone or brick, which don't have much strength under tension.

Arches of stone predate the lacy steel structures by at least a few thousand years. And the mere fact that an arch can be built in stone shows that all parts of the arch are held in compression. Or, as the poet Heinrich von Kleist put it: "Why, I thought, does the vault not collapse, though entirely without support? It stands, I replied, because all the stones want to fall down at the same time." Shown here is the Thomas Viaduct in Relay, Maryland, crossing the Patapsco River just south of Baltimore. It was built in the 1830s by Benjamin Latrobe and has been carrying railroad traffic ever since. (The nearby Carrollton Viaduct is a few years older but smaller.)

The Hell Gate Bridge is famous for its obscurity; among major New York City bridges, it's the one nobody can find. It hides out in the upper reaches of the East River, carrying four railroad tracks from Astoria Queens (at the right here) to Randall's Island. In terms of weight per unit length it is probably the heaviest steel arch bridge ever built. Some aesthetic critics suggest it may be heavier than it really needs to be.

Perhaps the lightest and airiest of all steel arch bridges is the New River Gorge bridge, near Fayetteville, West Virginia. When it was built in 1977, it was the longest arch bridge in the world, with a span of 1,700 feet. In 2003 it was surpassed by the Lupu bridge in Shanghai.

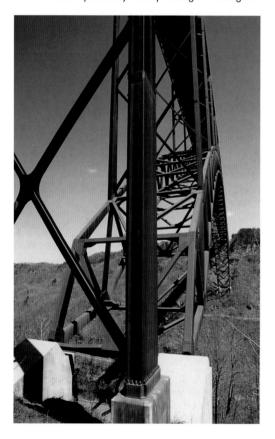

In the playroom you could build a toy arch bridge out of many small, wedge-shaped blocks. But if you try this experiment, you may well discover the two big problems of arch-bridge construction. First, the arch is a stable structure only when it's complete. As you build it up, working from the two ends toward the middle, you need to support all the blocks until the last one (the keystone) is dropped in place to lock the structure together. Second, there is an outward thrust at the base of the arch, which has to be resisted somehow or the arch will spread out and collapse. In bridge-building practice, the usual answer to the first problem is a scaffold called falsework or centering: You build a temporary bridge of timber to support a masonry arch until the masonry can support itself. As for the problem of horizontal thrust, an arch bridge needs massive abutments to push against. If the site of the bridge happens to be a rocky canyon, the vertical walls will serve this purpose. Otherwise, some kind of heavy anchorage will have to be built.

A trick for solving the problem of horizontal thrust is the tied arch, which works like a strung bow: a tension member across the bottom of the arch acts like a bow-string to restrain the ends. A bridge of this kind is generally built as a through arch (or rainbow arch), with the road deck itself serving as the bowstring.

Although the arch began as a trick for making heavy stone leap through the air, the form is so attractive and satisfying that arches are also built of other materials, including such modern stalwarts as steel and prestressed concrete. Whereas the ancient masonry arches were impressive for their mass, many of the modern ones are light and airy. The Sydney Harbour Bridge in Australia and the very similar Bayonne Bridge between Staten Island and New Jersey (two feet longer) are among the most famous examples. The Eads Bridge in St. Louis (the first crossing of the Mississippi

below its confluence with the Missouri) has three steel-and-iron arches that look massive today but were considered radically lightweight when they were built in 1874. And one of the most dramatic and elegant of all arch bridges is the Navaho Bridge over Marble Canyon in northern Arizona.

Suspension Bridges. The biggest load that a bridge ever has to support is itself. And so the key to stretching bridges across ever longer spans is to make the bridge as light as possible to reduce this self-burden. For the longest spans, the bridge design with the lowest weight is the suspension bridge.

A suspension bridge has the admirable property of functional transparency: how it looks is how it works. Tall towers—generally three or four times the height of the road-way—are built on opposite sides of the channel to be crossed. Stout cables are draped across the tops of the towers, then smaller wire ropes, called cable stays, are hung vertically from the main suspension cables. The bridge deck is attached to the cable stays.

The Wheeling Bridge, between West Virginia and Ohio, was the world's first bridge to span more than 1,000 feet and started a notable American tradition in suspension bridges. The first version was completed by Charles Ellet in 1849, but it was partially rebuilt a decade later by John Roebling, who made his fame with the Brooklyn Bridge.

Skeletal steel towers with a busy pattern of X-shaped braces are the distinguishing feature of the George Washington Bridge—but an accidental feature. When construction began in 1927, the plan was to cover the towers with a veneer of granite, but the Depression forced a postponement of this step. As the years went by, many people came to prefer the unadorned steel. The bridge, which crosses the Hudson River from Manhattan to Fort Lee, New Jersey, was designed by Othmar Ammann. The view is from the New Jersey side.

A suspension bridge both looks and works something like an upside-down arch. Whereas an arch is held in compression throughout its length and thrusts outward toward the bridge abutments, the main cables of a suspension bridge are in tension everywhere and pull inward on their anchorages.

The big cables are obviously the most vital part of the whole structure. They are woven in place out of much smaller steel cords, which are looped back and forth over the towers thousands of times. All the strands are carefully bundled together in a way that distributes stresses equally among them. In most cases a metal sheath is wrapped

THE BRIDGES OF THE FIVE BOROUGHS

No city in the world has a richer collection of bridges and tunnels than New York. Some 20 bridges and at least as many tunnels (plus sever-

al ferry lines and an aerial tramway) link Manhattan Island with the rest of the world; the outer boroughs also have some noteworthy specimens of bridge engineering. There are trusses, cantilevers, arches, and eight major suspension bridges. (All that's missing is a modern cable-stayed bridge.) Under the rivers, the ground is riddled with tunnels—enough to carry 18 lanes of automobile traffic and 36 railroad and mass-transit tracks.

Some of the great personalities of American civil engineering worked here. John Roebling and his son Washington, the builders of the Brooklyn Bridge, are probably the best-known figures. But the keystone of New York bridge building was Gustav Lindenthal, an autodidact immigrant from Brno (in what is now Slovakia), who built three of the major crossings. Lindenthal was also mentor to Othmar H. Ammann, who went on to install three monumental bridges of his own in the metropolitan area, and David B. Steinman, who built two more.

Suspense. On a cruise through New York waters, you can sail under a century of suspension-bridge history and evolution. Three early spans cross the East River between Lower Manhattan and Brooklyn. The Brooklyn Bridge *(see photo on page 392)*, with its distinctive gothic towers, was completed in 1883 by the

Roeblings. The Williamsburg Bridge came 20 years later, and the Manhattan Bridge *(see photo at left)* after another decade. Lindenthal supervised construction of the Williamsburg and was the principal engineer of the Manhattan.

Lindenthal's lifelong ambition was to bridge the Hudson River, but it was his assistant Ammann who accomplished this with the George Washington Bridge *(see photo on the opposite page)*.

Around the corner from the George Washington, on the upper reaches of the East River and the adjacent Long Island Sound, are three more suspension bridges: the Triborough, the Whitestone, and the Throgs Neck. For some reason, none of them have inspired the kind of affection lavished on the downtown East River crossings. The Triborough (which includes arch and lift bridges as well as the suspension span) was designed by Steinman.

The Verrazano Narrows Bridge *(see rainy-day photo above)*, which crosses New York Bay between Staten Island and Brooklyn, speaks

again of drama. It was Ammann's last major project. When it opened in 1964, it was the world's longest suspension span, and it remains the longest in the United States.

59th Street Bridge Song. The Queens-borough Bridge *(see photo on page 401)* is a sentimental favorite of New Yorkers. It's not the

biggest or the highest, but maybe it's the funkiest. As the songwriter Paul Simon put it, "Feelin' groovy!" The bridge, a cantilever truss, is another of Lindenthal's designs.

Three Arches. These bridges would be better known if they were not tucked away in corners of the city that few visit.

The Hell Gate Bridge *(see photo on page 404)* is one that millions have crossed but few have noticed; it carries the Northeast Corridor rail line across the East River at Astoria, Queens. The massive steel arch—some complain it is *too* massive—was built by Lindenthal and completed in 1917.

The Bayonne Bridge *(see photos above and on page 402)* is Ammann's homage (or rebuke?) to his master. Like the Hell Gate, it is a steel arch, but it is all spiderwebbery lightness where Lindenthal's bridge is ponderous. (In fairness, the Bayonne carries only highway traffic, whereas the Hell Gate groans under four railroad tracks.)

The Henry Hudson Bridge across Spuyten Duyvil at the northern tip of Manhattan is a steel arch 800 feet long. It is another David Steinman creation.

around the finished cable to prevent corrosion; unfortunately it also conceals the construction of the cable from view.

When the cables are first draped over the towers, they assume the same graceful curve as power wires hung from utility poles—or for that matter a clothesline in the backyard. The curve is known in mathematics as a catenary, a word that comes from *catena*, the Latin for "chain." A suspension bridge is often cited as the archetypal example of a catenary curve, but it turns out that once the bridge deck is hung from the cables, the shape of the curve changes. In case anyone ever asks you, it is closer to a parabola than to a true catenary.

The anchorages where the cables terminate serve the same function as tent pegs; they hold up the bridge. If an anchorage pulled out of the ground or broke like a failing tent peg, at least one of the bridge towers would likely fall inward toward the center of the span, and the bridge would collapse. You'll be reassured to know that the anchorages are massive structures dug deep into bedrock.

Try to take a close look at where the roadbed passes through the towers. In general, the bridge deck does not rest rigidly on the tower, but it may be connected by an arrangement of pivot arms or hinges or even rollers. (You can get the best view of these and other structural details on bridges that allow pedestrians. Some of the biggest and most famous bridges are also among the most accessible in this respect.

THE LEGEND OF GALLOPING GERTY

Like the dog that didn't bark in the Sherlock Holmes story, the bridge that isn't there anymore has a powerful presence in the minds of civil engineers. On November 7, 1940, the Tacoma Narrows suspension bridge, 30 miles south of Seattle, shook itself to pieces. It is certainly not the only bridge that ever failed, but it is the most famous. Part of the reason for its notoriety is that the collapse was caught on film. The best-known images were made by F. Burt Farquharson, a professor of engineering at the University of Washington. His presence at the bridge with a movie camera that morning was not a mere coincidence. He was there to study the alarming undulations that had been noticed during construction of the bridge and from the moment it opened to traffic, four months before. The violent flapping in the wind had earned the bridge the nickname "Galloping Gerty." Still, no one

expected it to crash into the water that day.

The causes of the disaster have been debated ever since. There was a strong wind that day—about 40 miles an hour—but the bridge

should have withstood much worse. One oft-cited culprit is "resonance," where a small force produces large effects by periodically reinforcing natural vibrational motions of the

structure. Another candidate is aerodynamic "flutter," a different kind of self-reinforcing behavior in which the twisting of the bridge, by altering the wind forces on the deck, causes more twisting. A mathematical theory focuses on the periodical tightening and slackening of the vertical cables from which the deck is hung. The one point on which all seem to agree is that the road deck was too light, slender, and flexible. The remedy adopted by designers of future bridges was to build heavier, thicker, and stiffer decks.

The pedestrian in the photo at left reached safety; a cocker spaniel did not.

The lost bridge at Tacoma Narrows was replaced a decade later with a bridge whose deck is braced by a very deep and stiff truss; the engineers were taking no chances.

(Photo credit: University of Washington Libraries, Special Collections, UW20731.)

The Mackinac Bridge *(left)*, which links the two pieces of the state of Michigan, is ranked by some bridge fanciers among the most graceful suspension spans, but others consider the deck truss to be obtrusively thick. The ultra-stiff deck was in part a reaction to the earlier failure of the Tacoma Narrows Bridge *(see box on opposite page)*. The Mackinac Bridge was designed by David B. Steinman. The ivory-and-green color scheme was his choice.

For example, the Brooklyn Bridge and the George Washington in New York and the Golden Gate in California all have walkways.)

Lightness is the great virtue of suspension bridges, but it has also been the main worry about them ever since the Tacoma Narrows Bridge in Washington twisted itself to ribbons in a high wind in 1940. Although the exact mechanism of that failure remains controversial, suspicion has focused mainly on the rather shallow girders of the bridge deck. Long-span suspension bridges built in the decades since have still been lightweight, but the deck has been stiffened by a deep truss, and the aerodynamics of the bridges have been looked at carefully through wind-tunnel studies of scale models or through computer simulations.

The Brooklyn, George Washington, Golden Gate, Mackinac, and Verrazano Narrows Bridges all in turn held the record for the world's longest bridge. More recent claimants of that title have been the Humber Bridge, in Britain, with a span between the towers of 4,626 feet, and the Great Belt East Bridge in Denmark, which was the first to pass the mile-long mark with a span of 5,328 feet. Both of those bridges have been far surpassed by the Akashi Kaikyo Bridge in Japan, which has a center span of 6,529 feet. The Akashi Kaikyo is part of a necklace of 18 ambitious bridges that will link the island of Shikoku with the main Japanese island of Honshu.

Cable-Stayed Bridges. There's a way to build a bridge even lighter than a suspension bridge: eliminate the suspension cables (which can be almost as heavy as the bridge deck they support). Instead of hanging cable stays vertically from the giant draped cables, let the stays fan out directly from the towers to the deck. The modified design is called a cable-stayed bridge.

Ever since the Roeblings built the Brooklyn Bridge, wire rope has been closely associated with the very idea of a suspension bridge, but many earlier bridges used chain rather than rope or cable. In the example below, the chains are more like bicycle chains than anchor chains; they consist of flat metal plates bolted together to form links that can flex only in the vertical plane. The bridge connects the cities of Buda and Pest in Hungary. It was designed by an English engineer, Adam Clark, and built in the 1840s.

If cable-stayed designs save the weight—and the cost—of the thick catenary cables, why aren't all suspension bridges built that way? There's a geometric problem that begins to get serious when the cable-stayed technique is adopted for really long spans. A cable stay supporting the deck near a tower is almost vertical and works much like one of the stays on an ordinary suspension bridge. But as the stays reach out farther from the towers, they become less efficient because they are pulling horizontally as well as vertically. When the angle gets beyond 45 degrees, the stays exert more horizontal force than vertical force and thus tend to buckle the roadway rather than hold it up. As the bridge span increases, keeping the stay angle in a reasonable range requires making the towers taller, but there are costs and technological limits of that strategy too.

Even lighter and more transparent than suspension bridges, cable-stayed bridges are the new passion of civil engineers. This one lies between the gloomy skies and the still waters of the Columbia River at Kennewick, Washington.

Diagonals are dynamic and dramatic in the geometry of cable-stayed bridges. At left, following a model introduced by the Sunshine Parkway Bridge in Tampa, Florida, the Varina-Enon Bridge near Richmond, Virginia, raises bright yellow triangular sails.

In Boston, the towers of the new cable-stayed bridge *(below)* echo the obelisk shape of the nearby Bunker Hill monument. It's not the longest such bridge, but it may be in contention for the longest name: the Leonard P. Zakim Bunker Hill Bridge.

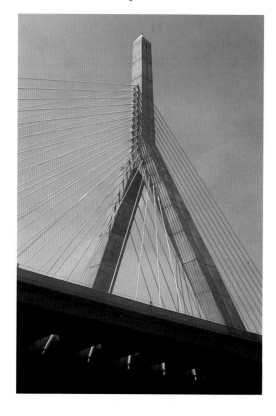

Until recently, cable-stayed bridges were seen as a kind of intermediate technology, filling a gap between the longer arches and cantilevered trusses and the shorter suspension bridges. Not many of them were being built. But engineers have developed a sudden infatuation with cable-stayed bridges, and the design has been chosen for a number of longer spans that would once have been considered the uncontested domain of the suspension bridge. One example is the Hale Boggs Bridge, which carries Interstate 310 across the Mississippi a few miles west of New Orleans. Completed in 1983, it has an unobstructed span of 1,222 feet. There was a special reason for choosing a lightweight design in this location: the soils are so soft that a heavier structure would have sunk into the riverbank.

Another spectacular cable-stayed bridge carries the Sunshine Parkway across Tampa Bay in Florida. The cables are wrapped in bright yellow plastic, so the bridge looks like a pair of glowing sails in the sunset. At least two more bridges of very similar design—right down to the yellow cable wrappers—have since been built.

The Clark Bridge, which crosses the Mississippi at Alton, Illinois, is the longest cable-stayed bridge in the United States at 1,360 feet, but builders elsewhere have gone much farther. In France, the Pont de Normandie is a cable-stayed design with a central span of 2,800 feet. It was built across the mouth of the Seine in 1995. The Tatara Bridge, one of two major cable-stayed bridges in the Japanese Shikoku-Honshu project, has a span of 2,920 feet.

It's worth noting that the Brooklyn Bridge is both a suspension bridge and a cable-stayed bridge. Although the weight of the deck is carried mainly by the four catenary cables, diagonal stays fan out from the towers to add stability. The criss-

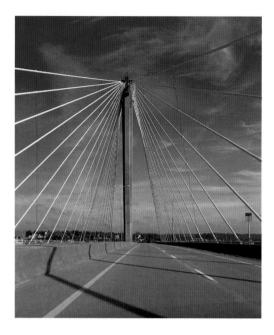

The Clark Bridge, which crosses the Mississippi at Alton, Illinois, north of St. Louis, spreads tentlike stays from central towers to the edges of the roadway. The visual effect is a little like that of the Brooklyn Bridge.

crossing of vertical and diagonal elements is one of the unusual features that has made that grand old bridge so popular with painters, photographers, and sightseers.

Floating Bridges. Most long bridges are humps in the road: you have to drive uphill to reach the level of the deck. Heading east out of Seattle there are two bridges, each more than a mile long, whose approaches are steep declines. They are floating causeways that carry major highways across Lake Washington. The larger of the bridges is supported by 25 concrete pontoons, each 350 feet long, 60 feet wide, and 14 feet deep. Cables and concrete anchors on the lake bottom hold the pontoons in place. A third floating bridge crosses the Hood Canal, west of Seattle. Two of the bridges have telescoping sections that can be retracted to let ships pass through.

The Seattle bridges are not quite unique, but they are rarities. Floating structures are vulnerable to currents, tides, floods, and ice. Sections of two of the Seattle bridges have sunk when some of the pontoons flooded.

Moveable Bridges. A waterway represents a barrier to road traffic, but a bridge obstructs ship movements. The standard compromise is a moveable bridge, which closes to let cars or trains through and opens for ships. There are three main types: the bascule bridge, the lift bridge, and the swing bridge. As machines go, they are fairly simple combinations of levers, gears, and bearings, but building them to survive the battering of both road traffic and the marine environment is a challenge.

The bascule bridge is what most people think of as the archetypal drawbridge. *Bascule* is French for "seesaw," and that pretty well describes how the bridge works. Part of the deck pivots upward, like an alligator's snout or the hood of a car; this represents the rising arm of the seesaw. A counterweight on the far side of the pivot point balances the weight of the deck and thereby reduces the amount of physical work done in opening and closing the bridge. Some bascule bridges have a single moving leaf; on larger bridges a pair of leaves meet in the middle.

Quite a variety of mechanisms can be seen at work in bascule bridges. (When I notice such variety, I always wonder if it means the "right" answer hasn't yet been

Another cable-stayed bridge over the Mississippi, at Burlington, Iowa, has a pronounced asymmetry: the western span (at the left in this photograph) is 40 percent longer than the eastern one. The number of cables is the same on both sides (indeed, they are the same cables, looped over the tower), and so some of them have to be bunched closely together at the eastern end.

At the mouth of the Pascataqua River, a lift bridge opens wide and high for a fishing boat. Note the massive counterweights inside the towers, which descend as the lift span rises. The bridge lies between Portsmouth, New Hampshire, and Kittery, Maine.

found.) Sometimes cables or chains attached to the deck are pulled up into towers, as in a medieval castle drawbridge. Or a large gear, attached to the axle around which the deck pivots, is slowly driven by a smaller pinion gear. The axle gear need not span a full circle, since the bridge is never tilted more than 90 degrees. Other bascule bridges are raised by a hydraulic piston. Almost always the ultimate source of power is electricity. Somewhere near the bridge you're likely to see a diesel generator for emergencies, since a stuck bridge is sure to annoy someone, either on land or on water.

The deck of a bascule bridge is often an open steel grid—which is murder on bicycle tires, not to mention high heels or bare feet. The advantage of steel over conventional paving is lightness.

A lift bridge is essentially an elevator for a section of road or railroad. Twin towers are the characteristic signature of the lift bridge. The deck between the towers is winched straight up out of the way of passing ships. The moving deck is guided by rails or channels and counterbalanced by weights on cables, which you may be able to see descending as the bridge rises.

Lift bridges are particularly common on railroads because they can support very heavy loads and because the problem of maintaining track alignment seemed easier to solve with a lift than with a bascule mechanism. A drawback of the design is that even in the raised position, it restricts the height of ships (whereas an open bascule is open to the sky). Also, if the lift mechanisms on the two ends get out of sync, the bridge can get jammed in an intermediate position that blocks both land and water traffic.

A swing bridge pivots horizontally, like a compass needle. This arrangement dispenses with counterweights and the struggle against gravity, but it opens two narrow

Proud bridges of yesteryear: a rolling bascule bridge in downtown Cleveland points permanently at the sky, leaving the waterway unimpeded but the rail line it once served blocked forever.

A swing bridge pivots like a compass needle to open a passage for boats. This one, seen here open for maintenance rather than for traffic, is on U.S. Route 90 in southern Louisiana.

channels for shipping rather than one wide channel. And the bridge itself is vulnerable to collision damage if a boat pilot miscalculates.

Moveable bridges of all kinds are out of fashion these days with state departments of transportation. The reason lies in the cost of operations, both for maintaining a finicky mechanism and for keeping a worker on duty to open the bridge. These costs come out of the state budget, whereas the capital for building a new, higher, stationary bridge often comes in part from federal funds. Under the circumstances, it's no surprise that drawbridges are disappearing.

The Underbelly of a Bridge. Even bridges that are not meant to move—the stationary ones, with no opening mechanism—do a surprising amount of squirming around. Walking across a bridge, you may be able to feel it flapping in the wind and bouncing under heavy truck traffic. But the most important movements are too slow to be perceived directly: they are the expansion and contraction of the deck as the temperature changes, a kind of daily and seasonal breathing. The structure has to accommodate these motions without breaking or pulling apart.

The most visible means of coping with thermal movements are expansion joints within the roadway. On smaller spans the joints are not much different from the tar strips in a concrete highway, but larger bridges can require more elaborate joints with meshing steel teeth or fingers so the deck can contract by several inches without creating a gap in the roadway.

Underneath the bridge there's more gear for dealing with movement, because the deck has to be able to grow and shrink while the supporting piers or towers or abutments remain still. The contraptions that allow the pieces of the bridge to move without falling apart are called bridge bearings. A tremendous variety of designs have been tried over the years—again, a tipoff that none of them work very well. There are steel plates that slide over each other, steel plates separated by lubricating pads of various kinds (porous bronze, lead, Teflon), "nests" of small steel rollers, single large

rollers, hinged bars that suspend the bridge deck like a porch swing, rack-and-pinion structures, and elastic rubber pads. All of these mechanisms work well enough when the bridge is first built. The trouble is that without regular maintenance, they tend to deteriorate. The moving parts get welded together by rust or clogged with

Between and beneath the piers of a cable-stayed bridge in St. Georges, Delaware, is a cathedral-like space. The bridge's central tower is visible through the gap between two approach roads. The triangular structures nested in the space between roadways are the cable anchorages that support the open-span portion of the bridge deck.

A sliding finger joint in a bridge deck allows the two sides to expand and contract independently without leaving a gap that vehicles or pedestrians might fall into. The joint is on a small bridge connecting Roosevelt Island with Queens in New York City.

dirt, and freeze up. Then something has to give, and cracks begin to appear either in the bridge deck or in the foundation. The rubber pads seem to be the least troublesome and are to be found in many newer bridges. To see any of these devices, you'll need to be under the bridge; look at the joint where the deck rests on the piers.

Many arch bridges have another kind of moveable joint. When the ends of an arch are attached rigidly to the abutments, motion of the bridge cause shearing and torsion stresses in the arch. These unwanted forces can be eliminated by installing a hinge joint at each end of the arch, allowing it to pivot at the abutments. Hinges are common in steel arches; they're harder to build in a masonry structure. The hinge itself looks something like a heavy-duty door hinge, with a steel pin several inches in diameter.

In earthquake country, the joints and bearings of bridges get still more complicated. The bridge deck has to be allowed to move freely over the foundations, but not so far that it falls off when the earth begins shaking. In California all new bridges are equipped with restraints meant to hold the joints together, and older bridges are being retrofitted with earthquake protection. From the underside of the bridge, look for stout steel cables, interlocking sills, and steel collars around concrete piers.

TUNNELS

Tunnels have their romance just as bridges do, but obviously there's less to see. Still, a tunnel is more than just a hole in the ground. Keeping the tunnel ventilated, drained, lighted, and safe from flood and fire calls for a fairly elaborate infrastructure.

Hinges allow an arch bridge to flex under both thermal cycles and changes in load, without causing stress or fatigue where the beams meet the bridge foundation. These hinges support one of a pair of matched arch bridges that take the New York State Thruway across the Mohawk River near Albany.

Bridge bearings make the connection between the girders of a highway overpass and the supporting foundation. A rigid connection between these parts would fail because of thermal expansion-contraction cycles or because of shock and vibrations from traffic. The bearings, which have a synthetic rubber pad inside a metal housing, provide a flexible cushion.

Shallow tunnels on land are usually built by the cut-and-cover technique: dig a trench, put the road or rail line in it, add a roof. Most city subways are excavated this way, and so was much of the recent "Big Dig" in Boston. Occasionally something similar is done for a river-crossing tunnel: a tube is assembled on the surface and sunk into a trench dredged into the river bottom, then it is covered over for protection from ship anchors and the like. But most tunnels are dug from the ends, either by blasting in hard rock or by a rotary boring machine that chews through the ground like an earthworm, grinding the rock at the advancing face of the tunnel and spewing out the digested debris behind it. Sometimes the work has to be done in pressurized air to keep groundwater from infiltrating.

Geometric Design. If you are a highway tunnel engineer, you want to make your tunnels as short as possible because clawing your way through solid earth costs more than any other kind of road building. But there are trade-offs to be made. Shortening a mountain tunnel usually means going higher up the mountain, which makes the approach roads longer or steeper or both. For a river crossing, minimizing the length of the tunnel is even trickier. The shortest routing would put the portal at the water's edge, but then the tunnel would have to descend too steeply to stay beneath the river bottom. Placing the portals farther back flattens the path but requires more excavation. Moreover, the strategy of moving farther back doesn't always work, since the riverbanks may rise steeply. One solution is to build a turn into the tunnel so that part of the descent can be made parallel to the riverbank. The Tyne Tunnel in Britain has a corkscrew ramp at one end, like the ramps that serve multilevel parking decks. The New Jersey approach to the Lincoln Tunnel also spirals into the ground.

For railroad tunnels the constraints on slope are even more severe because trains just don't climb steep hills. As a result, underwater rail crossings are uncommon. Notable exceptions are the tunnels that carry rail lines under the Hudson River and the East River in New York City. Those tunnels are practical only because the railroads remain

The New Jersey portal of the Lincoln Tunnel squeezes traffic into three tubes leading under the Hudson River and into Manhattan. The middle tube opened in 1937, the north (left) tube in 1945, and the south (right) tube in 1957.

underground all the way through Manhattan, and thus trains do not have to climb back to surface elevation until they reach Brooklyn or Queens.

Breathing Space. At many times in the history of tunneling, the limiting factor in going deeper and longer was not the difficulty of penetrating the earth but the challenge of supplying air for people passing through. In the first railroad tunnels, the problem was how to get rid of smoke and heat from coal-burning steam engines. In highway tunnels it's the exhaust fumes from thousands of car and truck engines.

The first tunnel where ventilation was seen as a matter of life and death was the Holland Tunnel, which passes under the Hudson River from Lower Manhattan to Jersey City, New Jersey. When it was built in the 1920s, it not only was the longest underwater highway tunnel (it runs for more than a mile) but also would carry very dense auto and truck traffic. The builders gave a lot of attention to renewing the air.

The tunnel consists of two circular tubes, each carrying two lanes of traffic. When you drive through the tunnel, however, the circular cross section is apparent only at the portals; inside the tubes, the roadway is housed within a flat rectangular box. That leaves space below the road deck and above the ceiling, most of which is used for ventilation. The supply duct runs under the road, pumping fresh air out of vents in the sidewalls at tailpipe height. The fouled air is withdrawn through ceiling vents into an overhead exhaust duct.

The fans for the Holland Tunnel are housed in four ventilation buildings. Two are on land near the portals; the other two are on piers built out into the river.

Drainage. Driving through an underriver tunnel, it's only natural to think about the tons of water flowing over your head and to wonder what it would be like if the tun-

nel were to spring a leak. As it happens, leaks are more of a problem in mountain tunnels, where the rock is fissured and porous. The strata underlying rivers tend to be naturally waterproof. (Think about it: if they weren't, the whole river would leak away!) Nevertheless, all tunnels need a drainage system. It has to handle not only leaks but also rainwater that comes in through the portals. And the largest volumes of water come from washing the tunnel walls. (The walls are scrubbed by a kind of inside-out car wash: a truck with big rotating brushes.)

Drains are installed at the curb, much like storm-water inlets in an ordinary roadway. From the inlets a drain pipe runs downhill to a low point in the tunnel profile, where it empties into a sump and is pumped back up to a discharge near the portals.

Fire. Although it runs counter to intuition, the big hazard of driving under a river is not water but fire. Tunnel fires are rare, but when they happen, they're doozies. The Holland Tunnel had a famous one in 1949, when a truck caught fire deep in one of the tubes. Smoke, fumes, and heat traveled uphill, igniting other vehicles. What eventually contained the blaze was the collapse of the ceiling, which created an oversize exhaust port so that the ventilation system was able to suck out the smoke. There were no deaths, but 66 people were hurt. Fifty years later the Holland Tunnel was fitted with a ceiling deliberately designed to collapse the way the first one did fortuitously.

In 1999, a fire in the Mont Blanc tunnel between Italy and France killed 39 people and closed the route for three years. Now the seven-mile tunnel has 37 emergency shelters, pressurized to keep out smoke.

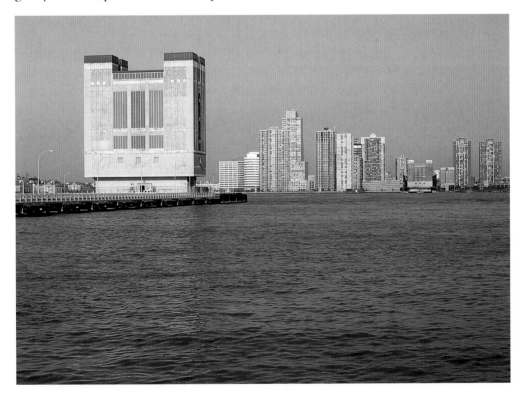

Ventilation towers for the Holland Tunnel, three miles south of the Lincoln Tunnel, are a conspicuous and yet strangely unnoticed part of the cityscape. The brick tower in the foreground, built on a pier reaching into the Hudson from Lower Manhattan, has a twin on the far bank; two more towers are farther inland, close to the tunnel portals. The louvers on the face of the tower are air intakes; exhaust air goes straight up.

THE FRENCH CONNECTION

Underground you go, and when you come up again half an hour later, you're in another country, where they speak a different language, use a different currency, and drive on the wrong side of the road. It's an extraordinary cultural journey (although some might say it's nothing compared to the trip through the Holland Tunnel from New York to New Jersey).

Officially the cross-channel tube is the Eurotunnel, but at the English-speaking end everyone calls it the Chunnel. It is a project with a history. Schemes for tunneling under the Channel go back to the Napoleonic era. By the 1970s, political and technological impediments had been overcome, but economic obstacles still proved insurmountable: the diggers had hardly gotten started when they went broke. The abandoned workings from that effort served as a base of operations when a new consortium started digging in the late 1980s. Eleven boring machines whirred away

through the chalk strata under the seabed, with the main shafts driven from both coasts toward the middle. The French and English teams met at the end of 1990. The first passengers made the undersea crossing in 1994.

The Chunnel route is more than 30 miles long, with three tubes running in parallel: a pair of operating tunnels and a smaller service tunnel that also provides emergency refuge. The link is by rail only; you can't drive through, although you can load your car onto "Le Shuttle" and drive off again on the farther shore. Heavy trucks go through the same way.

The most obvious challenges in such a project have to do with the sheer scale of the earth-moving effort—including where to put the 300 million cubic feet of spoil pulled out of the excavations. The logistics of operating the tunnel are also challenging. At maximum capacity (which, admittedly, has not yet been approached), 30 trains per hour could run through the tunnel in

each direction, and there might be 15,000 people underground at the same time. The trains could produce 100 megawatts of heat—enough to cook all those passengers unless an efficient cooling system intervenes. An early proposal called for dumping shaved ice from the back of trains, but in the end a more conventional chilled-water system was installed.

The worst fear was fire, under circumstances where the nearest help might be more than 10 miles away. In November 1996 the worst fear came true: a fire broke out on a train carrying trucks from France. Following instructions, the train driver tried to continue through to the far side but was forced to stop 10 miles from shore. The crew and the 31 truck drivers aboard the train got out safely, but it took another nine hours to extinguish the fire, and the tunnel was seriously damaged. Among policy changes made in the aftermath: Trains will now stop immediately when smoke is detected.

The Light at the Beginning of the Tunnel. The challenge to vision in tunnels is not that they're dark; people drive in the dark all the time. The problem is one of contrasts. Full daylight is about 4,000 times as bright as a well-lighted highway at night. On entering a tunnel, a driver has to make a transition between those levels in seconds.

The problem zone is not inside the tunnel but on the approach to the entrance portal, where the tunnel ahead appears as a black hole. Often, the first 50 feet or so of the tunnel are lighted much more brightly than the rest. Another strategy is to screen the portal with shades or louvers over the roadway so the outdoor light is dimmed gradually and the eye has a better chance to adapt.

At night the lighting problem is reversed. It's the tunnel exit that becomes a black hole. Again, there are two strategies for amelioration: dim the lights in the last few hundred feet of the tunnel, so the eye will begin adapting, or install brighter-than-normal streetlighting outside the portal.

Traffic Control. Riding through the Lincoln Tunnel as a child, I was fascinated by the guards stationed in glass booths along the narrow elevated walkway that parallels the road. It seemed such a strange and mysterious occupation.

The walkways and glass booths are still there in the Lincoln Tunnel, but they have not been occupied for years. Nowadays the tunnel staff keeps an eye on traffic with

television cameras. The images are displayed in a central control room, which also manages the tunnel infrastructure—ventilation, drainage, electric power, lighting.

In addition to the TV cameras, various sensors report their readings back to the control room. Carbon monoxide sensors provide the primary information needed to adjust the ventilation system. The sensor itself is likely to be out of sight in a utility cabinet, but near the ceiling you may spot a sampling port—a tube with a screen or filter where air is drawn in. A visibility sensor has two units mounted on the ceiling, pointing at each other across the width of the tunnel. An infrared beam bounces between them, its dimming being a measure of the murk in the air.

Air-velocity monitors were once spinning anemometers. Now airspeed is measured with ultrasonic sensors, which detect the Doppler shift in sound frequency caused by air motion. A pair of units is mounted high on the walls on opposite sides of the tunnel, pointed at a 45-degree angle across the roadway.

Perhaps the most important sensors in the tunnel are inductive-loop vehicle detectors buried in the pavement, which monitor the motion of the cars passing over them. These sensors work just like the ones that trigger traffic lights. The loop sensors are watched carefully because an accident or a disabled vehicle will halt traffic in at least one lane.

For all this surveillance to do any good, the control crew needs a way to communicate with drivers, especially in an emergency. Overhead lane markers (showing a red X or a green arrow) provide a means of communication with a pretty minimal vocabulary. Many tunnels now have more versatile signboards. Ordinarily they say something bland like "Stay in Lane," or "Maintain 45 MPH," but they can be reprogrammed instantly to warn "Accident Ahead" or even "Evacuate!"

The transition zone at the entrance to a tunnel *(above)* can become a black hole. At the Gran Sasso tunnel in central Italy, brighter lights are installed over the first 50 or 100 feet, and the walls are painted white. The white hole at the end of the tunnel *(below)* can be just as much an impediment to vision. This is the Tuscarora tunnel on the Pennsylvania Turnpike.

Another way of communicating with drivers is by radio. In some tunnels the car radio doesn't work at all; this should not come as a surprise because radio waves cannot penetrate far into the earth or water. In other tunnels, however, broadcast signals come through loud and clear. How does that happen? It's not an accident. The tunnel operators have put up an antenna to receive both the broadcast signals, which they retransmit through small antennas mounted in the tunnel ceiling. They don't go to all that trouble just to keep you entertained. They also have the ability to interrupt the broadcast on all frequencies with emergency announcements.

Drivers tend to be very cautious in tunnels, which keeps the accident rate low but also makes it a struggle to keep traffic moving. On the downgrade entering an underwater tunnel, a driver touches the brakes; the drivers behind overreact and slow down further. At the other end of the tunnel, drivers let their speed slack off on the upgrade. As cars bunch closer together, prudent drivers slow down still more. The result is a spontaneously generated traffic jam—the road is clogged for no good reason. These traffic instabilities also develop on surface roads, but they are particularly severe and long-lived in tunnels. The strategy for dealing with them is simply to stop traffic altogether at the entrance portal, not releasing a new "platoon" of cars until they can get through the entire tunnel without reaching the trailing end of the jam.

AVIATION

A S THE PILOT MANEUVERS into position for final approach, the aircraft banks to the right and you get your first clear view of the airport below. Lights leading up to the runway threshold flash in sequence like a rapid-fire movie marquee. Crossbars of bright red and white lights draw the eye toward the touchdown zone. The runway itself is outlined in multicolored lamps, with another row of lights down the center. As the airplane descends, you notice the rotating beacon atop the control tower, flashing alternately white and green. Once the aircraft is on the ground there are still more lights—a forest of dim blue ones mapping out the taxiways. What are all these lights and signals? And what about the stripes, bars, chevrons, numbers, and other markings painted on the pavement? What do they all mean?

Aviation breaks free of the earth—that's what it's all about—and yet it has left large marks on the landscape. Major airports are highly conspicuous; some of them are as big as cities, with populations to match. Away from the airport are other telltale signs of the airplane's influence, such as the flashing beacon lights and the red-and-white stripes or checkerboard patterns on water towers and the tall masts of radio and television transmitters. Another important part of the aviation infrastructure is almost invisible: the network of air routes that most aircraft follow from city to city. Even though the air routes themselves are out of reach far overhead, you may come upon some of the navigational beacons that act as signposts along the way.

THE AIRPORT

Airports are not among our best-beloved public places. Railroad terminals get preserved as historic landmarks even as the passenger railroads themselves wither away, but airports are at best tolerated as a necessary evil to be hurried through on the way

Like the minaret of a mosque, the control tower soars above the vaulted roofline of Reagan National Airport, just outside Washington, D.C. The tower and terminal were designed by architect Cesar Pelli. Controllers in the windowed "cab" at the top of the tower direct aircraft on the ground and while taking off or about to land; other flight operations are handled from a regional center. The "hat" atop the tower covers a radar antenna used to monitor aircraft movements on the ground. Since this photograph was made, that radar has been moved to another part of the airport.

New terminals at Charles de Gaulle airport in Paris reflect the "inside out" structure favored in many recent airport designs. Passengers arriving by road or by rail are deposited in the middle of the complex and then fan out toward terminals and gates, which in turn are surrounded by taxiways and runways. Seen here in a photograph made in September 2003 are terminal 2F (in the foreground) and 2E (in the background). On a Sunday morning in May 2004 a section of the roof of terminal 2E collapsed, killing four travelers.

from here to there. Passengers complain of the inhuman scale, the sterile architecture, the unwelcoming environment. It will be interesting to see whether the disdain for airports persists in generations to come. Who knows—maybe 50 years from now there will be a popular movement and fund-raising campaign to save O'Hare International from the wrecking ball.

Airport terminals have undergone a surprisingly complicated evolution since commercial air travel began in the 1930s. The prototypical passenger terminal was just a building that served as the buffer between ground transport and air transport. You drove up to the ground side of the terminal building, bought your ticket and checked your bags, then you walked through to the air side, going out the back door, across the tarmac, and up the stairs into a waiting airliner. At the end of the flight you passed through a similar building in the opposite direction.

As the volume of passenger traffic increased, terminals had to grow more elaborate. On the land side, the building is often split into two levels, separating arriving and departing passengers. The upper floor usually has the ticket counters; the lower floor, the baggage-claim carousels. On the air side, the main problem is finding space to park more and more aircraft, which grow larger and larger. As a rule of thumb, a parking space for a wide-body four-engine jet takes up 150,000 square feet—which would make a fair-sized lot for a suburban house. In response to the demand for more space on the air side, many terminals grew various protuberances and excrescences—long finger piers, Y-shaped or T-shaped branching structures, bulbous loops

on the ends of piers, and isolated satellite terminals reached by buses or underground passages. And passengers learned that to fly from New York to Chicago you have to walk as far as Cleveland.

The next major transformation of the airport was prompted by the spate of airplane hijackings and bombings in the 1960s and 1970s; it intensified after September 11, 2001. Security screening created a new set of zones in the passenger terminal. Now you pass not only from the land side to the air side, but also from the nonsecured to the secured zone. The secured zone is a lot like the sterile field of a surgical operating room: everything "outside" is officially classified as unsafe until it has passed through the metal detector or the x-ray machine at the security checkpoint; inside, everything is assumed to be safe. Note that the secured zone includes the interior of the aircraft, which means that the zone extends across whole continents. When you are cleared to board an airplane in Miami, you are still considered safe, and therefore you are admitted directly to the secured zone when you get off in Seattle. But if you stray a few steps back to the land side of the terminal, you are contaminated and must be inspected again for readmission.

Standard airport security practices actually create three zones. The outside world, where nothing can be trusted and where anyone may go, extends into the ground-side areas of the terminal as far as the security checkpoint. The secured zone, where all persons and carry-on bags have been examined, includes the air side of the terminal and the interior spaces of the gates and jetways and the aircraft. The third zone is the part of the airport outside the passenger lounges—the aprons, taxiways, run-

GETTING A LOOK

Once upon a time, my father would take me to the airport on Saturday afternoon. We weren't flying anywhere; we just parked the car at the end of the runway and watched the planes roaring overhead as they landed and took off.

Times have changed. Several years ago our spot near the runway threshold was posted with No Parking signs, and eventually the entire road around the airport perimeter was closed to traffic. (Another change is that ten-year-olds these days probably don't consider a trip to the airport as proper Saturday-afternoon entertainment.)

More than any other realm of daily life, aviation has been transformed by worries over security and the threat of terrorism. The airport is definitely not the place for casual trespass-

ing. Just opening the wrong door could land you in jail.

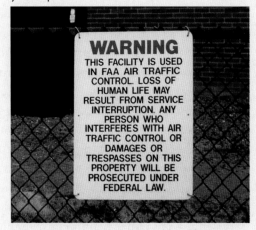

On the other hand, the public areas of airports remain very public indeed. Thousands of

travelers pass through them every day. Much of the aviation infrastructure is readily seen from the departure lounge, or from a window seat on any flight, or even from the upper levels of the parking deck. A few airports still have observation areas open to the public.

Navigational aids installed away from the airport are always fenced off and bear the kind of stern warning sign reproduced at left. Unlike most such signs, it not only tells you to keep out but also explains why it's important that you do so. The reasons are certainly good ones, but if you stay outside the fence, you'll do no harm—and come to no harm. (Note, however, that some countries consider air-traffic-control hardware to be militarily sensitive, and they frown on picture taking.)

Atlanta's Hartsfield Airport is the prototypical design for the age of hub-and-spoke airline operations. Ground-side facilities are at the far left, surrounded by parking lots and decks. Most of the boarding gates are in the five concourses in the middle of the field, which passengers can reach only via an underground rail link. Eliminating automotive connections to these concourses allows aircraft free access to all sides of them, maximizing the number of gates in a given area. The layout is ideal for passengers making connections between flights; it is less than optimal for those whose trips start or end in Atlanta. The airport has four parallel runways, which can all be active at the same time. In this aerial photograph, made as part of the U.S. Geological Survey Urban Areas series, the runways and taxiways form an elaborately perforated gasket-like pattern.

ways, and other areas with access to the exterior of airplanes, as well as to luggage holds. These places are off limits to everyone but employees with the right kind of identifying badge. (International terminals with customs and immigration inspections have an even more elaborate division into zones.)

Airport architecture has also been altered by the shift to hub-and-spoke operations by most scheduled airlines. In decades past, an airline that served 10 cities might have flown most of the 90 possible routes connecting pairs of those cities. Designating one of the cities a hub and flying only between the hub and each of the spoke cities reduces the number of routes to 9. This strategy simplifies operations for the airline, but it also means that most trips require two flights, with a change of planes at the hub. Accordingly, new demands are made on the hub airport. At a spoke airport, most passengers are either beginning or ending their journey, and the main imperative is to minimize the distance from the gate to the parking lot or the taxi stand. At hubs, many passengers are merely changing planes, and they never use the land-side facilities. For them the highest priority is to minimize the distance between gates.

Designers have responded to this new traffic pattern with a wholly new style of airport, one that would have seemed quite improbable 20 years ago. The new hub airports divorce the land side from the air side, isolating the two functions in separate buildings as much as a mile apart. This strategy allows cars free access to the entire perimeter of the land-side building, while aircraft can taxi up to gates all around the air-side concourses. Moreover, putting the air-side building out in the middle of the landing field reduces taxi distances to and from the runway, allowing quicker turnarounds for connecting passengers. The price is paid by those passengers whose trip begins or ends at the hub airport: they have a long trek between the land side and the air side. In the United States the two main examples of this design are Hartsfield-Jackson International Airport in Atlanta and the new Denver International Airport. In both cases an automated underground rail line carries passengers from the land-side building to the isolated gate concourses out in the middle of the field.

The largest metropolitan airports are hubs of economic activity as well as transportation. They employ tens of thousands, and at a busy hour their population can be that of a medium-size city. Some fraction of the purchase price of every airline ticket goes to support the airport's operation through landing fees and rents charged to the airlines. But this revenue stream is generously supplemented—and sometimes exceeded—by income from snack bars, newsstands, souvenir shops, and most of all from automobile parking. From an economic point of view, an airport is a device for persuading people to pay $50 to leave their car over the weekend or $5 for a warmed-over slice of pizza. It's no accident that some of the companies that build shopping malls are expanding into airport management, and vice versa.

THE TERMINAL APRON

Standing at the gate, you look out the window at the aircraft you're about to board. It stands on the terminal apron, the area where aircraft park while loading, unloading, and preparing for their next flight.

Parking a vehicle that's 200 feet long and almost as wide presents a considerable challenge. On the pavement you may see painted marks indicating the correct position for the nosewheel. Often there are several marks, labeled for different aircraft—Boeing 727, 737, Airbus 300, and so on. However, the pilot steering the airplane cannot see any of these marks when they are needed most because the nosewheel is under and behind the cockpit. Usually, the pilot follows the signals of a worker with flashlights or bright orange paddles. Some airports have a system of marker signs mounted on the wall of the terminal building that guide the pilot directly. Each sign is split into two pieces, with one piece mounted a few feet farther from the wall than the other piece. The two parts align correctly only when seen from the cockpit of a properly parked aircraft. Another system uses lights that change from red to green as the aircraft pulls into position.

At most airports the first item connected to an arriving aircraft is the telescoping jetway—the passage you walk through when you enter or leave the airplane. This strange articulated appendage is part building, part vehicle, and part bridge. It can nod up and down, extend and retract, sweep from side to side, and in many cases pivot at elbow-like and wrist-like joints. All of these motions are controlled with a joystick from a station next to where passengers board the plane.

Other connections to the parked aircraft are for utilities (fuel, water, electric power, air conditioning) and deliveries (luggage, meals).

Fuel. At most large airports today, fuel is delivered by a hydrant system. The driver of a small truck opens up an access panel in the pavement and connects a hose from the truck to a fitting in an underground vault; a second hose goes from the truck to a socket on the underside of the airplane wing. Actually, plugging in the hoses is not the first thing the driver does; before making any fuel connections, the driver clips on grounding wires that run from the airplane to the truck and to the underground vault to prevent sparks from static electricity. Once all the connections are double-checked, a valve is opened and fuel flows under pressure into the wing tanks.

The truck doesn't have to pump the fuel; the pressure is supplied by large, stationary pumps at the main fuel depot, which is generally on the periphery of the airport grounds. The machinery on the truck is mainly for filtering out air, water, or other contaminants in the fuel. The truck also measures the quantity of fuel delivered so that the airline can be billed. A few airports have filters and meters in each underground vault. This eliminates the need for the hydrant truck; the refueler just pulls a hose out of the vault and plugs it in.

Some airports have no underground fuel-distribution system at all. The fuel is delivered by tanker trucks—sometimes called bowsers—much like the trucks that bring gasoline to the corner gas station or fuel oil to homes. But the airport fuel trucks are larger, carrying as much as 8,000 gallons each. That makes them too big for the public roads, and yet they still can't carry nearly enough to quench the thirst of the biggest jetliners. A Boeing 747 holds roughly 50,000 gallons of fuel; in the long-range 747-400 model, the weight of the fuel is roughly equal to that of the airplane itself—more than 350,000 pounds. The refueling rate can reach 2,000 gallons per minute. If you had to fill up the tanks of a 747 from a gas-station pump, it would take the better part of a week.

Commercial jet aircraft burn a fuel called Jet-A, which is essentially kerosene and is similar to home-heating oil and diesel fuel. Piston-engined propeller planes burn high-octane gasoline (called avgas).

Electric Power. The heavy black cable that looks like a big extension cord is a big extension cord. It's usually draped along the underside of the jetway and plugs into a socket near the nose or on the underbelly of the airplane. If power isn't available from the terminal, the airline may roll up a generator cart that plugs into the same

The apron—the area just outside the terminal building where aircraft are serviced between flights—is the busiest part of an airport. At left, three aircraft nuzzle up to telescoping jetways at Miami International. On the opposite page, aircraft are guided and towed to parking places on the apron, and are connected to vehicles supplying air conditioning, fuel, and food. Below, deicing operations are seen from the point of view of one being deiced.

outlet. All larger aircraft have an on-board source of electricity—it's a small jet engine running an electric generator—but getting juice from the ground saves fuel and cuts down on noise.

Air, Water, Food, Baggage. A fat floppy hose about a foot in diameter, looking like a grown-up version of the hose that connects to the back of your clothes drier, carries chilled air to the plane and thereby takes the load off on-board air-conditioning equipment. As with the electricity, the chilled air could come from a source in the terminal or from a mobile cart.

Water comes in a tanker truck, often prominently marked "Potable Water." Perhaps that's to distinguish it from the *other* tanker truck, which pumps out the sewage.

The catering truck is a van whose body is mounted on a hydraulic lift, like an oversize version of the jack you would use for fixing a flat tire. The truck rises so that carts of meal trays can be rolled directly onto and off of the passenger deck.

For wide-body aircraft, baggage trucks use a similar lift mechanism. The bags are preloaded into aluminum or fiberglass containers, which are then hoisted up into the belly of the airplane. On smaller aircraft, bags are loaded individually, using an inclined conveyor belt. The bags are hauled around in trains of rubber-tired carts.

What's most notable about all these trucks and other vehicles that buzz around tending the airplane is their specialization: they would have no use off the airport grounds.

Tugs and Tractors. Although jet airplanes have thrust reversers, they are used only to slow the airplane right after touchdown; they can't be used as a reverse gear for backing out of a parking place. Thus, airplanes usually have to be pushed away from the gate. The machine that does the pushing is a low-slung tug or tractor with a pole

that links onto the nosewheel strut. Keeping enough tugs operating is a critical requirement of airport management, as a shortage can delay flight after flight.

Deicing. Accumulations of snow and ice can crush buildings and topple trees, so it's no surprise that they can also keep aircraft on the ground. (The problem isn't just the weight; a layer of ice can also change the shape and hence the aerodynamics of a wing.) At many airports deicing is done at the gate by workers in bucket-lift trucks. They spray warm antifreeze to wash the ice off. The trouble with this system is that the wings can gather a new coat of ice by the time an airplane taxies out to the runway; then it has to come back to the gate to be deiced again, a process that obviously could go on indefinitely. Some airports where icing is a frequent problem have a permanent deicing pad, something like an automated car wash, out near the departure end of the runway. Doing all the deicing in one spot also makes it easier to recover and recycle the antifreeze.

THE AIRFIELD

Beyond the gate and the terminal apron is the realm of runways and taxiways. Out in these wide-open spaces, aircraft enter their own element.

Runways. The runway is where the rubber *really* meets the road; it is the exact point of interface between air and ground transport. At first glance a runway looks just like a road—but it's a road writ large. A fully loaded Boeing 747 can weigh as much as 750,000 pounds (more than 10 times the weight of a big tractor-trailer rig), and it lumbers down the runway at 130 miles per hour or more. A surface built to highway specifications would not hold up for long under such treatment. Accordingly, the concrete slab of a runway may be as much as two feet thick, laid atop a carefully prepared foundation. Sometimes a thin layer of asphalt is laid down over the concrete to improve traction. (People are always talking about "the tarmac" in connection with airport paving, but actually there is very little of the tar macadam that properly goes by that name.)

A runway is not only thicker than a highway but also wider—up to 200 feet, or the equivalent of about 15 lanes of automobile traffic. All in all, a heavy-duty runway two miles long consumes enough concrete to build 60 miles of two-lane roadway.

The length of a runway depends mainly on the kind of aircraft an airport serves; obviously, big jets need more room than small private planes. The critical factor is take-off rather than landing, because an airplane needs more room to gather speed than it does to stop. Other factors also enter the equation: high elevation and high temperature both call for a longer takeoff roll. The reason is that heat and altitude reduce the density of the air, which means they also reduce the lift generated by an airplane's wings. The difference can be dramatic. A Boeing 727 that needs 4,000 feet

of runway at sea level on a cool day might require an 8,000-foot takeoff roll at an elevation of 5,000 feet in 100-degree weather. On an unusually hot day airlines may have to reduce load in order to take off safely, with the annoying result that your luggage may not get out of Denver with you.

Most airports that handle jet aircraft have at least one runway of 6,000 feet or more. For "heavy" jets, such as the Boeing 747 and the Airbus A-320, lengths of 8,000 to 10,000 feet are common. John F. Kennedy International Airport in New York has a runway 14,500 feet long, or nearly three miles. And there is an even longer runway at Doha International Airport in Qatar, where summer temperatures commonly reach 120 degrees Fahrenheit.

How can you determine the length of a runway? From the air, look for sets of white stripes, laid out in pairs on either side of the runway centerline. Typically there are single, double, and triple stripes, as well as broad white bars that mark the touchdown zone (the point that pilots are supposed to aim for when landing). Although there is some variation in the details of these markings, the distance from one set of stripes to the next is always 500 feet, and so you can use them as a ruler to estimate the total length.

Aircraft spend very little time on runways—takeoff and landing are the briefest phases of a flight—and yet those moments of transition between airborne and earthbound status are obviously crucial. On the opposite page, a U.S. Geological Survey aerial photo shows the markings on runways 34L and 34R at Fort Worth Alliance airport in Texas. The "34" designation means the runways are oriented with a magnetic bearing of approximately 340 degrees, or just west of due north. The stripe patterns are at intervals of 500 feet and thus can be used to estimate the length of the runway. At left, a runway and a taxiway are seen in the dawn light at Phoenix Sky Harbor Airport. A consistent marking scheme—white striping on runways, yellow on taxiways—helps avoid potential confusion.

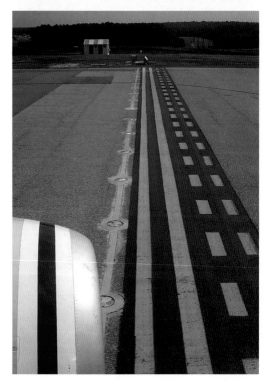

Red-and-white signs mark numbered runways; black-and-yellow signs point the way to lettered taxiways. At the bottom is a hold-short line, marking a point where aircraft must wait for clearance before entering or crossing a runway.

On the ground, during the takeoff roll, what matters most is not the total length of the runway but how much you have left ahead of you. For white-knuckle flyers who worry about such things, major airports have made the countdown easier. There are placards along the edge of the runway with the numbers 8, 7, 6, 5, and so on, marking off the remaining distance in thousands of feet. The placards have large white numbers on a black background. It's just like the Cape Canaveral countdown, but you want to reach liftoff *before* you get to 0.

Another runway marking is an identifying number, painted in numerals 20 feet high. The number represents the approximate magnetic compass direction of the runway, but there's a trick to reading it: the direction is rounded to the nearest 10 degrees and then the final zero is dropped. Thus, Runway 5 points toward 50 degrees, which is roughly northeast; Runway 18 has a bearing of 180 degrees, or south. Where an airport has two parallel runways, they are marked *L* and *R* for left and right. Runways can be used in either direction, so that Runway 9*R*, pointing east, would be marked at the opposite end as Runway 27*L*, heading west. Pilots learn to mentally calculate a "reciprocal bearing" to match up runways going in opposite directions. The arithmetic is simple: if the bearing is 18 or less, add 18; otherwise subtract 18. Because the bearings given are magnetic, they do not correspond exactly to geographic directions, although throughout the lower 48 states the difference between magnetic north and true north is no more than 20 degrees.

The centerline of a runway is marked with a dashed white line, and the edges of the paved area are outlined in solid white. These markings are similar to those of a highway, but they have a different meaning. They do *not* divide the runway into two lanes going in opposite directions; on the contrary, the pilot is expected to drive down the centerline, which would be unwise in an automobile.

Paved runway shoulders and paved areas beyond the runway threshold are painted in bold diagonal stripes or in a herringbone pattern. These are places where the pavement is not strong enough to bear the full weight of an aircraft, and the stripes warn pilots not to venture onto them. Why pave such places if you can't drive an airplane on them? The main reason is to prevent scouring of the soil by the blast from jet engines. There are various other strategies for dealing with the jet-blast problem, such as covering the area with stones or building a rampart of jagged concrete blocks. At the foot of a runway you may see a blast deflector, which looks like a giant immobilized snowplow, with a curved surface that directs the jet blast upward.

Taxiways. The taxiways that lead an aircraft to and from the runway are like frontage roads along a freeway. And as with freeways, some airports have high-speed exits that veer away from a runway at a shallow angle, allowing aircraft to clear the runway before they have slowed down to the usual taxiing speed of 20 or 30 miles an hour.

Taxiways are narrower than runways but are otherwise similar in construction and appearance. Indeed, one of the hazards of airport operation is that a confused pilot will try to take off from or land on a taxiway. To help dispel such confusion, all mark-

ings on taxiways are yellow, unlike the white markings on runways. Moreover, the taxiway centerline is not dashed, and the edges are marked by double yellow lines.

Whereas runways are numbered, taxiways are lettered. (Pilots and controllers pronounce the letters "alfa," "bravo," and so on, in the international phonetic alphabet.) Runway signs have white letters on a red background; taxiway signs are black and yellow.

Where a taxiway meets or crosses a runway, there is a "hold line": two solid stripes and two dashed stripes painted across the taxiway 100 feet short of the runway. Aircraft stop at the hold line until the control tower gives the pilot clearance. At various points around the airport you might also notice a hold line and a red-and-white sign with the legend "ILS." These mark positions where an aircraft would block the transmissions of an instrument landing system.

Layout of the Airfield. The mapmaker's symbol for an airport is three runways crossing in a triangular configuration. This design was once favored because it gives pilots six choices in selecting a runway aligned with the wind. (Both takeoff and landing are best done heading into the wind.) If the triangle is an equilateral one, there must always be a runway that deviates no more than 30 degrees from the wind direction.

The drawback of the triangular geometry is that only one runway can be active at a time. That limits the capacity of the airport during peak periods, as a single runway can accommodate only about 30 takeoffs or landings per hour even under the best of circumstances. To boost capacity, airport designers have turned to layouts with multiple parallel runways, spaced far enough apart that they can be in use simultaneously. One popular scheme puts the terminal in the middle, with parallel runways on either side.

A feature of parallel runways is that while coming in on final approach, you can sometimes look out the window and see another airplane performing the same maneuvers as your own. It's like having a mirror in the sky.

An airport's main parallel runways are lined up with the local prevailing winds. If space permits, an additional pair of cross runways can be built for those days when the wind shifts out of its usual quarter.

Airport Grading and Drainage. Airports tend to be *very* flat. The earthmoving needed to level the ground can be a major item in the construction budget, but there is not much choice about this investment. For each 1 percent increase in the slope of a runway, the length of the runway has to be increased by 10 percent for jet aircraft and by 20 percent for piston-engine aircraft. Extending the runway would cost even more than flattening the ground. This is one reason so many airports are built on marshes, meadows, and dry lake beds.

With vast expanses of flat, paved land, airports often have serious drainage problems. Runways and taxiways are crowned to help shed water, and often they are grooved to prevent hydroplaning, just as some highways are. Water drains into shal-

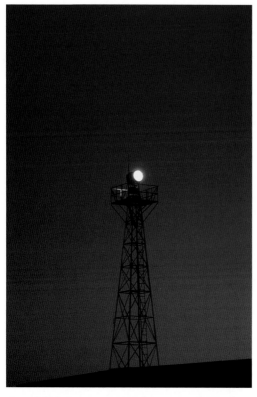

One of the longest-running traditions in aviation is the rotating airport beacon light, green on one side and white on the other. The one above beckons pilots in Hays, Kansas.

The classic triangular layout allows pilots plenty of options when choosing a runway headed into the wind, but only one of the runways can be active at a time. The landing field seen below is Flagler County Airport in Bunnell, Florida.

low ditches or collecting ponds and then is carried away either by open channels or by an underground storm-sewer system. The inlets to these sewers are not set at the curbline, as they would be on a city street, but are well back from the runway.

More Airfield Doodads. Here are a few more items you might spot as your plane taxies out to the departure runway.

The fuel depot. For a big airport, a month's supply of fuel is close to 100 million gallons, which means there will be a sizable tank farm. Finding the best place to put the tanks is a delicate problem. On the one hand, they should be as close to the terminal as possible to minimize pipeline and pumping costs; on the other hand, storing 100 million gallons of flammable hydrocarbons under the path of arriving and departing flights is not ideal.

The weather station. Pilots, like sailors, keep a close eye on the weather, and often the official weather-reporting site for a city is at the airport. The instruments are installed in a grassy area (the National Weather Service requires it) somewhere well away from jet blast. The standard package includes a thermometer, a barometer, and a hygrometer (for measuring humidity), all hidden inside a louvered shelter, as well as a funnel-like rain gauge and a spinning anemometer for measuring wind speed. But for a quick check on the wind, pilots prefer a glance at the wind sock, which indicates both speed and direction. Every airport has one, somewhere out on the field near the runways.

WHIRLYBIRDS

Igor Sikorsky, the helicopter pioneer, flew his prototype machines dressed in a suit and a Homburg. No crash helmet for this test pilot. In old films and photographs he sits in the open cockpit looking like a lawyer or accountant on his way to the office. And, in fact, Sikorsky's dream was that the helicopter would become the everyday conveyance that ordinary folks would fly to work or the grocery store. The family helicopter would replace the family car.

If you live anywhere near a public heliport—or even a landing pad where hospital or TV-station helicopters touch down from time to time—you are probably grateful that Sikorsky's vision never came true. Helicopters have improved a great deal since the early days, but they still make a fearsome noise. If you look around a large city and imagine replacing every car on the streets with a helicopter whomp-whomp-whomping through the skies, this is not a trade that most of us would welcome.

In principle, helicopters can land just about anywhere—a backyard, a supermarket parking lot, a rooftop. But in practice they often land at a facility specially built for just that purpose. The typical heliport is a fenced-off concrete pad with a big letter *H* painted inside a square. The *H* signifies a public heliport; private ones often substitute a company logo. A hospital landing pad may show a white cross with a red *H* in the middle. Still another kind of marking is seen on the roof of a tall office building or a high-rise hotel: a circle with a number in the middle. The rooftop structure labeled in this way is not meant for routine helicopter operations, but it can be used for emergency evacuations; the number indicates the maximum weight the platform will support, in thousands of pounds.

The FAA suggests that heliports be paved with concrete rather than asphalt. The reason is that the skids or wheels of a helicopter might sink into a softer material, especially in hot weather. Apparently there have been a few accidents where a helicopter tipped over on takeoff because it was stuck in its own ruts.

Although a helicopter doesn't need a runway, it does require an approach and takeoff zone cleared of tall structures. The direction of the recommended final-approach pattern is indicated by the alignment of the *H* or other marker. Wind is also important on takeoff and landing, and the heliport will have a wind sock somewhere near the touchdown zone.

Visibility measurements. Flying blind is routine these days; airline pilots seldom need to look out the window while they're aloft. On the ground, however, it's a different story. Airports sometimes have to close because the fog is too thick for pilots to feel their way to the terminal. For a long time, visibility was "measured" by looking out the window of the tower at some distant landmark. Instruments now give more consistent readings of RVR, or runway visual range. The instrument consists of two pods alongside a runway. A light source in one pod is beamed toward a reflector in the other, which bounces the beam back to a photocell in the first pod. The photocell measures how much of the emitted light comes back, which is an indicator of atmospheric clarity. A related instrument measures the height of the "ceiling," or the bottom of the lowest bank of clouds overhead. Again, the instrument consists of two pods, but they point upward rather than at each other. One pod shines a beam of light at a slight angle from the vertical; the other pod tilts until it detects the reflection of the beam. The angle of tilt indicates the height of the reflecting cloud layer.

Bird control. Flying machines and flying animals don't mix. A collision with a seagull can crack a jet transport's windshield; "ingesting" a flock of smaller birds, such as starlings, can destroy an engine. And the grassy expanses of an airport make an attractive habitat for birds. A wide variety of solutions have been offered—which is a pretty sure indicator that none of them work. There are noisemakers and scarecrows, phony owls, mechanical hawks, and recordings of bird-distress calls. One of the common devices is the gas cannon, a stout pipe a few inches in diameter connected to a canister of propane just like the one that fires up the backyard barbecue. Gas trickles into the base of the pipe, then a spark from a battery-operated controller ignites it. The boom is of chest-thumping intensity. Nevertheless, gulls quickly learn to ignore it. If bloodless measures fail, the airport authorities may seek permission to use deadly force. They might set out poison-baited feeders; introduce foxes, skunks, or other predators on birds and their eggs; bring out trained falcons; or hire hunters with shotguns. (But the hunters have to be careful not to shoot the big silver birds.)

BEACONS AND BEAMS

The crucial difference between aviation and other modes of transport is that an airplane can't pull off to the side of the road or drop anchor when the pilot gets sleepy or the weather turns ugly. In an airplane, once the fuel runs out, you *will* come back to earth, ready or not. Hence, aviation puts a lot of emphasis on guiding the pilot to a safe landing even through darkness or cloud cover.

Runway Lighting. Landing lights do not shine down on a runway the way streetlights illuminate a road. Instead, the lights are embedded in the concrete and point upward, guiding the pilot to a safe touchdown. The lights are mounted flush with the surface of the runway, powered by buried cables. Airplane tires run over them all the

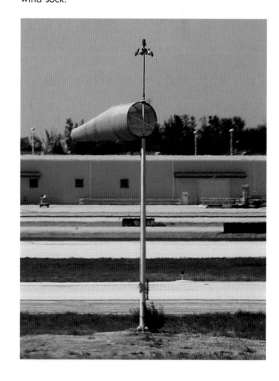

The twin pods staring at each other, above, measure runway visual range, or atmospheric clarity. Below is the most basic of all aviation weather instruments: the wind sock.

time; the fixtures are built to withstand this abuse. (But tire rubber has to be cleaned off the lenses from time to time.)

The lights outline the runway, giving the nighttime pilot the same clues to position and orientation that painted markings offer in daylight. Along the centerline, 200-watt lights are installed every 50 feet. At the arrival end of the runway, the centerline lights are white. Starting at 3,000 feet before the departure end, alternate lights are red and white; in the last 1,000 feet they are all red.

The centerline lights are only the beginning of the runway lighting system. There are also edge lights, which are white over most of the length of the runway but yellow in the last 2,000 feet. Also, at each end of the runway is a row of lights installed across the full width of the pavement and spaced closely enough that from a distance they merge into a continuous bar of light; this threshold marker is green at the arrival end of the runway for a pilot about to land and red at the departure end for a pilot about to run out of room for takeoff.

Near the arrival end, still more rows of lights extend perpendicular to the runway centerline. There are six lights in each row, and there are typically 30 rows spaced 100 feet apart. The main purpose of these lights is "roll guidance"—helping the pilot to keep the wings level in the last few seconds before landing.

On the main runways of large airports the lighting system extends a further 3,000 feet ahead of the runway threshold. These approach lights consist of "barrettes" of five lamps erected at 100-foot intervals. The barrettes also hold flash tubes, which are fired in sequence to create the effect of a ball of light rolling toward the runway. This eye-catching display leaves no ambiguity about where to land. On a cross-country flight at night you can often spot these beacons far below and at great distances.

For lighting taxiways, the current preference is for small, flush-mounted green centerline lights, like a necklace of emeralds, but many airports still conform to an older standard, with blue edge lights mounted on foot-high stalks.

Yet another system of landing lights is called VASIS, for *visual approach slope indicator system*. What VASIS tells the pilot is not where the airport is but rather where the aircraft is with respect to the runway. Specifically, it conveys information about the glide slope, the angle at which the aircraft is descending toward the ground. The customary glide slope is three degrees—which works out to a descent of 277 feet per mile of forward motion—and VASIS warns if the descent is either steeper than this angle or shallower. Horizontal bars of lights are mounted on each side of the runway at 500 feet and 700 feet from the arrival threshold. These specially rigged lights have lenses and colored filters arranged so that each light emits a split beam, white in the upper segment and red below. When the pilot is on the correct glide slope, the more distant bars (the "upwind bars") appear red, and the nearer ones (the "downwind bars") are white. If the aircraft is too high, both bars are white; if too low, both are red.

From a passenger seat in a jetliner I have never managed to glimpse the VASIS lights changing color as the pilot adjusted the landing trajectory, but you can watch them from a smaller aircraft by sneaking a peek over the pilot's shoulder. On the

Lights of the deepest cobalt blue, mounted on foot-high stalks, mark taxiways.

Approach lights held aloft on breakaway stanchions extend 3,000 feet ahead of the runway threshold. The airport is Raleigh-Durham in North Carolina.

ground, the light fixtures are easy to spot as your flight begins its takeoff roll. They are low structures off to either side of the runway, installed on concrete pads and painted—like nearly everything else on the airport grounds—bright red and white.

Instrument Landing Systems. Multicolored lights and painted markings aren't much use if the pilot can't see them because of fog or rain. The instrument landing system, or ILS, is meant to work in any weather. The pilot is guided by indicators on the control panel and looks up from these instruments only at the last minute, when the runway should be straight ahead and just below. It is an act of faith.

The ILS works much like VASIS, but instead of colored lights, the signals that guide the airplane are radio waves. The basic idea is to create a radio beam pointing upward at a shallow angle from the runway; the incoming aircraft slides right down the beam to the touchdown zone. But there's a problem with this idea. The radio beam has to come from an array of antennas, which are large metal structures that can't be buried in concrete the way a centerline light is. So how do you arrange to have the airplane fly toward the antennas but not crash into them?

The ingenious solution is to divide the ILS beam into two parts, for horizontal and vertical guidance, and build two antenna systems. The one that aligns the approaching airplane horizontally—keeping it from straying off to the left or right—has to be installed along the runway centerline, but longitudinally it doesn't have to be in the touchdown zone. In practice it's put beyond the far end of the runway. The beam for vertical guidance—keeping the aircraft descending at the correct rate—has to come from just the right distance down the runway, but it doesn't have to be centered laterally; the antenna is built off to one side of the runway at the touchdown zone.

The positioning of these antennas makes them easy to recognize. The horizontal beam, called the localizer, comes from an array of antennas that often stretches the full width of the runway, perpendicular to the centerline, several hundred feet beyond the threshold. The antenna elements vary in shape; some look like a ladder, some like a picket fence; most of the newer ones resemble a grape arbor, with irregularly spaced round bars poking out from both sides of a central beam.

The glide-slope antenna is mounted on a mast about 30 feet high, offset to the right or left of the runway far enough to be outside the wingspan of any aircraft. When your flight pulls into position for takeoff, the glide-slope antenna is up ahead a little ways, and so you pass it early in the takeoff roll; when you land, it ought to be pretty nearly abeam the aircraft at the moment the wheels thump down. The antenna elements are multiple bars or paperclip-shaped loops or bow ties, mounted in two or three groups at various heights along the mast, with mesh reflectors behind them.

To say that the ILS produces a beam for an airplane to follow is a bit of an oversimplification. Both the horizontal path and the vertical one are really defined by the intersection of two beams. The localizer antenna radiates two signals at the same radio frequency but with different audio tones impressed on them; the left side of the fan-shaped pattern is modulated at 90 hertz and the right side at 150 hertz. The aircraft is on course when the two signals are received at equal intensity. Similarly, the glide-slope antenna transmits waves modulated at 90 hertz above the correct slope and at 150 hertz below; again, the pilot follows the boundary line between the two regions. Of course the pilot doesn't have to do all this by ear; instruments automate the process.

The radio frequency of the localizer beam is in the band from 108 to 112 megahertz, which happens to lie immediately above the standard FM broadcast band. You might be able to hear the signals by tuning your car radio to the top of its range as you drive under the approach path. (Older radios work best for this. The new ones with digital tuning are just too good at rejecting off-frequency signals.)

The localizer and glide-slope signals keep an aircraft on track and headed toward the runway, but the pilot also needs another piece of information: an indication of how much distance remains to the touchdown point. This third dimension is added by another set of radio beacons, called the outer, middle, and inner markers.

Each of the three marker beacons comes from a transmitter directly below the flight path. They radiate upward in a fan-shaped pattern oriented perpendicular to the approach path. The outer marker is about four miles from the runway threshold, where the incoming aircraft is expected to cross at an altitude of 1,200 feet. The middle marker is 3,000 feet from the threshold, where the aircraft should be at an altitude of 300 feet. The inner marker—which many airports have not installed—is about 750 feet from the threshold, corresponding to an altitude of about 50 feet.

Many marker-beacon antennas have a distinctive Y shape; the upraised and outstretched arms create a flat, fanlike radiation pattern. Sometimes the design is simpler, just a few horizontal bars or loops on a vertical mast. Some beacon antennas could

A glide-slope antenna provides vertical guidance to pilots making a landing on instruments. It is mounted near the touchdown zone of a runway, but off to one side. The three trough-shaped reflector antennas on the tower project radio beams that help the pilot maintain a three-degree descent toward the runway. The installation shown is at Raleigh-Durham.

be mistaken for the kind of television aerial that used to decorate almost every rooftop in America. But there's a clue to the identity of the aviation markers: the antenna will always be painted either solid red or red and white—the only colors the FAA buys.

A localizer antenna provides the horizontal component of the instrument landing system. The antenna elements, which might be taken for some strange sort of gym equipment, point down the runway and keep an arriving aircraft on the centerline. (The aircraft above is departing rather than arriving and hence is not using the landing system.) The localizer in the photograph is at Reagan National Airport in Washington.

HIGHWAYS IN THE SKY

The instrument landing system guides an aircraft only in the last few minutes of its flight, but other aids to navigation provide a frame of reference all the way from take-off to landing. These electromagnetic landmarks create an invisible network of aerial routes that spans the North American continent and much of the rest of the world.

In the early years of aviation, aids to navigation were all visual—town names painted on factory roofs, and a network of beacon lights marking recommended routes through mountain passes. A remnant of the beacon-light system survives today.

Marker beacons are another component of the instrument landing system, which tell the pilot how much distance remains to the runway threshold. Above is an outer marker, four miles away from the touchdown point, on one of the approach paths to Dulles airport in Virginia. Below is a middle marker, 3,000 feet out from the threshold.

Most airports have a rotating beacon, green on one side and white on the other, so what you see from a distance is a sequence of alternating green and white flashes. At military airfields the white beam is split, creating a double white flash. At a seaplane airport the beams are yellow and white.

But visual signals have mostly been supplanted by radio-frequency "navaids." The simplest of these devices is called a nondirectional beacon, or NDB. As the name suggests, the beacon emits a signal that's the same in all directions. An instrument in the aircraft homes in on the beacon by turning a small loop antenna until the signal is strongest. From this information, the pilot can determine the bearing of the beacon with respect to the heading of the aircraft. For example, if you are flying straight toward the beacon, its bearing is 0 degrees; at a bearing of 90 degrees, the beacon is off your right wingtip. If you know the bearings to two or more beacons, you can draw intersecting lines on a chart to fix your own position. There are more than 1,500 NDBs operating in the United States.

The typical ground installation for a nondirectional beacon is a vertical tower about 30 feet tall, sometimes with an umbrella-like arrangement of ribs and spiral wires at the top. High-power beacons may use a horizontal dipole antenna—a wire or a set of parallel wires stretched between two poles. Any of these arrangements could be mistaken for the transmitting antenna of a small radio station, or perhaps the communication facilities of a police department or taxi company—if it weren't for the FAA's do-not-touch sign.

Nondirectional beacons broadcast at low frequencies—between 200 and 500 kilohertz, well below the AM broadcast band. But as you get close to the transmitter, you may be able to hear the signals anyway, even on an ordinary car radio. Spin the dial across the AM band and listen for a few letters of Morse code repeated over and over. The signal you pick up will be at double or triple the beacon's assigned frequency. The Morse code is an identifier for the beacon. Every navaid in the world is assigned a unique label, usually consisting of three letters; for a beacon at an airport, the letters are the same as those you find on your luggage tags.

The alternative to a nondirectional beacon surely ought to be called a directional beacon, but instead it's called a VOR or a VORTAC. Here we enter the dizzy world of aviation acronyms, which fill whole dictionaries. *VOR* stands for *VHF omnidirectional range*, where *VHF* in turn stands for *very high frequency*. A VORTAC combines a VOR with the military navaid called a TACAN, which is an abbreviation of *tactical air navigation*. Note that *VOR* is pronounced as three separate letters, but *VORTAC* is said as a word.

Whatever you call it, a VOR supplies directional information by sending out two carefully timed signals on the same carrier wave. To understand how it works, it's helpful to think in terms of light beams rather than radio waves. Imagine a lighthouse with a beam sweeping clockwise around the landscape, and a separate strobe light that sends out brief flashes in all directions at once. The strobe fires once per minute, and the rotating beam takes exactly a minute to sweep a full circle; furthermore, the

flashing and the rotating are coordinated so that the flash always comes at the instant when the rotating beam is pointing due north. With this system, determining your direction with respect to the lighthouse would require nothing more than a stopwatch. You start the watch ticking when you see the flash of the strobe light, and you stop the watch when the rotating beam sweeps past you. If the delay is 15 seconds, you are due east of the lighthouse; if 30 seconds, you are south; and so on.

In a VOR, the rotating lighthouse beacon is a radio beam that sweeps around the horizon 30 times a second, or 1,800 times per minute. The synchronizing signal—the conceptual equivalent of the strobe flash—is encoded in a warbling modulation of the same radio wave, with one cycle for each revolution. A special-purpose receiver in the aircraft disentangles the timed signals so that an instrument in the cockpit can simply point toward the VOR transmitter.

STRIPES AND CHECKS

Aviation leaves its mark on the land even miles from the nearest airport. Under regulations enforced by the Federal Aviation Administration, anything poking up into the sky far enough to snag an airplane gets painted red and white, or else it is decorated with blinking marker lights, or both. (Actually, the "red" paint has the official name aviation orange. It's the color of a ripe tomato.)

How high does a structure have to be before the FAA starts choosing your color scheme? That depends on where you are. On the airfield itself, an equipment shed the size of a doghouse or a knee-high lighting stanchion may get the red-paint treatment. At the airport boundary line, the height limit is usually 50 feet. Elsewhere, anything taller than 200 feet is a likely candidate for the paintbrush. This includes radio and television broadcast antennas, water towers, smokestacks, large oil and gas tanks, and high bridges. The standard treatment is to paint either the entire structure or the upper part of it in red-and-white stripes or checks. This pattern makes the object stand out in daylight. For night visibility, there are slowly winking red lights. In some cases only the lights are required; this is the common practice with tall buildings, which have enough bulk to be visible

by day without the garish paint scheme. (The New York City skyline would certainly be a different place if every building had to be decked out in FAA plaid.)

In recent years the FAA has introduced an alternative to the red-paint and red-light

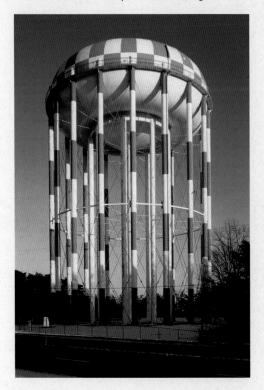

scheme. Many tall towers are now equipped with high-intensity strobe lights that emit brief but brilliant flashes of white light. The strobe flashes are easily visible by day as well as at night, and so there's no need for special paint. As a matter of fact, the strobes are so bright they have to be dimmed at night, or pilots might be dazzled by them.

A tall antenna tower will have three to five strobe units at roughly equal intervals along its height. They all flash simultaneously, at a rate of about 40 times a minute. A less common spectacle is a set of strobe lights timed to flash in rapid sequence from top to bottom so that a ball of light appears to be descending the tower. This distinctive marking is used on nearby pairs of towers to warn pilots that something is strung between them—usually a high-voltage power line—and so it wouldn't be a smart idea to fly through the middle.

At left is a water tower on the grounds of Andrews Air Force Base, just outside of Washington, decked out in gay FAA stripes and checks. Since this photograph was made, however, the tower has been repainted plain white. A flashing strobe light mounted at the top of the tank is all that remains to warn off low-flying pilots.

On the ground, a VOR installation looks like a broad-brimmed hat—usually a conical witch's hat, but sometimes a cylindrical, Abe Lincoln–style top hat. The brim is a circular reflector of solid metal or mesh that bounces radio signals away from the earth and toward the sky. The cone or cylinder in the middle—made of radio-transparent fiberglass—shields an antenna spinning at 1,800 revolutions per minute. If you can get close enough, you may be able to hear the motor whirring. At the edge of the hat brim there may be a dozen or so podlike attachments; they are antennas for monitoring and calibrating the VOR emissions. The whole apparatus usually sits atop a shed housing the transmitters and power supplies.

A newer style of VOR transmitter, called Doppler VOR, dispenses with the mechanical rotating parts (which are high-maintenance items). Instead of one spinning antenna, it has numerous stationary antennas arranged in a circle around a cen-

It's not a UFO that landed overnight in the Nevada desert. It's the Coaldale VORTAC, midway between Las Vegas and Reno, with the Silver Peak Mountains in the background. Inside the conical hat is an antenna twirling at 1,800 revolutions per minute, broadcasting a signal that aircraft instruments interpret as a directional beacon. The flat brim of the hat is a metal reflector, beaming the signal upward. Pods around the perimeter of the brim monitor and calibrate the signal.

tral emitter. The antenna in the middle radiates at all times, and the perimeter elements are activated in opposite pairs in a pattern that rotates at the standard 30-revolutions-per-second rate. For the airborne receiver, this electronic rotation is indistinguishable from a standard VOR signal. The circular array of antennas in a Doppler VOR would be hard to mistake for anything else. There are 52 antenna elements placed in a circle 44 feet in diameter, mounted on stalks above a metal reflector plane.

Much of the infrastructure for aerial navigation relies on machines that could qualify as antiques. The technology of the VOR dates back to the 1950s, and so does the ILS. Some of the radars and computers in air-traffic control centers are more than 30 years old. The FAA has been struggling to catch up, but technology has been moving faster than the agency can. For example, a replacement for the ILS, called the Microwave Landing System or MLS, was considered obsolete by the time the FAA finished testing it, and so the project was canceled. In the long run, all of the navaids described here will surely disappear. A long-range radio-navigation system called Omega has already been shut down. Another system called Loran (discussed in Chapter 12) is on the endangered-species list. According to one tentative plan, VOR and ILS services will be withdrawn sometime before 2010.

What will replace all these vintage technologies? The way of the future is satellite navigation. The reason becomes obvious when you consider that a passenger sitting in the back of an airplane, using a handheld GPS receiver that costs a few hundred dollars, can track the progress of a flight just as accurately as the pilots up front, who rely on multimillion-dollar avionics. All the thousands of VORs and NDBs worldwide can be replaced by a few dozen satellites in the Global Positioning System. Already the satellite signals are sometimes used for en route navigation. With just a little enhancement they'll be accurate enough to guide a plane right down to the runway threshold. (Some people worry about making transport so dependent on satellites, which are vulnerable to various kinds of attack and interference.)

Air Routes. The three-dimensional freedom of the skies is what's most alluring about aviation; you roam through the wild blue yonder, free as a bird. And yet, in practice, most flying is not so unfettered. Large tracts of airspace are regulated, and you may enter them only if you agree to abide by the instructions of air-traffic controllers. And most flights do not head out willy-nilly across the landscape. They follow well-worn tracks through the sky, invisible highways, where pilots are expected to stay in the correct lane, signal their turns, and obey traffic signs.

There are two systems of air routes over North America. The Victor airways are analogous to local blacktop roads. They get their name because they are designated V-21, V-44, and the like. *Victor* is pilot lingo for the letter *V*, as in VOR. The jet routes, with names such as J-14 or J-39, are the Interstate highways of the air, used by longer-range, higher-flying, and faster aircraft.

The Victor routes, at altitudes of 1,200 to 18,000 feet, run in straight lines from one VOR to another, so that you can lock onto a specific "radial" and then simply

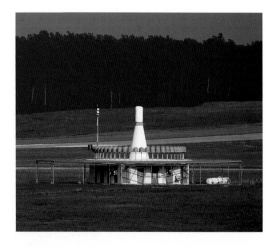

Doppler VOR *(above)* emits essentially the same signals but without the spinning antenna. The telltale mark of the Doppler VOR is a ring of 52 podlike antennas mounted above a metal reflector plane. The tall stalk next to the VOR is the antenna of a distance measuring equipment (DME) station. Whereas signals from the VOR tell the pilot what direction to steer to reach the site, those from the DME indicate how far away it is. A nondirectional beacon *(below)* is an even simpler navigational device. It merely broadcasts an identifying signal uniformly in all directions; a direction-finding antenna in the aircraft is needed to estimate the bearing.

The pagoda control tower at Dulles airport, designed by Eero Saarinen, is one of the most admired towers in aviation. Nevertheless, it has apparently outlived its usefulness. With ongoing expansion of the airport, Dulles has undertaken to build a new, higher tower. On the opposite page are some other distinctive towers: the bipedal design at Logan airport in Boston and a round glass tower at Charles de Gaulle in Paris, topped by a very French beret.

follow that course until the plane passes directly over the navaid. The jet routes are at altitudes of 18,000 feet and up. They also run from one VOR to another, but generally they make longer hops.

Air routes are multilane highways, but the lanes are not side by side as they are on a terrestrial freeway; they are stacked up one above the other. Furthermore, lanes going in opposite directions are interlaced. For example, a flight heading east on a certain air route might be assigned an altitude of 9,000 or 11,000 or 13,000 feet, whereas a westbound flight would have to choose 10,000 or 12,000 or 14,000 feet.

Although air routes are invisible, if you're patient you can sometimes notice aircraft tracing them out across the sky. Years ago I used to bicycle out to Farmington, Minnesota, south of the Twin Cities, where several air routes converge over a VOR at a major air-traffic-control facility. When the weather was right, high-altitude jets left condensation trails chalked against the dry midwestern sky. As the aircraft switched from an inbound to an outbound radial, I would see dozens of white trails pivoting on the same invisible pole directly overhead.

But aerial navigation is not the rigid, point-to-point discipline it used to be. For a long time pilots had to hop from one navaid to the next because it was too hard to calculate a direct course from departure to destination. With the simple cockpit instruments of earlier aircraft, you could steer straight toward a VOR or directly away from it, but more complicated routings required too much figuring. All that has changed with the advent of computers in the cockpit. Now a computerized instrument called a flight director or flight-management system continually monitors signals from several VORs and plots the aircraft's course with respect to all of them. The system is called RNAV, which is the aviation industry's strange way of abbreviating *area navigation* (or maybe it means *random navigation*—sources differ). When you hear your pilot announce, "We've been cleared direct from Baltimore to Cleveland," that word *direct* probably means you'll be following an RNAV route instead of skipping from one VOR to the next.

Air-Traffic Control. The controller, with a headset clamped over his ears, peers into the radar scope and calmly reassures the 12-year-old who has taken the controls of the airplane after the entire flight crew became stricken with ptomaine poisoning. How many times has this scene been played in the movies? Surely more than it has in real life. And yet it is not outside the job description of an air-traffic controller.

The airport landmark associated with air-traffic control is the tower, although only a small percentage of all controllers work in the glass-walled room at the top of the tower. This room is called the cab. It is where controllers handle flights that are just about to land or are ready to take off, as well as aircraft on the taxiways—basically all moving aircraft that are within sight of the tower. It's for this reason, of course, that the tower is a tower: the controllers have to be able to see everywhere on the airport grounds, which necessarily means that the tower itself can be seen from just about everywhere. The bigger the airport, the taller the tower. The one at the new Denver International Airport is 327 feet high.

Usually the windows of the cab are tilted inward at the bottom to avoid distracting reflections, or the glass may have a curved profile. (The same trick is sometimes employed with department-store display windows.) Sunscreens of various colors and degrees of transparency can be pulled down like window shades when needed.

All towers are tall, but apart from that they are anything but standardized. Airport architects have not been shy about making these structures distinctive as well as conspicuous. There is the two-legged Colossus of Logan in Boston, and a bright new minaret at National Airport in Washington; nearby Dulles International's tower is a kind of elevated pagoda; the new tower at Kennedy International in New York looks like an axe poised to fall; Chicago O'Hare has a golf tee covered with bright ceramic tiles. Some other designs look a bit too much like the guard towers of a prison.

The roof of the tower often has a small forest of antennas. These are mostly simple whip antennas used for radio communication with aircraft and with various ground vehicles (fuel trucks, snowplows, fire trucks). There may also be a rotating

radar antenna on the roof, perhaps enclosed in a protective radome. And the spinning green-and-white beacon light may be up there too. The tallest structure on the airport grounds is an obvious spot for installing all these items.

Most controllers track aircraft with radar rather than binoculars, and their surroundings are just the opposite of those in the tower cab; they live in a dim, windowless room, the better to see the blips on the screen. In some cases the radar room is on a tower floor below the cab, but it could also be in a nondescript building elsewhere on the airport grounds, or even some miles away. The controllers here work with aircraft at a somewhat greater range than their colleagues in the tower cab. They handle approach and departure patterns, making sure that all the flights converging on the airport and radiating from it don't get in each other's way. It is the approach controller who puts you in a holding pattern when traffic backs up.

There is still another cadre of controllers whose place of business is generally nowhere near the airport. They work in en route centers, directing traffic in the wide-open spaces at cruising altitude. There are 20 en route centers in the United States. They are mainly inconspicuous government buildings—there's no need for a tall tower—although some of them are located near VORs or other major navigational facilities. One distinctive feature you may spot near an en route center is a mast with a ladder-like array of long dipole antennas. Typically there are about 20 antenna elements, and the biggest of them is roughly 50 feet long. The whole contraption looks like a monstrous rooftop TV aerial. It is used for voice radio communications.

Controllers sometimes help with certain aspects of navigation—on request they will issue "vectors" telling a pilot which way to steer—but their main function is to

Still more airport control towers: at Venice (top), at O'Hare in Chicago (above), and at St. Louis (right).

SMALL-AIRPORT ETIQUETTE

Instrument landing systems and stroboscopic runway lights are standard fixtures at airports serving commercial airlines, but there are also hundreds of smaller airports that have no need for such high-tech facilities. At some of these airports, the most important maintenance task is mowing the runway.

Protocol at a small airport is very different from the rules that govern metropolitan flight operations. Here there is no control tower, and there are no air-traffic controllers keeping watch on radar scopes to prevent collisions. With no central authority to adjudicate the right-of-way, pilots must negotiate among themselves, much as drivers do at an intersection without traffic signals. If you watch the skies for an hour on a sunny Saturday afternoon, you may be able to figure out the etiquette of your local airstrip.

Most often, landing aircraft follow a three-legged, left-handed pattern. An approaching plane enters the pattern on the downwind leg, parallel to the active runway but in the direction opposite to the landing (and takeoff) direction. The aircraft makes a left turn, and then a second left turn lines it up with the runway cen-

terline for the final approach and landing.

How does the pilot know which runway is active and which way to land on it? Even airports without a control tower may have a radio service offering such information, but there is also a visual indicator. It is called the segmented circle (segmented because it is drawn with a dashed line). Look for it at a conspicuous spot on the landing field, such as near the intersection of two main runways. It's at least 100 feet in diameter, and the individual segments making up the circle are at least three feet wide and six feet long. Most often the segments are painted markings on a paved surface, but sometimes they are beds of white stone laid in a grassy field. (It can look a little like Stonehenge, and no doubt future archaeologists will speculate about its ritual significance.)

In the middle of the circle, mounted on a mast, is the wind sock. This device looks just like what you would expect from its name: a long cloth cone held open at one end by a hoop, and allowed to swivel so that the opening always faces into the wind. A sock is preferred over a simple weathervane in that it

indicates the force as well as the direction of the wind. As the wind grows stronger, air fills more of the sock and extends it to a greater length. A fully horizontal wind sock indicates a wind of about 30 miles an hour.

In the absence of any more definitive information, an arriving pilot will prefer to land into the wind and will look at the wind sock to choose a runway accordingly. But some airports have a device within the segmented circle to visually indicate the current active runway. It is a T-shaped or arrow-shaped sign made of wood or metal, big enough to be seen clearly from a few thousand feet up and mounted under the windsock. The arrow or the long arm of the T is turned so that it points along the active runway. If the wind shifts, someone has to go out on the field and spin the indicator around to establish a new traffic pattern.

Around the perimeter of the segmented circle, you may see a few L-shaped markers oriented in various ways. These tell the arriving pilot what kind of an approach pattern is customary for each runway—whether to make all right turns or all left turns when approaching the runway.

make sure that aircraft don't run into each other. Aircraft flying at the same altitude are kept at least three miles apart. When flight paths cross each other, the minimum vertical separation is 500 feet.

Radar. The revolving radar antenna is one of the most readily recognized features you're likely to spot at the airport. It's also one of those objects with the pleasing property that how it looks tells you what it does. You can learn a lot about a radar's purpose and operation just from the appearance of the antenna.

The idea of radar is simple, although getting it to work reliably was an engineering challenge that took the better part of a century. The twirling radar antenna sweeps a narrow beam of radio waves around the horizon, then listens for echoes reflected from aircraft or other "targets." From the echoes, the radar unit determines both the direction and the range of the target. (The term *radar* began as an acronym for *radio direction and range.*) The direction part is easy: it is the direction the antenna

Voice communication between pilots and controllers on the ground is the function of much radio hardware you might spot near airports. At top, a cluster of five towers in North Carolina provides a link between aircraft en route and a regional air-traffic control center. The structure in the middle photo looks much like a rooftop television antenna, but it is built on a larger scale; the longest of the crosspieces is about 50 feet long. At the bottom is a direction-finding antenna. Both of the latter devices are installed near MacArthur Airport in Islip, Long Island.

is pointing at the moment the echo arrives. To calculate the distance, the radar must measure the time needed for a radio pulse to make the round trip from antenna to target and back.

Two factors make the job of the radar difficult. First, only a minuscule fraction of the transmitted signal is reflected back to the antenna. Thus, the system has to emit megawatts and detect microwatts. Second, because the signals move at the speed of light, the timing of the radio pulses has to be controlled and measured with an accuracy of microseconds. (Bouncing a pulse off a target one nautical mile away takes 12.36 millionths of a second.)

When you look at a radar antenna, most of what you're seeing is a curved reflector that forms the radio signal into a narrow beam in much the same way as the reflector in a flashlight does. The shape of the reflector can tell you something about the shape of the radar beam. The larger the reflector in any one dimension, the more narrowly it focuses the beam in that direction. In most radars for air-traffic control, the antenna reflector is a low rectangle, longer than it is high. As a result the beam is strongly focused horizontally but not vertically: it is a thin wedge as seen from overhead but a broad fan when viewed from the side. This beam geometry allows the radar to locate aircraft precisely in the horizontal plane while sweeping up targets at all altitudes.

For some short-range radars the antenna is more a bar than a rectangle; it looks something like a rotating log. At airports these radars are often used for ground surveillance—that is, for keeping track of aircraft on runways and taxiways. Similar radar units are installed on boats and ships.

More often than not, the reflective surface of a radar antenna is a lattice or mesh—like a window screen or chicken wire fence—rather than a solid panel. A reflector full of holes wouldn't work very well for a flashlight, so how does it manage for radar? Doesn't the beam just leak through the sieve? The explanation lies in the difference between light and radio signals. For a reflector, holes and other irregularities don't matter much as long as they are smaller than the waves being reflected. For visible light, that means the surface has to be smoothly polished down to a scale of well under a millionth of an inch. But most radars operate at wavelengths of a few inches or even a foot or more, and so the openings in the mesh reflector cause no trouble. Roughly speaking, the waves are too big to pass through them. By reducing weight and wind resistance, the mesh eases the burden on the mechanical system that twirls the antenna.

The parabolic reflecting surface is the largest and most conspicuous part of a radar antenna, but the real business end of the system is a much smaller device called the feed horn. It is mounted in front of the reflector (that is, on the concave side) and rotates with it. If you can get a clear look—this is easier if the radar is shut down and not spinning madly around—you'll see a metal pipe connected to the feed horn. The pipe is called a wave guide, and it carries the radio-frequency signals from the transmitter to the antenna and then from the antenna back to the receiver.

One of the tricky parts of building a radar is connecting the rotating antenna to the stationary transmitter and receiver. The crucial component is called the rotary coupler joint; like something in a lawn sprinkler, it has to be loose enough to allow rotation but tight enough to prevent leaks. It tends to be a high-maintenance item.

The radar unit you are most likely to see at an airport is a short-range surveillance radar, also known as a terminal control radar. "Short range" means 50 to 100 miles; it covers all the traffic arriving at and departing from the airport, or perhaps from a

Airport surveillance radar keeps track of aircraft within 50 or 60 miles of an airport. The unit shown here is on the grounds of Miami International Airport. The most important components have helpfully been painted aviation orange for ease of identification. Atop the tower is the rotating antenna, which turns at 12 revolutions per minute. The curved, concave reflector is for the primary radar, which bounces microwave pulses off the surface of airplanes and detects the faint echoes. Note the two "feed horns" pointing into the reflector; they emit and receive the signals. The flat "picket fence" above the primary antenna interrogates a transponder installed on all larger aircraft; the signals returned by this device allow the radar display to identify individual aircraft and give their altitude. The two orange tubes installed along the near side of the tower are wave guides that carry microwave signals between the antenna and an electronics shed at ground level.

Long-range surveillance radar, reaching out several hundred miles, requires larger antennas, which are almost always protected from the weather by a "radome." The site shown here, on a hilltop near Tonopah, Nevada, was built for military purposes, but at the time this photograph was made, in 2000, it was being operated by the FAA.

cluster of nearby airports (as in New York, where there are three major airports within an area about 20 miles across). An important characteristic of a radar is the sweep rate, the speed at which the antenna rotates. A typical airport radar turns at roughly 12 revolutions per minute, or 5 seconds per sweep. (This means that the controller sees airplanes moving in jumps every 5 seconds.)

All radars, incidentally, rotate clockwise. As far as I can tell, there's no law of man or nature that requires them all to turn in the same direction, but that's the way it is. They go the same way in the Southern Hemisphere as in the Northern. So if you ever see one going counterclockwise, you'll know you have crossed over into a mirror universe.

Many airport radars have a second antenna—either a flat mesh screen or an elongated "hog trough"—mounted above the main reflector and rotating with it. This is part of a beacon-and-transponder system that allows aircraft to identify themselves individually on the controller's radar screen. A "passive" radar merely bounces signals off the skin of an aircraft and detects the faint echoes. With nothing more than this

to go on, a hang glider and a 747 look much the same. Furthermore, the radar echoes carry no indication of altitude. The beacon and transponder solve both of these problems. The transponder is installed in the aircraft. When the radar beam sweeps past it, the transponder broadcasts a coded message that identifies the aircraft and gives the current reading of the altimeter. With this information available, the controller's display can label each blip of light with a flight number and an altitude.

The radars that keep an eye on ground movements of aircraft need a range of only a few miles, but they have to give controllers a detailed map, accurate to within a few feet. The antennas are small, rapidly spinning, and often mounted atop the control tower, since the radar, like the controllers themselves, needs to have an unobstructed line of sight to all important areas of the airport.

Aircraft at high altitudes are tracked by long-range surveillance radars, which can reach a radius of 250 miles or more. It takes about 20 such radar installations to cover the lower 48 states. The antennas are larger and rotate slower: about 5 revolutions per minute instead of 12. Unfortunately, you are unlikely to see any details of the antenna because it is usually enclosed in a radome, a housing that protects the hardware from wind and weather. (A rotating antenna the size of a large billboard is not something you'd want to have out in a gale.) The radome looks like a giant golf ball or soccer ball and is made of fiberglass or a plastic transparent to radio waves. The entire structure has to be assembled without metal components, which would produce strong reflections bouncing around inside the dome.

In recent years, smaller radars have also been covered up by radomes, mostly to reduce maintenance costs. The latest thing is the spinning radome, often seen atop an airport control tower housing the ground-control radar. Surprisingly—and perhaps disappointingly—the spinning of the radome has nothing to do with the rotation of the antenna inside. The radome's rapid twirling is meant merely to shed rain, snow, and ice, which can distort the radar signals.

Some of the fanciest radars have no moving parts at all. They are called phase-array radars: the antenna is an array of hundreds of small antenna elements, and the beam is steered by electronically controlling the phase of the signals fed to these elements. A phased-array radar can flick its beam across the sky much faster than a mechanically steered antenna, but building big arrays is expensive. Some aircraft have a small phased-array antenna in the nose, hidden behind an aerodynamically shaped radome. On the ground, most phased-array radars are operated by the military. The big ones that stand on alert for missiles arcing over the North Pole look appropriately ominous and mysterious—pyramids in an arctic desert.

The wavelengths used by radar systems are the same as those of microwave ovens. (Indeed, an early brand of microwave oven was the Radar Range.) The power output of the radars is thousands of times greater than that of the kitchen appliance where you pop your popcorn. The hazard this suggests is quite real. If you stood directly in the beam emitted by a large radar feed horn, you too would pop. But the intensity falls off rapidly with distance. Stay outside fenced areas and you should be safe.

Radomes like this one in Pennsauken, New Jersey, also house antennas for weather radar.

SHIPPING

NOT SO VERY LONG AGO, the clipper ship sailing before the wind was the 747 and the Federal Express of transport. People and goods traveling from New York to San Francisco would pay a premium fare and go 10,000 miles out of their way, sailing all the way around South America, rather than travel over land. And they got there faster, powered only by the wind!

Shipping today has a very different complexion. It's not the fastest way to get anywhere. The passenger liners that once plied the Atlantic are either gone entirely or relegated to the cruise industry, making voyages to nowhere, like glorified amusement-park rides. And many of the waterfront neighborhoods where sailors and stevedores used to prowl have been transformed into fancy real estate. New York has its South Street Seaport and Baltimore has its Inner Harbor, but in neither of those places will you find a freighter or a tanker unloading its cargo. When those areas were working ports, few outsiders ventured into them; now tourists come to celebrate what's no longer there.

Yet the shipping industry has not sunk. Parts of it are thriving. The ports of New York and Baltimore handle far more cargo today than they ever did when merchant ships tied up at the finger wharves of Manhattan and spice boats anchored in the Inner Harbor. It's just that shipping, like so many other heavy industries, has fled from the high costs and constrictions of the central city for wide-open spaces elsewhere on the waterfront. In the New York area, for example, the biggest port facilities are across the Hudson River in Newark and Elizabeth, New Jersey—places where sightseers seldom go (if they can help it).

The new marine terminals would be alien terrain to a seaman stepping out of the pages of Herman Melville or Joseph Conrad. All the nautical kitsch has been left behind at the gentrified downtown harborfront. Near the working port, you won't find restaurants decorated with antique buoys and fishing nets and lobster pots. A

On the waterfront: the story is no longer about stevedores with cargo hooks toting sacks of coffee beans. In the new port all cargo comes prepackaged in steel containers, which are loaded onto ships and unloaded from them by distinctive cranes of a design seen nowhere else. The photograph on the opposite page looks out over the Seagirt cargo terminal in Baltimore from atop one of those cranes.

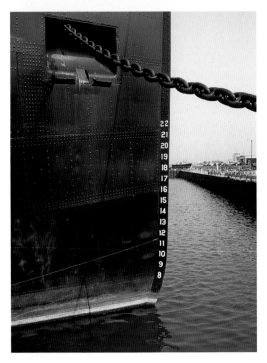

Load marks on the hull of a ship in Toledo, Ohio *(upper photograph)*, measure the ship's cargo capacity. If the hull is so deep in the water that the applicable load line is submerged, then the ship is overloaded. This particular set of load lines is used only on Great Lakes ships; oceangoing ships have slightly different markings. Draft marks *(lower photograph)* indicate what depth of water the ship needs to navigate safely. This ship, in Cleveland, is riding high in the water, drawing less than eight feet.

modern port is a vast and depopulated paved-over plain, as big as the parking lot of a megamall and almost as desolate. But there's a lot going on there for the aficionado of the industrial landscape.

SHIPS

As noted in earlier chapters, the machinery of every industry has its own distinctive technological style. No one would mistake a lightweight aluminum aircraft part from the sort of massive steel casting found on a railroad locomotive. Hardware for shipboard use is more like that of the locomotive—but even larger and heavier. Bolts and rivets are as thick as sausages. Light switches are mounted inside explosion-proof steel casings, and electrical outlets have screw-on brass caps. Watertight doors look like they could protect a bank vault. And perhaps most characteristic of all, everything in sight is slathered with paint; indeed, on an older ship there's not a sharp corner anywhere because every edge has been rounded over by years of painting and repainting.

Here are some more maritime idiosyncrasies to watch for when you're near the waterfront.

Hull Markings. I can never pass by a ship without checking its Plimsoll line. This is a painted mark—a circle or diamond with a horizontal slash through the middle, like a cartoon of a closed eye—somewhere on the hull midway between bow and stern. The mark is the maritime equivalent of the "maximum gross vehicle weight" that you see stenciled on the side of a truck—except there's no need for a ship to stop at a weigh station to make sure it's within legal limits. If the ship is overloaded, it will sink so low in the water that the Plimsoll mark drops below the waterline. Thus, the violation is immediately visible to anyone. (That's what's so clever about the laws of physics—you don't need the highway patrol to enforce them.)

Next to the Plimsoll mark there may be a little musical staff of other lines that give alternative loading levels for various conditions—freshwater, seawater, summer, winter. Why all the variations? Salt water is denser than freshwater, and cold water is denser than warm water. As a result, a ship that takes on cargo in New York in February will sink deeper in the water as it heads into the tropics on a voyage to Buenos Aires. The maximum safe load depends on when and where the ship will be sailing.

Who decides where to put the Plimsoll mark? Placing it higher on the hull would allow more cargo to be loaded, but it would also put the ship at greater risk in heavy seas. The regulations are enforced not by a government agency, but by insurance companies and the rating bureaus they have created for just this purpose. The most influential insurers are those who used to do business at a café called Lloyd's of London. Today you may still see the letters *LR* straddling the Plimsoll mark, indicating that the ship is listed in Lloyd's *Register of Ships*. American vessels carry the letters *AB*, for the American Bureau of Shipping.

The Plimsoll mark is meant to ensure that a vessel has enough "freeboard"—that it pokes far enough out of the water so the waves don't swamp it. Draft marks address the opposite worry, namely, that the ship may extend so far under the waterline that it scrapes bottom. The draft marks are like a ruler measuring distance from the lowest point on the keel. Sometimes there are two sets of draft marks, at the bow and the stern. Draft is usually measured in feet, not meters, even on ships registered in countries that otherwise use the metric system; occasionally both units are shown. For a long time, the marks were old-fashioned in another way as well: they were written in Roman numerals. Today most are in the more readable Arabic numerals. Each number is six inches high. If the water just laps the bottom of the number *20*, the draft is 20 feet; if the water covers the *20* mark, the ship is drawing 20½ feet.

Tonnage. How do you weigh a ship? Archimedes knew the answer more than two thousand years ago. Imagine a gigantic bathtub filled to the brim—at the very threshold of overflowing. Now gently float a battleship in this basin. If you catch all the water that sloshes over the sides of the tub, the weight of the displaced water is the ship's tonnage—also known as its displacement.

So where can you see ships being weighed in this way? Nowhere. Ship tonnage is never measured directly; it's calculated from blueprints. Furthermore, it turns out that displacement is only one of several kinds of tonnage. And how many pounds there are in a ton depends on what kind of tons you're talking about. And just to make things as confusing as they could possibly be, for some ships tonnage isn't a measure of weight at all, but a measure of volume.

The original ton was the tun—a big wine cask. The standard English tun was a barrel that held 250 gallons of wine, had a volume of 57 cubic feet, and weighed 2,240 pounds. That weight of 2,240 pounds is a unit of measure still in use today, called the long ton, but there are other kinds of tons as well. The usual American ton (known as the short ton) is 2,000 pounds, and the metric ton is 1,000 kilograms, equal to 2,200 pounds. For naval vessels and passenger liners, displacement weight is

GETTING A LOOK

The migration of port facilities away from downtown waterfront areas has taken the shipping industry out of the public eye. The new ports tend to be in out-of-the-way, little-visited, pedestrian-hostile, industrial areas. You may have to make an effort if you want to get a look at what's going on there. Sometimes there's a good spot on the opposite bank of a river or harbor. From a boat, of course, you can get an excellent view—but be careful maneuvering your kayak in the wake of a supertanker.

Many ports are operated by quasi-public agencies (such as the Port Authority of New York and New Jersey). In some places, group tours can be arranged.

Along inland waterways in the United States, navigational facilities such as locks are operated by the Army Corps of Engineers. Many of them have public observation areas.

Security has always been an issue at ports. In years past the major concerns were the pilfering of shipments and smuggling, but since September 11, 2001, new worries have come to the fore. Container ports are now high-security zones much like airports, where everyone entering must present identification.

expressed in short tons, so a 50,000-ton battleship weighs 100,000,000 pounds. But oil tankers are measured by a scheme called deadweight tons, using the long ton of 2,240 pounds. And for freighters, tonnage represents something else entirely, based not on the weight of a tun but on its size. Very roughly speaking, a freighter's tonnage is how many tuns it has room for in the cargo holds.

Lifeboats. In the movies, a lifeboat is a wooden rowboat in which shipwreck survivors suffer sun and thirst, and contemplate cannibalism. Real lifeboats are not so primitive. They come equipped with motors, sails, and enclosed cabins, and if the accommodations are not quite up to the standards of an oceangoing yacht, they would probably suffice for an admiral's motor launch.

These are the big lifeboats hanging from davits on the boat deck of a cruise ship, where the passengers assemble for the ritual lifeboat drill on the first day out of port. A ship may also have inflatable rafts or boats, housed in fiberglass pods lashed to the deck. They are designed to float free and inflate automatically if the ship sinks. An inflatable raft may sound a little impromptu for a disaster at sea, but these are no pool toys. They hold as many as 150 people.

Power Plants. *Steamship* is the generic term for any ship that doesn't hoist sails, but few of them today actually make steam. The power plant of choice is a big diesel engine. There's an interesting story behind this. Through the 1960s two kinds of power plants were widely used in large merchant ships. The diesel engine was one choice; the other was a steam turbine, a smaller version of the machines that drive

Mustard-yellow lifeboats hang from davits on the cruise ship *Navigator of the Seas*, photographed in port in Miami. The ship requires lifeboat capacity for 5,000 passengers and crew.

the generators in electric power plants. The diesel got better fuel mileage, but the boiler for the turbine could burn a cheaper grade of fuel, so the two choices came out about even economically. But then in the 1970s the price of both diesel fuel and boiler fuel shot up to 10 times their earlier levels, and the equation suddenly favored reducing fuel consumption. That was the end of the steam turbine. Diesels already had an advantage in efficiency, and they were soon made even more miserly in their consumption of fuel. Then they were adapted to burn the cheaper oils that used to power the steam turbines. These oils are the bottom of the barrel or resid fraction from oil-refinery operations (see Chapter 4). They look and smell more like tar than oil, and nobody else wants anything to do with them, so the price is right.

The diesel engines that propel a large ship work the same as the ones that power a truck or a locomotive, but on an entirely different scale. A marine engine towers two or three stories tall. The starter motor is bigger than an automobile engine. Whereas the pistons of a large truck engine might be the size of a gallon paint can, those of a marine engine are like 300-gallon drums. The parts of this hulking engine move at a stately, deliberate pace; cruising speed is about 100 revolutions per minute. (An automobile engine would stall if you tried to run it this slowly.)

FREIGHTERS AND BULK CARRIERS

For close to a century, the freighter was the mainstay and the jack-of-all-trades in the world's merchant shipping fleet. Freighters carried everything everywhere: sacks of sugar from Brazil, crates of shoes from New England, bottles of wine from Bordeaux, electrical machinery from Schenectady. It was all packed away in the cargo holds, or sometimes lashed to the decks, by the ship's crew and by an army of stevedores, or longshoremen.

This "break-bulk" freighter has not totally disappeared from the seas, but you may have to visit an out-of-the-way port to find one. In the prosperous harbors of Norfolk, Hong Kong, or Rotterdam, freighters have been displaced by container ships. Break-bulk freighters survive only by calling on the world's smaller ports, which lack the expensive shoreside equipment needed for loading and unloading containers. Even in this role they are being elbowed aside by other kinds of ships, such as the RoRo.

A break-bulk freighter carries its own equipment for loading and unloading: derricks or deck cranes. The classic derrick has changed little since the days of sail—indeed, it evolved from the rigging of sailing craft. A derrick has a vertical mast on the ship's centerline. Booms mounted near the base pivot for both up-and-down and left-to-right motion so they can swing out over the side of the ship to hoist cargo. Operated by a skilled crew, a derrick is a versatile instrument. But running it takes practice, coordination, and many hands. On newer ships the derricks are gone, replaced by cranes not much different from the ones you see at a construction site.

Derricks are the traditional instrument for loading and unloading cargo on freighters. These were photographed aboard the *Global Mariner*, a training ship operated by the International Transport Workers Federation. The photograph was made in 1998; two years later the *Global Mariner* sank after a collision in the Orinoco River. The bright yellow fittings on the deck are an addition allowing the ship to accommodate a few containers.

Cargo is loaded into a freighter's holds through deck hatches. On a pleasure boat, a hatch is a door you have to duck to get through, but the hatches on cargo ships can be the size of a tennis court. The hatch openings are not flush to the deck but are raised up by a waist-high rim called combing, which helps to keep any water that might wash over the deck from draining into the hold. The hatches are held closed by heavy bolts or clamps; tightening these bolts is what you do when you "batten down the hatches."

The classic freighter design, built anytime up until the 1970s, is the three-island freighter. The ship gets its name because it has three elevated structures—a forecastle (or fo'c'sle) at the bow, a bridge amidships, and a poop or afterhouse at the stern. Seen as a silhouette low on the horizon, the freighter looks like a chain of three islands. The three deckhouses reflected the social structure of the merchant marine. Traditionally, the navigation officers (including the captain) lived in the poop, the engineering officers lived in the midship bridge (which was built over the engines), and the seamen or deckhands occupied the forecastle. On newer ships, crews are dramatically smaller, and social distinctions have blurred. The three separate living quarters have been consolidated in a single deckhouse and bridge structure, which is usually toward the stern. That leaves a single uninterrupted expanse of deck for loading cargo.

A few commodities are shipped in such large volumes that it makes sense to build specialized ships and shoreside equipment that handle nothing else. On the Great Lakes, ore boats carry iron ore from Minnesota to the steelmaking centers of Ohio and Pennsylvania. The grain trade is also large enough to justify its own fleet of ships, which load up at ports on the Great Lakes or in New Orleans or Seattle and voyage to Europe, Asia, or Africa.

Roughly speaking, a bulk carrier is a large freighter stripped of the derricks or cranes. There's no need for on-board loading gear because the ports are equipped for those tasks. In the iron-shipping ports on Lake Superior, for example, ore boats pull up alongside high piers with rail spurs on top. Hopper cars from the mines dump their load of ore directly into chutes that lead to the hold of the waiting ship. At the other end of the voyage, in Cleveland or Erie or Toledo, the hatches are opened, and the ore is scooped out by gigantic cranes or excavators, which are not very different from the machines that were used to mine the ore in the first place. In the case of grain, the cargo is sometimes sucked out by vacuum hoses two feet in diameter.

I once watched an ore carrier being loaded in the northern Wisconsin city of Superior. Along the top of the deckhouse, red and green lights were blinking on and off at intervals of a minute or two. After I had puzzled over these signal lights for some minutes, it became clear what they are for and how they work. As chutes poured ore pellets into compartments on the starboard side of the ship, it gradually heeled over slightly toward that side. Although the tilt was imperceptible to an observer on shore, it was detected by a sensor on board the ship, which illuminated a red light on the starboard side. The chutes were then shifted to a port-side compartment to continue the loading and restore balance.

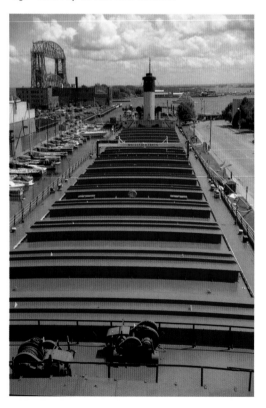

Cargo hatches cover the entire deck of the *William A. Irvin*, a retired ore ship berthed in Duluth, Minnesota, and maintained as a museum. As a bulk carrier, the ship has no on-board facilities for loading and unloading; it must rely on shoreside facilities.

TANKERS

Tankers are the largest of all moveable man-made objects. They dwarf even other large ships, including aircraft carriers and passenger liners. The largest tanker is the *Jahre Viking*, which is just over 1,500 feet long and about 225 feet wide. It needs water 90 feet deep, compared to 30 or 35 feet for a big freighter or container ship. When fully loaded, most of the tanker is below the waterline. It's an oilberg.

The *Jahre Viking* has an interesting history. When it was built in Japan and launched in 1979 as the *Seawise Giant*, it was not the largest ship afloat. It gained that distinction only three years later, when it was sent back to the shipyard and cut in half so that a new section of hull could be welded in, adding more than 500 feet of length. In other words, the tanker is a stretch limo. In 1988 the *Seawise Giant* became a victim of the Iran-Iraq war when it was hit by Exocet missiles. Although it was declared a total loss, the tanker was later salvaged and relaunched, and after several changes of name and ownership is sailing again.

Empty, the *Jahre Viking* weighs about 235,000 tons, but it can carry more than twice its own weight in cargo. Tanker capacity is measured in deadweight tons, equal to 2,240 pounds. One deadweight ton amounts to roughly seven barrels of crude oil, and each barrel is 42 gallons. If you do the math, you'll find that a ship loaded with half a million tons of crude is carrying almost 150 million gallons—enough to fill the tanks of about seven million cars. The cargo is worth well over $100 million—give or take a few tens of millions, depending on fluctuations in the market.

You won't see the *Jahre Viking* sailing into an American harbor. For most ports, receiving such ships is not even an option: the waterways leading to cities like Philadelphia and New Orleans are not nearly deep enough to float a supertanker. The few American ports that do have sufficient draft (such as Seattle) exclude the largest crude carriers because of fears of spills. As a result, some of the oil destined for the United States arrives via Canadian or Caribbean tank farms and refineries. Another option is to transfer the cargo of one big tanker into several smaller ones, a process called lightering. When I learned how lightering is done, I was taken by surprise. If you wanted to transfer cargo between two trucks, the first thing you'd do is park them. But that's not the best strategy with thousand-foot-long tankers at sea. It's easier to keep the vessels under control if they are making headway, so the ships come to parallel courses and keep moving while the oil is pumped through connecting hoses.

A modern tanker has a deckhouse or bridge at the stern of the ship; the engine room is also far astern, along with the large pumps that load and unload the cargo. The rest of the ship—the whole vast length and width and depth of it—is essentially hollow. There's nothing much in there but empty space for cargo and ballast. On the other hand, it's not just one big tank. The volume is subdivided into at least three tanks crossways and at least five longitudinally. There are several reasons for these partitions. They segregate different products or grades of oil; they help keep the ship upright when the

Three hinged booms called hard arms *(upper photograph)* swing out from a dock to plug into the piping manifold of a tanker. Aboard the ship *(lower photograph)*, the piping header has three corresponding connection points, numbered and color-coded to distinguish three sets of tanks: port, center, and starboard. The crane above the pipe header is used to handle hoses at ports that do not have hard-arm equipment. Note the draft marks and load lines painted on the hull. The ship extends another 52 feet below the surface; only about 25 feet of the hull are exposed. Water cannons both ashore and on the ship are for firefighting. The ship is the *Aldawha*, owned by the Arab Maritime Petroleum Tanker Company of Qatar; the port is Trieste in northern Italy. The *Aldawha*'s unloading at Trieste is further documented in the photographs on the next two pages.

Riding low in the water, the loaded *Aldawha (right)* is towed into the harbor by tugboats and then eased into the dock. The ship is 880 feet long and rated at a little more than 150,000 deadweight tons. Ships of this size are classified as Very Large Crude Carriers; there is a still grander class of Ultra Large Crude Carriers.

The bridge of the *Aldawha* is five stories high, the size of a small hotel. The engine room is aft of the bridge; the pumps are just forward of the bridge.

cargo starts sloshing in the tanks; and they limit the loss if the ship springs a leak. All the same, the tanks are enormous—like the inside of a Gothic cathedral.

Up on deck, racks of pipes run fore and aft, connecting every tank with the pump room in the stern and with a connecting header used for loading and unloading. When the tanker ties up in port, either the cargo is pumped through flexible hoses slung over the side of the ship, or else the connection is made through a "hard arm," a metal gantry with swivel joints that reaches from the dock to the pipe header on board.

What's that big electrical cord that also connects the ship to the dock? It's not a power cord to run machinery aboard ship; it's a grounding strap, to prevent sparks from static electricity. Something else you're sure to notice on a tanker is a big warning sign about the fire hazard. Frederick Allen, the editor of *American Heritage of Invention and Technology*, has remarked that all tankers seem to be named *No Smoking*.

The reason many people are aware of tankers—even people who are oblivious to the other kinds of ships that ply the world's oceans—is the hazard of oil spills. When the *Exxon Valdez* veered off course and ran aground in March of 1989, spilling 36,000 tons of crude oil into Prince William Sound on the southern coast of Alaska, the images of oil-soaked beaches and sea otters and waterfowl left a powerful imprint on public consciousness. By world standards, the *Exxon Valdez* spill wasn't even particularly big. The record loss was that of the *Atlantic Empress*, which dumped 257,000 tons after a collision in the West Indies in 1979.

Horrendous as those events were, most of the oil in the oceans doesn't come from tanker accidents. The largest share comes from air pollution that settles into the sea and from roadway runoff that gets washed into it. For a long time, there was another major source of oil contamination at sea. The tanker trade is a one-way proposition: When a ship carries crude from the Persian Gulf to the Gulf of Mexico, there's no paying cargo on the return voyage. But the ship can't sail completely empty, or it

would bob around like a cork; it needs ballast. The ballast is seawater, and until the 1970s it was carried in the same tanks used to hold the oil cargo. Furthermore, in the course of the voyage, the crew would use the ballast water to wash out the tanks, rinsing away a waxy sludge (they call it clingage) that builds up inside. Somewhere along the way, they'd pump the dirty water overboard. The result, for decades, was millions of tarballs washing up on Caribbean beaches.

Under regulations that began to take effect in 1973, tankers now segregate oil and water. There are separate ballast tanks that are never used to hold cargo. The ballast tanks are placed at the perimeter of the ship to provide some protection in collisions or groundings. Tank washing is done with the cargo oil itself. It is sprayed under pressure to dislodge the clingage, which winds up in the refinery rather than the ocean.

Laws passed in the aftermath of the *Exxon Valdez* spill require tankers entering U.S. waters to have a double hull—essentially a boat within a boat. In 2003 the European Union adopted similar regulations. The idea is that the inner shell will survive a grounding or collision and thereby prevent a spill.

The danger that tanker crews fear most is not a spill but a fire. The worry seems only natural when you're living atop a few hundred thousand tons of flammable liquid. But the fire hazard is not the liquid itself; it's the vapor above the liquid surface. Mixed with air in the "ullage"—the empty space at the top of a cargo tank—the vapors could be explosive. The solution is to fill the ullage with a gas that has too little oxygen to support combustion. And there's a handy source of such an inert gas: the exhaust from the ship's engine. The machinery for the inert-gas system is near the exhaust stacks above the deckhouse. A scrubber removes carbon monoxide, then large-bore pipes carry the gas to the cargo tanks. Some tankers have a separate inert-gas generator that burns fuel specifically to use up the oxygen in the air.

Does a tanker burn part of its cargo to fuel its engines? No, and for several reasons. In the first place, the cargo doesn't belong to the shipowner; it's being carried

During all tanker unloading operations, a boom boat— the small yellow vessel at left in the photograph above—patrols the harbor, ready to unreel a floating boom to confine any spills.

Twenty-four hours later, the *Aldawha* is nearly empty and riding much higher; the draft marks at the bow indicate 22 feet.

The height and the reach of the cargo cranes at a port, and the depth of the water, determine the maximum size of the ships the port can serve. The cranes at the Seagirt terminal in Baltimore can reach 140 feet outward from the dock and can load containers up to a height of 110 feet.

for hire, and the owner would not appreciate having part of it used as fuel. Second, crude oils are too variable to serve as good fuels. Finally, burning the cargo wouldn't make economic sense. Crude oil is worth more than the "bunker oil" that big tankers do burn—these are essentially waste products of the oil refinery, and so they actually cost less than the raw material from which they are made.

Years ago, a big tanker would have a crew of 25 or 30, roughly half of them seamen who work on deck and the rest "oilers" in the engine room. Newer ships make do with only half as many souls, and the traditional division between seamen and oilers has disappeared. All the crew members are listed as "mechanics"; in port they do deck work, loading and discharging the cargo, and at sea they maintain the power plant—although for the most part it runs unattended. The navigation officers have to stand watch around the clock in four-hour shifts, but the engineering crew works an eight-hour day, five days a week. Much of the time, the entire ship is controlled by one person, alone on the bridge. All ships are ghost ships these days.

Life on a tanker is said to be something like a long spaceship voyage to Mars and back. Tankers are not fast ships, and a round trip from Europe or North America

around the Cape of Good Hope to the the Persian Gulf takes two or three months. The crew are out of sight of land for almost all of that time; even when the ship reaches port, they stay only long enough to load or unload. There's no time to go ashore and tour the night spots of Bahrain.

CONTAINER SHIPS AND CONTAINER PORTS

The idea is so simple: put the cargo in boxes before you stow it aboard the ship. And the advantages are so obvious: faster loading and unloading, and better protection against breakage and pilferage. So why did the shipping industry need centuries to introduce such a simple and obvious concept? Not because shippers are hide-bound and backward. It took until the 1970s for containers to catch on because they work only if everybody agrees on how to use them. To make the containers stack neatly, they all have to be the same shape and size (or at least have only a few variations); ships have to be built to accommodate the containers; ports need special equipment to handle them; and on land there must be some convenient way to move them around by truck or rail. One component of the system is of no use without the others.

The container system that finally did catch on was invented in the 1950s by Malcolm McLean, who owned a trucking company that eventually became a major shipping line, the Sea-Land Corporation. And the connection with the trucking industry is no accident: the standard container that McLean introduced was basically a truck-trailer body without the wheels. This immediately solved the land-transport part of the problem: containers are driven to and from the loading dock by ordinary tractor-trailer rigs.

The first voyage of the first McLean container ship was in 1956. No one paid much attention for 10 years or more, but then suddenly the scheme took off in a big

Since the 1970s, great walls and embankments of stacked cargo containers have become a commonplace element of the industrial landscape. They are uniform in size and shape (or nearly so; note the differing heights in this stack at Seagirt) but highly variable in color, producing random checkerboard patterns.

Large ports handle more than 100,000 containers a year, so storing and retrieving them without waste motion is a significant challenge. When loading and unloading a ship or when dealing with stacks of containers on land, only the topmost container is immediately accessible. If the container you need is at the bottom of a stack, all those above it must be moved, one at a time. Organizing the task to minimize such movements is a kind of mathematical puzzle. At right, a straddle carrier at the port of Sorrento is shuffling containers in order to retrieve one further down in the stack, which will be loaded on a waiting truck. Below, a vehicle that looks like an overgrown forklift tends stacks of empty containers at a storage yard in Genoa.

way. By the 1980s, more than 80 percent of all freight was being shipped in containers. Today, in any port city you'll see great multicolor heaps of containers, often stacked six or seven high, like giant Lego bricks. You see them also on the highways and carried piggyback on railroad cars. Containers have become part of the everyday furniture of modern life.

Containers. To make sure that containers work with equipment the world over, their dimensions and the details of their construction are defined by international standards. The original idea was that all containers would be 8 feet wide and 8 feet high, and they would come in lengths of 10, 20, 30, and 40 feet. Because the lengths are all multiples of 10 feet, you can mix and match when you stack them up; a spot that would hold a 40-footer can equally well be filled by two 20s or four 10s. The capacity of container ships is measured in twenty-foot-equivalent units, or TEUs—in other words, the number of 20-foot containers that could be loaded on board.

Standardization has clear benefits: if everyone follows the rules, you can be sure that a container loaded in Shanghai will be compatible with the equipment in Seattle. On the other hand, there are also pressures to make exceptions to standards. Sometimes a cargo just won't fit in any of the standard boxes. As a result of these pressures, containers are not as standardized as shipping companies would like them to be. There are variations in height, as well as odd lengths such as 43 feet and 48 feet. But apparently there's still enough uniformity that the boxes pack together efficiently.

Every container has a label or stencil that tells who owns it. The owner code is four letters, but the last letter is always *U*. For example, containers marked SEAU belong to Sealand Services of New York, and those with the label HAMU come from Hapag

Lloyd of Hamburg. A six-digit serial number identifies the specific container, and then there's a check digit used for catching typographical errors. Most containers are also marked with their size and weight. The maximum gross weight is 67,200 pounds.

Out on the highway, the surest way to tell a seagoing container from an ordinary truck body is to look at the corners. A freight container has special reinforced fittings at all eight corners, with holes that are used as grab points to lift the box during loading operations. The same holes in the corner blocks engage with pins on the deck of a ship or on a truck or rail-car chassis to hold the containers in place. When multiple containers are stacked up, twist-lock keys clamp the stack together. (It really is like playing with Lego bricks.)

In addition to the common enclosed boxes, other kinds of freight containers are built to the same standards. Flatracks carry large and heavy items such as machine tools; tank containers hold liquids and gases; for perishables there are refrigerated containers, called reefers.

Imbalances in world trade can produce similar imbalances in container inventories. Because the United States buys more stuff from Asian nations than it sells there, containers tend to pile up on the West Coast, whereas they are sometimes scarce in Japan.

The Container Terminal. Containerization has forever altered the landscape and the working life of cargo ports. Fifty years ago, it took a lot of muscle to load a boat, and so the waterfront was crowded with longshoremen and other workers. Sacks and barrels and crates were piled up on the wharves. The streets nearby were lined with the offices of shipping agents, freight forwarders, and union hiring halls, not to mention the occasional tavern for sailors and stevedores.

The modern container port is a different kind of place. For one thing, it's bigger: a vast expanse of reinforced concrete where containers are stacked up in polychrome profusion. And it's a lot less crowded: apart from the guard at the gatehouse, the place seems deserted. The crane that loads and unloads the ships is run by a single operator in a cab high above the deck. A few truck drivers shuttle containers around the lot, and an office staff handles paperwork, but the crews of longshoremen are much depleted. Unloading a large ship used to keep a gang of 300 busy for a week; now it's the work of an afternoon for a couple of dozen.

The heart of any container port is the row of gantry cranes, specially built for lifting containers. The cranes are crucial to the operation of the port and are also the most conspicuous landmarks. Down on the waterfront, where the terrain tends to be flat, a massive crane towering 200 feet high really stands out. And—in a world where so much industrial equipment has been painted disappearing gray—I'm delighted to report that many container cranes are decked out in flamingo pink, robin's-egg blue, or other proud colors. In port cities like Baltimore and Oakland, the cranes are prominent monuments identifying industrial neighborhoods.

A container crane is different from a typical construction crane. For one thing, it's a lot beefier so it can handle heavier loads and move cargo in a hurry. On the other

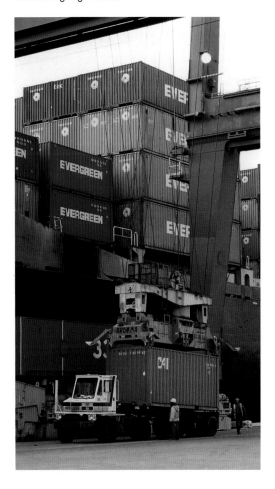

At Seagirt, a container unloaded from the ship *Ever Gaining* is gently nestled onto the frame of a truck known as a hustler. The spider-like device that grabs the containers from above is called a spreader; it engages fittings at the corners of the containers, which are also used to link the containers to one another like interlocking Lego bricks.

Idle cranes in Seattle *(top)* and Camden, New Jersey *(bottom)*, retract their booms to leave clearance for ships. With their long legs and crooked necks, they look like wading birds—cranes of another kind.

hand, it's also less versatile: the container crane has to deal with only one kind of load, and it only has to move containers in a straight line from the ship to the dock or vice versa; there is no need to swivel around to various parts of a building site.

When container cranes are idle, the boom that reaches out over the water is tilted up or retracted to keep it out of the way as ships maneuver into and out of their berth. When the crane is at work, the boom extends horizontally over the ship's deck. The "outreach" of a crane—the horizontal distance it can reach from the dock—and the maximum height to which it can lift a container determine the largest ship that a port can accommodate. (There's no point in building ships that are too big for any available crane, or in building cranes bigger than any expected ships.) Large cranes today have an outreach of about 140 feet and a maximum lifting height of over 100 feet; the biggest container ships use every inch of this capacity.

To operate the crane, you sit in a glass-bottomed booth that slides back and forth along the underside of the boom, trailing loops of a big electric-power cord. You look down between your knees at the container dangling below. Motions of the hoist are controlled by fingertip pressure on small joysticks, as in a video game. It's exacting work. You have to place 40-foot, 67,000-pound steel boxes with a precision of an inch or so, being careful not to bash anything or anybody nearby, and you have to do it quickly. An operator in good form moves two containers a minute.

When loading and unloading are going well, there's no wasted motion. A truck brings up an outbound container and waits on the dock below the crane. The crane operator lowers the spreader—a heavy steel platform that clamps onto the corner fittings of the container like some giant insect seizing its prey—and plucks the container off the truck bed. Lifting high enough to clear the deck of the ship and any other containers already aboard, the operator swings the container into its allotted space, and workers on board lock together the corner fittings. Then the crane picks up a container to be *un*loaded, carries it back to the dock, and places it on the bed of the truck still waiting below. This inbound container is hauled away, and another truck pulls up with the next outbound unit, so the whole cycle can begin again.

Watching this ballet high above the wharves, you can't help admiring the finesse of the crane operator and the other performers, but there is also a choreographer behind the scenes whose work deserves notice. For the process to go smoothly, the containers have to be stacked and unstacked in just the right order. When a ship departs for Seattle with further stops in Oakland and Los Angeles, you don't want the Seattle-bound containers buried under those going on to the later ports. And there are other constraints on container placement: differences in weight have to be taken into account to keep the ship in balance, and certain containers—such as those that need electric power to run a refrigeration unit—can be loaded only in certain spots on the ship. Planning and coordinating the loading of a container ship is like solving a difficult puzzle; these days it's done with computer assistance.

The trucks that shuttle containers around on the docks are called hustlers. And hustle they do: when I've visited container terminals, the hazard I've been warned

To serve places that have no container port, ships such as the *Tasmina* (left, photographed in Sorrento harbor) are equipped with on-board cranes capable of loading and unloading containers. The *Cartour Reggio Calabria* (below, also in Sorrento) is a roll-on/roll-off vessel, or RoRo, which transports entire trucks.

about most is not that I'll fall in the water or have a crane drop a container on my head, but that a hustler will run me down.

The hustlers move inbound containers to a staging area, where they can be loaded onto another truck for an over-the-road journey or onto rail cars for delivery to inland destinations. The loading is often done by an awkward-looking device called a straddle carrier, which pulls the container up into its belly and crawls around on widely outstretched legs, so that trucks can drive beneath. There are also forklifts with giraffe-like elongated necks to stack the containers in neat piles.

Container Ships. You'll have no trouble distinguishing container ships from other kinds of cargo-carrying, oceangoing vessels. Just look for the towering stacks of brightly colored boxes. Up close, it is the height of container ships that makes the most powerful impression. They are not as long or as wide as big tankers, but they tower over them, eclipsing the sun, far taller in fact than any of the "tall ships" that roamed the seas in the age of sail.

On most ships, containers are first stacked four or five deep inside the hull, then the holds are covered with watertight hatches, and another four or five tiers of containers are piled atop the hatches. But newer container ships have no hatches; the containers are stacked up continuously from the bottom of the hold to high above deck. This speeds loading and unloading, but it also means that any waves breaking over the deck will slosh right into the hold. To keep from filling and sinking, the ship has to rely on enormous high-speed bilge pumps.

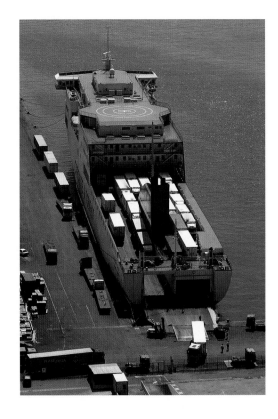

BARBIE OVERBOARD!

Anthropologists write of "cargo cults" among island peoples who supposedly venerate objects that wash up on their shores, like mysterious gifts from the gods. In an age of global commerce, when a single container lost at sea can launch 67,000 pounds of flotsam, the gods are offering many mysterious gifts. And the devotees of cargo cults now include a few oceanographers who trace the course of winds and ocean currents by gathering reports of lost merchandise found on the beach.

Perhaps the most famous such incident was the loss of 80,000 Nike sneakers when a container went overboard in the mid-Pacific in 1990. Over the next two years, thousands of the shoes came ashore on beaches in Canada and the western United States. Many of them were said to be in wearable condition, although finding a matched pair was a challenge. A few years later some of the same Pacific beaches were littered with hockey gloves (34,000 lost at sea) and another batch of 30,000 sneakers.

Accidents elsewhere have spilled rubber duckies and toy pieces described as "Barbie-doll butts." In 1997, some 62 containers en route from Rotterdam to New York were swept off the deck of the container ship *Tokio Express*. One of the lost boxes held 4,756,940 plastic Lego pieces, a few of which soon found their way to British beaches. (The prevalence of sports gear and toys in these incidents is curious, and probably reflects a certain amount of reporting bias.)

Curt Ebbesmeyer, an oceanographer in Seattle, has been gathering beachcomber reports of recovered cargo for more than 15 years as a means of studying ocean circulation. He can apply his knowledge in the other direction as well: He correctly predicted that the hockey gloves would come ashore weeks ahead of the sneakers that went into the drink at the same time. The reason is that hockey gloves float with one finger poking up into the wind.

The largest container ship so far is the *Sovereign Maersk*, launched in 1998, at 1,138 feet long. It carries 6,600 TEUs.

A container ship of the kind described here is sometimes called a LoLo, for *lift on/lift off*. The alternative is a RoRo, which stands for *roll on/roll off*. Think of a RoRo as a floating parking lot, with multiple parking levels, and spiraling ramps for moving between them. Instead of standardized containers, the RoRo carries truck trailers, wheels and all. At the departure port, a tractor hauls each trailer up a ramp to its assigned deck, then parks it for the voyage. At the other end, another tractor drives the trailer off. This scheme eliminates the need for cranes and all the other dockside hardware of the container port; the price is that the ship can't be packed as densely.

In some cases a RoRo carries entire trucks—tractors as well as trailers—and the truck drivers too. Essentially it is a ferry for trucks.

Some cargoes come with their own wheels attached—namely, automobiles. Special car-carrier ships are variants of the RoRo. The biggest of them transport more than 6,000 automobiles on 13 decks. At the ports where they unload, a common sight is a vast parking lot full of identical new cars, protected by a security fence festooned with the meanest-looking razorwire.

PORTS AND HARBORS

The basic problem that a port has to solve is how to get a big ship, which needs lots of deep water, right up against the land so people and cargo can go back and forth. There are two solutions. One is to extend the land into the water; the finger pier jut-

ting out perpendicular to the shoreline is the standard example of this approach. Alternatively, the water can be brought to the land by scooping out a deep pool right next to a retaining wall along the embankment of a harbor. In this case the ships must do parallel parking, like cars along a curb line.

Two tools are indispensable for port building: the pile driver and the dredge. Piles are long shafts, looking very much like telephone poles, pounded deep into the soft muck at the bottom of a body of water. Most piles are wood, heavily treated with creosote or some other nasty goo to prevent rot; there are steel and concrete piles too. The pile driver is essentially a hammer: it repeatedly lifts a big weight to the top of a frame and then drops it on the end of the pile.

Dredging is needed in many ports and harbors to make channels and basins deep enough for ships to enter without risk of running aground. The dredge is a barge with a large pump and engine. The intake end of the pump is connected to a pipe a foot or two in diameter that's lowered over the side to Hoover up sediment from the bottom. The other end of the pump connects to a much longer pipe, which carries the dredged mud and debris to a dumping site. Where to dump it is the whole problem with dredging. For a long time, the usual practice was to pick a nearby marsh to fill in; a few years later, the "reclaimed" land would sprout a strip mall and a housing development. This is less common today. A dredging project in New York Harbor has been delayed for decades because the spoil is considered hazardous waste.

One of the all-time championship dredging projects created the Port of Houston 50 miles inland from the Gulf of Mexico. Ships reach the port through the Houston Ship Channel, which was once a narrow swampy bayou a few feet deep. A hundred and fifty years of dredging has transformed it into a maritime thoroughfare 40 feet deep and more than 100 feet wide.

A dredge at work off the coast of Santa Barbara is employed in "beach nourishment"—moving sand from offshore onto the beach, at least until the next storm blows in. The dredge itself is mounted on the swiveling boom at the near side of the barge, with the pump mounted in the red enclosure at the top of the arm. Dredged material is carried ashore through the floating pipe at the stern of the barge.

A bright orange tug (above), backing the tanker *Aldawha* into a berth in Trieste, looks like such a jolly vessel it could be a bathtub toy. A tug of an older generation (below) at South Street Seaport in New York has a front bumper crafted from old rope. But the two tugs have something in common: old tires as fenders.

It's worth noting that dredging is one of those areas of life where averages don't count; it's the *minimum* depth that matters. Major waterways are maintained at a specified "controlling depth." In the Great Lakes and the St. Lawrence Seaway, the controlling depth is 27 feet; on the Mississippi, it's just 9 feet.

Jetties, Breakwaters, Seawalls. The port is a place where land and water come together—but it's also important to keep them apart. Many harbors are guarded by various kinds of barriers to keep out stormy seas.

A typical jetty or breakwater looks like nothing more than a pile of rock and rubble dumped into the water, but there may be more structure than meets the eye. Often there's a core of more carefully compacted crushed stone and gravel, or even concrete; the boulders strewn on the surface are armor meant to protect the core and dissipate the energy of the waves. Fancier jetties are armored not with mere rocks but with manufactured blocks in special shapes that lock elbows so as to better resist the force of the surf. Seawalls are built of concrete or of sheet piling—interlocking steel staves driven into the harbor floor.

Port Furniture. All the miscellaneous fittings and fixtures on wharves and piers and elsewhere in nautical neighborhoods are known by the charming term *port furniture*.

Fenders are meant to protect both the ship and the wharf. In the simplest case, for a fishing boat, a fender might be an old tire slung over the side of a wood dock. For bigger wharves and bigger vessels, the fenders become elaborate structures built out of timber, springs, and rubber bearings. They don't appear to have any give at all—if you ran into one of these bumpers with your car, you'd find them quite unyielding—but they bend readily enough when nudged by a ship that weighs 300,000 tons.

A dolphin is a big post out in the middle of the water that a ship can use as a bumper, mooring post, or pivot point—and that a pelican can use as a roosting spot. The typical dolphin is a cluster of either 7 or 19 pilings lashed together with wire rope. (Why 7 or 19? If you want to make something that's approximately round, those are good numbers to choose. Try it with a handful of pencils.) Bigger dolphins are made of steel or concrete.

The name *dolphin* is also sometimes used for the post on a wharf where mooring lines from the ship are attached, but this item of port furniture is more frequently called a bollard. Modern bollards are cast steel, weigh several hundred pounds, and come in a variety of shapes, all designed to hold a loop of hawser. They have to be incredibly strong—but not too strong. If a ship cannot be held fast to the dock, you want the bollard to give way, not the whole structure of the pier.

Dry Docks. A dry dock is the maritime counterpart of the hydraulic lift that raises your car so that the mechanic can get underneath to change the oil and fix the muffler. The traditional dry dock (also called a graving dock) is a huge concrete basin, connected at one end to a river or harbor. A ship needing work on the underside sails into the dock, then a watertight gate is closed behind it. As the water is pumped out, the ship settles onto a cradle erected on the bottom of the basin. Few graving docks are still in use. They have been made obsolete by ship-lifting technology. The new procedure is to float the ship into position over a submerged cradle, or dolly. Then electric winches pull the cradle—and the ship—out of the water onto a broad, sloping deck. The system is called a synchrolift, because the action of many winches has to be carefully synchronized. The big advantage of the new scheme is that one lifting dock can be used to bring many ships on land, where they can all be repaired at once, whereas the old graving dock could accommodate just one vessel at a time.

Tugboats. Tugboats are among those few industrial artifacts that are looked on as cute, charming, or endearing. The origin of this cultural fondness is mildly mysterious—but it has produced a succession of songs, stories, and even a movie and TV series.

The classic tugboat of 50 years ago—in the era of *Tugboat Annie*—had the look of a lion or a buffalo: at the bow was a great golden mane—a bumper or fender made from worn-out hawsers teased into bristly fibers. The bumper provided both traction and cushioning when the tug was nudging a ship into place. Nowadays the same

Port furniture: a dolphin in Baltimore *(top)*, a bollard in San Diego *(middle)*, a bollard of another type in Philadelphia *(bottom)*.

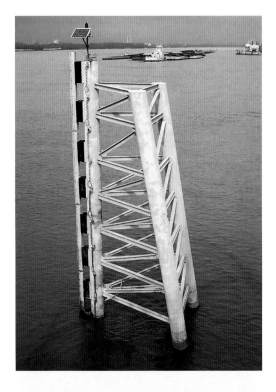

function is performed by old tires lashed to the hull, or by specially made rubber cushions.

There is no question that tugboats are all brawn, with an engine strong enough to move a supertanker crammed into a craft the size of a fishing boat. Yet many of them also have a certain geometric grace. They have an upswept sheer line—the curve traced out by the upper edge of the hull when seen in profile—that would be at home on a classic racing yacht.

The most common and familiar tugboats are harbor tugs. Their main job is to assist in docking large ships, helping them maneuver in close quarters. In many places, the presence of a harbor tug is required by law, whether or not it's actually needed. There are also oceangoing tugs, which are much larger craft that usually pull their cargo on a long tow rope rather than pushing or nudging it with their nose.

INLAND NAVIGATION

Some of the world's busiest ports are in cities hundreds of miles from the nearest salt water—in Buffalo, Pittsburgh, Minneapolis. In some cases, oceangoing vessels sail upriver all this distance, but more commonly the inland traffic is carried by distinctive boats or barges specially adapted to river and canal navigation.

Barges. *Barge* was once a term for a very classy vessel. It was a barge that carried Cleopatra down the Nile; later, in the British navy, a barge was a motorboat reserved for the use of an admiral. But what we know as a barge today is utterly utilitarian.

A modern barge is about as plain as a watercraft can be: it has no pointed bow to cut through the waves, no keel for stability in the wind, no motor, no rudder, certainly no polished brass railings or mahogany decks—it's just a rectangular box, flat-bottomed, square-sided, tough. Hopper barges haul heaps of coal, ore, crushed stone, scrap metal, and other stuff that doesn't need protection from the weather. Dry-cargo barges have covered compartments; most of them transport grain. There are also tank barges for petroleum and other liquids.

American barges come in two standard sizes. The Pittsburgh barge is 175 by 26 feet and can be loaded with about a thousand tons of coal. The jumbo barge is 195 by 35 feet and holds 1,500 tons. But a single barge seldom travels alone. Multiple barges are lashed together with steel cables to make a "tow" or "hitch." On some parts of the Mississippi, a tow can be seven barges long and six or seven barges wide, which makes it roughly the same size as the largest supertanker. And whereas big tankers stay out in the open seas, the barge train has to slip between the piers of bridges while contending with river currents and shifting sandbars.

More port furniture: a steel bollard for tying up barge trains in Baton Rouge *(top)*, a fender made of wood, steel, and rubber in New York *(bottom).*

Towboats. The boat that pushes a tow of barges is called a towboat. The tugboat and the towboat have a lot in common and are often confused: they are both small craft

with big engines, and they both make their living by pushing bigger vessels around. But it's easy to tell them apart. The tugboat has a a pointed prow, whereas the towboat is squared off in front so that it can be lashed solidly to a train of barges. At the bow of a towboat are two upright triangular structures called knees that make contact with the tow; they may have a staircase built in to make it easy to reach the deck of the barges. The cabin of the towboat is three or four stories high, with the navigation bridge at the top so the pilot can see over those five acres of barges.

Canals. In the early years of the nineteenth century, before the railroads were built, water transport was so much faster, cheaper, and easier than overland travel that there was a great spurt of canal building, taking boats and barges where the rivers don't go. In the United States the most famous of these canals was the Erie Canal through Upstate New York, connecting the Hudson River (and thus the city of New York) with Buffalo and the Great Lakes. Proposed in 1809 and completed in 1825, it was the first great civil engineering project in the nation. The original channel was 40 feet wide and 4 feet deep; the route covered 363 miles. Barges specially designed for the canal were pulled by teams of mules treading a tow path along the banks.

The heyday of the Erie Canal lasted just a few decades, until it was supplanted by the railroads. The waterway fell into disuse and disrepair. Then, a century after its original construction, it was rebuilt as the New York State Barge Canal, with an enlarged channel. In this form it is still navigable, although it's now used mainly by cruising pleasure craft, and the tow paths are the territory of joggers and bicyclists.

Europe also has a long tradition of canal building, as well as one remarkable recent addition to the world's navigable waterways. Tributaries of Europe's two greatest rivers, the Rhine and the Danube, arise just a few miles apart in the German province of Bavaria. The idea of connecting them goes back to the time of Charlemagne, who tried digging a trench across the watershed in AD 793. That dream was finally realized in 1992, with the completion of the Rhine-Main-Danube Canal, also known as the Europa Canal, along a route from Bamberg through Nuremberg to Regensburg. With

The towboat *Dennis C. Bottorff* pushes a tow upriver on the Mississippi at Baton Rouge.

this link in place, ships and barges can sail overland from the Dutch ports of Rotterdam or Amsterdam all the way to the Black Sea, a distance of 2,200 miles. In the process they climb and descend a hill whose peak is some 1,300 feet above sea level.

Of course, the most famous canals of all are the ones in Suez and Panama, which offer a shortcut between oceans. Technologically and commercially, both have proved successful. At times, ships line up for days waiting their turn to pass through. A tanker or container ship might pay a toll of more than $100,000 for a single passage—an amount calculated to be just a little less than the cost of sailing the long way around Africa or South America. But in describing the canals' success, I should not neglect to mention that their construction was a saga of human hardship, as well as a source of bitter political and social turmoil. The canals were built by colonial powers; wars have been fought in both places over control of the route. Ownership of the canals has now been returned to the countries they cross.

Most canals are less contentious—and less ambitious—than those of Panama and Suez. They are the nautical equivalent of a highway bypass. For example, the Welland Canal carries shipping traffic around Niagara Falls. Along the Atlantic and Gulf coasts of the United States, dozens of canals link various natural bays and channels to form the Intracoastal Waterway, which allows vessels to sail most of the way from New Jersey to Mexico without venturing into the open ocean.

Locks. The canal builder's dream is a level route from start to finish, so the canal is essentially just a big ditch filled with water. This design is often feasible in flat coastal areas; for example, the Cape Cod Canal cuts all the way across the base of that Massachusetts peninsula at sea level. (If you take a boat trip through the canal, you see a complete cross section of the Cape.) The Suez Canal is also at sea level.

Steeper terrain presents a challenge. A canal can't just run up and down mountainsides the way a road would; the water would drain out of the high spots and overflow the low spots. Going around or cutting through a hill is not always practical. If the two ends of the canal are at different elevations—in the case of the Erie Canal, the western end is more than 200 feet higher—then a level route is simply not an option. The solution is a lock, a device that looks like a giant stairstep in the course of the canal but that works more like an elevator.

A lock sits between two sections of a canal at different elevations. The lock itself is a basin with watertight gates at each end. Suppose initially that the lower gate is open and the upper one is closed; thus, the water level inside the basin is equal to that of the downstream canal. A ship heading upstream enters the lock, and the downstream gate is closed behind it (so that both gates are closed). Now, passages called sluices are opened up between the upstream canal and the lock chamber, allowing water to fill the basin and raise the ship to the level of the upper canal. Once the water levels equalize, the upper gate can be opened, and the ship proceeds on its way. At this point a ship heading downstream can enter the chamber from the upper canal, and the upstream gate is closed again. The water inside the chamber drains

through another set of sluices into the lower canal, the lower gate opens, and the downstream ship continues on.

A lock is like an elevator because ships are raised and lowered vertically. But note that there's no need for motors or hoists or even pumps to do the lifting: the system is gravity-powered, allowing the water in the upper canal to do the work of lifting as it is released to flow downhill, one lock full at a time. (It's crucial that the two gates of the lock never be open at the same time; if they were, the lock would become a waterfall.)

Locks are generally the narrowest choke point on a canal, and so they limit the size of the ships that can pass. A class of ships known as Panamax are built so that they can just barely squeeze through the Panama Canal. The narrowest lock in Panama is 108 feet wide, and some of the ships have a 106-foot beam. The lock length is 1,000 feet, and Panamax vessels leave less than 25 feet of room at each end. The Panama locks were designed in 1909 to accommodate the largest ship then afloat, the U.S. battleship *Pennsylvania*, as well as a somewhat larger vessel that was then under construction in Ireland. As it turned out, the latter ship never passed through the canal; it was the White Star liner *Titanic*.

You get the best closeup view of a lock by "locking through" on a boat. If you get a chance to do this, make note of the system for mooring vessels within the lock chamber. In small locks, this may be no more elaborate than a series of ropes hanging down the walls. Boaters are instructed to hold onto a rope by hand but not to tie it fast to the structure of the boat—this would be a particularly embarrassing mistake on the downstream transit, when the water level is falling. Locks that accommodate larger craft have floating bitts—tie-off points that rise and fall along with the water level.

Locks are found not only on canals but also on rivers that have been dammed. On the upper Mississippi there are locks at each of the 27 dams built from Minneapolis down to St. Louis.

AIDS TO NAVIGATION

On land we can find our way around by looking out for landmarks, but on the water there are no watermarks. The blankness of the sea has made marine navigation a serious art and science for centuries. Inventions like the magnetic compass, the sextant, and the chronometer have altered the course of history by making possible long-range navigation—and thereby long-range commerce and conquest. Mariners today still learn to use those instruments, but they also have other devices to rely on.

Buoys and Channel Markers. The road signs and guardrails of the sea, buoys and channel markers delineate safe routes, warn sailors away from rocks and shoals, and regulate traffic in crowded ports and rivers. As with highway signs and signals, the

Locking through: On the opposite page a towboat with a small hitch of barges comes upriver on the Mississippi through Lock and Dam No. 16 at Muscatine, Iowa. Below is a floating tie-off in the wall of a lock at Bonneville Dam on the Columbia River near Portland. The device allows a vessel to stay firmly tethered to the wall even as the water level rises or falls.

Channel markers at Gulfport, Mississippi, provide guidance to mariners and a perch for herons. According to the rule "red right returning," it appears we are not returning.

nautical markers rely on combinations of color, shape, and lights to convey meaning. Sometimes sounds get into the act too.

Rules of the road are a bit more complicated at sea because the road itself is invisible. Instead of the simple keep-right rule of the (non-British) highways, mariners are taught the formula "red right returning." When entering a bay or river from the sea—or in general when going upstream—you keep red signs and lights on your right. The other boundary of the safe channel, on the left, is indicated by green markers. When you're going downstream, of course, you have to remember that red right returning means "red left when not returning." And to further complicate matters, it's not always obvious which direction is "returning." The Intracoastal Waterway, for example, runs parallel to the shore so neither direction is clearly upstream. There's a special set of signals for just this case—yellow triangles and squares added to the usual red and green signage.

It gets even worse if you sail your boat across the pond. All the above rules apply in North and South America, and in Japan, South Korea, and the Philippines, but the rest of the world has adopted exactly the opposite convention: when you are motoring up the Seine toward Paris, it's "rouge a gauche revenant."

Color isn't the only clue to what a channel marker means. The markers are also numbered in sequence, much like houses, with odd numbers on the green side and even numbers on the red side. Some markers have lights (also red and green) that blink in distinctive patterns. Finally, port and starboard buoys are differently shaped: The green ones are cans, which are essentially cylindrical; the red ones are nuns, tapered at both ends to form a double cone.

Some of the busiest waterways now have traffic-control facilities much like those of the air-transport industry, with control towers and radars. At Harwich on the English Channel, a ship cannot enter the harbor without advance permission from the vessel traffic system, just as an airline pilot cannot land at Chicago O'Hare without clearance from air-traffic control. Similar rules govern parts of the Intracoastal Waterway in the United States. Sections of this channel on the Gulf Coast sport TV cameras on masts, just like the ones that monitor expressways at rush hour.

Lighthouses. Few of us have warm-and-fuzzy feelings about container ports or supertankers, but when it comes to lighthouses, we go positively gushy. A typical lighthouse is roughly the same size and shape as a factory smokestack—but you don't find many smokestacks on souvenir postcards or calendar art. Lighthouses are seen as romantic monuments and emblems of local identity. A few years ago when the Coast Guard threatened to abandon the Cape Hatteras Light in North Carolina to encroaching seas, there was tremendous public outcry; eventually the lighthouse was moved inland a few hundred yards.

The technology is ancient, the most famous example in antiquity being the lighthouse at Alexandria, Egypt, built in the third century BC. The Alexandria light was on the island of Pharos, and words derived from this place name are today the generic terms for a lighthouse in French, Italian, and Spanish. According to some reports, the Pharos tower was more than 400 feet high (twice as tall as any modern light). The light itself was an open wood fire with a curved focusing mirror behind it.

The Pharos was a beacon meant to guide ships crossing the Mediterranean into the harbor of Alexandria. Later lighthouses have the opposite function, warning sailors away from hazardous shoals or rocks.

Nineteenth-century lighthouses burned lamp oil or acetylene gas. Modern lights, naturally, are electric. The beacon typically uses a 1,000-watt quartz-halogen light bulb, much like the one in the projector at your local cinema.

The big innovation in modern lighthouse design is not the bulb that provides the source of illumination but the lens that forms it into a narrow beam sweeping across the sea. The lens needs to be several feet in diameter, and casting or grinding it as a single piece of glass would make it too thick to be practical. The solution to this problem came from Augustin-Jean Fresnel, a French engineer, who showed that light could be formed into a beam by a carefully placed array of many small rings and prisms of glass. Fresnel lenses the size of a garden gazebo became standard equipment of lighthouses by 1850, and some of the original glass elements remain in use today.

In response to the public fondness for lighthouses—and perhaps also because of their declining importance to navigators in this age of radar, sonar, and global positioning satellites—many lighthouses have been opened to tourists. If you have the stamina, you can climb the spiral staircase.

Radio Aids. Even the brightest and tallest lighthouses fade from view once you get more than a few miles from shore. Thousands of miles of featureless ocean lie beyond. How do ships find their way? For two or three millennia, the answer was celestial navigation—estimating your position by observing the stars, the sun, and the moon. The invention of radio a century ago opened up some new possibilities.

Ordinary commercial radio broadcasts can provide a little navigational help. In principle, you can get a fix on your position at sea by making a cross-comparison of heavy metal with country-and-western. All that's needed is a radio receiver with a directional antenna—one that gives a stronger signal when it's pointing toward the

The color-coding of nautical markers is totally different in European waters. These red and yellow hazard markers, warning of submerged rocks, are in Sardinia.

transmitter. By twirling the antenna, you can find the compass bearings of two stations whose transmitters are at known locations. On a chart, draw lines through the site of each station oriented along the measured directions; wherever the two lines cross, that's where you must be. This scheme works, but it's crude and inaccurate.

A better system, called Loran—an abbreviation for *long-range navigation*—was invented during World War II. The version in use since the 1950s is Loran-C. A Loran receiver doesn't measure the direction to a transmitting station; instead, it measures the difference in the time of arrival between signals from three or more stations.

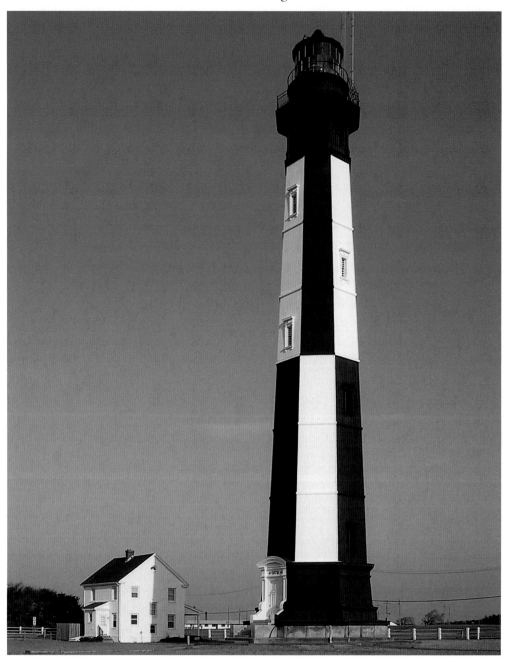

The Cape Henry Light in Virginia marks the entrance to Chesapeake Bay. The black and white pattern is meant to make the tower visible during the day, and distinguishable from other lighthouses along the coast. The cast-iron structure was put up in 1881. Officially, it is the New Cape Henry Light; the old one, from 1791, is still standing, though not lit.

Think of two people standing at the opposite goalposts of a football field and slowly clapping in perfect unison. You are somewhere on the field between them. If claps from the two ends reach your ears at exactly the same instant, then you know you must be on the 50-yard line. If one clap arrives ahead of the other, you are nearer one end. With just two people doing the rhythmic clapping, you can determine your position only along a single dimension, in this case the length of the field. If a third person claps in synchrony from the sidelines, you can also estimate your location along the cross-field dimension. The Loran-C system relies on the same principle on a much larger scale. Transmitters send out radio pulses (analogous to clapping hands), and receivers measure the differences in time of arrival. If at least three transmitters are in range, the receiver can calculate its position on the earth's surface.

There are 29 Loran-C transmitters in North America, each with a range of roughly a thousand miles. With help from a couple of Russian stations, they cover all of the continent and its coastal waters except for the remote north. Stations elsewhere cover much of Europe, the Mideast, and the Pacific Rim.

The typical Loran range of a thousand miles allows the system to reach well off the coastline, but it leaves vast blanks in the open oceans. The need for navigation in those areas was addressed by an alternative set of radio transmitters called Omega. The basic idea was the same—measure relative time of arrival at the receiver—but the signals were at lower radio frequencies, which carry farther. Omega covered the globe with just eight stations. They were placed in North Dakota, Hawaii, Argentina, Norway, Liberia, Australia, Japan, and the French island of Réunion in the Indian Ocean. Because of the low frequencies and long wavelengths, the transmitter stations required very tall antennas; the towers in Liberia and Argentina were the tallest structures in Africa and South America, respectively. They became tourist destinations.

Omega was launched in 1983 and was a technological success, but it didn't last long; in 1997 the system was shut down permanently. The North Dakota station was in the tiny town of La Moure, 100 miles from Fargo. When I visited, I found an Omega Motel, an Omega Cafe, and an Omega Mall, but no Omega transmitter.

The future of Loran-C is also in jeopardy, challenged by the Global Positioning System, or GPS, which relies on orbiting satellites instead of tall-masted earth stations. From the user's point of view, the change is equally revolutionary: an Omega receiver was a boatload of equipment costing thousands of dollars, but a GPS receiver is something you can slip in your pocket when you go canoeing. And yet GPS uses the same underlying mechanism as the older systems. Each of the satellites broadcasts precisely timed signals, and the receiver compares their time of arrival to calculate the distances to three or four satellites, thus pinning down its own position on the earth.

Ordinarily there are 24 operational GPS satellites (plus a few spares), all at an altitude of about 11,000 miles. From any point on the earth with a clear view of the sky, at least five of the satellites should be in range at all times. How interesting that after all these years we have returned to celestial navigation, but with artificial stars.

This Coast Guard base is 200 miles from the nearest coast in need of guarding. It is one of 29 North American transmitter stations for the Loran-C navigational system. Two of the towers that support the antennas are visible.

WASTES AND RECYCLING

THE STORY TOLD IN THIS BOOK began with the extraction of raw materials from deep in the earth. We end by closing the cycle, looking at what becomes of the leftovers, the wastes, the stuff nobody wants anymore. More often than not, the ultimate fate of these materials is to go back into the earth. It's mining in reverse.

When we enter the world of refuse and waste, we cross over into a mirror-image economy. In the "normal" world, we pay to acquire things; on the other side of the looking glass, we pay to get rid of them. Junk isn't merely worthless; it has negative value. A chemical engineer once told me about a recent improvement in a manufacturing process; by fine-tuning a chemical synthesis he had increased the yield of a certain commodity from 98 percent to 99 percent. I congratulated him, but I couldn't help remarking that this seemed like a rather small improvement. "Ah, you miss the important point," he said. "The amount of waste goes from 2 percent down to 1 percent. It's cut in half. We save tremendously on disposal costs."

RUBBISH

English has some fine, forthright words for it: *trash, rubbish, refuse, garbage.* For waste professionals, shades of meaning distinguish these terms. Trash is dry stuff (junk mail, empty cans, ashes, gum wrappers) and garbage is wet stuff (moldy cheese, orange peels, used chewing gum). Refuse is a broader term, encompassing both wet and dry; it's everything you'd ever put in the trash can or the recycling bin, including grass trimmings and other yard scraps. Rubbish means about the same as refuse, but sometimes it also includes debris from construction sites, such as broken concrete.

These are all fine words, but sanitation workers I've met talk about MSW. The initials stand for *municipal solid waste*.

Household refuse collected from the streets of New York City (*opposite page*) tumbles out of a truck and plops into a barge waiting below. When the photograph was made, New York's trash was being sent by barge to the Fresh Kills landfill on Staten Island. The landfill was closed at the end of 2001, and so was the system of truck-to-barge transfer stations. The city's refuse is now trucked some hundreds of miles to out-of-state disposal facilities.

The Environmental Protection Agency (EPA) keeps tabs on just how much MSW Americans produce, as well as what's in it and where it goes. The latest figures are for 2001. Total collections that year were just under 230 million tons, which works out to 4.4 pounds a day for every person in the nation. Is that a lot? Is it too much? Are we scandalously wasteful people? A throwaway society?

The historical context is hard to interpret. Over the 40 years that the EPA has been keeping records, the total annual tonnage has more than doubled. Although part of that increase can be explained away as a consequence of population growth, the per capita discard rate has also risen sharply, from about 2.7 pounds a day in 1960 to the 4.4 pounds of 2001. It's interesting to note, however, that all of the increase in rubbish per person was recorded before 1990; the rate has been steady since then. And, looking farther back, it seems likely that our forebears a century ago were even trashier than we are. In the era of wood-burning and coal-burning stoves, ashes alone amounted to more than 3 pounds per person per day, according to some estimates.

Collecting It. In many cities the agency charged with collecting rubbish is called the Department of Sanitation, a name that takes us back into the nineteenth century. When I looked into the history of these departments in a few cities, I was surprised to learn that emptying trash bins was not given much emphasis in the early days. Instead, the principal task of sanitation workers was cleaning the streets. Of course, this task still needs to be done, and most cities have specialized equipment for it—sweeper trucks with rotating brushes and water sprays—but street cleaning no longer claims a big share of the sanitation budget and labor force. I wondered why it was so much more important in those days, and then I caught on: in the nineteenth century, street cleaning *was* trash collection. No one put the rubbish out neatly packaged in a black plastic bag closed with a twist-tie or a bright yellow drawstring. There were no 40-gallon roll-out bins. Much of a city's refuse wound up directly in the gutter, and so that's where it had to be collected. And it wasn't just household waste and litter that needed sweeping up. New York City had more than 100,000 horses in 1900, each producing 20 pounds of manure per day.

Sanitation workers who cleaned the streets circa 1900 were equipped with a cart, a can, and a broom. Working conditions have improved since then, or at least the tools have. The key innovation was the compactor truck, invented in the late 1930s and widely adopted by the 1950s. The classic version has a low, round-bottomed hopper at the rear, where workers toss bags or dump cans. When the hopper is full, a steel blade driven by hydraulic rams descends from overhead in a scooping motion and stuffs the trash up into the body of the truck, squeezing it down to less than half its original volume.

I have had an opportunity to spend a morning working alongside the crew of a compactor truck, and I must confess I was awestruck by its power. Seeing bags of ordinary household refuse disappear into the maw of the machine was impressive but not surprising; I could not remain so blasé, however, when we heaved a three-

A decendant of the Dempster Dumpster is emptied by a truck equipped with hydraulic lift arms. This mechanism was probably the first successful example of automation in trash collection, and it remains in common use.

A hydraulic lift on the back of a compactor truck dumps the contents of a 96-gallon roll-out cart, designed explicitly for this system of collection. The trash being collected in this photograph is the author's own.

cushion, blue velour sofa into the hopper, and the compactor swallowed it whole without so much as a burp or a hiccup.

Hydraulics help to *un*load the truck as well as load it. The entire rear section of the body, including the hopper, the compacting blade, and the mechanism that drives it, swivels upward and out of the way like the rear door of a hatchback car. Then another hydraulic ram pushes the load of refuse out of the bin.

The introduction of the compactor made a big difference in trash collecting. A truck without this device would fill up quickly and spend much of the day shuttling back and forth to the unloading site. A compactor that can squeeze the trash down to half its original volume also cuts the number of trips in half.

In recent years, collector trucks have seen a number of further experiments and innovations, mostly directed to improving worker productivity. (Or, to look at it from the other side, mostly aimed at squeezing more tons per hour out of each worker.) The traditional collecting crew consists of a driver (who does nothing else) and two loaders (who can ride standing on small platforms at the rear of the truck, but who in practice walk most of the route since they are busy lifting bags or cans into the hopper). One approach to reducing the size of the crew is the "standup truck," designed to be driven from a standing posture so the driver can get in and out fast enough to help with loading. If the driver is to have loading duties, it also helps to redesign the truck so that the loading hopper is at the side, just behind the cab, rather than in the rear.

Another strategy is to replace ordinary cans or bags with standardized roll-out carts that are hoisted into the hopper by a hydraulic lift. The benefit in this case is not just productivity but also lower risk of injury. The ultimate in garbage automa-

On the tipping floor of the Tinker Creek Transfer Station in Roanoke, Virginia *(right)*, local collection trucks disgorge their loads; then the rubbish is pushed through an opening in the concrete floor into rail cars waiting below. Tampers compact the trash. Before the rail cars are hauled the 30 miles to the county landfill, they are fitted with lids *(below)*.

tion is a truck equipped with a robotic arm that reaches out to the curb to empty specially designed carts; a driver alone can thus handle an entire route without ever leaving the cab of the truck. This sounds like a futuristic vision—and attempts to make it work for residential collections are still at the experimental stage—but there is a well-established precedent going back to the 1950s called the Dempster Dumpster. Some of the large steel trash bins used by commercial buildings are emptied by a truck that lifts the bin high overhead and upends it.

Transfer Stations. The logistics of trash collection get more complicated when the ultimate destination of the trash is more than a few miles away. Driving farther to unload obviously leaves less time for loading. This is the rationale for building refuse transfer stations, which have become very common in the past decade or two, as more of our trash is hauled to outlying disposal sites.

At first glance, the idea of a transfer station seems inefficient, or even futile. What good does it do to transfer trash from one truck to another if that second truck is just going to have to drive to the same remote disposal site anyway? But the second truck can be much larger because it doesn't have to maneuver through city streets to make collections. Furthermore, the long-haul truck has only a driver, not a crew of loaders. And sometimes the second truck is actually a rail car or a barge.

A transfer station usually has two levels. The collection trucks drive to the upper level, called the tipping floor, where they dump their loads on a concrete deck. Meanwhile, the long-haul trailers line up below a slot or chute in the tipping floor. Bulldozers or front-end loaders push the trash into the slot, where it falls into the waiting trailers. Crane-like hydraulic arms pound the trash to compact it.

At the Roanoke Area Resource Authority in western Virginia, the landfill is about 30 miles from the city—not an extreme distance by any means, but compactor trucks would spend 45 minutes each way driving out there. The city found it advantageous to build a downtown transfer station, where the trash is loaded into rail cars for the trip to the landfill. The cars are specially designed for hauling trash, with tall bodies to increase the volume and removable lids so they can be loaded from overhead. At the unloading station, they are turned upside down by a giant rotary table.

The tipping floor of a transfer station is a challenging environment. Noise and dust make an immediate impression—both made more intense because the space is enclosed. The fine gray dust billows up from the floor as each load spills out of a truck; some transfer stations have a misting system to help suppress it. As for the noise, the most insistent and annoying sound is the constant beeping of trucks and other vehicles as they back up. Also, when a collection truck is emptied, some of the drivers bang the hydraulic ram the way you would pound on a can to shake the last bits out. The falling garbage makes only squishy, well-cushioned noises. The space is enclosed to keep birds, rodents, and insects out, and to keep smells in. Since transfer stations are usually close to town, these are likely to be nonnegotiable demands of the neighbors.

BURYING IT

Don't call it a dump, or you'll get a stern lecture on the distinguishing traits of the modern sanitary landfill. In a nutshell, a dump leaves garbage exposed to the weather and to scavengers. At a landfill, each day's deposits are covered with a layer of soil.

GARBOLOGY

William Rathje and his students at the University of Arizona in Tucson know more than anyone else about America's garbage. They learn it the hard way: by sorting through the contents of trash cans, and listing what they find, piece by piece. They have also drilled and dug into landfills to see what happens to the garbage after it's buried. (The answer, usually, is not much. Often they can still read newspapers buried 20 or 30 years ago.)

Rathje is an archaeologist, and he applies to landfills and to fresh garbage the systematic techniques that would be used in excavating an ancient city or burial site. Most of the garbage sorting is done by students at the university. They have a secret for making the work a little less daunting than it might otherwise be: they freeze the garbage before they pick through it, and they work fast so it doesn't thaw before they finish.

What the Garbage Project finds in the trash is not at all what most people expect. They once asked a group to guess how much of the trash in a landfill would be made up of fast-food wrappers, disposable diapers, and Styrofoam cups. People estimated that these three kinds of things together would amount to at least two-thirds of the landfill contents. Actually, they make up only about 3 percent, so even if we got rid of all of them, it wouldn't make much difference in how fast landfills fill up. What takes up most of the space in landfills is paper (especially newspapers), grass clippings and other yard waste, and construction debris, such as concrete and lumber from buildings being torn down.

The Garbage Project also reports some stranger results. For example, after Halloween they find lots of candy wrappers in the trash, but not much candy. After Valentine's Day people throw away the candy too.

Trash deposited in a landfill is dumped in a small, sloping region called the active face. A day's worth of fresh refuse is spread over the face of the slope, compacted, and finally covered with soil, to form the substrate for the next day's active face. The load about to be dumped in this photograph is at Fresh Kills in 1997.

Dumps are ancient, even prehistoric, but the sanitary landfill is a modern idea, apparently unknown before the twentieth century. Exactly where and by whom it was invented is a slightly contentious issue. Some factions claim a British origin in the 1920s, but Martin V. Melosi, in a history of urban waste disposal, argues that operations much like sanitary landfills were already running in Champaign, Illinois, in 1904; in Dayton, Ohio, in 1906; and in Davenport, Iowa, in 1916. The term *sanitary landfill* is attributed to Jean Vincenz, commissioner of public works in Fresno, California, in the early 1930s. In any case, the idea really caught on during World War II, when the Army Corps of Engineers adopted the landfill as the disposal method for military bases. By 1945 a hundred American cities had landfills.

In years past, both landfills and dumps often had the dual purpose of getting rid of trash and creating something new at the same time—namely, land. Rubbish was used to fill in swamps and marshes, with the intent of leveling the surface and building on the new land. In the New York City area, for example, most of the land for all three major airports—LaGuardia, Kennedy, and Newark—was created by landfilling. Those projects would not get regulatory approval today.

The typical landfill of recent years begins as a broad, flat-bottomed pit about 25 feet deep. Excavating this depression has two purposes: it increases the capacity of the site and it provides a stockpile of earth for use as daily cover when the landfill begins operating.

In an earlier generation, digging the pit was the extent of site preparation, but the process has become much more elaborate. All the further steps concern the problem of water that falls on the landfill and percolates through the layers of rubbish, pick-

ing up various contaminants along the way. The fluid that drips down to the bottom is known as leachate, since it has leached chemicals from the strata above. The nightmare of the landfill operator is that leachate will continue infiltrating the soil below, contaminating groundwater and showing up in the neighbors' wells or the public water supply.

The strategy of landfill builders now is to confine and collect leachate. That means sealing the bottom of the pit; it is built like a bathtub, with an impervious shell and a drain at the low point. The shell consists of several layers. On the bottom comes a thick liner of clay, which resists water penetration. But the clay is actually the second line of defense; above it is a complete membrane of "geo-textile," a heavy waterproof fabric or plastic. The textile comes in rolls 20 feet wide. It is unfurled in strips oriented vertically down the sides of the pit. Adjacent strips overlap a few inches and are then either sewn together or joined by heat sealing (depending on the type of

A fleet of bulldozers driving on toothy steel wheels drove over the day's load to compact it at Fresh Kills. Seagulls were a constant presence.

The daily cover of soil was spread over each day's receipts at Fresh Kills, so that the internal structure of the landfill is finely striated, with alternating layers of refuse and soil. Securing an adequate supply of cover soil can be a major challenge of landfill operation.

fabric). Placing, sealing, and inspecting the geomembrane is a delicate operation, and workers have to be careful not to rip or puncture it in subsequent construction steps. The membrane is covered with a few feet of crushed stone and gravel, both to protect the fabric and to create a permeable zone where the leachate can circulate. Perforated plastic pipes installed within this layer collect the leachate and carry it to the low point of the structure. There a sump pump brings it back to the surface.

What becomes of the leachate? One possibility is to spray it over the surface of the landfill so it circulates continuously and keeps the refuse saturated with moisture. This strategy encourages biodegradation of the organic components in the refuse—in other words, rotting. There is a faction that favors this bio-reactor approach to landfill operation, but the EPA recommends keeping the interior of landfills as dry as possible. Accordingly, the leachate has to go somewhere else. In some cases it can be pumped to an existing sewage-treatment plant. Otherwise, the landfill will need a small treatment plant of its own. As with sewage, the treatment is mainly biological, meant to digest the organic matter dissolved and suspended in the fluid. But the treatment does not work on leachate heavily contaminated with old crankcase oil, paint, insecticides, battery acid, and all those other things we are told not to put in the trash. Enforcement is problematic.

When the landfill goes into operation, the refuse is not spread out in a uniform horizontal layer over the bottom of the pit. Instead, it is formed into an embankment with a steep slope. Trucks drive to the top of the embankment and dump their load over the edge so it tumbles down the slope. The dumping is confined to a small area, called the active face, which is typically no more than a hundred yards wide even in

THE HIGH POINT OF THE EASTERN SEABOARD

It seemed only fitting that a city of skyscrapers would have a high-rise garbage dump, a mound of refuse touted as the highest point of land on the eastern seaboard and the largest landfill in the world. This was Fresh Kills, on the western shore of Staten Island.

Fresh Kills opened in 1948. It was such a success that several landfills and incinerators in other boroughs were closed one by one, until by the 1990s Fresh Kills was receiving all the residential rubbish of the entire city. Staten Island residents did not feel blessed by this distinction, and at a moment of political advantage they extracted a promise that Fresh Kills would be shut down by the end of 2001. To everyone's surprise, the promise was kept.

I first visited Fresh Kills in 1997, soon after the plan to close it was announced, and I went back again two years later. A number of photographs made during those visits appear elsewhere in this chapter.

In those days, Fresh Kills was receiving 13,000 tons of rubbish every day, six days a week. Almost all of this material was delivered by barge. Seeing the barges heading downriver doubtless helped perpetuate the belief that New York dumped its garbage at sea. In fact, that practice was outlawed in 1888 and actually stopped, for the most part, in 1934.

At the far end of the voyage, a hydraulic crane lifted the rubbish out of the barge, and it was loaded into giant off-road dump trucks, called pay-haulers, for a trip to the top of the hill. The trucks were much like the ore haulers at an open-pit mine, reinforcing the sensation that a landfill is a mine working in reverse.

Atop the growing heap, the trucks dumped their loads, which tumbled down the active face and were immediately attacked by bulldozers, graders, compactors, and other heavy equipment, not to mention sea gulls. At the end of the day the new layer of refuse was sprayed with a deodorant and covered with about a foot of earth.

What surprised me most about Fresh Kills was not any aspect of the actual garbage-handling operation, but rather the amount of infrastructure needed to support it. There were garages and maintenance yards for all the varieties of heavy equipment. Refueling trucks circulated throughout the site 24 hours a day to keep the cranes, trucks, and bulldozers running. A stone-crushing plant supplied material for the internal network of roads. The landfill had its own fire department.

At the time of my visits, the tallest of four mounds at Fresh Kills had reached an elevation of about 150 feet. If waste shipments to Staten Island had continued for another 20

years, the hills would have been approaching 500 feet, which was considered the safe limit. It was not to happen.

When the closure was announced, I asked a supervisor at Fresh Kills what New York would be doing with its trash in the future. "Oh," he scoffed, clearly not believing that the landfill would really close, "they'll just load it in a rocket and send it to the moon." As far as I can tell, the moon shot has not been given serious consideration, but there *was* talk at one point of a semisubmersible ship that would ferry the garbage to the Caribbean. The official plan is to compress and "containerize" the rubbish for easy shipment by rail or sea, sending it wherever it might find a low-cost welcome. A necessary first step is to rebuild the transfer stations that used to send refuse to Staten Island as container-packing stations. But three years after Fresh Kills closed, work on the transfer stations has not yet begun. So all of New York's residential rubbish (the total is down to about 11,000 tons per day) leaves town by truck—some 500,000 trips per year, half of them by tractor-trailers and half by ordinary sanitation collection trucks. The city trash is reloaded at transfer stations in New Jersey and then hauled over the highways to landfills in Virginia, Pennsylvania, and Ohio.

Meanwhile, back at Fresh Kills, life is calmer and quieter, but work at the landfill is far from done. First, the mounds must be fitted with a multilayer "cap" structure. Directly above the garbage itself is a porous layer of stones threaded with perforated pipes that carry away gases generated in the decomposing organic matter. Another porous layer drains rainwater from the hill. Between these two layers are a series of impervious membranes, liners, and clay fillers meant to keep the gas from percolating upward and water from infiltrating downward. Finally, the whole layer cake is iced with two feet of earth and six inches of topsoil, on which tall grass is growing. (The photograph at left was made in 1999, on the first section of the landfill to be capped and reclaimed.)

Eventually, Fresh Kills is to become New York City's largest park. But the key word is *eventually*. Although some areas can be opened for public use soon, it could be 30 years before the hills are completely stable and ready for skiers.

a large landfill. Bulldozers and other heavy vehicles neaten up the pile around the edges and repeatedly drive over the rubbish to compact and consolidate it. Then at the end of the day they spread the daily cover. The layer of rubbish might be a foot or two thick, the daily cover a little less. If you could cut a cross section through a landfill that had been operating for some years, what you would see would be alternating stripes at an angle.

The active face, as the name suggests, is a hectic place. The haul trucks, compactors, and bulldozers wallow through the rubbish like awkward grazing animals. Again, there is the constant bleating of vehicles in reverse gear. But there's something else as well, at least at the landfills I've had a chance to visit: the air is full of sea gulls—soaring, cawing, diving, quarreling. People worry about rats and roaches when it comes to garbage, but the most visible scavengers are birds.

Unless you get very close, the rubbish itself is just gray mush in black bags. You can't recognize individual banana peels or Publishers Clearing House sweepstakes envelopes. Up close, what attracts the eye are those things that glint in the sun or flutter in the wind. A few years ago, at one landfill, what I noticed most was magnetic recording tape, both the narrow audio kind and the wider ribbons of video tape; it was tangled in all the equipment, blowing across the surface of the landfill and decorating the fences like tinsel on a Christmas tree. More recently, at another location, I caught sight of the bright, multicolor iridescence of a CD. Presumably, this will become more common, and the tape rarer, as the years go by.

What about smell? If it were a wine, I would describe it as fruity, sour, grassy, a little overripe. Workers tell me that some days are worse than others, but in general what bothers the nose is not stench but dust.

BURNING IT

The advantages seem obvious: Take a large heap of untidy and perhaps smelly rubbish and reduce it to a much smaller pile of inert ash. Along the way, produce a quantity of heat that can be converted into electrical energy. So why aren't we burning more trash? In 2001 only about 15 percent of the MSW collected was incinerated, and the percentage has been declining. There are two big reasons. First, burning trash isn't as easy as you might think; many of the facilities that do it have high operating costs, so that burning turns out to be more expensive, even when some of the costs are recovered by sales of electricity. Second, incinerators or waste-to-energy plants are no more welcome in most neighborhoods than landfills are. Residents worry in particular about what might come out of the smokestack.

The Covanta Fairfax facility in Lorton, Virginia, is one of a newer generation of waste-to-energy plants, commissioned in 1990. One end of the plant looks just like a transfer station or any other waste facility: there's a large tipping floor where trucks disgorge their loads. The other end is just like a conventional steam power plant, with

A deep storage pit next to the tipping floor at the Covanta Fairfax waste-to-energy plant holds a stockpile of refuse and is also used as a mixing basin. Loads of refuse differ in fuel value (especially because of differences in moisture content). Cranes grapple through the refuse to mix it before loading it into the furnaces.

A clamshell grapple drops a load of refuse at the top of an incline leading into one of the four furnaces at the Fairfax plant. The rubbish—which from this point on can be considered fuel—will slowly be drawn into the furnace, dried, heated, ignited, and eventually reduced to ash.

turbines, generators, and a high-voltage switchyard. It's where these two realms meet in the middle that things get complicated.

MSW is not a fuel that a power-plant engineer could love. For one thing, it's just too lumpy: one batch is all paper and plastic, which burn fiercely, but the next is all wads of wet grass, which will barely smolder. Some components are totally non-combustible (ashes and cinders, glass, plaster). To reduce variations in fuel value, the refuse is thoroughly mixed in a huge pit adjacent to the tipping floor. The process relies on the judgment and experience of a crane operator, who grapples the fuel out of the hopper and feeds it into the furnace.

What about aerosol spray cans, batteries, and other products that carry a "Do not incinerate" label? In the large and very rugged furnaces of a waste-to-energy plant, the explosion of such small items is of no consequence. What does cause trouble is the occasional automobile engine or large tree stump that can get jammed in the furnace grate. Crane operators are supposed to spot such artifacts and set them aside, but sometimes a large item slips by. (It says something about the volume of rubbish that an automobile engine can escape notice.)

At a coal-fired power plant, the fuel is blown into the furnace as a fine powder, which burns much like a gas or an aerosol spray. There's no hope of turning rubbish into such a well-behaved fuel. In Fairfax the MSW is loaded onto the upper end of a long, sloping grate, which has a mechanical agitator. The fuel slowly makes its way into the flame zone of the furnace, gradually drying, then igniting and burning for some 45 minutes or an hour before the last of it is reduced to ash.

Variability in the fuel is unwelcome because it leads to variations in power output, but that's not the worst of it. To ensure complete combustion of the waste

The stoker grate furnaces at Fairfax *(upper photograph)* burn the refuse as it travels down an inclined grate. The sloping feed chute can be seen where it enters the furnace at the top of this image. To ensure complete combustion, the fuel is mechanically agitated, and air is blown in both over and under the grate. A small glass port offers a view of the flame zone *(lower photograph)*.

The turbines and generators of the Covanta Fairfax plant recover 79 megawatts from the heat of burning rubbish. One of two generator units is shown here. The turbines are in the background; the generator is the large blue oblong casing; the smaller device at the near end of the generator is the exciter generator, which supplies the magnetic field needed by the generator itself.

stream, the furnace is required to maintain a temperature above 1,000 degrees Fahrenheit. A wet mattress traveling down the grate could soak up enough heat to momentarily depress the temperature below the limit. Temperatures that drift too high are also a problem. When overheated, the ash can begin to melt, or at least turn gooey, and clog the grate or other parts of the machinery. Aluminum in the fuel causes a similar problem even at normal operating temperatures: it melts and needs to be cleaned off the grate.

The Fairfax plant has four furnaces, each of which burns 750 tons of refuse per day. This is essentially all of the trash from residents of the surrounding community. From this fuel, the plant generates 79 megawatts of electricity.

An alternative to burning raw MSW is to upgrade it to a slightly enriched fuel called RDF *(Refuse-Derived Fuel)*. Processing machinery breaks open the garbage bags, then workers at a manual picking station remove noncombustible and non-shreddable items (including recyclables and especially metals). What's left goes to a

Pollution-control equipment of a waste-to-energy plant is illustrated by the RESCO facility in Baltimore. Flue gases from the plant's three furnaces pass through scrubbers, which are the tall cylinders mounted on an upper level of the building. The scrubbers treat the gases with lime to neutralize acids. The scrubbed gases then pass on to electrostatic precipitators (the large, white rectangular structures), which remove particulates.

shredder. The chopped-up trash can be much more thoroughly mixed, making the burn more uniform, and the mere fact that it's reduced to smaller pieces also helps. But the drawback is the labor-intensive preliminary processing. Also, shredding trash is more exciting than one might wish: with aerosol cans and butane lighters in the mix, explosions and fires inside the shredder are routine.

RECYCLING IT

The blue recycling bin out for curbside collection; the neighborhood dropoff point for newspapers, phone books, and cardboard, with separate receptacles for clear, brown, and green glass; the nickel deposit on soda bottles in some states—these have all become familiar social institutions. We associate them with the flowering of environmental consciousness that began in the 1970s. But recycling was not invented on Earth Day.

The Scrap Trade. Not many junkyards are called junkyards anymore. They have renamed themselves salvage yards or dismantlers. Outside Portland, Oregon, I saw one whose sign advertises "auto recycling."

The nature of the business has changed along with the name. The old automobile graveyard had acres of rusting hulks that would be kept on the lot for decades. Customers would buy a taillight from one car and a door from another. Only after many years of this slow nibbling at the carcass would the skeleton be sold as scrap

iron. Today the turnover rate is much faster. Valuable and easily saleable parts are removed right away: radios, generators, air bags. Those parts go into a warehouse and a computerized inventory. The rest of the car is then flattened or baled or shredded, and the metal is sold back to the mill.

What transformed the junkyard was a change in the economics and structure of the steel industry. When most steel was made from iron ore reduced in the blast furnaces of Pittsburgh and Cleveland, there wasn't much of a market for recycled steel.

THE WORST WASTES

Ordinary household rubbish, the sort of thing collected once or twice a week at curbside, can be assigned an economic value somewhere in the range of –$10 to –$100 a ton. The dollar amounts are what it costs to get rid of the refuse by incineration or burying in a landfill—in other words, it's what you have to pay someone to take the stuff off your hands. The range is wide because these costs vary quite a lot from place to place. (New York City is now paying more than $250 a ton.) As a point of reference, the price of coal is generally around $20 a ton; the price of steel, from $200 to $300 a ton.

The worst wastes are hazardous materials that have negative values of much greater magnitude. For routine household hazardous wastes—half-full containers of paint, pesticides, drain cleaners, and so on—the cost of disposal could be in the range of $1,000 a ton. Biomedical wastes—known as red-bag wastes—could run as much as $10,000 a ton. And then there are high-level radioactive wastes, whose negative value remains incalculable because at this point there is no one who can dispose of them for you.

Landfills remain an option for some hazardous wastes, but red-bag wastes are mostly incinerated. Much of the cost of disposal is really a payment for assuming risks and legal liability, but there are also higher direct costs of operating the incinerator. Hazardous-waste incinerators are smaller than those for run-of-the-mill refuse, and they do not rely on the

waste itself to serve as fuel; they are gas-fired burners. The wastes must reach a temperature of at least 1,800 degrees Fahrenheit and remain above that level for a second or two. The challenge to the operator is that the temperature is very close to the point where ash begins to soften and form slags that clog the hearth of the furnace.

The largest hazardous-waste incinerators are being constructed by the U.S. Army, which has stockpiles of chemical weapons to be destroyed. The first facilities are in Tooele, Utah, and Anniston, Alabama.

The most problematic of all wastes are radwastes, especially the leftover radioactive materials from dismantled nuclear weapons and the spent fuel rods from the nation's nuclear power reactors. The ultimate disposal of these materials has been a contentious issue for at least 40 years. Deep burial has been the basic plan since 1956, but the first site selected—a salt mine in Kansas—was very quickly shown to be unsuitable. It took another 30 years, until 1987, before the Department of Energy settled on a new site, at Yucca Mountain, Nevada, about a hundred miles from Las Vegas. The facility to be built there was legally obligated to start accepting waste in January 1998. That date came and went. The projected date of opening is now 2010, but many legal and political challenges to the plan are still pending, and it is far from certain that Yucca Mountain will ever become a radwaste repository.

If an ordinary landfill is an open-pit mine in reverse, then the plan for the Yucca Mountain repository would create a deep underground mine in reverse. It has all the same equipment you'd see at hard-rock mines throughout the surrounding Nevada region—hoists, ventilation shafts, and fans. The difference is that the cavern is being excavated so that we can put things in it rather than take them out.

The radwastes are hot both in the thermal sense and in the radioactive sense. They present some unique challenges in waste disposal. For example, it's imperative to keep fissionable materials in separated, small parcels, lest they form a critical mass and start a nuclear chain reaction. And the canisters holding the waste will need to remain intact for 10,000 years or so to prevent contamination of air or water. It's worth noting that few man-made artifacts beyond the simplest stone tools have survived so long.

Proponents of the Yucca Mountain plan have endeavored to show that the repository can meet the requirements, and opponents have been diligent in pointing out flaws in the analysis. When it comes to making a decision, however, the question in the background is always, What's the alternative? The reactor wastes are now stored at more than 100 power-plant sites scattered throughout the country, mostly in "swimming-pool" tanks where they will remain safe only as long as they are submerged in water. The survival of these facilities for 10,000 years is extremely unlikely.

Composting—the recycling of organic materials such as lawn clippings—can be done on a municipal scale. Here a specialized vehicle called a scarab (named for the dung-burying beetle) turns a row of compost at a Staten Island facility.

But for the new steel-making furnaces of the minimills, scrap is the primary raw ingredient. Suddenly it became a commodity in great demand, and perhaps the most impressive example of a successful recycling program in modern times.

The technological challenge of turning an automobile back into a raw material is not to be taken lightly. You might think that the hardest part would be grinding up the heavy items such as the engine block, but in fact the most troublesome problem is "fluff"—the industry's derisive term for all the nonmetallic components of a car, the upholstery, the carpets, the plastic steering wheel, perhaps even the burled-walnut dashboard. Years ago, hulks of cars were burned to remove the fluff, but pollution-control authorities did not smile on this practice. The current solution is to shred the entire car, reducing it to chunks an inch or two across, and then sort the resulting stream of mixed material.

Before a car goes into the shredder, the dismantler is supposed to remove all fluids (gasoline, motor oil, antifreeze, refrigerant from the air-conditioner). Often the body is either flattened or squeezed down to a hay bale–size brick for convenience of shipping. Flattening was once done by dropping a heavy steel plate onto the car with a crane. Then hydraulic flatteners were introduced: they pressed a beam down onto the roof with rams at both ends. But now there's a machine that eats the car the way a snake swallows a rat: opening the jaws wide and chomping down, crushing it gradually from end to end. After flattening, a dozen or so cars fit on a flatbed truck.

Baled cars are even more dense and compact. Hydraulic rams press in from three directions. (The same machines are also used for recycled paper.)

The first successful shredders for automobiles were built in the late 1950s. The heart of the shredder is a hammer mill. The hammers are bell-shaped pendulums

mounted on a shaft driven by a huge motor (typically 1,000 horsepower). As the rotor turns at 500 or 600 revolutions per minute, the hammers clang against a stationary breaker bar, or anvil, battering into small pieces anything that gets in their way. All this mechanized mayhem happens inside a protective steel housing with walls as much as four inches thick. Whole flattened automobiles are fed into the maw of the shredder, and they come out as popcorn-size lumps. This includes the engine, transmission, drive train, axle, frame, and springs.

Shredding is the brute-force step in automobile recycling; the rest of the process calls for more subtlety and ingenuity. The aim is to sort all the materials that went into building the car into separate streams for recycling. The steel is easier to isolate than other metals because it is strongly magnetic. The nonmetallic fluff is also relatively easy to separate, because it is light; a blast of air will remove most of the fluff while leaving the metal behind. Separating metals such as copper, zinc, and aluminum is more difficult.

Tires. If automobiles and steel are the poster children of modern recycling, then tires and rubber are the *problem* children. Roughly 250 million new tires a year are sold in the United States, which means that 250 million old ones need to go somewhere. Yet more than half the states ban tires from landfills while offering no alternative disposal or recycling options. As a result, vast numbers of worn-out tires—perhaps as many as a billion—are marooned in illegal dumps. These towering heaps regularly become the sites of spectacular fires.

The car was once somebody's pride and joy, but now it's a pancaked hulk, ready to be recycled, treated without ceremony by a grappling crane. The auto recycling operation is at Atlantic Scrap and Processing in Kernersville, North Carolina.

An automobile at the end of its life slides into the jaws of the shredder at Atlantic Scrap, to be reduced to metallic confetti in a matter of seconds. The hammer mill at the heart of the shredder is inside the heavy steel housing flanked by yellow-painted stairways. The apparatus just above the mill is a feed roll that forces the scrap into the mill. Metal fragments emerge on the conveyor belt at the bottom of the mill; the large-diameter pipe carries the "fluff" on a current of air.

The tires are not without uses. As a fuel, they have a heat value slightly better than that of coal, with less sulfur. Both whole tires and shredded ones have been used to fire cement kilns.

Household Recycling. Thirty percent of the MSW collected, according to the EPA, is diverted from the waste stream by recycling. In terms of the 4.4 pounds of waste per person per day, that's 1.3 pounds that doesn't go to the landfill or the incinerator. Breaking it down further, 0.3 pound consists of yard trimmings and other organic wastes that are composted and subsequently reused as mulch; the remaining 1 pound is refuse that's successfully recycled. Most of this consists of collections from curbside or dropoff recycling programs for newspapers and cardboard, glass and plastic bottles, and aluminum cans.

Recycling is an information-intensive process. The value added by information is illustrated most clearly in the case of glass for making beverage bottles. The bulk raw material, consisting of broken glass, is called cullet. It is classified in three color categories: clear, green, and brown. Mixed cullet, containing all three colors, is essentially worthless; glassmakers have no use for it. But exactly the same material, separated into three separate bins, is a saleable commodity—it can be used to make new bottles. In other words, the value lies not just in the material itself, but in the knowledge or information that goes into keeping it sorted by color.

The principal product of the shredding operation is steel scrap, which tumbles off the end of a conveyor belt after various other metals have been separated from it.

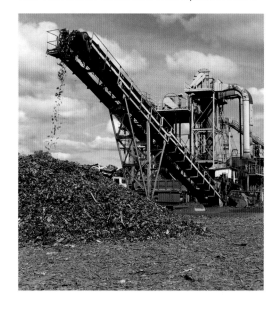

Corrugated cardboard is loaded into a baling machine for recycling at Asheville Waste Paper Co. in North Carolina.

The same principle applies more broadly to recycling as a whole. The key step that converts a waste stream of zero or negative value into a commodity with positive value is separation, classification, unmixing. A physicist would describe it as reducing the entropy, or disorder, of the waste stream.

The separation can be done at many different points in the handling of the materials. At one extreme, the responsibility to classify discards is put entirely in the hands of the individual consumer. This approach has been tried in some dropoff recycling centers, which provide separate receptacles for various kinds of paper, plastic, glass, steel, aluminum, and other materials. The other extreme allows all solid wastes to be collected in a mixed stream, and then relies on postprocessing to extract recyclable materials. In recent years no programs of this kind have been attempted in the United States, but in some parts of the world, communities support themselves by scavenging saleable materials from rubbish dumps. Of course there are many possible middle courses between the two extreme policies.

The value of glass for recycling depends critically on keeping it sorted according to color. In the photograph at right, brown, green, and clear glass bottles are stockpiled at Asheville Waste Paper. Close examination shows that the separation by color is not perfect. On the opposite page, various recycled commodities are baled for shipping: three kinds of plastic bottles, steel cans, egg cartons, mixed paper, cardboard cartons.

Most of the blue bin curbside recycling programs take an intermediate approach to sorting. The consumer is expected to separate recyclables from trash, and in some cases to do a little preliminary classification, such as separating paper from bottles, but the bulk of the sorting is done at a central materials recovery facility, or MRF (pronounced "murf").

The automated classification of recyclables remains a fond dream; at MRFs today sorting is a hands-on, labor-intensive process. Typically there is a tipping floor, where loads are dumped and given a preliminary inspection; then a conveyor system spreads out the materials and carries them along a sorting table, where they are classified by the skilled hands and eyes of workers who can instantly tell the difference between an HDPE bottle and a PTE one.

NIMBY, BANANA, AND NOPE

There is a widespread sense that waste disposal is one of the pressing problems of modern times, that we risk going down in history as the civilization that was buried by its own waste. No one could reasonably argue that we are *literally* running out of room for garbage; even in the most densely populated states, landfills have never occupied as much as one-tenth of 1 percent of the land area. Still, as a practical matter, there's no question that finding a place to put a new landfill or a rubbish burner, and getting the approval to build it, has become fiercely difficult and contentious.

Waste-disposal facilities of all kinds—landfills, incinerators, even transfer stations—are sure bets for generating the NIMBY response: *not in my backyard*. In its most cynical form, NIMBY is the attitude of citizens who acknowledge the need for a facility, somewhere, but who oppose a plan for building it simply because the selected site is too close to their own property. But opposition to landfills and many other kinds of development goes well beyond cynical NIMBY. The new catch phrase for this phenomenon is BANANA: *build absolutely nothing anywhere near anybody*. Or else it's NOPE: *not on planet earth*.

For many of those alarmed about the disposable society, the motives are not self-interest and property values; the debate is about affluence, materialism, sustainability, culture, lifestyle. To reduce waste, they are prepared to reduce consumption. This is a crucial test.

In the long run, a society's inputs and outputs are necessarily in equilibrium. If you go on acquiring new stuff and never throw anything away, then eventually the closets and cupboards will be stuffed so tight that you can't add even one more item. Conversely, and even more obviously, if you keep throwing things away and never replace them, then at some point there's nothing left. Neither way of life can be maintained indefinitely. Like it or not, over some archaeological period, our wastes will balance our acquisitions. We'll put back into the earth exactly as much as we take out: dust to dust, ashes to ashes.

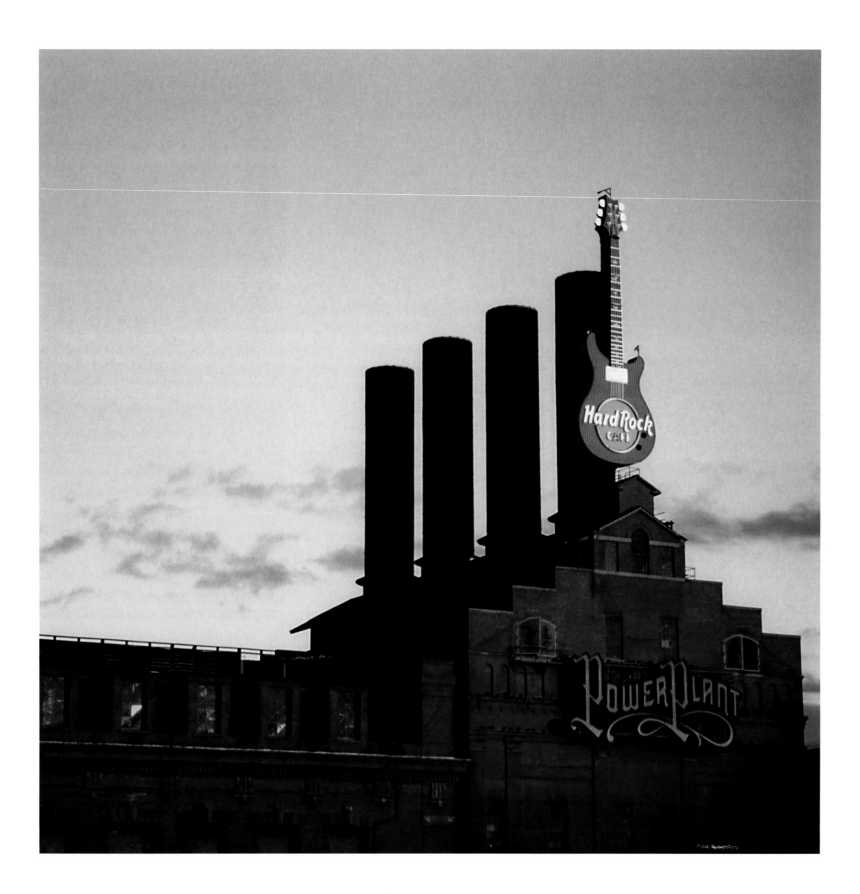

THE POSTINDUSTRIAL LANDSCAPE

A FEW YEARS AGO I had an epiphany in a parking lot. I was visiting a railroad yard where freight cars are sorted according to their destinations and assembled into trains. It was a big place, a hub of the national rail network, and when I drove through the gate, I wasn't surprised to find a parking lot with space for 200 cars. But the lot was empty except for a dozen cars huddled near the entrance to the main building. The superintendent who was showing me around soon explained. At one time, the yard employed a large number of brakemen, who rode along on each of the freight cars to control their speed during the sorting process. But the role of the brakemen has been taken over by mechanical "retarders" installed in the track and operated by computer control. In the old days there was also a large room filled with clerks, who handled the paperwork that accompanied every freight car on its journey across the country. But the routing of cars is now accomplished by electronic communication from one computer to another, and so the sorting yard is paperless and clerkless. The room where the clerks had their desks is as empty as the parking lot.

What struck me that morning was just how lonely a place the industrial landscape has become. It's not just railroad freight yards; I found the same haunting depopulation almost everywhere I looked. On the docks of a cargo port, gangs of longshoremen used to swarm over a ship to load or unload it; now most of the work is done by one artful crane operator, perched high overhead, lifting 60,000-pound containers at the rate of two a minute. Where miners used to toil underground, drilling and blasting, the earth is now ripped open by gargantuan shovels and draglines; these machines, too, are controlled by one worker in a high glass booth. Telephone switching centers, once filled with the voices of hundreds of operators, are silent, dark, and deserted. On the high plains of Kansas, a solitary farmer in a tractor plows and plants a thousand acres of land. At an oil refinery, the rows of tall distilling towers and chemical reactors give the place the look of a city of skyscrapers, but it is a vacant city,

An unwritten rule of urban development says: Always celebrate what's no longer there. In the Inner Harbor area of Baltimore, what's no longer there is a neighborhood of wharves, warehouses, and heavy industry. One end of the harbor is dominated by the brick building in the photograph on the opposite page. In case the carefully preserved smokestacks on the roof are not a sufficient clue, the owners have put up a helpful sign identifying the building's former function: Power Plant. The other sign, the guitar attached to one of the smokestacks, advertises the new function: Hard Rock Café.

with no one on the streets; everything is watched over by a few engineers and technicians inside a windowless control room.

Fifty years ago, "automation" was a matter of considerable public interest, a subject for academic white papers, newspaper editorials, and congressional hearings. The prospect of replacing human labor with machines seemed both attractive and forbidding at the same time. According to one faction, automation would liberate us all from drudgery, giving us the time and economic freedom to cultivate higher callings; we would be a society of poets and scholars at leisure. The other side asked: If our jobs are taken by sleepless machines, how shall we live? At the time when these competing visions of the future were being debated, most people probably believed neither of them. The idea that automation might either displace or liberate some large fraction of the work force was one of those fantasies that would always remain just beyond the horizon, like the nuclear-powered flying automobile. Whether it was a threat or a promise, automation was for the future, not the present. But now automation is here, even though the word itself is seldom spoken anymore. Machines have insinuated themselves into our lives in ways that the futurists of the 1950s could not have anticipated, and as a result whole categories of jobs have all but disappeared. Elevator operators, typesetters, and airplane navigators have followed milkmaids and lamplighters into oblivion.

The social and economic consequences of these developments are not yet fully understood. So far, automation has not made us a nation of poets and scholars. So far, armies of the dispossessed and unemployed are not roaming the streets. I would not attempt to predict how the important questions of work, livelihood, and income

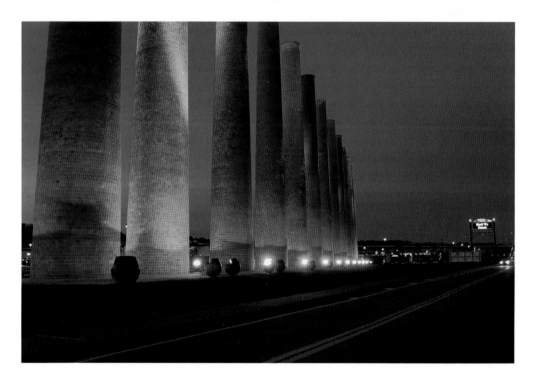

Where the Homestead mill of United States Steel once stood on the banks of the Monongahela River in Pittsburgh, a dozen brick smokestacks were left as ornaments to the shopping mall that now occupies the site.

In Akron, Ohio, the storage silos of the Quaker Oats grain elevator are converted into novelty hotel rooms.

distribution will ultimately be settled. But I do feel I can say something about the effect of all these changes on public perceptions of the industrial landscape.

Now that so few of us spend our daily working lives in agriculture, manufacturing, and tending the industrial infrastructure, those occupations have begun to seem more exotic and specialized, and the locales where they are practiced are more alien and mysterious. In a sense, that's what makes a book like this one possible. When most of us were out there getting our hands dirty in the industrial landscape—mining coal, making steel, growing crops, loading ships—a field guide to that landscape was hardly needed. We were producers as well as consumers of industrial goods, and so we knew something about where the raw materials of life came from. Now, we spend our working days in offices, and we never see the inside of a mine, a mill, a factory, a power plant. The facilities that support our way of life have become invisible. Your home is probably connected to an electric-power substation, a telephone switching office, a water filtration plant, a sewage treatment plant, and a natural-gas distribution depot. Have you ever been inside any of those facilities? Do you know where they are in your community or what they look like, even from the outside?

Having lost contact with industry on a day-to-day basis, one common response is to romanticize or sentimentalize what we have left behind. We don't want to tear down the old water-powered mills along New England's rivers, where generations of factory workers mass-produced shoes and textiles; instead we turn them into restaurants or shops where we buy expensive craft goods, made by hand. The brickfront warehouses and factories of urban industrial districts become artists' lofts or condominium apart-

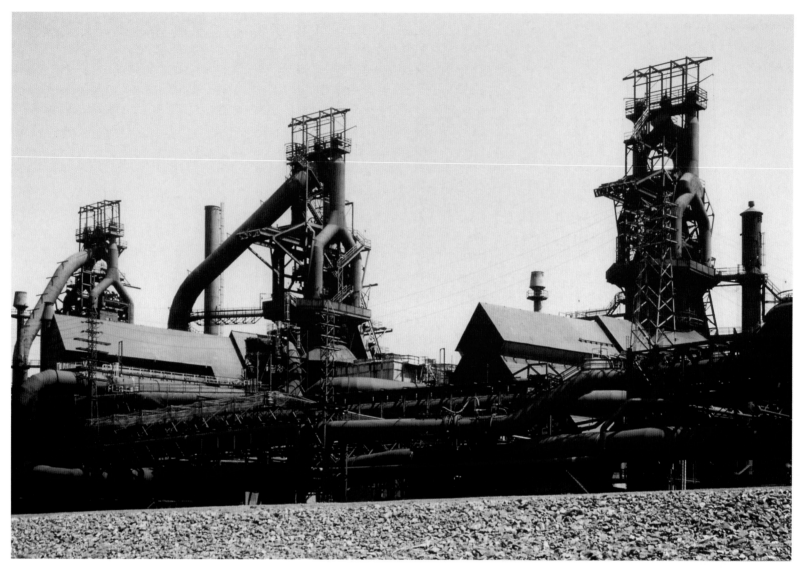

In Bethlehem, Pennsylvania, a row of blast furnaces has stood idle since 1995. Part of the site may become a National Museum of Industrial History.

ments. A steel mill in Duisburg, Germany, has been converted into a "landscape park," where children play among the ruins of blast furnaces. As the photographs that accompany this essay attest, there are many other projects for turning industrial artifacts into something else—something more approachable and friendly.

But the kinder-and-gentler treatment seems to work only for bygone industries. An old mill with a waterwheel is charming; a modern power plant is merely menacing. When it comes to industrial operations still running today, the effect of our alienation and unfamiliarity is to make the industrial infrastructure seem all the more sinister. At a refinery or a petrochemicals plant, we don't know what goes on behind the chain-link fence or what comes out of the smokestacks, and therefore we suspect the worst. At a hog farm or a poultry farm, we can't see inside the sheds where the animals are kept, and we imagine horrors of inhumanity. The operators of these facilities, feeling besieged by a hostile and uncomprehending public, respond by closing

the gates and building the fences higher. Their secrecy, naturally, tends to confirm public suspicion that they must have something to hide. It is a spiral of distrust and animosity.

This estrangement from industrial enterprise is not going to be reversed anytime soon. Our children and grandchildren will live in a world where nearly everyone works in an office or a classroom or a retail outlet or some other sort of service establishment; only a tiny minority of workers will be needed to keep factories, mills, mines, and farms running. Already, most of us work more with "bits," the fundamental units of information, than with atoms, the units of matter. Manufacturing tangible, ponderable objects is now secondary to the creation and management of "intellectual property"—words, numbers, data, images, accounts, programs—information in all its manifestations.

There is something of a paradox here. On the one hand, people today deal with machines on a much more frequent and intimate basis than earlier generations did. We pump our own gas; we get cash from the ATM instead of from a bank teller; we check out our own groceries at the supermarket and our own books at the library; we make our own airline reservations over the Internet instead of consulting a travel agent. But most of us know less and less about how all these machines work. We know how to use them, but not how to build or fix them. As for the more remote machinery—the turbines, pumps, generators, transformers, switches, amplifiers, transmitters, and all the rest of the apparatus that keeps an industrial economy humming—all that is quite out of sight. The technological infrastructure is someone else's responsibility; we just want the lights and the phones to work when we need them.

Is this situation something to be worried over? After all, most of us will never need to know how to run a nuclear power plant or how to operate a strip-mining dragline. Yet there is something sad about a society in which large numbers of people don't understand the basic substrate of their own world. In the case of the natural world, everyone ought to have at least a rudimentary grasp of the laws of physics and those of biology, such as Darwin's principle of evolution by natural selection. The same imperative applies to the world of technology. Without a sense of how materials and energy flow through an industrial economy, you miss something basic about the world you live in.

And cutting people off from the industrial infrastructure has practical consequences, too. Sooner or later, decisions about the direction of important technologies have to be made by a democratic process. People who have never seen a power plant, who know nothing of how it works, who have never met anyone who works there, are poorly equipped to judge the relative merits of nuclear and coal-fired technologies, or to seek alternatives that might allow us to dispense with both. To make good decisions about such issues, citizens need to get better acquainted with the technological underpinnings of their own communities. To allow that to happen, those who own and operate the various elements of the infrastructure will have to open up the gates and invite the people in.

Before and after: The image above is an unaltered and unadjusted scan of a photograph of the Ravenswood power plant, made August 24, 1997. The image at right is the corrected and adjusted version of the same photograph; it also appears on page 199. The changes made include slight adjustments to color and contrast; in addition, the geometry has been altered somewhat. In the original image the smokestacks and the walls of the building seem to lean inward; in the final version, this apparent tilt has been removed.

A NOTE ON THE PHOTOGRAPHS

Most of the photographs in this book were made by the author between 1992 and 2004. (Photos made by others are identified in the captions.)

In some cases the photography was done with the permission and assistance of the owners of the site being photographed. In those cases, a representative of the company was usually standing beside me as the photograph was made. Thus, the company had some control over what I could and couldn't see, although in practice those companies that allowed me to photograph their operations put few constraints on my movements. (Many other companies refused to allow me access at all.)

The majority of the photographs were made without invitation or authorization, by searching out a vantage point on the public side of the fence where I could get a clear view of the operations. In one instance I chartered an airplane to make aerial photographs, but most of the aerial photos were made through the windows of commercial aircraft during ordinary airline flights. A few photographs were made from behind the wheel of a moving car.

Until 2000, the images were recorded on 35-millimeter color film and later digitized by scanning a print or the negative. More recent photographs were recorded with a digital camera.

The fact that the photos have been digitally processed raises questions about verisimilitude. Have I altered or enhanced the images? I have not abided by the strict regimen of some documentary photographers, who contend that the camera captures truth, and any change made outside the camera is a falsification. I disagree. I would argue, on the contrary, that sometimes it takes a good deal of hard work to undo the camera's lies and restore some semblance of truth.

In preparing the photos for publication, I have made changes of the following kinds:

• In images scanned from a film original, I have removed dust, scratches, fingerprints, and other blemishes.

• I have corrected colors, contrast, and brightness. Neither film nor the electronic sensors in digital cameras detect light in the same way as the human eye and brain. To make a picture "look right"—in the sense that it matches what the photographer saw through the viewfinder—it is almost always necessary to adjust the color balance of an image and its brightness or contrast.

• Sharpening. The process of creating a digital image—reducing a continuous scene to discrete dots of color—always sacrifices information about the finest details in the image. Sharpening restores some of that information.

• Geometric correction. Just as film does not always render colors in the same way that the eye and brain perceive them, camera lenses do not always record the shape of objects and their geometric relations in the same way that people see them. Indeed, different lenses can give very different views of the same scene. Some of the geometric effects are simply flaws that would not be present in an ideal lens; an example is the barrel or pincushion distortion that transforms straight lines into curves. Another geometric issue—the treatment of perspective—is not so clear-cut. Some cameras can be set up so that parallel lines in a scene remain parallel in the image. But with most cameras, a photograph of a tall building will show the walls converging toward the top. Which is the truth? I don't believe there is a definitive, universal answer. Aesthetic judgment has to be applied on a case-by-case basis. I have tried to do just that, although I admit to a strong bias in favor of rectilinear forms. The geometric corrections were made not by using a perspective-altering camera but by transforming the digitized image with computer software.

Another contentious issue is the "doctoring" of photographs. If you are about to click the shutter and you notice a distracting beer can in the foreground, is it acceptable to remove the can, or would that amount to tampering with the scene? If you notice the beer can later, in the photograph, is it acceptable to remove it with the electronic equivalent of an airbrush? Making such changes soon puts you on the slippery slope that leads to pictures of aliens embracing movie stars on the front page of supermarket tabloids. I have not been a purist in these matters, but I have exercised restraint. There are neither aliens nor movie stars in these pictures, and I believe they convey a faithful impression of what the photographer was looking at when they were made.

S OME OF THE LITERATURE of the industrial landscape is more exotic and obscure than the landscape itself. Textbooks for the power-plant operator and trade magazines for dairy farmers or landfill managers are not items that turn up on many reading lists. Any library will offer information on the general principles or theories behind the operation of the electric-power grid or the telephone network, but it's harder to come by the nitty-gritty, hands-on, inside dope about what particular machines look like and how they work. The best resources belong to the blue-collar literature, such as repair manuals and training guides for technicians and operators.

What follows is not a comprehensive or scholarly bibliography but merely a selection of readings that may be useful for those who would like to pursue a topic further. Items followed by the symbol **KIDS** are suitable for younger readers; the notation **GEEKS** flags material that is technical or specialized but may interest enthusiasts.

The Internet also offers boundless resources on many of the themes in this book, but Web pages come and go too quickly for citation here.

GENERAL AND MISCELLANEOUS WORKS

Axelrod, Karen, and Bruce Brumberg. *Watch It Made in the U.S.A.: A Visitor's Guide to the Companies That Make Your Favorite Products.* 2nd edition. Foreword by Richard S. Gurin. Santa Fe, NM: John Muir, 1997. **KIDS**

Brain, Marshall, and the staff at HowStuffWorks.com. *Marshall Brain's How Stuff Works.* New York: Hungry Minds, 2001. **KIDS**

Feldman, Anthony, and Bill Gunston. *Technology at Work.* New York: Facts on File, 1980.

Graedel, T. E., and B. R. Allenby. *Industrial Ecology.* Englewood Cliffs, NJ: Prentice Hall, 1995.

Jones, Pamela. *Under the City Streets*. New York: Holt, Rinehart and Winston, 1978.

Macaulay, David. *The Way Things Work*. Boston: Houghton Mifflin, 1988.

McKibben, Bill. "A Reporter at Large: Apartment." *New Yorker*, March 17, 1986, pages 43–91.

Stilgoe, John R. *Common Landscape of America, 1580 to 1845*. New Haven: Yale University Press, 1982.

———. *Outside Lies Magic: Regaining History and Awareness in Everyday Places*. New York: Walker, 1998.

Sullivan, Robert. *The Meadowlands: Wilderness Adventures at the Edge of a City*. New York: Scribner, 1998.

Zukin, Sharon. *Landscapes of Power: From Detroit to Disney World*. Berkeley: University of California Press, 1991.

PICTORIAL WORKS

Arthus-Bertrand, Yann, with text by the editors of the annual *L'etat du monde*. *Earth from Above*. New York: Harry N. Abrams, 1999.

Becher, Bernd, and Hilla Becher, with text by Armin Zweite. *Typologies of Industrial Buildings*. Cambridge: MIT Press, 2004.

Corner, James. *Taking Measures: Across the American Landscape*. Photographs by Alex S. MacLean. Foreword by Michael Van Valkenburgh. New Haven: Yale University Press, 1996.

Greenberg, Stanley. *Invisible New York: The Hidden Infrastructure of the City*. Introduction by Thomas H. Garver. Baltimore: Johns Hopkins University Press, 1998.

Higgins, James Jeffrey. *Images of the Rust Belt*. Kent, OH: Kent State University Press, 1999.

Jackson, John Brinckerhoff. *American Space: The Centennial Years, 1865–1876*. New York: W. W. Norton, 1972.

Levy, Builder. *Images of Appalachian Coalfields*. Introduction by Helen Matthews Lewis. Foreword by Cornell Capa. Philadelphia: Temple University Press, 1989.

Plowden, David. *Industrial Landscape*. New York: Chicago Historical Society in association with W. W. Norton, 1985.

CHAPTER 1: OUT OF THE EARTH

Bartlett, Robert W. *Solution Mining: Leaching and Fluid Recovery of Materials*. Philadelphia: Gordon and Breach Science, 1992. **GEEKS**

Davis, E. W. *Pioneering with Taconite*. St. Paul: Minnesota Historical Society, 1964.

E. I. du Pont de Nemours & Co. *Blasters' Handbook*. 16th edition. Wilmington, DE: E. I. du Pont de Nemours, 1980. **GEEKS**

Francaviglia, Richard V. *Hard Places: Reading the Landscape of America's Historic Mining Districts*. Foreword by Wayne Franklin. Iowa City: University of Iowa Press, 1991.

Gani, M. S. J. *Cement and Concrete*. New York: Chapman and Hall, 1997.

Gregory, Cedric E. *A Concise History of Mining*. New York: Pergamon Press, 1980.

Gunn, John M. *Restoration and Recovery of an Industrial Region: Progress in Restoring the Smelter-Damaged Landscape near Sudbury, Canada*. New York: Springer-Verlag, 1995.

Hartman, Howard L. (editor). *SME Mining Engineering Handbook*. 2nd edition. Littleton, CO: Society for Mining, Metallurgy, and Exploration, 1992. **GEEKS**

Husband, Joseph. *A Year in a Coal Mine*. New York: Arno Press, 1977. (Original publication: Houghton Mifflin, 1911.)

Isalski, W. H. *Separation of Gases*. Oxford: Clarendon Press, 1989. **GEEKS**

Kroll-Smith, J. Stephen, and Stephen Robert Couch. *The Real Disaster Is above Ground: A Mine Fire and Social Conflict*. Lexington: University Press of Kentucky, 1990.

Stack, Barbara. *Handbook of Mining and Tunnelling Machinery*. New York: Wiley, 1982.

Thomas, Richard (editor). *E/MJ Operating Handbook of Mineral Processing: Concentrating, Agglomerating, Smelting, Refining, Extractive Metallurgy*. New York: McGraw-Hill, 1977.

Voynick, Stephen M. *The Making of a Hardrock Miner*. Berkeley: Howell-North Books, 1978.

CHAPTER 2: WATERWORKS

Berk, Sharon G., and John H. Gunderson. *Wastewater Organisms: A Color Atlas*. Boca Raton, FL: Lewis, 1993. **GEEKS**

Davis, Margaret Leslie. *Rivers in the Desert: William Mulholland and the Inventing of Los Angeles*. New York: HarperCollins, 1993.

Fair, Gordon Maskew, John Charles Geyer, and Daniel Alexander Okun. *Elements of Water Supply and Wastewater Disposal*. 2nd edition. New York: Wiley, 1971.

Golzé, Alfred R. (editor). *Handbook of Dam Engineering*. New York: Van Nostrand Reinhold, 1977.

Hammer, Mark J. *Water and Wastewater Technology*. SI version. 2nd edition. New York: John Wiley, 1986.

Hundley, Norris, Jr. *The Great Thirst: Californians and Water, 1770–1990s*. Berkeley: University of California Press, 1992.

Kahrl, William L. (editor). *The California Water Atlas*. Sacramento: Governor's Office of Planning and Research, distributed by William Kaufmann, Inc., 1979.

McGhee, Terence J. *Water Supply and Sewerage*. 6th edition. New York: McGraw-Hill, 1991.

McPhee, John. "Atchafalaya." In *The Control of Nature*. New York: Farrar, Straus and Giroux, 1989, pages 3–92.

Melnick, Mimi. *Manhole Covers*. Photographs by Robert A. Melnick. Foreword by Alan Sekula. Cambridge: MIT Press, 1994.

Metcalf & Eddy, Inc. *Wastewater Engineering: Treatment, Disposal and Reuse*. 3rd edition. Revised by George Tchobanoglous and Franklin L. Burton. New York: McGraw-Hill, 1991.

Pattison, Kermit. "Why Did the Dam Burst?" *Invention and Technology*, Summer 1998, pages 23–31.

Petersen, Margaret S. *River Engineering*. Englewood Cliffs, NJ: Prentice Hall, 1986.

Petroski, Henry. "Hoover Dam." *American Scientist*, Vol. 81, No. 6, November–December 1993, pages 517–521.

Smalley, Ian. "The Teton Dam: Rhyolite Foundation + Loess Core = Distaster." *Geology Today*, January–February 1992, pages 19–22. **GEEKS**

Van Veen, Johan. *Dredge, Drain, Reclaim: The Art of a Nation*. 5th edition. The Hague: Martinus Nijhoff, 1962.

CHAPTER 3: FOOD AND FARMING

Bell, Brian. *Farm Machinery*. 3rd edition. Ipswich, UK: Farming Press, 1989.

Ewing, Sherm. *The Ranch: A Modern History of the North American Cattle Industry*. Missoula, MT: Mountain Press, 1995.

Finner, Marshall F. *Farm Machinery Fundamentals*. Madison, WI: American Publishing, 1978.

Frazier, Ian. *Great Plains*. New York: Farrar, Straus and Giroux, 1989.

Gardner, Bruce L. *American Agriculture in the Twentieth Century: How It Flourished and What It Cost*. Cambridge: Harvard University Press, 2002.

Gillespie, James R. *Modern Livestock & Poultry Production*. 7th edition. Clifton Park, NY: Thomson/Delmar Learning, 2004.

Glass, Joseph W. *The Pennsylvania Culture Region: A View from the Barn*. Ann Arbor, MI: UMI Research Press, 1971, 1986.

Hart, John Fraser. *The Rural Landscape*. Baltimore: Johns Hopkins University Press, 1998.

Hoy, James F. *The Cattle Guard: Its History and Lore*. Foreword by Jimmy M. Skaggs. Lawrence: University Press of Kansas, 1982.

Jacobs, Frank. *Cattle and Us, Frankly Speaking, or, Cattle Come in Five Sexes*. Calgary, Alberta, Canada: Detselig Enterprises, 1993.

Jensen, M. E. (editor). *Design and Operation of Farm Irrigation Systems*. St. Joseph, MI: American Society of Agricultural Engineers, 1980.

Johansen, Harley E., and Glenn V. Fuguitt. *The Changing Rural Village in America: Demographic and Economic Trends since 1950*. Cambridge, MA: Ballinger, 1984.

Klamkin, Charles. *Barns: Their History, Preservation, and Restoration*. New York: Hawthorn Books, 1973.

Lovenheim, Peter. *Portrait of a Burger as a Young Calf: The True Story of One Man, Two Cows, and the Feeding of a Nation*. New York: Harmony Books, 2002.

Mallin, Michael A. "Impacts of Industrial Animal Production on Rivers and Estuaries." *American Scientist*, Vol. 88, No. 1, January–February 2000, pages 26–37.

Noble, Allen G., and Richard K. Cleek. *The Old Barn Book: A Field Guide to North American Barns and Other Farm Structures*. Illustrations by M. Margaret Geib. New Brunswick, NJ: Rutgers University Press, 1995.

Rath, Sara. *About Cows*. Minocqua, WI: Heartland Press, 1987. **KIDS**

Razac, Olivier. *Barbed Wire: A Political History*. Translated from the French by Jonathan Kneight. New York: New Press, 2002.

Sheldrake, Rupert. "Cattle Fooled by Phoney Grids." *New Scientist*, February 11, 1988, page 65.

Sloane, Eric. *Eric Sloane's an Age of Barns*. Des Moines: American Museum of Natural History, 1967, 1976.

Splinter, William E. "Center-Pivot Irrigation." *Scientific American*, June 1976, pages 90–99.

United States Department of Agriculture, Office of Communications. *Agriculture Fact Book 2001–2002*. Washington, D.C.: U. S. Government Printing Office, 2002.

CHAPTER 4: OIL AND GAS

Alvarez, A. *Offshore: A North Sea Journey*. Boston: Houghton Mifflin, 1986.

An Overview of the Alaska Highway Gas Pipeline: The World's Largest Project. Papers presented at the National Convention of the American Society of Civil Engineers in Pittsburgh, PA, April 1978. New York: American Society of Civil Engineers, 1978. **GEEKS**

Baker, Ron. *A Primer of Oilwell Drilling: A Basic Text of Oil and Gas Drilling*. 5th edition. Austin: Petroleum Extension Service, University of Texas at Austin, 1994.

Berger, Bill D., and Kenneth E. Anderson. *Modern Petroleum: A Basic Primer of the Industry*. 3rd edition. Tulsa: PennWell Books, 1992.

Boyd, Dan T. "Oklahoma Oil: Past, Present, and Future." *Oklahoma Geology Notes*, Vol. 62, No. 3, Fall 2002, pages 97–106.

Deffeyes, Kenneth S. *Hubbert's Peak: The Impending World Oil Shortage*. Princeton: Princeton University Press, 2001.

Evans, Frank L., Jr. *Equipment Design Handbook for Refineries and Chemical Plants*. Vol. 2. Houston: Gulf Publishing, 1974. **GEEKS**

Gary, James H., and Glenn E. Handwerk. *Petroleum Refining: Technology and Economics*. 3rd edition. New York: Marcel Dekker, 1994.

Jakle, John A., and Keith A. Sculle. *The Gas Station in America*. Baltimore: Johns Hopkins University Press, 1994.

Kennedy, John L. *Oil and Gas Pipeline Fundamentals*. 2nd edition. Tulsa: PennWell Books, 1993.

Langston, Leslie V. *The Lease Pumper's Handbook*. Norman, OK: The Commission on Marginally Producing Oil and Gas Wells of Oklahoma, 2003.

Leffler, William L. *Petroleum Refining for the Non-Technical Person*. 2nd Edition. Tulsa: PennWell Books, 1979, 1985.

Mead, Robert Douglas. *Journeys down the Line: Building the Trans-Alaska Pipeline*. Garden City, NY: Doubleday, 1978.

Petroleum Extension Service, University of Texas at Austin. *Oil Pipe Line Pumping Station Operation*. Austin: University of Texas at Austin, 1956, 1976.

———. *Field Handling of Natural Gas*. 3rd edition. Austin: Petroleum Extension Service, University of Texas at Austin, 1972.

———. *Oil Pipe Line Construction and Maintenance*. 2nd edition. Austin: University of Texas at Austin, 1973.

———. *Introduction to the Oil Pipeline Industry*. Austin: University of Texas at Austin, 1973.

———. *Plant Processing of Natural Gas*. University of Texas at Austin, 1974.

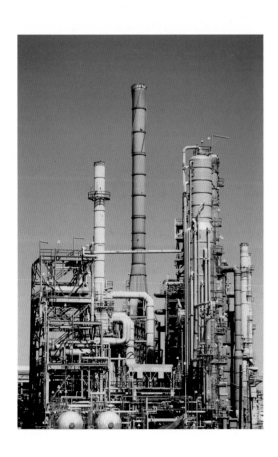

Witzel, Michael Karl. "Gas Pumps." *American Heritage of Invention and Technology*, Winter 1997, pages 58–63.

CHAPTER 5: POWER PLANTS

Baker, T. Lindsay. *A Field Guide to American Windmills*. Foreword by Donald E. Green. Norman: University of Oklahoma Press, 1985.

Elliott, Thomas C., Kao Chen, and Robert C. Swanekamp. *Standard Handbook of Powerplant Engineering*. 2nd edition. New York: McGraw-Hill, 1998. **GEEKS**

Elonka, Stephen Michael. *Standard Plant Operators' Manual*. 2nd edition. New York: McGraw-Hill, 1975. **GEEKS**

El-Wakil, M. M. *Powerplant Technology*. New York: McGraw-Hill, 1984. **GEEKS**

Gipe, Paul. *Wind Energy Comes of Age*. New York: John Wiley, 1995.

Lish, Kenneth C. *Nuclear Power Plant Systems and Equipment*. New York: Industrial Press, 1972. **GEEKS**

Pansini, A. J., and K. D. Smalling. *Guide to Electric Power Generation*. Lilburn, GA: Fairmont Press, 1994.

Rogovin, Mitchell, and George T. Frampton Jr. *Three Mile Island: A Report to the Commissioners and to the Public*. Vol. II, Part 2. Washington, D.C.: Nuclear Regulatory Commission, Special Inquiry Group, 1980.

Smil, Vaclav. *Energy at the Crossroads: Global Perspectives and Uncertainties*. Cambridge: MIT Press, 2003.

Weisman, Joel, and L. E. Eckart. *Modern Power Plant Engineering*. Englewood Cliffs, NJ: Prentice Hall, 1985.

Winter, C.-J., R. L. Sizmann, and L. L. Vant-Hull. *Solar Power Plants: Fundamentals, Technology, Systems, Economics*. Berlin: Springer-Verlag, 1991.

Woodson, Riley D. "Cooling Towers." *Scientific American*, May 1971, pages 70–78.

CHAPTER 6: THE POWER GRID

Burke, James J. *Distribution Engineering: Fundamentals and Applications*. New York: Marcel Dekker, 1994.

Casper, Barry M., and Paul David Wellstone. *Powerline: The First Battle of America's Energy War*. Amherst: University of Massachusetts Press, 1981.

Chiles, James R. "Learning from the Big Blackouts." *American Heritage of Invention and Technology*, Vol. 1, No. 2, Fall 1985, pages 27–30.

Crowe, Sylvia. *The Landscape of Power*. Diagrams by Michael Laurie. London: Architectural Press, 1958.

Doocy, E. S., A. R. Hard, C. B. Rawlins, and R. Ikegami. *Transmission Line Reference Book: Wind-Induced Conductor Motion*. Palo Alto, CA: Electric Power Research Institute, 1979. **GEEKS**

Eaton, J. Robert. *Electric Power Transmission Systems*. Englewood Cliffs, NJ: Prentice Hall, 1972.

Electric Power Research Institute. *Transmission Line Reference Book, 345 kV and Above*. Palo Alto, CA: Electric Power Research Institute, 1975.

Fardo, Stephen W., and Dale R. Patrick. *Electrical Power Systems Technology*. 2nd edition. Boston: Newnes, 1997.

Giles, R. L. *Layout of E.H.V. Substations*. Cambridge, UK: Cambridge University Press, 1970. **GEEKS**

Goulty, George A. *Visual Amenity Aspects of High Voltage Transmission*. Taunton, Somerset, UK: Research Studies Press, 1990.

Graneau, Peter. *Underground Power Transmission: The Science, Technology and Economics of High Voltage Cables*. New York: John Wiley, 1979.

Hughes, Thomas P. *Networks of Power: Electrification in Western Society 1880–1930*. Baltimore: Johns Hopkins University Press, 1983.

Kurtz, Edwin B., and Thomas M. Shoemaker. *The Lineman's and Cableman's Handbook.* 8th edition. New York: McGraw-Hill, 1955, 1992.

Myers, William A. *Iron Men and Copper Wires: A Centennial History of the Southern California Edison Company.* Glendale, CA: Trans-Anglo Books, 1984.

Nye, David E. *Electrifying America: Social Meanings of a New Technology, 1880–1940.* Cambridge: MIT Press, 1990.

Pansini, Anthony J. *Electrical Distribution Engineering.* New York: McGraw-Hill, 1983.

Rieder, Werner. "Circuit Breakers." *Scientific American,* January 1971, pages 76–84.

Stockbridge, G. H. "Overcoming Vibration in Transmission Cables." *Electrical World,* Vol. 86, No. 26, 1925, pages 1304–1305.

Weedy, B. M. *Electric Power Systems.* 3rd edition. New York: John Wiley, 1979.

CHAPTER 7: COMMUNICATIONS

American Telephone and Telegraph Company. *Things Worth Knowing about the Telephone.* New York: American Telephone and Telegraph Company, 1923.

———. Special issue on TAT-1. *Bell System Technical Journal,* Vol. 36, No. 1, 1957. **GEEKS**

Balston, D. M., and R. C.V. Macario (editors). *Cellular Radio Systems.* Boston: Artech House, 1993. **GEEKS**

Bell Telephone Laboratories. *Engineering and Operations in the Bell System.* New York: Technical Publication Department, Bell Laboratories, 1977.

Briley, Bruce E. *Introduction to Telephone Switching.* Reading, MA: Addison Wesley, 1983.

Carr, Joseph J. *Practical Antenna Handbook.* New York: Tab Books, 1994.

Chaffee, C. David. *The Rewiring of America: The Fiber Optics Revolution.* Orlando: Academic Press, 1988.

Chomycz, Bob. *Fiber Optic Installations: A Practical Guide.* New York: McGraw-Hill, 1996.

Coe, Lewis. *The Telegraph: A History of Morse's Invention and Its Predecessors in the United States.* Jefferson, NC: McFarland, 1993.

Fischer, Claude S. *America Calling: A Social History of the Telephone to 1940.* Berkeley: University of California Press, 1992.

Goff, David R. *Fiber Optic Reference Guide: A Practical Guide to the Technology.* Boston: Focal Press, 1996.

Hayes, Brian. "The Numbering Crisis in World Zone 1." *The Sciences,* November–December 1992, pages 12–15.

———. "Computing Science: The Infrastructure of the Information Infrastructure." *American Scientist,* Vol. 85, No. 3, May–June 1997, pages 214–218.

Hayes, Jim. *Fiber Optics Technician's Manual.* Albany: Delmar, 1996.

Johnson, Richard C. (editor). *Antenna Engineering Handbook.* 3rd edition. New York: McGraw-Hill, 1961, 1993. **GEEKS**

Noll, A. Michael. *Introduction to Telephones and Telephone Systems.* 2nd edition. Boston: Artech House, 1991.

Parsons, Patrick R., and Robert M. Frieden. *The Cable and Satellite Television Industries.* Boston: Allyn and Bacon, 1998.

Reeve, Whitham D. *Subscriber Loop Signaling and Transmission Handbook: Analog.* New York: IEEE Press, 1992. **GEEKS**

————. *Subscriber Loop Signaling and Transmission Handbook: Digital.* New York: IEEE Press, 1995. **GEEKS**

Wang, Michael. *Metropolitan Microwave Network: Design and Implementation.* Englewood Cliffs, NJ: Prentice Hall, 1990.

CHAPTER 8: ON THE ROAD

Crowe, Sylvia. *The Landscape of Roads.* Diagrams by John Brookes. London: Architectural Press, 1960.

Giblin, James M., Walter H. Kraft, James Rudden, and Robert Sands. *Traffic Signal Installation and Maintenance Manual.* Englewood Cliffs, NJ: Prentice Hall, 1991.

Helbing, Dirk. "Traffic and Related Self-Driven Many-Particle Systems." *Reviews of Modern Physics*, October 2001, Vol. 73, pages 1067–1141. **GEEKS**

Kell, James H., and Iris J. Fullerton. *Manual of Traffic Signal Design.* 2nd edition. Englewood Cliffs, NJ: Prentice Hall, 1991.

Lay, M. G. *Ways of the World: A History of the World's Roads and of the Vehicles That Used Them.* New Brunswick, NJ: Rutgers University Press, 1992.

A Policy on Geometric Design of Highways and Streets, 1994. Washington, D.C.: American Association of State Highway and Transportation Officials, 1995.

Schlereth, Thomas J. *U.S. 40: A Roadscape of the American Experience.* Indianapolis: Indiana Historical Society, 1985.

Stewart, George R. *U.S. 40: Cross Section of the United States of America.* Boston: Houghton Mifflin, 1953.

Transportation Research Board. *Highway Research Program, Synthesis of Highway Practice*, Vol. 41. Washington, D.C.: Transportation Research Board, 1977.

Wixom, Charles W. *Pictorial History of Roadbuilding.* Washington, D.C.: American Road Builders' Association, 1975.

CHAPTER 9: THE RAILROAD

Adams, Ramon F. *The Language of the Railroader.* Norman: University of Oklahoma Press, 1977.

Archdeacon, H. C. (editor). *The Track Cyclopedia.* 9th edition. Omaha: Simmons-Boardman Books, 1978. **GEEKS**

Barry, Michael. *Through the Cities: The Revolution in Light Rail*. Dublin: Frankfort Press, 1991.

Brignano, Mary, and Hax McCullough. *The Search for Safety: A History of Railroad Signals and the People Who Made Them*. Graphic design by Sid Navratil. New York: American Standard, Union Switch & Signal Division, 1981.

Fisher, Ralph E. *Vanishing Markers: Memories of Boston and Maine Railroading, 1946–1952*. Brattleboro, VT: Stephen Greene Press, 1976.

Foster, Gerald. *A Field Guide to Trains of North America*. Boston: Houghton Mifflin, 1996.

General American Transportation Corporation. *GATX Tank Car Manual*. 4th edition. Chicago: General American Transportation Corporation, 1979. **GEEKS**

General Railway Signal. *Elements of Railway Signaling*. Rochester, NY: General Railway Signal, 1979.

Profillidis, V. A. *Railway Engineering*. Aldershot, UK: Avebury Technical, 1995.

Seki, Nagaomi. *Shinkansen*. Translated by Don Kenny. Osaka, Japan: Hoikusha, 1983.

St. Clair, David J. *The Motorization of American Cities*. New York: Praeger, 1986.

Stilgoe, John R. *Metropolitan Corridor: Railroads and the American Scene*. New Haven: Yale University Press, 1983.

Stover, John F. "One Gauge." *American Heritage of Invention and Technology*, Winter 1993, pages 54–61.

The Track Data Handbook. Omaha: Simmons-Boardman Books, 1982. **GEEKS**

U.S. Department of Transportation. *Rules, Standards and Instructions Governing the Installation, Inspection, Maintenance, and Repair of Signal and Train Control Systems, Devices and Appliances*. The Code of Federal Regulations, 49 CFR 236. Washington, D.C.: U.S. Government Printing Office, 1996. **GEEKS**

White, John H., Jr. *The American Railroad Freight Car: From the Wood-Car Era to the Coming of Steel*. Baltimore: Johns Hopkins University Press, 1993.

———. "Changing Trains." *American Heritage of Invention and Technology*, Spring–Summer 1991, pages 35–41.

CHAPTER 10: BRIDGES AND TUNNELS

Anderson, Graham, and Ben Roskrow. *The Channel Tunnel Story*. London: E & FN Spon, 1994.

Bickel, John O., Thomas R. Kuesel, and Elwyn H. King (editors). *Tunnel Engineering Handbook*. 2nd edition. New York: Chapman and Hall, 1996.

Billah, K. Yusuf, and Robert H. Scanlan. "Resonance, Tacoma Narrows Bridge Failure, and Undergraduate Physics Textbooks." *American Journal of Physics,* Vol. 59, No. 2, February 1991, pages 118–124. **GEEKS**

Brown, David J. *Bridges*. New York: Macmillan, 1993.

Buckley, Tom. "A Reporter at Large: The Eighth Bridge." *New Yorker*, January 14, 1991, pages 37–59.

DeLony, Eric. "The Golden Age of the Iron Bridge." *American Heritage of Invention and Technology*, Fall 1994, pages 8–22.

Jackson, Donald C. *Great American Bridges and Dams*. Foreword by David McCullough. New York: John Wiley, 1988.

Megaw, T. M., and J. V. Bartlett. *Tunnels: Planning, Design and Construction*. Two volumes. Chichester, UK: Ellis Horwood, 1981.

Moran, Barbara. "A Bridge That Didn't Collapse." *American Heritage of Invention and Technology*, Fall 1999, pages 10–18.

Normile, Dennis. "A Short Course in Modern Bridges." *Technology Review*, November–December 1994, pages 52–59.

Petroski, Henry. "Still Twisting." *American Scientist*, Vol. 79, No. 5, September–October 1991, pages 398–401.

———. "Bridges of America." *American Scientist*, Vol. 84, No. 3, May–June 1996, pages 215–219.

Steinman, David B. "Bridges." *Scientific American*, November 1954, pages 60–71.

Transportation Research Board, National Research Council. *Highway Tunnel Operations*. Washington, D.C.: Transportation Research Board, 1975.

———. *Bridge Bearings*. Washington, D.C.: Transportation Research Board, 1977.

Troitsky, M. S. *Planning and Design of Bridges*. New York: John Wiley, 1994.

CHAPTER 11: AVIATION

Ashford, Norman, and Paul H. Wright. *Airport Engineering*. 2nd edition. New York: John Wiley, 1984.

Coombs, Charles. *Cleared for Takeoff: Behind the Scenes at an Airport*. New York: Morrow, 1969.

Dempsey, Paul Stephen, Andrew R. Goetz, and Joseph S. Szyliowicz. *Denver International Airport: Lessons Learned*. New York: McGraw-Hill, 1997.

Federal Aviation Administration, Office of System Safety, Safety Promotion Staff. *Airport Markings, Signs, and Selected Surface Lighting*. Washington, D.C.: U.S. Department of Transportation, 1995. **GEEKS**

Gould, Frederick L. *Radar for Technicians: Installation, Maintenance and Repair*. New York: Tab Books, 1995. **GEEKS**

Hart, Walter. *The Airport Passenger Terminal*. New York: John Wiley, 1985.

Horonjeff, Robert. *The Planning and Design of Airports*. New York: McGraw-Hill, 1962.

Kaplan, James. *The Airport: Terminal Nights and Runway Days at John F. Kennedy International*. New York: William Morrow, 1994.

Kayton, Myron, and Walter R. Fried (editors). *Avionics Navigation Systems*. 2nd edition. New York: John Wiley, 1997.

CHAPTER 12: SHIPPING

Angelucci, Enzo, and Attilio Cucari. *Ships*. New York: Greenwich House, 1983.

Basnight, Bobby L. *What Ship Is That? A Field Guide to Boats and Ships*. New York: Lyons and Burford, 1996.

Berteaux, H. O. *Buoy Engineering*. New York: John Wiley, 1976. **GEEKS**

Bruun, Per. *Port Engineering*. Houston: Gulf Publishing, 1973.

Cockcroft, A. N., and J. N. F. Lameijer. *A Guide to the Collision Avoidance Rules: International Regulations for Preventing Collisions at Sea*. Oxford: B. H. Newnes, 1996.

Committee on Tank Vessel Design, Marine Board, Commission on Engineering and Technical Systems, National Research Council. *Tanker Spills: Prevention by Design*. Washington, D.C.: National Academy Press, 1991.

Eyres, D. J. *Ship Construction*. 4th edition. Oxford: B. H. Newnes, 1994.

Gilman, Roger H. "Cargo Handling." *Scientific American*, October 1968, pages 80–88.

Hershman, Marc J. (editor). *Urban Ports and Harbor Management: Responding to Change along U.S. Waterfronts*. New York: Taylor and Francis, 1988.

International Organization for Standardization. *Freight Containers*. ISO Standards Handbook 34. Geneva: International Organization for Standardization, 1989. **GEEKS**

Karmon, Yehuda. *Ports around the World*. New York: Crown, 1980.

Mostert, Noël. *Supership*. New York: Alfred A. Knopf, 1974.

Nordbok. *The Lore of Ships.* New York: Crescent Books, 1975.

Osborne, Walter. *Tug, Tow, and Barge.* New York: Van Nostrand Reinhold, 1969.

Sauerbier, Charles L., and Robert J. Meurn. *Marine Cargo Operations.* 2nd edition. New York: John Wiley, 1985.

Tusiani, Michael D. *The Petroleum Shipping Industry.* Vol. 1: *A Nontechnical Overview.* Vol. 2: *Operations and Practices.* Tulsa: PennWell Books, 1996.

CHAPTER 13: WASTES AND RECYCLING

Ahearne, John F. (editor). Radioactive Waste: Special Issue. *Physics Today,* June 1997.

Bagchi, Amalendu. *Design, Construction, and Monitoring of Landfills.* 2nd edition. New York: John Wiley, 1994.

Bouverie, Jasper. "Recycling in Cairo: A Tale of Rags to Riches." *New Scientist,* June 29, 1991, pages 52–55.

Brunner, Calvin R. *Handbook of Incineration Systems.* New York: McGraw-Hill, 1991.

Crowther, Richard. "Space Junk—Protecting Space for Future Generations." *Science,* Vol. 296, May 17, 2002, pages 1241–1242.

Darlington, Arnold. *Ecology of Refuse Tips.* London: Heinemann Educational, 1969.

Franklin Associates. *Characterization of Municipal Solid Waste in the United States: 1997 Update.* Washington, D.C.: Office of Solid Waste, Municipal and Industrial Solid Waste Division, U.S. Environmental Protection Agency, 1998.

Freeman, Harry M. (editor). *Standard Handbook of Hazardous Waste Treatment and Disposal.* New York: McGraw-Hill, 1989.

Hayes, Brian. "Garbage." *Muse,* February 1999, pages 10–16. **KIDS**

Lund, Herbert F. (editor). *The McGraw-Hill Recycling Handbook.* 2nd edition. New York: McGraw-Hill, 2001.

McPhee, John. "A Reporter at Large: Duty of Care." *New Yorker,* June 28, 1993, pages 72–80.

Miller, Benjamin. *Fat of the Land: Garbage in New York, the Last Two Hundred Years.* New York: Four Walls Eight Windows, 2000.

Nijkerk, Alfred A. *Handbook of Recycling Techniques.* The Hague: Nijkerk Consultancy; distributed in the English-language countries by American Metal Market, New York, 1996.

Palmisano, Anna C., and Morton A. Barlaz (editors). *Microbiology of Solid Waste.* Boca Raton, FL: CRC Press, 1996. **GEEKS**

Pfeffer, John T. *Solid Waste Management Engineering.* Englewood Cliffs, NJ: Prentice Hall, 1992.

Rathje, William, and Cullen Murphy. *Rubbish! The Archeology of Garbage.* New York: HarperCollins, 1992.

Rhyner, Charles R. *Waste Management and Resource Recovery.* Boca Raton, FL: Lewis, 1995.

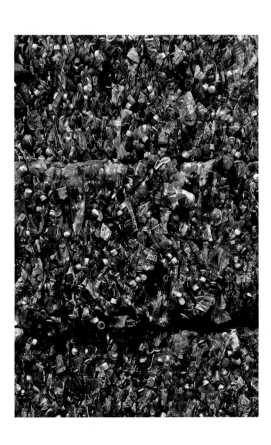

Stessel, Richard Ian. *Recycling and Resource Recovery Engineering: Principles of Waste Processing*. New York: Springer-Verlag, 1996.

Tierney, John. "Recycling Is Garbage." *New York Times Magazine*, June 30, 1996, pages 24–53.

U.S. Environmental Protection Agency. *Municipal Solid Waste in the United States: 2001 Facts and Figures*. Washington, D.C.: Office of Solid Waste and Emergency Response, 2003.

AFTERWORD: THE POSTINDUSTRIAL LANDSCAPE

Bellamy, Edward. *Looking Backward*. New York: Ticknor, 1888.

Fox, Nicols. *Against the Machine: The Hidden Luddite Tradition in Literature, Art, and Individual Lives*. Washington, D.C.: Island Press/Shearwater Books, 2002.

Goudie, Andrew. *The Human Impact on the Natural Environment*. 5th edition. Cambridge: MIT Press, 2000.

MacKenzie, Donald, and Judy Wajcman. *The Social Shaping of Technology*. 2nd edition. Philadelphia: Open University Press, 1985, 1999.

Mann, Floyd C., and L. Richard Hoffman. *Automation and the Worker: A Study of Social Change in Power Plants*. New York: Holt, 1960.

Marsh, George Perkins. *Man and Nature*. Edited by David Lowenthal. Cambridge: Belknap Press of Harvard University Press, 1864. (Reprinted 1965.)

Marx, Leo. *The Machine in the Garden: Technology and the Pastoral Ideal in America*. New York: Oxford University Press, 1964.

Nye, David E. *American Technological Sublime*. Cambridge: MIT Press, 1994.

Philipson, Morris. *Automation: Implications for the Future*. New York: Vintage Books, 1962.

Ruskin, John. *Unto This Last: Four Essays on the First Principles of Political Economy*. Edited with an introduction by Lloyd J. Hubenka. Lincoln: University of Nebraska Press, 1862.

Sale, Kirkpatrick. *Rebels against the Future: The Luddites and Their War on the Industrial Revolution: Lessons for the Computer Age*. Reading, MA: Addison Wesley, 1995.

Tunnard, Christopher, and Boris Pushkarev. *Man-Made America: Chaos or Control? An Inquiry into Selected Problems of Design in the Urbanized Landscape*. New Haven: Yale University Press, 1963.

INDEX

Page numbers in *italics* refer to illustrations.